Helmut Klausing

Politics and Innovation in the Development of Radar Technology

in Germany and Great Britain up to 1945

VERLAG KARL ALBER

Die Deutsche Nationalbibliothek verzeichnet diese Publikation in der Deutschen Nationalbibliografie; detaillierte bibliografische Daten sind im Internet über http://dnb.d-nb.de abrufbar.

ISBN 978-3-495-99048-3 (Print)
ISBN 978-3-495-99049-0 (ePDF)

Onlineversion
Inlibra

1. Auflage 2026

Besuchen Sie uns im Internet
verlag-alber.de

Foreword

This book offers a comprehensive comparison of British and German radar development up to the end of World War II. It examines both technological progress and the organisational frameworks within which research and development took place. At its core is the question of how the two countries' different innovation systems shaped their ability to advance. The central argument is that the outcome of the war was not determined by technological competition alone. In Britain in particular, a strategic vision at the leadership level, stable organisational structures, and forward-looking personnel and resource planning proved decisive in sustaining a dynamic process of innovation. Beyond the technological race, the study highlights complementary measures that had a profound impact, including British signals intelligence and codebreaking at Bletchley Park, which played a crucial role in the decryption of German communications and was only fully acknowledged in the mid-1970s. Equally important was the pioneering British concept of operational research, a novel method for analysing military operations that was not recognised in Germany until after the war.

Given the wide range of actors involved in both countries, the study focuses on those who played decisive roles in politics, the military, science, and industry. Its findings draw on an extensive range of national and international sources, including the Federal Archives in Germany, the National Archives in the United Kingdom, and influential post-war publications. Particularly valuable are the contributions of the Defence Electronics History Society (DEHS), whose work on British radar sheds important light on technological progress. For Germany, this material is complemented by sources from the Federal Foreign Office and the Wehrmacht's war diaries.

Through detailed archival research and discussions with military historians from Great Britain, France, and the Netherlands, this study reassesses key events and developments that shaped the course of the Anglo-German conflict. It not only makes a significant contribution to the history of warfare but also opens new perspectives on the role of technological innovation within national systems of innovation in their historical context.

Professor Dr. Sönke Neitzel
Chair of War Studies, Department of History
University of Potsdam

Acknowledgements

This scientific publication is the result of an extensive research process that covered both technical and historical aspects. It soon became clear that a purely technological comparison of radar developments in Germany and Great Britain is not sufficient to fully understand the events and their impact during World War II. Therefore, this work focuses on embedding the technical developments in their political and military context to enable a comprehensive and differentiated understanding.

My research took me to important historical sites such as Cornwall, Bawdsey, Malvern, Bletchley Park, and London in the United Kingdom, and Cape Cod in the United States, where Guglielmo Marconi made significant contributions to early wireless communication around 1900.

These visits provided valuable insights into British radar technology and operational innovations during World War II. In addition, numerous discussions with military historians in the United Kingdom, France, and the Netherlands enabled me to reevaluate important events and developments in an international context.

My special thanks go to Dr Phil Judkins from the United Kingdom and Arthur O. Bauer from the Netherlands, whose tireless support and expertise were essential to my work. Their contributions have inspired me time and again.

Dr Phil Judkins, chairman of the Defence Electronics History Society (DEHS) and of the Communications and Electronics Museum Trust, and patron of the Bawdsey Radar Trust, is a leading expert on the history of radar. His expertise and dedication to preserving this important part of military history have been invaluable to my research, while his patient explanations of the political and technical background of British radar development have greatly enhanced the scientific quality of this work.

Arthur O. Bauer, who sadly passed away in February 2025, was one of the most renowned technical and military historians, particularly in the field of radar development. His extensive publications have contributed significantly to the documentation of technical advances in the 20th century. In particular, his website CDVandT.org offers a wealth of original historical documents and publications and serves as an indispensable resource for experts and technology enthusiasts. His unique radio and radar museum near Amsterdam, which houses an extraordinary collection of functioning devices and offers fascinating insights into the development of radar and communication technology, was another highlight of my research. Arthur Bauer's unwavering support and motivational words, especially during difficult phases of this work, have given me the strength to continue my research. This personal support and constant exchange of ideas have contributed significantly to the progress of my work.

I am grateful to Paul Reuvers and Marc Simons from the Crypto Museum in Eindhoven for their spontaneous support, which helped me to make the complex subject of cryptology more accessible.

I would like to thank Diane Clouting and Peter Wain of the Bawdsey Radar Trust for their guidance and insights into the historical background of Bawdsey.

My gratitude also goes to Simon Watts for his invaluable guidance on airborne radar in Great Britain, to Mike Burstow, former Chairman of the Malvern Radar and Technology History Society (MRATHS), for his insightful suggestions, as well as to Yves Blanchard for his contributions from a French perspective.

Finally, I am grateful to all those I have met at radar history events whose kind and open discussions, even on controversial topics, have enriched my understanding.

I would like to express my special gratitude to Wolfgang Schlaffer for his technical and editorial support, which significantly advanced the progress of this work.

I am also grateful to Lukas Trabert and Maik Sühr of Nomos Verlag for their professional guidance during the preparation of this book, and to Ella Shvets, whose careful and competent oversight of the production process was particularly valuable.

Helmut Klausing

Technology and Responsibility

When comparing German and British radar development during World War II, I am struck by how it is often portrayed as a contest between rival scientists and engineers, as if it were a sporting competition. This image highlights creativity and ingenuity, yet it can obscure the human lives that stand behind the technology.

Radar development saved countless lives through the detection of German U-boats and the protection of merchant ships. At the same time, it was used in destructive ways, as seen in the air raids on Hamburg in July 1943. These examples show that technological progress always carries both benefits and human consequences.

I am convinced that technology is never neutral. Every innovation carries ethical responsibility, and decisions about development, deployment, or limitation directly affect lives. Each loss was a personal tragedy; human suffering should never be reduced to numbers or strategic gains.

Through this work, I aim to compare Germany's and Britain's radar advancements while emphasising that technological developments must always be assessed in light of their ethical implications. Behind every triumph are real people, and every life matters.

Ultimately, radar development in World War II is a story of both brilliance and responsibility. To understand history and guide the future, we must remember the human consequences behind every technological achievement.

Table of Contents

1 Introduction

The principle of radar technology[1] was recognised and experimentally proven in several countries almost simultaneously in the 1930s. The first successful attempts to measure reflected electromagnetic waves were made in Germany as early as 1904 by the Düsseldorf engineer Christian Hülsmeyer, who first had the idea of using electromagnetic waves to locate ships. However, there was no targeted further development for practical use and Hülsmeyer's findings were forgotten.

From 1933 onwards Germany intensively promoted the development of radar technology. Once its military potential became apparent, Great Britain, alongside other countries such as France, Italy, Japan, the Netherlands, the Soviet Union and the USA began focusing on the technology in 1934.

In the early years, Germany had a significant technical lead over Great Britain. The key to further progress in the technology, development, and application of radar was the interlocking of political leaders with the military, science, and industry before and during the war in both countries. It turned out that, despite a high level of technological expertise and mature manufacturing capabilities, even a major technical lead would not result in success without a long-term, top-down strategy initiated by the highest levels of government[2] and the military. The present work aims to expand the existing studies on the development and application of radar technology on the German and British side up to the end of World War II by analysing additional influencing factors such

---

1 The term Radar is an acronym from the English language and stands for Radio Detection and Ranging. It was coined by the US Navy in 1940. For reasons of simplification, the term radar is used instead of German Funkmeß or English Radio Direction Finding, except in quotations.

2 In the context of this work, the term government is used as an overarching term to refer to both political institutions and ministries. This simplification serves readability and clarity, without fundamentally sacrificing the differentiation where it is relevant for understanding.

as signals intelligence, radio direction finding, jamming measures, operational research and considering their interaction.

Operational research was unknown in Germany during the war and for a long time afterwards. Today, there is a consensus that these additional factors and their interaction had a major influence on the duration and outcome of World War II and ultimately contributed to the Allies' advantage.

To this end, the respective innovation systems in both countries are analysed in depth.[3] To assess their performance, it is necessary to examine political actions and technological developments in detail, as seemingly insignificant decisions often had a significant impact on the course of the war. Another important factor in comparing the two innovation systems is a thorough analysis of key political events following the signing of the Treaty of Versailles. In particular, British efforts to maintain peace during the interwar period require closer examination within the context of Great Britain's defence capabilities.

Promoters play a key part in driving innovation processes across all fields and levels. In this context, promoters are understood to be individuals who, due to their hierarchical position, expertise or function, actively shape and advance the innovation process, overcoming or circumventing barriers to innovation. Recognising opponents with their specific goals at an early stage requires a constructive approach to overcome these barriers.

This necessitates a closer analysis of both external and internal influence groups, their key players, and their differing interests and objectives. Another critical aspect of the innovation process is identifying innovation steps and barriers. To this end, the activities of key individuals in various roles as promoters or opponents during the period from 1933 to 1945 are analysed. These individuals played very different roles during this period. The promoter model, based on Hauschildt[4], is used to assess their activities in connection with groundbreaking innovation steps and barriers.

---

3 The Austrian economist Joseph Schumpeter (1883–1950) used the term Innovation for the first time in his two-volume work "Business Cycles: A Theoretical, Historical, and Statistical Analysis of the Capitalist Process", published in 1939.

4 Hauschildt et al. 2016.

Due to the vast number of influencing groups and actors in both countries, the present analysis focuses on those involved in decision-making processes related to key research projects in radar development and application from the beginning to the end of the war.[5]

During the course of the study, it became apparent that clear objectives and the consistent overcoming of innovation barriers had a strong impact on the success and duration of the respective projects. Radar development demonstrates how the British side was able to compensate for the initial backlog in research and development through the effective cooperation of key promoters, ultimately surpassing the German technical lead and significantly influencing the course and outcome of the war. The alternation between defence and offence on both sides throughout the war is also discussed.

A notable example of the translation of scientific findings into practical applications was the invention and further development of the magnetron at the University of Birmingham in 1940. With the help of this magnetron, it was possible for the first time on the British side during World War II to install high-resolution radar equipment on high-performance aircraft. From the beginning of 1943 onwards, this led to both the large-scale bombing of German cities and the decisive advantage in submarine warfare in the Atlantic. Close cooperation between Great Britain and the USA as part of the Tizard Mission in 1940 led to significant progress in the development and mass production of key components of radar technology and other essential technologies. In Germany, the establishment of the Arbeitsgemeinschaft Rotterdam in 1943 marked the first successful attempt to catch up with the Allies' technological advancements through trusted cooperation and the unhindered flow of information. However, by the end of the war, these advances no longer played a decisive role.

5 Numerous discussions with German and British historians have highlighted that a certain degree of uncertainty always remains when analysing and evaluating the respective roles, as the descriptions of the activities and biographies of individual actors are often shaped subjectively and remain the subject of controversial debate, particularly in the UK.

Navigating the Chapters

Following the introduction in Chapter 1, Chapter 2 provides a historical overview of the key political, economic, and diplomatic developments that shaped Germany and Great Britain during the interwar period. It begins with the Treaty of Versailles and its impact on Germany, which paved the way for the rise of the National Socialist regime. The chapter then examines German foreign policy and strategic aims towards Great Britain before World War II.

It also addresses Germany's efforts to achieve economic self-sufficiency and expand its military capacity through intensified resource mobilisation and a shift towards a war economy, with a focus on the Air Force, the Navy, and the industrial sectors.

On the British side, the chapter reviews the policy of appeasement, discussing disarmament efforts and diplomatic initiatives aimed at avoiding conflict. It concludes with a critical reflection on the effectiveness of this policy in the years leading up to the war.

Chapter 3 describes the development of radar technology from the early 1930s up to the decisive turning point in 1943 in both countries. Initially, Germany led in radar development, particularly at longer wavelengths, but from around late 1942 research on shorter wavelengths was largely reduced. Britain, early aware of the threat from enemy bombers and the limits of geographic protection, continued to advance and achieved significant breakthroughs. These differences in strategy and support contributed to Germany losing its early advantage.

This development had a major impact on the course of the war and shows how technological progress, combined with political and military decisions, can decisively shape the outcome of a conflict.

Chapter 4 introduces the fundamental principles of radar technology. It begins with an explanation of the radar principle itself and an overview of various radar types, providing the technical foundation necessary to understand the challenges of radar development and its wide-ranging applications. The chapter also covers key performance parameters, highlighting both the decisive advantages and the limitations of this technology, factors that in many historical situations could determine the difference between victory and defeat. This chapter thus lays the groundwork for understanding the technical aspects and practical applications of radar.

Chapter 5 traces the origins of modern radar technology, beginning with the pioneering experiments of Christian Hülsmeyer in May 1904, whose work was largely unrecognised at the time. It highlights key developments in the 1920s in the United States, where experiments with reflected electromagnetic waves provided insights that helped shape the advancement of radar. From 1933 onwards, Germany, Great Britain, other European countries and the United States intensified research and experimentation. The chapter focuses particularly on Germany and Great Britain, which played central roles in the practical and military application of radar. This period marked a shift towards more practical and increasingly military-oriented uses, reflecting the growing strategic importance of radar in the lead-up to and during World War II. Understanding this development is essential to appreciate radar's transformative impact on both technology and warfare.

Chapter 6 addresses the generation of the shortest wavelengths as a decisive technological breakthrough in radar development. The focus is on Germany and Great Britain, detailing the challenges and progress made around 1942 and 1943. The chapter examines the innovative technologies and engineering solutions that enabled these advances, as well as the key scientists and engineers involved. Particular emphasis is placed on the British development of the cavity magnetron, which allowed the production of high-power microwaves at short wavelengths and profoundly influenced the course of the war. The chapter also considers German and British research in centimetre-wave radar within their respective political and military contexts. These developments transformed radar from a primarily defensive tool into a powerful offensive weapon and include a timeline of key events and contributions from other countries.

Chapter 7 examines the development of radar technology in Germany and Great Britain from 1933 until the end of World War II. Germany initially held a technical lead but focused mainly on improving existing systems, while Britain built an integrated air defence network with effective coordination. Over the war, Germany became primarily defensive, and Britain increasingly adopted an offensive stance.

The chapter emphasises cooperation between political leaders, government institutions, the military, scientists, and industry, which was critical for achieving strategic radar objectives. It covers early developments such as the rise of GEMA, Britain's defence committees, and key contributions from the Watson-Watt memorandum and the Daventry Experiment.

It also explores Britain's comprehensive radar system, including the Chain Home network, airborne radar, and key facilities, as well as major disputes and the evolution of radar use across the army, navy, and air force.

Chapter 8 assesses the political influence on the development of radar technology in Germany and Great Britain. It explores how the arms race and the prevailing culture of innovation in both countries shaped the progress of this technology. In Germany, persistent neglect of radar development by political and military leadership until 1943 had a lasting negative impact on its operational capabilities. This part delves deeper into the German political environment and its effects on radar advancement.

By contrast, Great Britain benefitted from close cooperation between political leaders, the military, scientific institutions, and industry, along with the allocation of key human and financial resources, which significantly contributed to its success. The chapter also focuses on the integrated political and military support that underpinned British radar developments.

Chapter 9 describes further technological and operational measures such as signals intelligence, radio direction finding, jamming, and operational research, as well as the interplay between these elements. These factors, alongside radar development, played a vital role in shaping the outcome of World War II.

Signals intelligence, particularly the work carried out at Bletchley Park, was instrumental in decrypting the radio communications of the German armed forces. This breakthrough significantly aided the Allies in securing victory during the Battle of the Atlantic by allowing them to anticipate and counter German naval operations effectively.

The chapter also addresses the Oslo Report as a notable example of human intelligence. This report documents the brave actions of an individual who resisted the Nazi regime by providing crucial information to the Allies, demonstrating the profound impact that human courage and espionage had on the war effort.

Together, these technological and human factors illustrate the multifaceted nature of intelligence and warfare during the conflict.

Chapter 10 is central to understanding the interplay between politics, the military, science, and industry, and the factors that enabled or constrained technological breakthroughs such as radar. It examines the innovation systems in Germany and Great Britain, highlighting the roles of key actors, interest groups, and promoters who either facilitated or hindered progress. The chapter analyses essential innovation steps and barriers to shed light on the political and technical decisions that influenced the course of the war.

Innovation partnerships were crucial for progress in radar development. In 1940, close cooperation between British and American researchers as part of the Tizard Mission led to significant technological advancements and expanded production capacities for radar equipment, which proved decisive in the war against Germany. In Germany, the Arbeitsgemeinschaft Rotterdam, established in 1943, brought together government institutions, the armed forces, industry, and research organisations for targeted collaboration, following recognition of the British advantage in radar development, particularly in the centimetre-wave range.

In the appendices of Chapter 11, a list of biographies of key figures from politics, the military, science, and industry in Germany and Great Britain is provided first. This is followed by descriptions of key pioneers in radio technology, offering background information to deepen the understanding of the different paths that led to radar technology as we know it today. Included in the appendices is a supplementary article titled *Microwave Generation Into A Key Military Technology in World War 2: A Comparative Study*, which explores the historical, technical, and scientific aspects of magnetron technology, drawing on numerous British sources.

Note on Terminology

For reasons of accuracy and historical context, the original designations of ministries, offices, working groups, and official titles are retained in their respective languages throughout this book. To assist the reader, dedicated glossaries have been compiled that explain these terms and provide translations where necessary.

2 Historical Background

An assessment of German and British radar developments before and during World War II from the perspective of the history of innovation requires a detailed analysis of political history, especially concerning the effects of the Treaty of Versailles. German-British relations from 1933 until the outbreak of the war were shaped by two main factors: Great Britain's Appeasement Policy[6] and Hitler's efforts to secure British non-intervention in his aggressive expansion plans to the East. In exchange for tolerating German hegemonic ambitions on the European continent, Hitler sought to respect British interests in overseas territories and sea routes.[7]

The following sections outline key developments that had significant political and economic impacts on the domestic and foreign policies of both countries up to the declaration of war by Great Britain and France on Germany on 3 September 1939.

2.1 Treaty of Versailles

The Paris Peace Conference convened on 18 January 1919, shortly after the Armistice of Compiègne ended active hostilities on 11 November 1918. Its purpose was to negotiate a comprehensive settlement to formally conclude World War I and establish a new international order. Separate treaties were negotiated with each of the defeated Central Powers. Among them, the Treaty of Versailles[8]

6 Appeasement (English: appeasement, reassurance). In international relations, appeasement refers to a policy of restraint, concession, and accommodation towards aggressive states in foreign policy (Schubert und Klein 2021, p. 26). A notable example is the British foreign policy under Prime Minister Neville Chamberlain towards the Third Reich from 1937 to 1939. The connection between British radar development and the policy of appeasement is discussed in Kaiser 1996a, pp. 29–52; Kaiser 1996b, pp. 145–160.

7 Graml 2010, p. 16.

8 RGBl. 1919, Nr. 140, p. 687.

was the principal agreement concerning Germany.[9] The negotiations were mainly influenced by the so-called Big Three, which were France, Great Britain, and the United States. France, having suffered immense destruction during the war, demanded harsh reparations and strong security guarantees. Great Britain aimed to preserve imperial stability and naval supremacy. The United States, which had entered the war relatively late in April 1917 and was considered an associated rather than a full Allied power, played a distinctive role through President Woodrow Wilson's international vision.

Already on 8 January 1918, more than a year before the Paris Peace Conference, Wilson presented his Fourteen Points to the United States Congress, outlining principles for a just and lasting peace.[10] These included open diplomacy, freedom of the seas, reduction of armaments, national self-determination, and, notably, the creation of a League of Nations to ensure peace and security.[11] In the last point of his Fourteen Points, Wilson emphasised:

> "A general association of nations must be formed under specific covenants for the purpose of affording mutual guarantees of political independence and territorial integrity to great and small states alike."[12]

This idea, while central to Wilson's vision, was only partially realised in the final treaty. By late September 1918, the German military leadership recognised that the war was lost.[13] After Prince Max von Baden was appointed Reichskanzler by Kaiser Wilhelm II on 3 October 1918, Germany began a transition from a semi-authoritarian monarchy towards a parliamentary system.[14] On 4 October, the German government sent a diplomatic note to Woodrow Wilson, indicating its willingness to negotiate an armistice based on the principles outlined in his Fourteen Points.[15]

9 MacMillan 2003; Neiberg 2017; Krüger 1993; Boemeke et al. 1998; Schwarz 1973, pp. 11–26; Neitzel 2014, pp. 59–62.

10 MacMillan 2003, pp. 495–496; Schönbrunn 1979, pp. 104–106.

11 Kraus 2014, p. 17.

12 MacMillan 2003, p. 496.

13 Kraus 2014, p. 16.

14 Ibid., p. 18.

15 Ibid., pp. 16–17.

Wilson initially hesitated and demanded internal political reforms in Germany as a prerequisite. It was not until 5 November that Germany was officially informed that the Allies were prepared to negotiate an armistice as well as to enter into subsequent peace negotiations aimed at formally ending the war.[16]

The Armistice of Compiègne was signed on 11 November 1918 by Matthias Erzberger[17], head of the German Armistice Commission, marking the end of active combat in World War I. Two days earlier, revolutionary developments in Germany had led to the abdication of Kaiser Wilhelm II on 9 November and the proclamation of a republic. On the same day, Friedrich Ebert[18], a Social Democrat, was entrusted with the leadership of the Reich government by Prince Max of Baden.

This provisional government laid the foundation for the Weimarer Republik, which was formally established with the adoption of the Weimar Constitution in August 1919.

The Paris Peace Conference, chaired by French Prime Minister Georges Clemenceau and lasting from January 1919 to January 1920, brought together the principal Allied leaders and led to the founding of the League of Nations, which was formally established on 10 January 1920, the same day the Treaty of Versailles came into force.[19]

The German delegation was presented with the draft treaty on 7 May 1919 but was excluded from the negotiations.[20] After Germany's requests for revisions were rejected, the final version of the treaty was delivered on 16 June and signed under protest by German representatives on 28 June 1919 in the Hall of Mirrors at Versailles. In total, 27 Allied and Associated Powers signed the treaty.[21] The Treaty of Versailles consisted of 15 parts and 440 articles. It imposed harsh conditions on Germany, including significant territorial losses, strict military limitations, and the obligation to pay reparations. Germany lost 13 percent of its territory and about 10

16 Ibid., p. 19.

17 Epstein 1976.

18 Mühlhausen 2025.

19 RGBl. 1920, Nr. 35, p. 237.

20 Schwarz 1973, pp. 12–13.

21 Kraus 2014, pp. 25–27; Schwarz 1973, p. 14.

percent of its population, including all overseas colonies.[22] The treaty also prohibited unification with Austria, abolished conscription, limited the army to 100,000 men and the navy to 15,000. The air force was disbanded, and armaments were strictly regulated, with many weapons completely banned.[23] In addition to these military limitations, Germany lost resource-rich areas such as Alsace-Lorraine to France, parts of Upper Silesia to Poland, and the Saar Basin was placed under League of Nations administration, all of which significantly weakened its industry.

The Danube was placed under international control to ensure free navigation via the Danube Commission, whose authority was strengthened after 1919, while the Rhine, Elbe, and Oder remained under German sovereignty but were regulated internationally. The Memel Territory was similarly administered by the League of Nations before being annexed by Lithuania in 1923. Furthermore, Germany was required to hand over large parts of its merchant fleet and railway equipment as reparations, significantly hampering its economic recovery and capacity for international trade.

A central and highly controversial element of the treaty was Article 231, the so-called Kriegsschuldparagraph. It stated:

> „Die alliierten und assoziierten Regierungen erklären, und Deutschland erkennt an, daß Deutschland und seine Verbündeten als Urheber für alle Verluste und Schäden verantwortlich sind, die die alliierten und assoziierten Regierungen und ihre Staatsangehörigen infolge des ihnen durch den Angriff Deutschlands und seiner Verbündeten aufgezwungenen Krieges erlitten haben.“[24]

This clause served as the legal basis for demanding reparations and became a symbol of national humiliation in Germany. It triggered widespread resentment and was instrumentalised by nationalist and anti-democratic forces, who portrayed the treaty as a betrayal and the assignment of war guilt as unjust.

22 Sturm 2003, p. 19; Piper 2021, p. 21; Schwarz 1973, pp. 14–17.

23 MacMillan 2003, p. 176; Elz 2008, p. 31.

24 RGBl. 1919, Nr. 140, p. 985. "The Allied and Associated Governments affirm, and Germany accepts the responsibility of Germany and her allies for causing all the loss and damage to which the Allied and Associated Governments and their nationals have been subjected as a consequence of the war imposed upon them by the aggression of Germany and her allies." (translated by the author).

The harsh terms of the treaty caused political instability and social unrest. Right-wing groups condemned the treaty as a Diktat[25], or imposed peace. Many Germans, having been exposed to years of victory propaganda, were unprepared for defeat and its consequences. The economic burden of reparations, high inflation, and growing unemployment further deepened dissatisfaction.

As a result, there was little room for critical engagement with Germany's role in the war. Instead, the myth of the Dolchstoßlegende[26] gained traction, accusing democratic politicians and revolutionaries of betraying the supposedly undefeated army. Nevertheless, Germany gradually regained diplomatic recognition. Following diplomatic efforts and changing international attitudes, Germany was admitted to the League of Nations on 8 September 1926; the formal ceremony and an address by Gustav Stresemann[27] who had briefly served as Reichskanzler in 1923 before becoming Außenminister took place two days later, on 10 September.[28]

However, the fragile Weimarer Republik soon faced increasing political polarisation and the delegitimisation of democratic institutions. This climate contributed to the rise of radical political movements, including right-wing extremism and antisemitism. The Republik was further destabilised between 1930 and 1933, paving the way for the Nazi seizure of power.[29]

In the context of the global economic crisis, the Lausanne Conference of 1932 effectively suspended German reparation payments, although by then they had already been significantly reduced through earlier agreements. Nevertheless, the political damage caused by the treaty had already taken root. While the treaty aimed to ensure lasting peace, it ultimately sowed the seeds of future conflict.

---

25 The term Diktat refers to a peace settlement imposed without negotiation or the consent of the defeated party. In post-World War I Germany, it became a symbol of national humiliation and resentment towards the Treaty of Versailles.

26 The Dolchstoßlegende was a false narrative claiming that Germany was not defeated on the battlefield but betrayed from within by politicians and civilians. It was later used by nationalist groups to delegitimise the Weimarer Republik.

27 Eschenburg und Frank-Planitz 1978.

28 Schwarz 1973, pp. 123–124.

29 Wildt 2012, pp. 22–27; Thamer 2003, pp. 5–19.

2.2 The Rise of the National Socialists in Germany

With the appointment of Adolf Hitler[30] as Reichskanzler by Reichspräsident Paul von Hindenburg on 30 January 1933, Hitler assumed leadership of a coalition government consisting primarily of the NSDAP[31], the DNVP[32], and other conservative forces. This event was later stylised by National Socialist propaganda as the Tag der Machtergreifung[33], although it was in fact a constitutionally legitimate appointment conducted within the legal framework of the Weimar Constitution, one that nonetheless paved the way for the rapid dismantling of the democratic system. Upon taking office, Hitler swore the traditional oath without explicit reference to the constitution. In the early days of his chancellorship, he sought to present the change in government not as a revolutionary seizure of power, but as a national renewal under the leadership of Reichspräsident Hindenburg. Nevertheless, Hitler quickly moved to consolidate his authority. Through the systematic weakening of both Hindenburg's position and the constitutional order of the Weimarer Republik, he laid the foundations for the establishment of a dictatorship. The actual seizure of power, that is the transformation of the Weimarer Republik into a one-party state and ultimately into a totalitarian regime under Hitler's absolute control, took place in the weeks and months following his appointment.

On 1 February 1933, Hitler succeeded in having the Reichstag dissolved by Reichspräsident Hindenburg in accordance with Article 25 of the Weimar Constitution. He justified this step by claiming that there was no functioning majority in the Reichstag, despite the presence of parliamentary opposition.

New elections were subsequently scheduled for 5 March.[34] Hitler's entire foreign and domestic policy, including his economic

---

30 Kershaw 1998; Kershaw 2000; Kershaw 2009; Pyta 2024.

31 NSDAP: Nationalsozialistische Deutsche Arbeiterpartei.

32 DNVP: Deutschnationale Volkspartei.

33 According to Norbert Frei, the formation of the government was the result of negotiations between Hitler, Franz von Papen and Alfred Hugenberg. Both were prepared to join a cabinet and wanted Hitler at the head of the cabinet. In this respect, according to Frei, it was not a seizure of power, but rather a transfer of power (Frei 1983, pp. 136–145; Hildebrand 2009, pp. 1–2).

34 Hildebrand 2009, pp. 2–3; Tofahrn 2003, pp. 20–21; Hofer 1979, pp. 16–17.

measures and cooperation with the Reichswehr, aimed at strengthening Germany's political and military power. The Reichswehr played a central role in these plans. Hitler's demands for rearmament and revision of the military provisions of the Treaty of Versailles were well received by the military leadership, as the Reichswehr had already pursued secret rearmament programmes in 1928, 1929, and 1932 with the goal of initiating negotiations to revise the treaty's military clauses.[35]

On 3 February 1933, just four days after his appointment as Reichskanzler, Hitler delivered a secret policy address to the commanders of the Reichswehr.[36] In this speech, he outlined the core elements of his domestic and foreign agenda: the elimination of parliamentary democracy, the suppression of political opposition, the revision of the Treaty of Versailles, and above all the reintroduction of universal conscription as a prerequisite for rearmament.

Hitler used the period leading up to the new elections on 5 March 1933 to pursue his goal of securing an absolute majority for the NSDAP. This was achieved through terror against political opponents, the use of emergency decrees, and the gradual suspension of basic constitutional rights. Despite widespread terror and demagogic propaganda, the NSDAP failed to secure an absolute majority in the German Reichstag, which was only made possible through the mandates of the DNVP.[37]

A few days prior to the Reichstag election, on 28 February 1933, the Verordnung des Reichspräsidenten zum Schutz von Volk und Staat, commonly referred to as the Reichstagsbrandverordnung, was promulgated.[38] This decree marked the beginning of the systematic suspension of fundamental civil liberties in the Weimarer Republik. The Reichstag fire of 27 February 1933 served as a justification for this emergency measure, which effectively nullified key

35 Thamer 2004, p. 32.

36 Hitler's two-and-a-half-hour speech was summarised in bullet points by Lieutenant General Curt Liebmann, in the so-called Liebmann-Aufzeichnung; Hofer 2011b, pp. 180–181; Wirsching 2001, pp. 517–550; Michalka 2002, pp. 17–18; Studt 2002, p. 10.

37 Hofer 1979, pp. 18.

38 RGBl. 1933, Teil I, Nr. 17, p. 83; Raithel und Strenge 2000, pp. 413–460; Hofer 2011b, pp. 55–56; Hildebrand 2009, p. 3; Tofahrn 2003, pp. 19–20.

constitutional rights, including the freedoms of expression, assembly, and the press. Subsequently, on 24 March 1933, the Gesetz zur Behebung der Not von Volk und Reich, commonly referred to as the Ermächtigungsgesetz, was enacted.[39] This law granted the government authority to enact legislation without parliamentary consent, even if such laws contravened the constitution. It thus abolished the separation of powers, suspended democratic oversight, and dismantled the constitutional foundations of the Weimarer Republik. Together, the Reichstagsbrandverordnung and the Ermächtigungsgesetz legitimised the National Socialist dictatorship and paved the way for a totalitarian regime.

On 7 April 1933, the Gesetz zur Wiederherstellung des Berufsbeamtentums was passed.[40] This law included the Arierparagraph, which excluded Jews and other undesirable civil servants, such as political leftists, from the civil service or forced them into retirement. Many of those dismissed were replaced with individuals loyal to the Nazi regime. This policy also had a profound impact on universities, where a large number of professors lost their positions. The student body was similarly subjected to alignment with National Socialist ideology.

On 14 July, the Gesetz gegen die Neubildung von Parteien was enacted, effectively eliminating democratic parliamentarianism in the German Reich.[41]

Following the death of Reichspräsident Hindenburg on 2 August 1934, Adolf Hitler immediately merged the office of the Reichspräsident with that of the Reichskanzler, assuming the combined title of Führer und Reichskanzler. This centralisation of executive authority was subsequently legitimised by a referendum held on 19 August 1934, which reported an approval rate of approximately 90 percent. However, the plebiscite took place under conditions

39 RGBl. 1933, Teil I, Nr. 25, p. 141; Hofer 2011b, pp. 58–60; Hildebrand 2009, pp. 4–5; Tofahrn 2003, p. 22.

40 RGBl. 1933, Teil I, Nr. 34, pp. 175–177; Hildebrand 2009, p. 3; Tofahrn 2003, p. 24.

41 RGBl. 1933, Teil I, Nr. 81, p. 479; Hofer 2011b, p. 62; Tofahrn 2003, p. 27.

shaped by intense National Socialist propaganda and widespread intimidation.[42]

This consolidation of power was codified in the Gesetz über das Staatsoberhaupt des Deutschen Reichs, enacted on 1 August 1934, which further entrenched the totalitarian nature of Hitler's regime.[43] On the same day, the Reichswehr swore a personal oath of loyalty to Hitler, who now held supreme command authority over the Reichswehr. Notably, this oath was sworn not to the constitution or to the German state, but directly to Hitler himself, demanding unconditional loyalty and obedience. Subsequently, on 22 August 1934, the German government enacted legislation requiring all civil servants, including university professors, to swear an oath of loyalty to Hitler personally.[44] This development drew international attention and concern.

European countries and the United States expressed alarm at the open persecution of communists and social democrats, the increasing oppression of Jewish populations, and the regime's anticlerical policies. In Britain, public opinion became increasingly critical of the German government's actions; on 27 April 1933, German ambassador Leopold von Hoesch reported that Germany had lost an extraordinary amount of sympathy in recent weeks.[45]

By late 1934, Hitler's domestic political objectives and authoritarian course were clearly evident. At the same time, the Nazi regime sought to overcome Germany's international isolation, which had resulted from its role in World War I and the Treaty of Versailles. Although open military conflict was not yet imminent, the regime began to pursue a more assertive foreign policy aimed at revising the postwar order.

In pursuing this aim, Hitler relied heavily on propaganda and sought to exploit the European powers' desire for peace. In this context, he delivered his so-called Friedensrede to the Reich-

42 Tofahrn 2003, p. 36; on 18 July, the British government under Prime Minister MacDonald decided to increase the number of Royal Air Force squadrons from 94 to 131.

43 RGBl. 1934, Teil I, Nr. 89, p. 747; Hofer 2011b, p. 72; Tofahrn 2003, p. 36.

44 Tofahrn 2003, pp. 36–37.

45 Hoesch von (1933), ADAP 1971, Serie C, Band I,1, Nr. 193, pp. 351–352; Hauser 1972, pp. 261–265.

stag on 17 May 1933, which was earlier than 1934, portraying National Socialism as a movement committed to peace.[46] The speech was primarily aimed at reassuring foreign governments and shaping international opinion, while concealing the regime's ongoing, largely secret rearmament efforts and repressive domestic policies. This speech, his first major government address on foreign policy, served as a calculated propaganda measure, especially aimed at Great Britain. For the first time, Hitler transferred his techniques of demagogic rhetoric from domestic policy to foreign affairs, effectively replacing traditional diplomacy with ideological messaging. He later claimed that only by persistently emphasising Germany's peaceful intentions was he able to carry out the military rearmament necessary for the next stage of his plans.

As part of Germany's open defiance of the Treaty of Versailles and its remilitarisation efforts, the Gesetz für den Aufbau der Wehrmacht, enacted on 16 March 1935, laid the foundation for transforming and expanding the Reichswehr into the Wehrmacht.[47] Public support for these measures was strong, as the Treaty of Versailles was widely regarded by the German population as unjust and humiliating. Even parts of the victorious powers viewed this sentiment with a degree of understanding.

To complement this structural reorganisation, the Wehrgesetz, promulgated on 21 May 1935, reintroduced general conscription for German men and regulated military service obligations, while women were exempt from conscription but could serve in auxiliary roles. Meanwhile, on the same day, the Reichsverteidigungsgesetz was passed, granting the government extensive powers to prioritise resources and coordinate arms production. Although publicly announced, many details of the economic mobilisation were carried out discreetly to avoid scrutiny. Behind the scenes, the economy

---

46 Reichstagsprotokoll 1933, 3. Sitzung, pp. 47–54; Studt 2002, p. 19; Frei 2013, pp. 80–81.

47 The Gesetz für den Aufbau der Wehrmacht, which reintroduced compulsory military service and formally established the Wehrmacht. This marked a clear violation of the Treaty of Versailles, which had limited Germany's army to 100,000 men and prohibited conscription, an air force, and a general staff (RGBl. 1935, Teil I, Nr. 28, p. 375); Tofahrn 2003, p. 40; Studt 2002, p. 48; Hildebrand 2009, p. 22).

had already been committed to arms production under this law, ensuring the rapid militarisation of German industry.[48]

In a second major foreign policy speech on 21 May 1935, often referred to as Hitler's second Friedensrede, he once again emphasised Germany's desire for peace and promised to respect international treaties provided they were concluded voluntarily.[49] The speech was intended to reassure the Western powers, particularly Britain and France, of Germany's peaceful intentions even as rearmament efforts continued unabated behind the scenes. Delivered shortly after the passage of the Wehrgesetz and Reichsverteidigungsgesetz, the speech marked another calculated use of propaganda to mask the regime's true objectives.

In a further address on 7 March 1936, Hitler once again proposed non-aggression pacts with the Western nations. This offer served as a diplomatic tactic to reduce concerns about Germany's rearmament and territorial ambitions even while those efforts continued without interruption.[50]

2.3 German Foreign Policy up to the Outbreak of World War II

Already during the Weimarer Republik, efforts were undertaken both to alleviate Germany from the obligations imposed by the Treaty of Versailles and subsequently to integrate the country into the League of Nations, thereby reducing its international isolation.

Through a series of primarily bilateral treaties, attempts were made to improve relations with the Western European powers, par-

48 Tofahrn 2003, p. 42; Studt 2002, p. 50.

49 Reactions abroad to the speech on 21 May were mixed. In Britain and France in particular, the speech was interpreted very differently. Britain showed understanding for Germany's need to be a free, equal, and strong nation rather than a people subjugated by the dictates of Versailles (The Times, 22 May 1935). In contrast, in France, the speech was understood as an expression of Hitler's policy outlined in Mein Kampf, aiming at political and military supremacy (Le Journal des Débats politiques et littéraires, 23 May 1935). Hitler's speech on 21 May 1935 is one of the best examples of the art of misleading and deceiving (Hofer 1979, p. 48).

50 Hofer 1979, p. 53.

ticularly France. The conclusion of these agreements, under the leadership of Stresemann, significantly contributed to restoring the Weimarer Republik's international standing.

Moreover, Germany's decision to join the League of Nations marked a crucial step in overcoming the division between victors and vanquished of World War I, fostering a climate of détente. Germany joined the League of Nations on 8 September 1926, gaining a permanent seat on the Council. Two days later, on 10 September, Außenminister Stresemann delivered a speech to the League's Assembly in Geneva, outlining German foreign policy and emphasising the significance of the country's admission.[51] Clarifying the reparations issue arising from the Treaty of Versailles was crucial for German foreign policy, as it constrained Germany's economic flexibility and became a point of contention for forces hostile to the Republic.

The period from 1933 until the declaration of war by Great Britain and France on 3 September 1939 was marked by several bilateral agreements between Germany and various partners. Hitler's strategy in concluding these treaties was based on the assumption that the consequences of violating bilateral agreements would be less severe or more predictable than breaching multilateral treaties involving multiple states.

A significant milestone in Germany's rearmament occurred on 26 February 1935, when the Reichskabinett was formally informed of plans to establish the Luftwaffe[52] as the third branch of the German armed forces. This decision was officially communicated to foreign governments on 9 March. Shortly thereafter, on 16 March, the introduction of general conscription was announced, and Adolf Hitler unequivocally repudiated the military restrictions imposed by the Treaty of Versailles.[53]

On 25 and 26 March 1935, British Foreign Secretary John Simon and Lord Privy Seal Anthony Eden conducted an extensive exchange of views with Hitler. Their visit was widely interpreted as

51 Stresemann 1926.

52 The term Luftwaffe refers to the German Air Force, established in 1935 as a crucial branch of the Wehrmacht.

53 Churchill 2005, pp. 128–131; Churchill 1954, pp. 179–182; Tofahrn 2003, p. 40; Studt 2002, p. 48; Hofer 1979, p. 47.

indicative of a shift in British policy, signalling that Great Britain no longer regarded itself as bound to enforce the military limitations stipulated by the Treaty of Versailles.[54] This visit can be interpreted as an early example of the appeasement policy, which was later criticised as a sign of blatant weakness on the part of the British government. At this time, however, Germany remained largely isolated in terms of foreign policy due to the consequences of its previous actions.

On 18 June 1935, the Anglo-German Naval Agreement was signed, aiming to regulate the naval forces of Great Britain and Germany. The core provision stipulated that the size of the German navy could not exceed 35 per cent of the British Royal Navy. Initially, the German submarine fleet was limited to 45 per cent of British submarine strength, but this limit was later raised to 100 per cent following further negotiations.[55]

This agreement represented a clear departure from the Treaty of Versailles, which had imposed strict limitations on the German armed forces after World War I. By permitting Germany to expand its navy, the agreement played a key role in Anglo-German relations in the pre-war period. While both sides viewed the agreement as an attempt at détente and mutual understanding, it ultimately enabled Germany to further strengthen its military presence at sea, contributing to rising tensions in Europe.

Great Britain negotiated with Germany without involving the League of Nations or its French ally, aiming instead to secure strategic room for manoeuvre to protect its interests as a colonial power. In doing so, however, the naval limitations imposed by the Treaty of Versailles were effectively undermined. Meanwhile, Hitler continued his dual strategy of revision and expansion, despite facing domestic political resistance.

Hitler's territorial ambitions in Eastern Europe were outlined on 5 November 1937 during a secret meeting with the commanders-in-chief of the German Wehrmacht and the Foreign Secretary. The content of this meeting was later recorded in the document known as the Hoßbach-Protokoll. This memorandum clearly re-

54 Graml 2010, p. 9.

55 Deutsch-Britisches Flottenabkommen; ADAP 1975, Serie C, Band IV,1, Nr. 156, pp. 315–320; Dönitz 2018, pp. 14–15; Deist 1989, p. 547; Kershaw 2009, p. 365.

vealed Hitler's plans for a foreign policy based on violent expansion and is considered a key source on the prehistory of World War II.[56]

The realisation of Hitler's aims was facilitated by the attitude of Great Britain, which regarded a potential conflict with Japan as the main threat to the British Empire. As a result, Prime Minister Neville Chamberlain and his government viewed a policy of conciliation with Germany as a pragmatic course of action. Britain sought an understanding with Germany and was willing to support Hitler's demands for a revision of the Treaty of Versailles, particularly on points it considered justified. This policy of appeasement came to define Anglo-German relations until Britain and France declared war on Germany in 1939.[57]

On 12 March 1938, German troops marched into Austria.[58] The following day, the Gesetz über die Wiedervereinigung Österreichs mit dem Deutschen Reich, commonly referred to as the Anschluss, was enacted.[59] Shortly afterward, the Austrian army was incorporated into the Wehrmacht. In a referendum held on 10 April, a large majority of Austrians and Germans voted in favour of Austria's annexation to the German Reich.[60] This vote, held under National Socialist control, served to legitimise the annexation after the fact. The incorporation of Austria's institutions was part of the broader National Socialist plan to bring all state, social and political structures in Austria into line following the Anschluss.

In the context of the Sudetenkrise, Hitler began to implement his plans to dismantle Czechoslovakia. He used the demands of the approximately 3.5 million Sudeten Germans to justify the incor-

---

56 The meeting was recorded in key points and essential sentences by Colonel Friedrich Hoßbach, who was serving as Hitler's military adjutant. A few days later, he compiled a written account that detailed Hitler's speech and summarised the subsequent discussion. In this sense, the Hoßbach document is not a protocol in the strict sense, but rather a transcript or memorandum (ADAP 1950, Serie D, Band I, Nr. 19, pp. 25–32; Hoßbach 1937; Hoßbach 1965, pp. 181–192; Hofer 2011b, pp. 193–196; Tofahrn 2003, p. 57; Bussmann 1968, pp. 373–384).

57 Kaiser 1996a, pp. 29–52; Kaiser 1996b, pp. 145–160.

58 Tofahrn 2003, p. 59; Hildebrand 2009, p. 38; Hofer 2011b, pp. 197–199; Hofer 1979, p. 138.

59 Hofer 2011b, pp. 198–199; Studt 2002, p. 79.

60 Tofahrn 2003, p. 60; Thamer 2004, p. 40.

poration of the predominantly German-speaking border regions of Czechoslovakia into the German Reich. These demands were deliberately manipulated to exert political pressure on the Czechoslovak government and to gain international support for the annexation.[61] As tensions escalated and international concern grew, diplomatic efforts were launched to resolve the crisis peacefully. After two meetings between Chamberlain and Hitler, the first on 15 September 1938 in Berchtesgaden and the second from 22 to 24 September in Bad Godesberg, a conference of the major powers took place in Munich on 29 September 1938 at the initiative of Chamberlain and Mussolini.[62]

In the Münchener Abkommen, signed during the night of 29 to 30 September 1938 by Adolf Hitler, Neville Chamberlain, Benito Mussolini, and Édouard Daladier, Germany, Great Britain, Italy, and France agreed on the gradual evacuation of the Sudeten German territories by Czechoslovakia between 1 and 10 October, and their subsequent cession to the German Reich[63]. The agreement was perceived by the Czechoslovak government as a betrayal. For Hitler, it marked the peak of his popularity and further cleared the path towards war, while Chamberlain, still committed to his policy of appeasement, remained convinced that he had secured lasting peace.[64]

Hitler's next goal, the complete destruction of Czechoslovakia, was realised with the invasion of Bohemia and Moravia by German troops on 15 March 1939. On 16 March Hitler proclaimed the establishment of the Reichsprotektorat Böhmen und Mähren[65], effectively dissolving Czechoslovakia. Bohemia and Moravia became part of the German Reich as the Protectorate of Bohemia and Moravia, while Slovakia was set up as a formally independent state, though it remained under German control.[66] On 22 March Nazi Germany took control of the Memel Territory, which belonged to Lithuania, under pressure and the threat of military force.[67] Lithua-

61 Hofer 1979, pp. 141–143; Thamer 2004, p. 41.

62 Studt 2002, pp. 87–88.

63 Munich Agreement; Hofer 2011a, pp. 205–207; Studt 2002, p. 88; Thamer 2004, pp. 42–43; Hildebrand 2009, p. 40; Hofer 1979, pp. 145–146.

64 Chamberlain 1999a, p. 170.

65 Bohemia and Moravia.

66 Studt 2002, pp. 98–99; Hofer 2011b, pp. 223–225.

67 Studt 2002, pp. 99–100; Hofer 1979, p. 158.

nia was forced to cede the territory, which was effectively an annexation, although it was not the same formal incorporation as Austria or the Sudetenland.

In response to the failure of the appeasement policy, Great Britain issued a declaration on 31 March 1939, guaranteeing Poland's independence and promising to support the country in the event of an attack.[68] A few days later, France joined this guarantee.

On 7 April 1939, Italy, under Mussolini, invaded Albania. In reaction to this act of aggression and the rising tensions in Europe, Great Britain and France issued separate declarations on 13 April, guaranteeing the independence of Greece and Romania.[69] On 28 April 1939, Adolf Hitler unilaterally declared the termination of both the German-Polish Non-Aggression Pact, concluded on 26 January 1934, and the Anglo-German Naval Agreement, signed on 18 June 1935, thereby marking a decisive turning point in the Third Reich's foreign policy towards open confrontation and the abandonment of diplomatic constraint.[70] A declaration of guarantee was also issued by Great Britain to Turkey on 12 May, followed by a similar declaration from France on 23 June.[71]

On 1 September 1939, Nazi Germany invaded Poland. When the ultimatum issued by Great Britain and France demanding the immediate withdrawal of German forces expired, both countries declared war on Germany on 3 September 1939.[72]

2.4 Hitler's Diplomatic Strategy Towards Great Britain

Since coming to power in 1933, Adolf Hitler actively sought to win over Great Britain often referred to by him as England either as an ally or at least as a neutral power. His primary objective was to secure strategic freedom for his territorial ambitions in the East and to avoid the danger of a two-front war. Hitler regarded Great Britain as a strategically vital country respecting its global

---

68 Hildebrand 2009, pp. 46–47; Studt 2002, pp. 100–101.

69 Studt 2002, p. 102.

70 Ibid., p. 103.

71 Studt 2002, p. 104; Hofer 1979, p. 161.

72 Studt 2002, p. 113; Frei 2009, pp. 3–5; Michaelis 1979, pp. 3–7.

supremacy and viewing its support or neutrality as essential for the success of his war plans.

These efforts manifested in diplomatic initiatives aimed at rapprochement including agreements such as the Anglo German Naval Agreement of 18 June 1935. However British scepticism toward Nazi Germany grew steadily particularly following German violations of the Treaty of Versailles and aggressive actions in Europe.

Until the war against Great Britain, Hitler's aims regarding his relationship with the country shaped by both power political and ideological considerations were characterised by the concepts of cooperation, neutrality, and, finally, opposition.[73]

He considered a war against Great Britain only as a last resort an emergency measure necessary to secure the military freedom of action required for his Eastern ambitions. Should attempts at understanding fail Hitler contemplated an early military campaign against the Soviet Union to achieve decisive victories in the East thereby isolating Great Britain and compelling it to capitulate.

Already in the 1920s Hitler openly expressed admiration for the British Empire emphasising what he perceived as a shared Aryan racial heritage and envisioning a global order dominated by Aryan powers including Britain and Germany. He aimed to avoid conflicts in regions of British interest believing that Britain would tolerate German territorial expansion as long as its own imperial possessions remained unaffected. Moreover, Hitler assumed that a rapprochement with Great Britain would weaken France's position in Europe and further isolate his western rival. In addition to his British policy Hitler sought friendship with Italy recognising its crucial role in facilitating the annexation of Austria into the German Reich. Austria's geographic location made it a strategic base for Hitler's hegemonic ambitions toward Eastern and Southeastern Europe. Nevertheless, without Italy's consent to the Anschluss Hitler considered it impossible to realise his goals.[74]

73 Graml 2010, pp. 33–68; Henke 1973, pp. 35–108; Hauser 1982, pp. 42–67.
74 Graml 2010, pp. 19–20.

2.5 Hitler's Autarky and Rearmament Policies

2.5.1 Economic Mobilisation

2.5.1.1 Vierjahresplan

An important step in preparing for the coming war was the implementation of the Vierjahresplan.[75] Although a first phase of economic measures aimed at reducing unemployment had been pursued earlier, the preliminary objectives of what would later be formalised as the Vierjahresplan were announced during the Reichsparteitag der Ehre in Nuremberg on 9 September 1936.[76] This initial presentation emphasised Germany's aim to achieve war readiness within four years by fostering autarky in key sectors such as raw materials and food production, alongside a substantial expansion of armaments manufacturing. The Vierjahresplan was subsequently formally enacted and endowed with legal authority through the publication of the Verordnung zur Durchführung des Vierjahresplanes on 18 October 1936.[77] Responsibility for implementing the plan was transferred from Hitler to Hermann Göring, who was thereby granted extensive powers over economic and armaments planning. In its initial phase up to 1938, economic control primarily rested with military personnel from the Reichsluftfahrtministerium (RLM) and prominent representatives of key industries. Notably, engineers were excluded from decision-making processes at all levels.[78] During this phase, the focus was on securing independence from foreign raw materials to prepare Germany economically and strategically for a future war.

After 1938, emphasis increasingly shifted towards direct war preparations, leading to accelerated rearmament and the relative deprioritisation of self-sufficiency policies. Despite these changes, Göring retained centralised control over economic and armaments planning. In addition to securing raw materials, the plan sought self-sufficiency in food production to reduce Germany's depen-

75 Petzina 1968; Benz et al. 2001, pp. 782–783.

76 Hofer 1979, p. 131; Bührer 2001, pp. 114–119; Tofahrn 2003, p. 52.

77 RGBl. 1936, Teil I, Nr. 96, p. 887; Hofer 1979, p. 131; Studt 2002, p. 65.

78 Ludwig 1974, pp. 160–161.

dence on imports in the event of war.[79] Earlier, a secret memorandum drafted by Hitler in August 1936 had marked a significant turning point, underscoring the regime's explicit shift towards rearmament and military readiness. Once the global economic crisis had essentially been overcome, Hitler's focus was clearly on preparing for a war of aggression within a fixed four-year timeframe. Civilian needs were subordinated to military priorities throughout this period.[80]

In August 1936, Hitler issued a secret memorandum outlining the objectives of the Vierjahresplan, in which he set the following tasks:

> „I. Die deutsche Armee muß in 4 Jahren einsatzfähig sein.
> II. Die deutsche Wirtschaft muß in 4 Jahren kriegsfähig sein.“[81]

As Beauftragter für den Vierjahresplan, Hermann Göring was granted extensive authority, which he exercised from the Prussian State Ministry. At a Council of Ministers meeting on the implementation of the Vierjahresplan on 4 September, Göring concluded his remarks on Hitler's memorandum with the following statement:

> „...Alle Maßnahmen haben so zu erfolgen, als ob wir uns im Stadium der drohenden Kriegsgefahr befänden.“[82]

Furthermore, on 11 October Rudolf Hess appealed to the population to accept supply shortages and to adopt wartime restraint.[83]

It should be noted that Germany's dependence on foreign countries for essential raw materials remained very high, including

79 Bührer 2001, p. 114; Hofer 1979, p. 131.

80 Hitler 1936, p. 31; Hofer 2011b, pp. 84–86; Weinberg 1988, pp. 133–135; Treue 1955, pp. 184–210; Hofer 1979, pp. 131–132; Michalka 1985, pp. 188–190; Studt 2002, p. 64; this secret memorandum is one of the key documents of the Third Reich.

81 "I. The German army must be operational within four years. II. The German economy must be ready for war within 4 years." (translated by the author); Hitler 1936, p. 31.

82 Hofer 1979, p. 132; "...All measures are to be carried out as if we were in a state of imminent danger of war." (translated by the author).

83 Michalka 1985, pp. 191–192.

iron ore, mineral oils, rubber, textiles, and leather. The annexation of Austria in March 1938 and the occupation of Czechoslovakia in March 1939 provided some relief, as did the swift military victories during the early years of the war.[84]

From a military rather than purely economic perspective, the implementation of the Vierjahresplan shifted after about two years to focus increasingly on preparations for the coming war. Although various competing directives, plans, and priorities existed at the war's outset, the Vierjahresplan continued to function remarkably well.[85]

In 1940, Hitler extended Göring's assignment for the Vierjahresplan by another four years, which by then also encompassed the occupied territories.[86]

Due to the initial military successes, there was no urgent need for a radical shift toward a full war economy. Between 1938 and 1942, industrial production saw only slight growth compared to the steep increase in armaments output, although not all potential was fully exploited.

Thus, at the beginning of the war, a complete mobilisation of all economic resources had not yet been achieved. Nevertheless, the economy was ultimately ill-prepared for a prolonged war.[87]

2.5.1.2 War Economy

Although the German economy was officially in a state of war on 3 September 1939, historians still have doubts as to whether the term war economy applies to the first two years of the war. Only from 1942 onwards, with Albert Speer[88] as Rüstungsminister, can we speak of a war economy. Although a preparatory defence economy was in place before the start of the war, conditions after the war began were very similar to those in peacetime.[89]

As a comprehensive analysis of armaments production during World War II cannot be carried out within the scope of this work,

84 Hofer 1979, p. 132.
85 Ibid., p. 133.
86 Benz et al. 2001, pp. 782–783.
87 Bührer 2001, p. 119.
88 Herbst 2010; Speer 2013; Klee 2011, p. 590.
89 Müller 1988, p. 349.

and against the background of ongoing discussions on the so-called Rüstungswunder[90], the following section briefly examines the considerable growth in labour productivity in the armaments industry from the time Speer took office in 1942 until 1944.

In April 1942, shortly after Speer took office, an institution of central planning was established. To this end, the so-called Planungsamt was established on 4 September 1943 within the Rüstungsministerium, acting as the executive arm of the newly formed Central Planning authority.[91] Its aim was to obtain a realistic overview of the available resources and to identify impending bottlenecks at an early stage. The Planungsamt was tasked with specifically controlling and monitoring individual programmes and priorities in order to prepare decisions by the central planning department on tasks essential to the war effort.[92]

The figures given for 1943/44 are based on a report on Germany's economic situation 1943/44[93] from the Reichsministerium für Rüstung und Kriegsproduktion, which was distributed in hundreds of copies to Reich ministries, Reich leaders and Gau leaders with a cover letter signed by Speer, dated 24 August 1944.[94] According to this letter, the production of armaments had increased by almost two-thirds from 1942 to 1943 and continued in 1944, where twice as much war material was produced as in 1942, but this included armaments production in the occupied territories, especially in France.[95] Speer's power increased even further from 1943, when he was also responsible for the control of civilian products and as a result the Vierjahresplan authority, the Ministry of Economics and the Wehrmacht lost influence, although the require-

---

90 Schroeter 2019, pp. 42–52.

91 Scherner 2007, p. 501.

92 Scherner 2007, pp. 501–502; Müller 1999, pp. 303–307, pp. 343–347; Speer 2013, pp. 235–236; Milward 1966a, pp. 77–78.

93 The 2007 study by Jonas Scherner from the Institut für Zeitgeschichte analyses this key document by Albert Speer from August 1944 on the economic situation in the second half of the war across all sectors of the German economy (Scherner 2007, pp. 499–546). However, Scherner explicitly points out that four different versions of the report were found in various archives, some of which differ noticeably from the version cited here (Scherner 2007, p. 499).

94 Scherner 2007, pp. 511–546.

95 Ibid., p. 524.

ments of the Wehrmacht had absolute priority.[96] In recent historical research, the role of Speer, known as the Mythos Speer, has been critically scrutinised. The discontinuity between the Rüstungswunder after Speer took office and the still largely peacetime war economy in the period from 1939 to 1941 is also discussed.[97] Year after year, the Allies doubled the number of bombs they dropped on Germany. However, until 1943/44, industrial plants and transport routes were only hit sporadically, and the damage was quickly repaired. In 1943, air raids targeted companies such as Telefunken and Lorenz, along with other radio technology firms.

Although these attacks caused delays in production in several areas, there was no serious crisis. However, a bomb strike at Telefunken caused a two-month delay in the development of the critical Berlin radar series. These devices, operating in the centimetre wave range, were urgently needed at the front. Even the 100 ground panorama radars for air surveillance against enemy aircraft and night fighters, which Siemens had committed to delivering by June 1944, were reduced to an estimated 25 units.[98] By January 1944, 75,000 people were employed in the radar industry, with 37,500 working in Berlin-based factories.[99]

On 28 June 1943, Hitler issued a decree ordering the relocation of armaments production essential to the war effort from particularly vulnerable areas. Some of this production was to be moved to underground caves. By late 1944, a plan called Weingut I foresaw the commissioning of an underground armaments facility by July 1945, but its implementation remained incomplete and was not achieved before the end of the war.[100]

From spring 1944 onwards, the Allies systematically targeted aircraft manufacturers and other armaments companies crucial to the war effort. In the following period, the Allies intensified their bombing raids, and from July 1944, the bomb load reached around

---

96 Bührer 2001, p. 119.

97 Scherner und Streb 2006, pp. 172–196; see also the British estimate of the German war economy, as discussed in Webster's analysis (Webster and Frankland 2006, pp. 271–283).

98 Reuter 1971, p. 201.

99 Ibid., p. 200.

100 Ibid., p. 201.

60 percent of the total bombs dropped during the war. Additionally, production facilities of raw materials, supply companies, and transport equipment were also attacked.[101] Despite this, the relocation of operations to areas less threatened by air raids meant that production losses were short-lived, lasting around one month. By December 1944, the radar industry had reached its highest production figures of the entire war. According to Leo Brandt[102], 400 Würzburg radar devices were manufactured in December 1944. The total number of centimetre-wave devices completed by April 1945 was 500.[103]

2.5.2 Rearmament

2.5.2.1 Air Force

According to Article 198 of the Treaty of Versailles, Germany was prohibited from maintaining any military air forces, whether on land or at sea. In addition, civil aviation and aircraft manufacturing were subject to strict regulations designed to prevent any potential military application. Following the gradual relaxation of restrictions on civil aviation beginning in 1922, the resulting technological stagnation was progressively addressed, leading to a revival of the German aviation industry.

As early as the 1920s, the Reichswehr had begun clandestine preparations for the creation of a future air force. These efforts included the covert training of pilots under the guise of civilian, sport, and commercial aviation programs. Such activities were carried out incrementally and with great discretion, as military aviation remained explicitly prohibited under the terms of the Treaty.[104]

After the National Socialist regime came to power, these preparatory measures were significantly expanded. In 1935, the Air Force was officially established as an independent branch of

101 Müller 1999, p. 755.

102 Mittermaier und Rusinek 2009; Brandt 1962b; Mauel 1983a, pp. 25–27; Weinreich 2001d, pp. 109–110.

103 Brandt 1962a, p. 106.

104 Völker 1967, pp. 28–70.

the newly formed Wehrmacht, alongside the Army and Navy.[105] This development gave rise to internal tensions within the state and military apparatus, particularly between the Oberkommando des Heeres (OKH) and the newly established Air Force, concerning strategic priorities, resource allocation, and jurisdictional authority.

After Hermann Göring had already held the position of Reichskommissar für den Luftverkehr since 30 January 1933[106], he was appointed head of the newly established RLM on 28 April 1933. The official decree establishing the Ministry was issued on 5 May 1933.[107]

On 11 March 1935, the German government publicly announced the existence of the Air Force thereby formally breaking the restrictions imposed by the Treaty of Versailles. It was not until 1 May 1935 that Hermann Göring was officially appointed Commander-in-Chief of the Air Force assuming responsibility for both military and to some extent civil aviation.[108]

At the time of the announcement, the German Air Force already comprised more than 900 aviation officers and 200 antiaircraft officers commanding approximately 17,000 non-commissioned officers and enlisted personnel. By August 1939, the force had expanded to over 15,000 officers and around 370,000 non-commissioned officers and enlisted men. Similarly, employment in the aircraft industry, excluding the workforce of companies involved in equipment and repairs, increased from about 4,000 in January 1933 to 204,100 by 1 October 1938.

At the outbreak of World War II, Allied blockades and trade restrictions caused significant shortages of raw materials in Germany, which also affected the German Air Force. The air force had to operate under constrained supplies of fuel, spare parts, and ammunition, negatively impacting its operational readiness and effectiveness. Despite these limitations, waste of valuable raw materials remained widespread within the aircraft industry until 1942.

105 Lemke 2001, pp. 300–301; Neitzel 1995, p. 16; for aviation research under National Socialism, see Maier 2007b, pp. 104–122.

106 RGBl. 1933, Teil I, Nr. 8, p. 35.

107 RGBl. 1933, Teil I, Nr. 47, p. 241; Benz et al. 2001, pp. 682–683; Studt 2002, p. 18.

108 Benz et al. 2001, p. 682.

In that year, Staatssekretär der Luftfahrt in the RLM and Generalluftzeugmeister (GL) Erhard Milch was informed that supplies of aluminium and copper were critically limited and that inefficient manufacturing processes urgently required reform.

Against the backdrop of increasing air raids on its production facilities, the German aircraft industry was able to maintain production levels until its infrastructure was ultimately destroyed by the end of the war. In 1939, the first year of the war, and again in 1944, the last full year, Germany temporarily surpassed British aircraft production. Despite sustained Allied bombing, German fighter aircraft production peaked in 1944. However, overall German output remained far below that of the Allies. The main manufacturers of combat aircraft on the German side were listed in Table 1.

For example, the Messerschmitt Bf 109, the Heinkel He 111, and the Junkers Ju 87 were successfully tested by the Condor Legion during the Spanish Civil War (1936–1939). The Bf 109 established air superiority and escorted bombers such as the He 111, which conducted precise attacks on strategic targets. The Ju 87 Stuka demonstrated its destructive power in dive-bombing missions against ground targets, including both military positions and civilian infrastructure. Their combined effectiveness highlighted the potential of modern air warfare.

A detailed overview of the development of the operational readiness of the German Air Force's front-line units from 1939 to 1945, covering all types of combat aircraft, can be found in Kroener.[114] The main manufacturers of combat aircraft on the British side were listed in Table 2.

On the British side, the Supermarine Spitfire and the Hawker Hurricane were among the combat aircraft that played decisive roles during the Battle of Britain. The Spitfire was renowned for its speed, manoeuvrability, and overall performance, while the Hurri-

109 Kosin 1990, pp. 103–106.
110 Ibid., pp. 128–131.
111 Ibid., p. 99, p. 134.
112 Ibid., pp. 163–165.
113 Ibid., p. 134–135.
114 Kroener 1988, p. 963.

Messerschmitt AG, Augsburg: Bf 109 (in service from 1937 to 1945)[109]
- The Bf 109 G was the standard fighter and the most frequently built version with around 25,000 aircraft; a total of around 33,000 aircraft including Bf 109 variants were produced.
Focke-Wulf Flugzeugbau AG, Bremen: Fw 190 (in service from 1941 to 1945)[110]
- The Focke-Wulf Fw 190 was built in various versions and was a robust standard fighter and fighter-bomber; a total of around 19,500 were produced.
Ernst-Heinkel-Flugzeugwerke GmbH, Rostock and Oranienburg: He 111 (in service from 1936 to 1944)[111]
- The HE 111 was a medium standard bomber; a total of around 7,600 were produced.
Junkers Flugzeug und Motorenwerke AG, Dessau: Ju 88 (in service from 1936 to 1945)[112]
- The Ju 88 was one of the Air Force's standard fighter aircraft (dive bombers, night fighters and fast bombers); a total of around 15,000 aircraft were produced.
Junkers Flugzeug und Motorenwerke AG, Dessau: Ju 87 (in service from 1937 to 1945)[113]
- The Ju 87 was a dive bomber (Stuka) that could hit targets with pinpoint accuracy; fewer than 6,000 were produced in total.

Table 1: German combat aircraft

cane, also a fighter aircraft, was particularly crucial in defending Britain against German air raids, accounting for the majority of enemy aircraft shot down during the battle. Together, these combat aircraft significantly contributed to the British victory.

The Spitfire continued to demonstrate its exceptional qualities on key fronts in the following years. Other notable combat air-

115 Thetford 1979, pp. 481–496.
116 Ibid., pp. 339–346.
117 Ibid., pp. 132–135.
118 Ibid., pp. 140–145.
119 Ibid., p. 140.
120 Ibid., pp. 118–120.

Supermarine Aviation Works, Ltd., Southampton: Supermarine Spitfire (in service from 1938 to 1948)[115]
- The Supermarine Spitfire was a highly manoeuvrable fighter aircraft; a total of around 20,300 were produced.
Hawker Aircraft Ltd, Kingston: Hawker Hurricane (in service from 1937 to 1944)[116]
- The Hawker Hurricane was a successful fighter aircraft; a total of around 14,500 were produced.
Bristol Aeroplane Company, Bristol: Bristol Blenheim (in service from 1937 to 1943)[117]
- the Bristol Blenheim was a light and for a long time the fastest bomber of the Royal Air Force, but was also used as a fighter aircraft; a total of around 4,400 were produced.
Bristol Aeroplane Company, Bristol: Bristol Beaufighter (in service from 1940 to 1945)[118]
- The Bristol Beaufighter was a fighter-bomber and was widely used as a night fighter, torpedo bomber and against ships. It was the first night fighter with sufficient performance to be equipped with the top-secret AI (Airborne Interception) radar during the German air raids in the winter of 1940–1941[119]; a total of around 5,500 were produced.
Boulton Paul Aircraft Ltd, Wolverhampton: Boulton Paul Defiant (in service from 1939 to 1943[120]
- The Boulton Paul Defiant with its special armament was mainly used to escort bombers; a total of just over 1000 were produced.

Table 2: British combat aircraft

craft included the Bristol Blenheim, originally developed as a light bomber but also employed as a fighter, particularly during the early stages of the air war. The Boulton Paul Defiant was primarily used for bomber escort missions.[121]

The numbers of German and British combat aircraft are estimates and can vary considerably depending on the source.[122] It

121 A further detailed overview of Royal Air Force aircraft from 1918 onwards can also be found in Thetford 1979 and even more clearly illustrated in Gunston 1982.

122 For the status of British fighter aircraft at the beginning of the war, see Price 2009, pp. 2–5; Price 1978, pp. 9–16.

is important to note that these figures include only the main aircraft models produced by the respective companies during World War II; other models and variants were also developed by these and other manufacturers.

2.5.2.2 Navy[123]

In addition to the air supremacy of the Allied Air Forces, the course of the U-boat war and the Allies' use of new technologies such as radar and radio direction finding played a central role in this development. These advances were crucial in bringing the U-boat campaign to a standstill by 1943.[124] Starting in 1936, the German Navy began systematically expanding its submarine fleet, made possible by the 1935 Anglo-German Naval Agreement. This agreement permitted Germany to rebuild a limited fleet, including a specified number of submarines. However, within the naval leadership, opinions on the strategic importance of submarines were divided.

Admiral Erich Raeder[125], Commander-in-Chief of the Navy, regarded submarines primarily as a complementary component of a traditional fleet, the core of which consisted of capital ships. Captain Karl Dönitz[126], on the other hand, recognised the strategic potential of submarines. After being entrusted with the development of the submarine force on 1 October 1935 and promoted to captain, Dönitz did everything in his power to convince naval leadership of their capabilities. Driven by his conviction that war with Great Britain was inevitable, he worked tirelessly to position the submarine arm as a central element of the Kriegsmarine's overall strategy.[127]

It was not until early summer 1939 that Dönitz finally succeeded in persuading Raeder of the submarines' strategic impor-

123 The term Navy in this context refers to the Kriegsmarine, which was commonly referred to as Marine in German.

124 Schramm 1963, p. 1497.

125 Salewski 2003, pp. 104–106; Fischer 2011, pp. 185–194; Raeder 1957; Klee 2011, p. 476.

126 Hillmann 2022; Kraus 2011, pp. 316–325; Dönitz 2018; after the resignation of Erich Raeder on 30 January 1943, Hitler appointed Karl Dönitz as the new Commander-in-Chief of the Navy.

127 Kraus 2011, p. 317.

tance, leading to an acceleration of fleet expansion. Nevertheless, submarine construction never became a top priority for the Navy.[128] Until shortly before the outbreak of war, the German Navy continued to prioritise the construction of a surface fleet over the development of a large submarine force.

It was only with the outbreak of the war and the early successes of submarines in the Atlantic that their strategic importance was fully recognised. On 1 September 1939, the German Navy had 57 submarines.

Between 1 September 1939 and 8 May 1945, the German Navy commissioned an additional 1,113 submarines. Of these, 1,099 were newly built by German shipyards, 4 were newly built by foreign shipyards, and 10 were captured foreign submarines.[129]

The primary task of German submarines was to attack and sink enemy merchant ships and convoys, aiming to cut off the supply of essential goods and war materials to Great Britain. This strategy caused significant losses on the Allied side during the early phase of unrestricted submarine warfare around the British Isles in 1940.

By the end of 1940, British forces had begun to understand German operational tactics, particularly the so-called Wolfsrudeltaktik, which involved coordinated attacks by multiple submarines on single convoys. As a result, they recognised the value of radio direction finding and radar, leading to their increasing use in locating German submarines.

Initially, direction-finding stations along the English coast were used to determine approximate submarine positions. However, it was not until November 1942 that shortwave direction finders became standard equipment on escort vessels, marking a decisive improvement in convoy protection.[130] Between autumn 1942 and May 1943, German submarines achieved their greatest successes in the Battle of the Atlantic. In March 1943 alone, 140 Allied ships were lost.[131] However, from May 1943 onwards, there was a noticeable

128 Brennecke 1994, p. 13.

129 Dönitz 2018, p. 483.

130 Dönitz 2018, p. 485; Reuter 1971, p. 150.

131 Schramm 1961, p. 1439; Burns 2013, pp. 275–277; Reuter 1971, p. 158.

decline in successes against Allied convoys.[132] The Allies' superior detection systems and weaponry resulted in such heavy losses that most U-boats were redeployed away from the North Atlantic. They continued to engage convoys there only when conditions were favourable.[133] Exactly half of the U-boats were sunk by aircraft, some in joint operations with escort ships. Of the 1,170 submarines commissioned between 1939 and 1945, 863 were deployed to the front and carried out one or more combat missions. Of these, 630 were lost, with 603 of them sunk due to enemy action.[134]

Increasing Allied successes in military operations from 1940 onwards were largely based on powerful radar systems used in aircraft and on ships. By the end of 1940, the British had successfully deployed airborne radar equipment for the first time to combat submarines. These early devices operated at a wavelength of 1.5 metres and were primarily used to detect surfaced U-boats during poor visibility or at night.

In response to these developments, the German Navy introduced the Metox warning receiver in the summer of 1942.[135] This device allowed U-boats to detect the radar emissions of Allied aircraft operating in the metre wavelength range, giving them time to dive before being attacked. As a result, U-boats continued to surface at night to recharge batteries and ventilate the boat, assuming relative safety in the dark.

However, the introduction of centimetric radar by the British in early 1943 fundamentally changed the situation. The newly developed ASV Mk. III radar[136], based on cavity magnetron technology, operated in the centimetre wavelength range and was undetectable by Metox. From spring 1943 onwards, Coastal Command aircraft were increasingly equipped with this radar, which provided significantly improved detection range and resolution. This technological advance enabled aircraft to locate U-boats even at night, precisely when the submarines were most vulnerable during surface operations.

---

132 Burns 2013, pp. 275–277; Dönitz 2018, pp. 397–422; Reuter 1971, pp. 158–159.

133 Schramm 1961, p. 1440; Rahn 1990, p. 84.

134 Dönitz 2018, p. 483; Brennecke 1994, p. 9; according to Brennecke, 63 per cent of the 603 German submarines lost in action were destroyed by air attacks.

135 Reuter 1971, p. 153; Bauer 1997a, pp. 194–203.

136 ASV: Air to Surface Vessel.

In the following months, more and more aircraft, both British and American, were fitted with centimetric radar. This significantly improved the accuracy of submarine detection and increased the effectiveness of combat operations against German submarines.

In addition to radar and high-frequency direction-finding systems (HF/DF), British cryptanalysts at Bletchley Park achieved major breakthroughs by deciphering secret German radio messages. This included the successful decryption of the ENIGMA M4, a more complex version of the cipher machine introduced in early 1942 for U-boat communications.[137] Initially considered unbreakable due to its fourth rotor, the M4 was eventually broken after British forces captured vital key material from German vessels in late 1942. From early 1943 onwards, Bletchley Park was increasingly able to read M4-encrypted messages, which provided valuable intelligence and significantly supported Allied anti-submarine efforts.

As a result of the growing crisis in submarine warfare, Karl Dönitz, as Commander-in-Chief of the Navy, ordered a reorganisation and expansion of the Navy's entire technical intelligence apparatus in late 1943.

In December of that year, the Wissenschaftlicher Führungsstab beim Oberkommando der Kriegsmarine (WFM) was established. Its purpose was to consolidate naval research capacities in the field of high-frequency technology and to issue directives to affiliated institutions and companies. This step marked a systematic, though belated, attempt to counter the increasing technological superiority of the Allies.[138]

2.5.2.3 Industry

Although Germany had a great deal of experience in the field of tube and transmitter development and production, particularly through the Telefunken company, which was founded in 1903[139], it was only in the following years that this lead was effectively utilised for emerging radar development. This became especially

137 Dönitz 2018, pp. 485–486.

138 Reuter 1971, p. 159.

139 Thiele 2003.

evident in the failure to cooperate with the Nachrichtenmittel-Versuchsanstalt der Marine (NVA) at the end of 1933, which led to the founding of GEMA[140] in January 1934. GEMA had to build up the necessary technological knowledge and manufacturing capabilities from scratch. This was a missed opportunity for Germany to take a leading role in the development of powerful high-frequency tubes which were essential for effective radar technology. The history of Telefunken's introduction of tube transmitters in Germany dates back to the period between 1915 and 1918.[141] Initially, Telefunken's development focused on radio and television technology.[142] In addition, numerous experiments with radio relay at a wavelength of 50 centimetres were conducted in the early 1930s.[143] This gave Telefunken a great deal of practical experience, which remained unutilised for radar development for a long time.[144] In 1935, however, Wilhelm Runge[145] began his own initial reflection measurements at Telefunken, which formed the basis for extensive developments in future radar technology.[146]

Telefunken played an important role in the course of World War II, particularly with the development of anti-aircraft guns. In 1939, Telefunken employed around 23,500 people. By the height of the war, the number had risen to approximately 40,000, including a significant number of forced labourers. About half of the employees worked in eight production facilities for high-frequency devices and systems.[147] In 1941, AEG[148] took over the Telefunken shares from Siemens & Halske and continued to run the company as a fully owned subsidiary until the end of 1966. On 1 January 1967, the business of Telefunken AG was transferred to the parent company

140 GEMA: Gesellschaft für elektroakustische und mechanische Apparate m.b.H. (Kroge 2000; Kroge 1998); see also Chapter 7.1.1.4.

141 Bredow 1928, p. 23.

142 Rukop 1953, pp. 165–176; Bredow 1953, pp. 177–180.

143 Runge 1953, p. 183.

144 Runge 1934a, pp. 24–33.

145 Rangnick 2005; Jäger und Heilbronner 2010, pp. 371–372; Nestel 1983, pp. 151–153; Weinreich 2001e, pp. 105–106.

146 Runge 1971, pp. 42–43; Reuter 1971, p. 32; Bekker 1980, pp. 45–46.

147 Thiele 2003, pp. 34–35.

148 AEG: Allgemeine Electricitäts-Gesellschaft; Strunk 1999.

AEG and renamed Allgemeine Eletricitäts-Gesellschaft AEG-Telefunken.[149]

GEMA also played a central role in the development of long-range radar systems for air warning and contributed to the overall radar infrastructure used in anti-aircraft defence, complementing systems developed by Telefunken.

As a pioneering company in the field of radar technology, GEMA developed numerous radar and radio navigation systems up until the end of the war.[150]

The C. Lorenz AG[151], founded in 1880, initially specialised in telegraphy and railway signalling systems. Over time, the company expanded into teleprinters and telephony and became a global leader in the development of air navigation and landing systems. In connection with wireless telegraphy, Lorenz also emerged as a pioneer in radio technology. Between 1939 and 1944, Lorenz was tasked with producing radar equipment for diverse military applications, encompassing air defence, installation on naval vessels and submarines, and airborne search radar for maritime target detection.

Another key contribution from Lorenz was the development of the Lorenz-Schlüsselmaschine[152], a cipher machine used for top-secret communications. It is renowned today for its complexity and for having been ultimately broken by Allied codebreakers at Bletchley Park.

2.6 British Appeasement Policy

2.6.1 Challenges of the Appeasement Policy

The appeasement policy, meaning the consistent pursuit of peace through restraint, accommodation, and the search for solutions to existing tensions with aggressors, characterised British foreign pol-

149 Strunk 1999, p. 235.

150 Watts and Klausing 2025, pp. 24–33; Willisen ca. 1984, pp. 1–27.

151 C. Lorenz AG 1955; C. Lorenz AG also employed forced labour during the war.

152 The Lorenz-Schlüsselmaschine (cipher machine) included the SZ42, a key component officially designated by the German Army.

icy during the period between the world wars. This policy has been the subject of historical controversy, particularly from the Anglo-German Naval Agreement in June 1935 to the Münchener Abkommen in September 1938. Specifically, the appeasement policy of Neville Chamberlain, who served as British Prime Minister from 1937 to 1940, is often viewed negatively. Opponents of this policy, including the future Prime Minister Winston Churchill[153], argued that pragmatic politics require a position of military strength. The appeasement policy had already been shaped before Chamberlain by his predecessors Stanley Baldwin and Ramsay MacDonald.[154]

MacDonald played a significant role in promoting a peaceful foreign policy for Great Britain during the 1920s and 1930s. Even before Hitler came to power, MacDonald supported Germany's economic recovery and international reintegration. At the Lausanne Conference in 1932, amid the Great Depression, the Allies agreed to suspend Germany's reparation payments. This suspension effectively amounted to a de facto cancellation, taking into account the economic situation of Germany and the other countries involved. However, France and other signatory states insisted that the territorial provisions and military restrictions of the Treaty of Versailles would remain binding. This compromise reflected both a willingness to stabilise the international economy and a continued mistrust of Germany's political direction.[155]

Although the agreement was never formally ratified due to the U.S. Congress's refusal to cancel inter-Allied debts, it effectively came into force and provided substantial economic relief to Germany. Nevertheless, conservative circles in Germany, supporters of the monarchy, and the National Socialists were strictly opposed to the treaty.

To understand British appeasement policy, a more in-depth analysis is required that does not focus solely on foreign policy but also considers social and domestic factors. British foreign policy was characterised by the hope of resolving conflicts through peaceful means, either within the framework of the League of Nations or through a system of collective security based on various agree-

153 Addison 2018; Roberts 2019.
154 Marquand 2015; Marquand 1997.
155 Graml 2010, p. 34.

ments between the nations involved. At the same time, however, the appeasement policy reflected widespread war-weariness among the population, coupled with fundamental doubts about the necessity of military conflict.[156]

In the 1920s and 1930s, Great Britain faced increasing political, economic, and military challenges. Additionally, the burdens of maintaining the British Empire complicated any comprehensive rearmament efforts. At the same time, the country was losing influence in the Mediterranean where Italy expanded under Mussolini and in the Far East where Japan was growing in power. These geopolitical pressures, combined with internal modernisation processes, made a sustainable military build-up nearly impossible, forcing Great Britain to rely on a policy of diplomatic conflict resolution.[157]

Moreover, beyond the peaceful resolution of conflicts, this policy of understanding aimed to morally discredit states pursuing aggressive domestic and foreign policies. Great Britain also had to consider the stance of the Dominions such as Canada, Australia, and South Africa, which gained greater foreign policy independence through the Statute of Westminster in 1931. Prior to this, these countries had been bound to Great Britain in matters of foreign policy. The Statute granted the Dominions the ability to conduct independent foreign policies and enter into their own diplomatic relations and alliances. Although they remained formally part of the British Commonwealth, their legal ties to Great Britain became less binding. Consequently, Great Britain could intervene in foreign policy matters only if requested by the Dominions or in cases of joint defence. These changes allowed the Dominions greater autonomy while preserving their cultural and historical connections with Great Britain.[158]

However, Britain's efforts to reintegrate Germany into a system of collective security were hindered by the stringent French policy. Developments in Germany particularly the erosion of parliamentary rights and the persecution of political and ethnic groups placed clear limits on those advocating for a policy of understanding. It became

156 Hauser 1972, pp. 9–10.

157 Tyrell 1987, p. 2.

158 Doherty 1972, pp. 209–234; Hauser 1972, p. 10.

evident early on that Germany's visible rearmament efforts would have to be met with a sustained strengthening of Great Britain's own armed forces. In addition to diplomatic negotiations aimed at securing peace, Britain's goal was to keep armament costs low and to delay military conflict for as long as possible. Furthermore, Germany was also viewed as a bulwark against communism.[159]

2.6.2 World Disarmament Conference

The World Disarmament Conference of the League of Nations, the second major international disarmament conference, commenced in Geneva on 2 February 1932 and continued until 11 June 1934, albeit with several interruptions.[160] Its primary objective was to reduce military armaments worldwide and to establish a common security framework among member states. Key participants included major European powers such as France, Britain, and Italy, alongside Japan and the Soviet Union. Although the United States was not a member of the League of Nations, it was represented by observers. Germany took part in the conference from its beginning, although it was still subject to the unilateral disarmament obligations imposed by the Treaty of Versailles. However, Germany later sought to bypass these restrictions through secret rearmament programmes and reintroduced compulsory military service in 1935. The conference ultimately failed due to conflicting security interests and disagreements over specific disarmament measures.

On 11 December 1932, the Five-Power Declaration was signed by the United States, Great Britain, France, Italy and Germany. This declaration formally recognised Germany as an equal negotiating partner in the disarmament talks, paving the way for Germany's gradual reintegration into the international disarmament process. This was a significant milestone in Germany's foreign policy, as revising the Treaty of Versailles was a central aim, which Hitler later pursued after coming to power in 1933. Great Britain faced a difficult position: on one hand, there was sympathy for Germany's

159 Hauser 1972, p. 10.
160 Hofer 1979, p. 38.

demand for military equality, while on the other hand, the security concerns of France, insisting on strict adherence to the Treaty's disarmament provisions, had to be carefully considered.

2.6.3 MacDonald Plan

On 16 March 1933, British Prime Minister Ramsay MacDonald presented a plan that became known as the MacDonald Plan.[161] The plan proposed allowing Germany to increase its troop strength to 200,000 men, while France would reduce its forces to the same level. According to the Treaty of Versailles, the German Army was limited to 100,000 men and general conscription was forbidden. As part of this plan, the Reichswehr was to be transformed from a professional army into a conscript army with a service period of eight months, over a transitional period of five years. The plan also included disarmament proposals for other countries, such as the reduction of large-calibre guns and tanks in terms of calibre, number, and weight. Significantly, Germany, unlike other nations, was still prohibited from building or possessing military aircraft.

Although the MacDonald Plan was controversial within the British government, it was also seen as a means of preventing uncontrolled German rearmament in the event that the disarmament negotiations failed. No sanctions were foreseen in the case of a breach of the plan's provisions, but a Permanent Disarmament Commission, consisting of representatives from the signatory states, was to monitor compliance.[162]

After MacDonald presented his plan in Geneva, he sought contact with Mussolini, who, as the leader of Fascist Italy, sympathised with National Socialist Germany, in order to promote an approach in which Great Britain and Italy could act as mediators between France and Germany. At Mussolini's initiative, the foundation was laid for a pact between Great Britain, France, Italy, and Germany, aimed at fostering understanding and cooperation between the powers to resolve European disputes through peaceful

161 Hauser 1972, pp. 15–25, pp. 258–261.
162 Ibid., p. 19.

means. This so-called Four-Power Pact was seen as a replacement for the Disarmament Conference, with its inconclusive negotiations, and aimed to create a new foundation for a stable European policy while respecting collective security and acknowledging the League of Nations. It was clear from the text of the treaty that Germany was regarded as an equal partner, which granted international recognition to National Socialist Germany and provided Hitler with a significant boost in prestige.[163] The Four-Power Pact was finally signed on 15 July 1933, after an amendment to the treaty.[164] However, the pact did not come into effect due to Germany's withdrawal from the League of Nations. France maintained its demand for a period of five years during which German troop strength was to remain fixed at 100,000 men, and German efforts towards peaceful cooperation were to be put to the test.

French resistance to German rearmament in accordance with the MacDonald Plan, as well as reservations about Hitler's policies and his persona, were repeatedly reinforced by events in Germany, such as the Ermächtigungsgesetz of 24 March 1933, the organised boycott of Jewish businesses, doctors, and lawyers from 1 April 1933[165], the dissolution of trade unions on 2 May 1933[166], and the Gesetz gegen die Neubildung von Parteien on 14 July 1933.[167] Germany's political and, above all, moral isolation had now become unmistakably evident.

In this atmosphere of tension and isolation, Hitler presented National Socialism as a movement committed to peace in his first foreign policy government statement to the Reichstag on 17 May 1933.[168] Initially, Hitler gained ground with his speech, but everyone involved knew that Germany's seriousness in striving for a viable solution to the armaments issue would only be proven at the Geneva Disarmament Conference.

163 Graml 2010, pp. 42–43.
164 Hauser 1972, pp. 19–20; Tofahrn 2003, p. 27.
165 Studt 2002, p. 16.
166 Ibid., p. 18.
167 Ibid., p. 24.
168 Ibid., p. 19.

2.6.4 Simon Plan

In the period that followed, the attitudes of Germany and France revealed that the MacDonald Plan had failed to achieve its goal of securing dialogue partners for a voluntary agreement. However, in response to this failure and growing French concerns, the British Foreign Secretary, John Simon, presented a revised proposal in October 1933. The declaration, known as the Simon Plan[169], was finalised around 14 October 1933, the same day that Germany announced its withdrawal from both the League of Nations and the Disarmament Conference.

This plan was the result of British-French negotiations at the Office of the Disarmament Conference in Geneva and aimed at a partial revision of the MacDonald Plan. The key elements of the proposed changes included extending the implementation period of the treaty to eight years. At the same time, the Simon Plan outlined a programme that, at the end of a period of eight to ten years, would ensure two main objectives: first, that actual disarmament of the heavily armed states had taken place, and second, that the principle of equal rights, in line with the Five Power Declaration of 1932, would be realised by all participants in the Disarmament Conference. The British government emphasised that the successful implementation of this plan for the equal rights of all states would depend on sincerity and loyalty.

During the preliminary negotiations on the Simon Plan, Great Britain made it clear from the beginning that it was strictly opposed to the use of sanctions in the event of treaty violations, particularly by Germany. In the case of a German breach of the proposed agreements, the signatory countries would regain their freedom of action and would no longer be bound by the planned armament limits or the elimination of weapons.

During the negotiations in the summer of 1933, in which Italy and the United States also participated, the discussions were dominated by Germany's demands for defensive weapons, which were permitted to other states. France agreed to the immediate reduction of its troop strength but categorically rejected the reduction of heavy equipment as outlined in the MacDonald Plan unless strict

169 Hauser 1972, pp. 26–38, pp. 266–268.

controls were introduced to ensure compliance with the armament restrictions designed to preserve the defensive character of the respective armies.

In contrast to the MacDonald Plan, which had proposed that artillery with a calibre over 155 mm and tanks weighing more than 16 tons should be destroyed by one third within the first twelve months and the remaining two thirds within three years, the French government took a much more restrictive stance. It insisted that the heaviest military equipment be stored on French territory under the supervision of the League of Nations. Furthermore, France demanded that only artillery with calibres exceeding 220 mm be dismantled by the end of the eighth year. At no point was there any mention of aligning Germany's armament with the levels maintained by the other powers during the two implementation phases.[170]

2.6.5 Further Developments Towards the Münchener Abkommen

Thereafter, Germany was to remain bound by the restrictions of the peace treaties, meaning that no alignment of German armaments with those of other states was planned for about four years. The Reich government, which received the proposal from the President of the Disarmament Conference on 17 July 1933, accepted the renunciation and monitoring of offensive weapons but demanded defensive weapons similar to those of other states, albeit in smaller numbers. Shortly afterwards, Hitler categorically rejected any trial controls, casting doubt on the outcome of the negotiations. While Italy showed some readiness to authorise sample weapons for Germany, France and Great Britain rejected this outright. This position remained firm even after German Foreign Minister Konstantin von Neurath reiterated the demand for prototypes, such as guns up to 155 millimetres calibre. Everyone involved recognised that the issue of German sample weapons could very likely cause the failure of the Geneva negotiations. Von Neurath therefore warned early on that the refusal to accommodate Germany's claims would risk undermining the entire negotiation process.

170 Ibid., p. 27.

The United States attempted to mediate by proposing to allow the Reichswehr a certain number of 155-millimetre guns, tanks and reconnaissance aircraft per division. However, this conflicted with France's demand for rearmament without heavy equipment. Neither France nor Germany was willing to compromise. Germany would view sanctions as aggression and insisted on model weapons authorisation, while France remained equally firm against them. Great Britain ultimately sided with France, signalling that its alliance took precedence over the risk of the conference failing. Hitler nevertheless claimed to consider the MacDonald Plan as a basis but renewed his demand for weapons exceeding the 105-millimetre calibre limit set by Article 167 of the Treaty of Versailles. He insisted that Germany be treated equally, called for cancellation of the Versailles weapon limits, and rejected any unilateral testing period.

Hitler's declaration was handed to British Foreign Secretary John Simon on 6 October 1933, reiterating Germany's demands and rejecting unilateral trial controls. The following days brought growing scepticism in Britain regarding the chances of successful negotiations.[171]

On 11 October 1933, Foreign Minister von Neurath submitted a supplementary communication clarifying Germany's position and urging continuation of talks.[172] He warned that Germany could not accept less in 1934 than had been promised for 1938 and explicitly indicated that Germany might withdraw from the Disarmament Conference if its demands continued to be rejected.[173] Despite these efforts, France and Britain remained firm on the issue of sample weapons.

Although there were plans to grant disarmed states the same number and size of guns, tanks and aircraft as other states during the second stage, exceeding Hitler's own demands, these plans failed due to persistent disagreements.

France's refusal to grant Germany increased military strength gave Hitler a justification for his subsequent actions. On 14 October 1933, Germany left the Geneva Disarmament Conference and declared its withdrawal from the League of Nations. This marked

171 Ibid., p. 36.

172 Neurath (1933). ADAP 1971, Serie C, Band I,2, Nr. 489, pp. 890–891.

173 Hauser 1972, p. 34.

the ultimate failure of the MacDonald Plan, which had sought to bridge differences and restore national pride and dignity, especially to Germany.

2.6.6 Münchener Abkommen

The Münchener Abkommen of September 1938 is often seen as a symbol of the failed appeasement policy of the Western democracies in the face of the National Socialist threat of violence.[174] It was described as "A total and unmitigated defeat"[175], as Churchill titled his speech in the British House of Commons on 5 October 1938. He argued that the policy of appeasement towards Hitler would bring neither peace nor security. Instead, he saw the agreement as a sign of weakness that could encourage Hitler to pursue further aggression.[176]

Despite being one of Churchill's most famous speeches, the House of Commons voted 366 to 144 in favour of a motion supporting the government's approval of the agreement. Although they opposed the agreement, Churchill and his supporters in the Conservative Party chose to abstain, thereby not voting against the motion.

On 31 March 1939, Great Britain and France declared their guarantee of Poland's independence.[177] With the declaration of war by Great Britain and France against Germany on 3 September 1939, following Germany's invasion of Poland on 1 September, the policy of appeasement can be regarded as having ended in failure.

At 11:45 a. m. on Sunday, 3 September 1939, Neville Chamberlain gave a sombre radio address from the Cabinet Room at Number 10 Downing Street:

> "This morning the British Ambassador in Berlin handed the German Government a final Note stating that, unless we heard from them by 11 o'clock that they were prepared at once to withdraw their troops from Poland, a state of war would exist between us.

174 Churchill 2005, pp. 268–288; Churchill 1954, pp. 364–391.

175 Churchill 1999a, pp. 172–175.

176 Churchill's speech of 5 October 1938 marked one of his earliest prominent warnings about the threat posed by Nazi Germany and remains a key moment in his opposition to Prime Minister Neville Chamberlain's policy of appeasement.

177 Studt 2002, pp. 100–101; Hildebrand 2009, pp. 46–47.

> I have to tell you now that no such undertaking has been received, and that consequently this country is at war with Germany.
>
> You can imagine what a bitter blow it is to me that all my long struggle to win peace has failed. Yet I cannot believe that there is anything more or anything different that I could have done and that would have been more successful ....”[178]

Chamberlain resigned as Prime Minister on 10 May 1940 and was succeeded by Churchill on the same day. Churchill also took on the role of Minister of Defence.[179]

Churchill had been warning about Germany’s rearmament since the mid-1930s and had repeatedly called for massive armament efforts to counter Germany from a position of strength. On the day of his appointment as Prime Minister, 10 May 1940, Germany invaded the Netherlands, Belgium, Luxembourg, and France, thereby opening the Western campaign of World War II. The aim of the German offensive was to force France to surrender, to reach a settlement with Great Britain, and subsequently to turn eastwards against the Soviet Union.[180]

Churchill was convinced that war between Great Britain and Germany was inevitable. In his first address to the House of Commons as Prime Minister, delivered on 13 May 1940, he declared: “I have nothing to offer but blood, toil, tears, and sweat”.[181] With this speech, Churchill made clear that only unconditional victory over Nazi tyranny would be acceptable. It was followed by further defining speeches, including “Be Ye Men of Valour”[182] on 19 May 1940 and “This Was Their Finest Hour”[183] on 18 June 1940. In May 1940, Churchill formed a streamlined War Cabinet to coordinate the British war effort. While Neville Chamberlain had already established a larger War Cabinet at the beginning of the war in 1939, Churchill reduced it to a smaller group of key ministers

178 Chamberlain 1999b, pp. 179–180; Chamberlain 2008; Price 2009, p. 1; Price 1978, pp. 9–10.

179 Addison 2018, pp. 32–33; Tyrell 1987, pp. 13–23.

180 Studt 2002, p. 129.

181 Churchill 1999c, pp. 187–188; Addison 2018, p. 34.

182 Churchill 1999b, pp. 188–190.

183 Churchill 1999d, pp. 190–191.

who worked in close consultation with him.[184] This War Cabinet remained in place until the end of the war in 1945 and was dissolved after the general election of that year, when Churchill was succeeded by the newly elected Labour government.

Construction of a secret and heavily fortified underground command centre beneath the Treasury building in Whitehall, London, began in June 1938 (Fig. 1).[185] Initially referred to as the Central War Room, it was renamed the Cabinet War Rooms at the end of 1939 and served as the central command facility of the British War Cabinet under Prime Minister Churchill until 1945.[186] It provided protection from aerial bombardment and served as the location for planning and overseeing strategic operations during the war. Churchill reportedly said shortly after becoming Prime Minister:

> "This is the room from which I will direct the war."[187]

Fig. 1: Cabinet War Room[185]

184 Tyrell 1987, p. 14.
185 Image credit: IWM
186 Holmes 2011.
187 The exact wording of the sentence may not be fully documented, but it reflects the significance and symbolic value of the room from which Churchill addressed the nation.

After the war, the rooms were decommissioned in 1945 and later restored as a museum.

Although Great Britain and France declared war on Germany following the invasion of Poland on 3 September 1939, this initially had no direct military consequences for Germany. The Western powers were divided on how to proceed, as their preparations for war were still incomplete and strategic thinking remained strongly shaped by the defensive experiences of World War I. France, in particular, relied on its fortifications along the Maginot Line and adopted a defensive posture.

Neither the German invasion of Poland on 1 September nor the Soviet advance into eastern Poland on 17 September 1939 prompted a military response from the Western powers. These events unfolded despite the Anglo-French guarantee of Polish independence declared on 31 March 1939, which ultimately failed to prevent the attack and could not stop the collapse of Poland by early October.

Following the Polish defeat, Hitler felt under pressure to begin the war in the West. He considered Great Britain his principal adversary, while France remained its most important continental ally.

2.6.7 Reflections on the Appeasement Policy

The historical assessment of British appeasement policy should not be reduced to its failure. Rather, it must be acknowledged that the period between the Münchener Abkommen in autumn 1938 and the Battle of Britain in summer and autumn 1940, followed by the subsequent Blitz into early 1941, provided Great Britain with valuable time to prepare militarily for an armed conflict, particularly in the area of air defence.[188]

The development of radar technology illustrates how, alongside Chamberlain's appeasement strategy, there was a targeted interconnection of politics, the military, science and industry, all focused on the defence of the British Isles. This indicates that preparations for a potential war with Germany were a key element of British strategic planning throughout this period.[189] The question

188 McKercher 2008, pp. 391–442.

189 Kaiser 1996a, pp. 43–44, p. 47; Kaiser 1996b, pp. 156–158.

of why the European powers allowed Hitler so much room to consolidate his power, despite the clear programmatic aims expressed in Mein Kampf[190], remains a subject of controversial debate and diverse interpretation.[191] The well-known sentence in Mein Kampf, "Deutschland wird entweder Weltmacht oder überhaupt nicht sein", encapsulated Hitler's programmatic objectives.[192]

Just as the appeasement period was shaped by Germany's internal political developments, including the dismantling of democratic institutions, the suppression of press freedom and the establishment of concentration camps, public opinion in Britain gradually shifted. The persecution of entire population groups, especially Jews, further intensified negative attitudes toward the Nazi regime after Hitler's appointment as Chancellor. While some conservative circles in Britain expressed a certain level of understanding for developments in Germany, this did not translate into broader support or sympathy for Nazism.

On 5 April 1940, just over a month before the German offensive against France, Joseph Goebbels held a confidential briefing for selected German press representatives on the regime's foreign policy tactics. In it, he stated (excerpt):

> „...Bis jetzt ist es uns gelungen, den Gegner über die eigentlichen Ziele Deutschlands im unklaren zu lassen, genauso wie unsere innenpolitischen Gegner bis 1932 gar nicht gemerkt haben, wohin wir steuerten, daß der Schwur auf die Legalität nur ein Kunstgriff war. Wir wollten legal an die Macht kommen, aber wir wollten sie doch nicht legal gebrauchen. Wir wollten nicht Parteien dulden, die mit uns nach einem Jahre etwa so verfahren würden, wie wir mit ihnen zu verfahren beabsichtigten. Man hätte uns ja erdrücken können, so sehr schwer war das nicht. Aber man tat es nicht. Man hätte 1925 ein paar von uns in Haft nehmen können und alles wäre aus und zu Ende gewesen. Nein, man hat uns durch die Gefahrenzone hindurch gelassen. Genauso war das in der Außenpolitik...“.[193]

190 Hitler 1938.

191 Metzmacher 1966, pp. 369–412.

192 Hitler 1938, p. 742. "Germany will either become a world power or cease to exist." (translated by the author).

193 Jacobsen 1965, p. 181; Wette 1989, p. 134.
"...So far we have succeeded in leaving our opponents unclear about Germany's real objectives, just as our domestic opponents did not realise until 1932

Recent assessments of the appeasement policy, including those by historian Frank McDonough, recognise it as the only viable option at the time, yet criticise its hesitant and delayed implementation. McDonough argues that an earlier confrontation with Hitler might have prevented the outbreak of war.

An examination of the development of British radar technology for air defence from as early as 1935 reveals a deep mistrust of Germany's ambitions. The policy of appeasement granted Britain crucial time to strengthen its military capabilities in preparation for a possible conflict.

Importantly, the advancement of radar and other military measures was not merely defensive, but part of a broader strategic approach. While diplomatic efforts to avoid war continued, the British government simultaneously prepared for armed conflict, enhancing its defences in anticipation of a future confrontation with Nazi Germany.[194]

where we were heading and that the oath to legality was merely a stratagem. We wanted to come to power legally, but we did not intend to use that power legally. We did not want to tolerate parties that, after a year, might treat us in the same way we intended to treat them. We could have been crushed; it really would not have been that difficult. But it was not done. They could have arrested a few of us in 1925, and everything would have been over and finished. No, we were allowed to pass through the danger zone. It was the same in foreign policy..." (translated by the author).

194 McDonough 2002; McDonough 1998.

3 Radar Technology as a Core Element of Warfare

3.1 Race Against Time

Aeroplanes became increasingly important during World War I, particularly in German air raids on the British Isles.[195] The first German bombing raid using Zeppelins took place on 19 January 1915 against targets on the East Coast of England, followed by more intensive attacks, including the first raid on London on 31 May. In 1916, an additional 123 missions were carried out. From 1917 onwards, the first bombings of the British capital by Gotha bombers began, culminating in the heaviest attack on 13 June 1917, which claimed 162 lives.[196] At the beginning of the bombing campaign, the German attackers faced an unprepared enemy.[197] The lack of effective early warning systems often led to widespread panic among London's civilian population.[198] For the first time since the Norman Conquest in 1066, British civilians experienced direct attacks on home soil.

In addition to bombing raids, Zeppelins were employed by Germany during the war primarily for reconnaissance and the monitoring of enemy troop movements. These large airships provided valuable intelligence due to their long range and ability to operate at high altitudes. While other nations, such as Britain, did not deploy rigid Zeppelins, they relied instead on fixed-wing aircraft for aerial reconnaissance and bombing missions.

Their ability to stay airborne for extended periods and to cover long distances gave them a tactical advantage. Zeppelins could operate at altitudes initially beyond the reach of enemy aircraft. How-

195 Judkins 2007, pp. 82–90.

196 Süß 2011, pp. 28–30; Zimmerman 2013, p. 18; Judkins and Bauer 2014, pp. 3–11.

197 Price 2009, p. 3; Price 1978, p. 12.

198 Süß 2011, pp. 29–30.

ever, by 1916, aircraft had developed to the point that they could reach these altitudes and effectively counter the Zeppelins.

In 1917, the German Army abandoned its Zeppelin fleet due to heavy losses, shifting to more powerful aircraft instead. From that point on, the most advanced Zeppelins were used primarily by the Navy.[199] This shift foreshadowed a broader trend: in the years that followed, the military importance of the aeroplane steadily increased. With its growing range and payload capacity, the aeroplane could penetrate deeper into enemy territory.

From the early 1930s, this development led to an increased threat from enemy aircraft. The extension of their operational range and bomb load became particularly significant for Great Britain. The island's geographic position no longer guaranteed the population lasting protection from attack and destruction.

On 10 November 1932, Stanley Baldwin delivered his sombre speech "A Fear for the Future" in the House of Commons, a speech remembered above all for his warning that "the bomber will always get through."[200] At that time, Baldwin was serving as Lord President of the Council in Ramsay MacDonald's National Government. He had previously served two terms as Prime Minister and would later return for a third term from June 1935 to May 1937.

In his speech, Baldwin warned that Britain's geographical insularity no longer guaranteed protection from air raids. He expressed serious concern about the destructive potential of enemy bombers, emphasising that despite advances in air defence, there was little that could be done to fully prevent such attacks.

This apprehension was partly informed by memories of World War I, during which Germany conducted bombing campaigns over Britain using Zeppelins and early aircraft. Baldwin's assessment reflected the widespread anxiety of the period regarding the increasing threat posed by aerial warfare, which had developed considerably since the previous conflict.

His statement highlighted a sense of vulnerability and the pressing need for Britain to rethink its air defence strategy in response to technological advancements in aviation and the changing nature of warfare.

---

199 Zimmermann 1992, pp. 342–343.

200 Baldwin 1999, pp. 124–127; Ball 2019, p. 21.

His startling statement was:

> "...I think it is well also for the man in the street to realise that there is no power on earth that can protect him from being bombed. Whatever people may tell him, the bomber will always get through, and it is very easy to understand that, if you realise the area of space. I said that any town within reach of an aerodrome could be bombed..."

He argued that offensive action was the only way to ensure their survival and went on to say:

> "..The only defence is in offence, which means that you have to kill more women and children more quickly than the enemy if you want to save yourselves."[201]

Later, on 30 July 1934, he said:

> "When you think of the defence of England you no longer think of the chalk-cliffs of Dover; you think of the Rhine. That is where our frontier lies."[202]

The message conveyed by this statement was that, owing to the absence of effective technical means of defence, the only viable response to aerial attacks was a counteroffensive, which ultimately would not spare the civilian population either. Baldwin's speech left Members of Parliament feeling uneasy and uncertain, particularly when he unequivocally asserted that the bomber would always get through. This statement was regarded as discouraging by some and prompted debates concerning the efficacy of disarmament and rearmament initiatives.[203]

Another key aspect of early twentieth-century warfare was the advancement of submarine technology and its associated operational capabilities. The invention of the submarine dates back to the nineteenth century. In Germany, however, submarines were at first assigned only a subordinate role in comparison with the

201 Baldwin 1999, p. 125; Burns 2013, p. 95; Zimmerman 2013, p. 61; Buderi 1997, p. 53; White 2007b, p. 1.
202 Messerschmidt 1989, pp. 708–709, footnote 43; Ball 2019, p. 21.
203 Baldwin 1999, p. 127; Fisher 1989, pp. 81–83.

Hochseeflotte.[204] At the time, Germany possessed the second largest battle fleet in terms of tonnage after Great Britain, although naval powers such as the United States and France were also significantly expanding their fleets. German military planners only gradually began to recognise the submarine as a viable and strategically valuable weapon against enemy warships.

Although submarine warfare proved surprisingly effective from 1915 onwards, challenging the dominance of the British Royal Navy, German decision-makers never established a clear and consistent strategic vision for future naval armament before the war's end in 1918. Despite growing doubts about its practical effectiveness, the German Hochseeflotte[205], remained a powerful symbol of national strength and maritime prestige.[206]

In retrospect, the initial underestimation of submarine warfare proved short sighted, as submarines would later become a central element of naval strategy in World War II. Meanwhile, the foundational principles of radar technology in Germany were already demonstrated publicly before World War I.

On 18 May 1904, Christian Hülsmeyer[207] presented his Telemobiloskop beneath the Dom Bridge in Cologne as a device to detect ships on the Rhine. This early invention can be regarded as a precursor to later radar systems.[208] Unlike the successful development of airplanes and submarines, Hülsmeyer's invention received little support and soon fell into obscurity. The concept of using reflected electromagnetic waves for military detection purposes was not seriously pursued again until the early 1930s, when multiple countries independently resumed research into radar technology.[209]

Leo Brandt reported later:

> „Die Tatsache, daß bereits im Jahre 1904 der deutsche Ingenieur *Hülsmeyer* in Deutschland und England Patente für ein Gerät erwirkt hat-

---

204 The first German submarine, U-1, was put into service in 1906.

205 The German Hochseeflotte was the main battle fleet of the German Navy before and during World War I.

206 Epkenhans 1997, p. 334.

207 Jäger und Heilbronner 2010, pp. 205–206; Weiher 1983c, pp. 83–85; Weinreich 2001b, pp. 102–103; see also Chapter 5.2.

208 See also Chapter 5.2.

209 Klausing und Holpp 2022, pp. 22–33; Watson 2009, pp. 37–378.

> te, mit dem er die von einem Schiff zurückgestrahlten Funkwellen aufnehmen konnte, war in beiden Ländern vergessen worden. Wenn man heute diese Patentschriften betrachtet, so erscheint es völlig unverständlich, warum dieser Weg drei Jahrzehnte lang nicht begangen wurde…“[210]

In Germany, radar development from 1933 onwards was primarily driven by the German Navy's demand for a detection system superior to acoustic methods. Naval artillery required an all-weather targeting solution, while anti-aircraft artillery needed a weather-independent tracking device. In Great Britain, the concept of using electromagnetic waves to detect incoming aircraft emerged in 1934 from a theoretical analysis of various positioning methods across the electromagnetic spectrum. Ultimately, radar emerged as the most effective option.

In both countries, it was increasingly recognised that conventional methods such as acoustic detection systems and optical observation posts were of limited use for the early warning of enemy aircraft, which were becoming faster and capable of reaching higher altitudes. At night, powerful searchlights, guided approximately by sound detectors, were deployed and could reach distances of up to 10 kilometres. In poor visibility conditions such as rain, fog, or clouds, only the imprecise acoustic systems remained operational.[211] Nonetheless, both Germany and Great Britain continued to rely on traditional observation posts equipped with optical devices as well as acoustic listening methods throughout World War II to assist in the reconnaissance of approaching aircraft. Despite this, radar quickly became the dominant technology for aerial detection.

In autumn 1933, the first attempts to use reflected electromagnetic waves instead of sound waves for target detection were undertaken at the NVA in Kiel under the direction of Rudolf Kühn-

210 Brandt 1962c, p. 53. "The fact that as early as 1904 the German engineer Hülsmeyer had obtained patents in both Germany and England for a device capable of detecting radio waves reflected from a ship had been forgotten in both countries. Looking at these patent documents today, it seems utterly incomprehensible why this path was not pursued for three decades…" (translated by the author).

211 Trenkle 1986, p. 36.

hold.[212] These early efforts initially focused on locating ships but were soon expanded to include air surveillance and fighter direction.[213]

In Great Britain, similar developments were initiated in June 1934 by Albert Rowe[214] at the British Air Ministry, where an evaluation of existing detection technologies for air defence was conducted. There, too, it was firmly concluded that conventional systems, especially acoustic detection, were no longer sufficient for early warning.

Rowe explicitly rejected the idea of continuous airborne patrols, as they would have required far more aircraft and personnel than were available. In a formal memorandum, he called for scientific research into a new technological solution to support national air defence.[215] Shortly thereafter, several approaches were investigated, including infrared detection, barrage balloons, aerial mines, and parachute bombs.[216]

As in Germany, it ultimately became evident that the analysis of reflected electromagnetic waves offered the most promising and effective basis for detecting and repelling enemy air attacks.

In the period that followed, it became apparent that, in addition to technological progress, the ability to implement innovations swiftly and without bureaucratic obstacles also played a decisive role.

A comparison of the major innovation steps in Germany and Great Britain before and during World War II reveals that both countries developed new technologies and applications in parallel. However, Great Britain also introduced entirely new innovations in the areas of strategic and organisational planning, underpinned by a clearly structured innovation process.

This process was largely in place in Britain from the outset. In contrast, Germany lacked key elements of such a process un-

---

212 Reuter 1971, p. 18, pp. 235–238; Weinreich 2001g, pp. 108–109; Surhone 2010, pp. 1–2.

213 See also Chapter. 7.1.1.1.

214 Watson 2009, pp. 60–62; Putley 2009, pp. 45–48; Swords 2008, p. 84; Lovell 2004a; Latham and Stobbs 1999, pp. 31–33.

215 See also Chapter 7.1.2.1.

216 Latham and Stobbs 1999, p. 8.

til the end of the war. Moreover, the early successes of Blitzkrieg led to serious errors in both technological and personnel planning, which ultimately contributed to Germany's weakening and the Allies' growing advantage from the middle of the war onwards.

Technical innovations can be broadly analysed in terms of their drivers and the scope of their impact. They may be prompted by a technological push, or a demand pull.[217] Innovations tend to succeed when these two forces converge. The impact of an innovation can be categorised as either radical or incremental.
Radical innovations typically arise from major technical or scientific breakthroughs and are often based on new key technologies. They lead to fundamental changes in existing systems or practices. Incremental innovations, by contrast, focus on improving existing products by adding new features, properties, or functionalities.[218]

The invention of the cavity magnetron at the University of Birmingham in 1940 serves as a prime example of a radical innovation. It marked a major technological advance that significantly enhanced the performance of airborne radar systems, giving the Allies a decisive advantage during World War II.[219]

One element for assessing the performance and limits of existing technologies is the Technology S-Curve, a building block of strategic innovation management. The S-Curve concept, as proposed by Richard Foster at McKinsey and based on the Arthur D. Little technology life cycle model, shows the progression of a technology's performance as a function of research and development expenditure during each phase of its life cycle.[220] Technological development then asymptotically reaches its performance limit. In this phase, only incremental increases in performance can be achieved. The technology is mature, and even increased expenditure on research and development will only bring a small improvement. At this point, a decision must be made as to which substitution technology should replace the existing one. As part

217 Hauschildt et al. 2016, pp. 4–5; Vahs und Brem 2013, pp. 111–114.

218 Vahs und Brem 2013, p. 67.

219 See also Chapter 6.2.2.

220 Vahs und Brem 2013, pp. 123–124; Gausemeier et al. 2009, pp. 154–156; Foster 2006, pp. 39–47, pp. 111–116; Foster 1986, pp. 31–32; Krubasik 1982, pp. 28–33.

of research and development planning, both the right time for the transition to the substitution technology must be identified and the question of which technological trends will shape the future and should therefore be pursued must be answered. Once this question has been resolved, a comprehensive concept for technology management must be developed as part of strategic planning, and all necessary resources must be allocated.

This strategic understanding of the Technology S-Curve also played a significant role in military innovation during World War II. For instance, in radar development, the transition from low-frequency systems to high-frequency cavity magnetron-based systems marked a shift from incremental improvement to a radical technological leap. Such breakthroughs illustrate the importance of recognising the limits of existing technologies and timely transitioning to superior alternatives within a structured innovation process.

A comparison of the S-Curves between Germany and Great Britain regarding the status of existing technologies shows that in a relatively short period in 1933/34, the two countries independently recognised, although from different perspectives, that conventional means such as optical instruments, underwater sound detection or listening devices could no longer be relied upon to counter future threats. Fig. 2 and 3 illustrate the trajectories of the respective S-Curves in both countries with regard to radar technology.

The comparison of significant innovations in Germany and Great Britain before and during World War II should not be limited to the development of technical equipment, even though technological advances and their applications progressed simultaneously. In Great Britain, however, it was primarily strategic and organisational innovations, consistently implemented even before reaching full practical maturity, that substantially contributed to the country's growing advantage as the conflict unfolded.

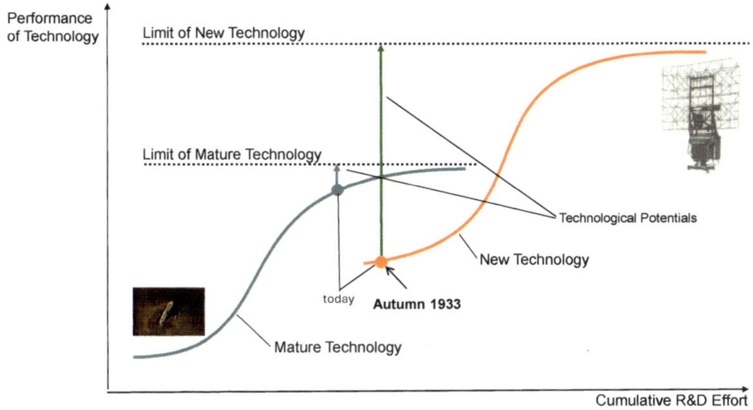

Fig. 2: S-Curve for Germany[221]

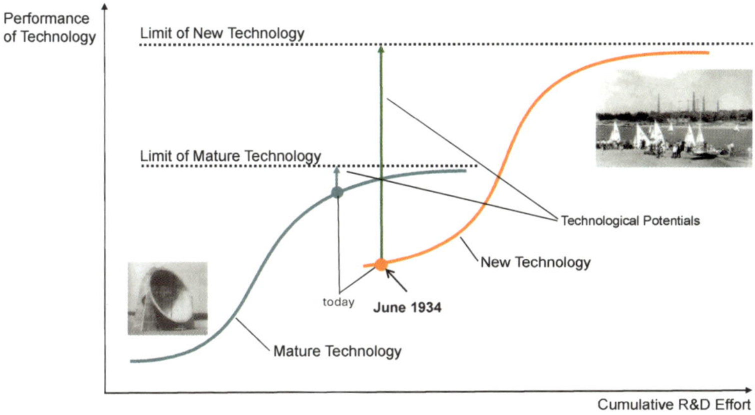

Fig. 3: S-Curve for Great Britain[222]

3.2 Milestones

Major radar development in Germany began in 1937/38 with GEMA's work on the Seetakt[223] radar system for the German Navy,

221 Author's own illustration.

222 Author's own illustration.

223 German naval radar used during World War II to detect targets typically operated at ranges between 10 and 25 kilometres, depending on environmental conditions and the size of the target.

building on earlier experimental work from 1934/35 to 1936. Initially, a small series of about five units was produced in 1937/38 for installation on ships. In 1938, the armoured cruiser Admiral Graf Spee became one of the first vessel to be equipped with a Seetakt device from this series.

In the same year, GEMA received an order for a pre-production series of 31 devices, employing a longer wavelength compared to the initial models, which was delivered from October 1939 to December 1940. At the request of the German Air Force, the Freya[224] radar family was subsequently developed.

With its longer wavelength and greater range, it was initially used for air surveillance and early warning of approaching aircraft. Later, it was also employed to direct anti-aircraft artillery and achieved ranges of up to 150 kilometres. By the end of 1939, following the outbreak of World War II, GEMA had delivered seven Freya devices and four Seetakt devices. The first Freya radar was installed in the spring of 1939 in the Riesengebirge[225], shortly after the German occupation of Czechoslovakia. The seven Freya units were deployed on various islands along the North Sea coast.
By the end of the war, more than 2,000 Freya devices had been manufactured, incorporating various further developments.

GEMA subsequently focused on the further development of existing devices as well as the creation of new systems for air warning, with ranges of up to 300 kilometres.

In March 1939, the Telefunken company presented the first prototype of the Würzburg-A[226] fire-control radar, which had a range of up to 30 kilometres. Testing took place later that year, including at the Rechlin testing centre and with anti-aircraft units. Although development was largely complete by late 1939, serial production and deployment of the Würzburg-A began only in early 1940. The main German radar systems were thus the Freya de-

224 Watts 2025a, p. 1, pp. 3–26, p. 28; German air warning radar used during World War II to detect enemy aircraft typically had a range of up to 120 kilometres, and under optimal conditions could reach about 150 kilometres.

225 Riesengebirge: Giant Mountains (translated by the author).

226 First operational German fire-control radar, developed by Telefunken and first presented in 1939; serial production and deployment began in early 1940. It was used to direct anti-aircraft artillery and had a range of up to 30 km.

vice, used for air warning with a range of over 100 kilometres, and the Würzburg device, which offered higher resolution capabilities and an initial range of 30 kilometres. From 1941 onwards, the Würzburg-Riese was introduced as a target-tracking radar to support fighter guidance. It featured a parabolic reflector with a diameter of 7.4 metres and a range of up to 70 kilometres against aerial targets. By the end of the war, around 4,000 Würzburg radar devices had been produced.[227]

While Germany concentrated on the continuous refinement of existing radar systems, the United Kingdom commenced its initial experiments with radar technology for air warning in 1935, primarily along the East Coast.

The fixed Chain Home (CH)[228] radar stations can be regarded as a rough British counterpart to the German Freya radar. Operating at longer wavelengths than Freya, CH offered lower resolution and was initially less technologically advanced in terms of target precision. Nevertheless, it provided effective long-range early warning of incoming aircraft.

However, CH radars were capable of detecting high-flying aircraft at ranges of up to 190 kilometres under favourable conditions, providing vital early warning despite their technical limitations.

Nonetheless, the United Kingdom integrated the CH radar stations into a comprehensive air defence network at an early stage. This holistic approach to countering German air attacks fostered not only technical innovations but also substantial advancements in organisational structures and operational procedures. Beyond radar, radio direction finding and jamming techniques, signals intelligence efforts at Bletchley Park[229], and the introduction of Op-

227 See also Chapter 7.2.1.1.

228 UK's first operational early-warning radar network, established in the late 1930s. Operating at longer wavelengths than German systems like Freya, it had lower resolution but played a crucial role in Britain's comprehensive air defence during World War II.

229 Main British codebreaking centre during World War II, where German encrypted communications were decrypted, significantly aiding the Allied war effort.

erational Research[230] a concept largely absent in Germany played particularly significant roles.

The formulation of an effective and comprehensive air defence strategy was made possible by the generally close and efficient collaboration among political leaders, military authorities, scientific experts, and industry stakeholders. Nevertheless, a closer look at this cooperation reveals various conflicts over military priorities. While these tensions undoubtedly caused delays in the development of radar technology, they did not lead to critical misjudgements, as occurred in Germany, that fundamentally weakened technological progress or ultimately the UK's defensive capabilities.

At the time of the declaration of war on Germany on 3 September 1939, 18 Chain Home radar stations were already operational across Great Britain and integrated into a networked information system with clear communication and command structures. Two additional stations in Scotland were operational but not yet connected to this system. Furthermore, Great Britain began developing radar equipment for locating air and sea targets aboard aircraft earlier than Germany.

By the end of 1937, the two projects, ship detection and air interception, were designated with the abbreviations ASV[231] and AI.[232] In the following year, the focus shifted to installing radar equipment in aircraft of the Royal Air Force (RAF) and the Royal Navy.

By April 1941, around 110 British aircraft had been fitted with early airborne radar systems, most of which were still in development or testing, operating at a wavelength of approximately 1.5 metres. Experimental systems, such as the AI Mk. III developed between 1939 and 1941, had detection ranges of about 3 to 5 kilometres, enabling night fighters to intercept enemy bombers more effectively.

230 Multidisciplinary approach developed in the UK during World War II, using mathematics, statistics, and scientific analysis to optimise military operations and improve strategic decision-making.

231 Air to Surface Vessel (ASV): Radar system developed for detecting ships and submarines from aircraft, primarily used by the Royal Navy.

232 Airborne Interception (AI): Radar system designed for detecting and tracking enemy aircraft, mainly used by the Royal Air Force for fighter guidance.

In April 1942, the Halifax V9977 undertook its first flight with a radar system operating in the centimetric wave range at a wavelength of approximately 10 centimetres, offering all-round visibility, although its performance was only moderately successful.

In Germany, the Telefunken company brought the first operational airborne radar, the Lichtenstein BC[233], into service in autumn 1941. Its performance was broadly comparable to that of the British AI Mk. III, with a similar detection range of around 3 to 5 kilometres. The radar operated at a wavelength of approximately 61 centimetres and required a bulky antenna array mounted on the aircraft's nose.

Subsequently, Telefunken developed the Berlin A radar set for night fighters, operating at a much shorter wavelength of approximately 10 centimetres. Technically comparable to the British AI Mk. VIII, which had been under development since 1941 and became operational in 1943 with detection ranges of up to 9 to 10 kilometres, the Berlin A saw only very limited operational use. A further improved version, known as the Berlin N1A, was developed for night fighter operations but did not enter mass production before the end of the war. As the war progressed, the dynamic between Germany and Great Britain changed significantly, both strategically and technologically.

While Germany initially focused on offensive operations, it was increasingly pushed onto the defensive. In contrast, Great Britain's early focus on defence against enemy air attacks gradually shifted towards offensive operations. This transition was particularly driven by major technological advances, notably the development of high-resolution centimetric radar systems that were installed in aircraft. For a long time, Germany was unable to counter this technological breakthrough. In addition, Great Britain had a well-developed concept for the centralised networking and analysis of all information gathered through air warning.

It is undisputed that radar played a significant role in the advantage enjoyed by the Allied forces during World War II. However, the often-cited claim that radar technology was the main reason for

233 First operational airborne radar of the German Luftwaffe, introduced in autumn 1941. It enabled night fighters to detect enemy aircraft at ranges of approximately 3 to 5 kilometres.

the Allied victory must be critically examined in light of current knowledge.

The frequently cited statement by many radar specialists, "The bomb may have ended the war, but radar won it"[234], which was often repeated after the atomic bomb was dropped in August 1945, referred only to specific missions. It should be noted that at the time, the work of the cryptologists at Bletchley Park and the findings of Operational Research were still largely unknown.

This assessment should also be seen in the context of the fact that the highly successful signals intelligence operations at Bletchley Park remained secret for decades after the end of the war. It was the integration of radar with signals intelligence, radio direction finding, jamming, and operational research that laid the foundation for the Allies' growing advantage throughout the conflict.

Innovative cooperation played a decisive role in the progress made in the development and production of technologically advanced radar equipment. The collaboration between Great Britain and the United States in the field of centimetre wave technology was particularly significant in this regard. The journey of British scientists to North America in 1940, led by Henry Tizard[235] and known as the Tizard Mission[236], is an outstanding example of successful cooperation between scientists, the military, and industry in both countries, despite occasional serious tensions between the parties involved. The framework for this collaboration had been agreed in advance at the highest political level by Winston Churchill and US President Franklin D. Roosevelt.

In Germany, close and targeted cooperation between state institutions, the Wehrmacht, industry, and research bodies began under industrial leadership with the Arbeitsgemeinschaft Rotterdam (AGR)[237], after a radar device recovered from a British bomber shot down near Rotterdam on 3 February 1943 was found to operate in the centimetre wave range. Despite considerable progress, the Allies' technological lead could no longer be overtaken by the end of the war.

---

234 Brown 1999, p. x.

235 Piller 2008; Clark 1965; Watson 2009, pp. 42–43.

236 See also Chapter 10.5.2.1.

237 See also Chapter 10.5.1.1.

3.3 Turnaround

In Germany, radar technology focused on the continuous refinement of existing devices, resulting in steady incremental improvements in performance. At the start of the war, Germany possessed technically advanced equipment operating at wavelengths of 2.4 metres, 80 centimetres, and 50 centimetres. At that time, there was little knowledge of Allied developments in the shorter wavelength range, providing no apparent incentive to pursue further research into other wavelength bands. The existing systems delivered excellent performance according to their specifications. Consequently, further technological advancements, such as those in centimetre wave technology or antenna frequency agility, were abandoned, despite valuable research into centimetre wave technologies having been conducted as early as the 1920s.[238] The limited research capacity in this field was ultimately discontinued in November 1942. These significant misjudgements resulted in a developmental backlog that Germany was unable to overcome.

According to the war diary of the Oberkommando der Wehrmacht (OKW), 1943 marked a turning point in Germany's offensive behaviour:

> „Der Scheitelpunkt einer militärischen Angriffshandlung ist dann erreicht, wenn es nicht mehr gelingt, den Angriff weiter zu führen, wenn die Kraft nicht mehr ausreicht, um dem Gegner das Gesetz des Handelns vorzuschreiben. In der Taktik erfolgt sodann in der Regel der Übergang zur Verteidigung, es kommt der Moment des Stillhaltens, des zähen Abringens der Kräfte oder auch die Umgruppierung, ja die Zurücknahme der eigenen Truppe in die Ausgangsstellung."[239]

---

238 Although centimetre waves were successfully generated and studied in laboratories in the 1920s, practical measurement tests and radar experiments using these wavelengths were not undertaken in Germany until the autumn of 1933 by Abraham Esau. He began practical research on short centimetre waves from 1933 onwards, while systematic investigations of millimetre waves only started around late 1943, which later contributed to developments in radar and high-frequency technology. In contrast, the first British experimental devices operating at a wavelength of approximately 9 centimetres were developed around 1939–1940, representing a major step forward in radar technology.

239 Schramm 1963, p. 1489; "The culmination point of a military attack is reached when it is no longer possible to continue the offensive, when the force is insuf-

This fundamental assessment by the German Wehrmacht leadership in 1943 is further corroborated by the trajectory of research and development in German radar technology up to the end of the war, particularly following the acute realisation in February 1943 of the advanced capabilities of British radar systems operating within the centimetre-wave spectrum. In addition to severe shortages of personnel and material resources, the OKW identified two pivotal events in the summer of 1943 as critical factors:

> „....die in ihren Auswirkungen Katastrophen befürchten ließen: die Luftherrschaft über dem Reichsgebiet fiel der alliierten Luftwaffe zu, die nun ohne wesentliche Störungen eine Stadt nach der anderen mit schwersten Angriffen, bald auch bei Tage, heimsuchte. Und schlagartig hörte infolge der Überlegenheit der gegnerischen Ortungs- und Waffenwirkung der Unterseebootkrieg auf...“ [240]

The bombing raid on Hamburg during the night of 30/31 January 1943 marked the first instance in which British Pathfinder[241] aircraft, equipped with airborne centimetre-wave radar, supported the operation.[242] Further devastating attacks on Hamburg occurred between 24 July and 3 August 1943 as part of Operation Gomorrah. The initial major raid by the Royal Air Force took place on 24 July and continued on 25 July. Alongside the British assaults, United States Army Air Forces also participated in the offensive. Particularly catastrophic was the second major RAF raid on the night

ficient to dictate the course of action to the enemy. At this stage, a transition to defence usually occurs, followed by a period of holding ground, the tenacious depletion of forces, regrouping, or even the withdrawal of one's own troops to their starting positions." (translated by the author).

240 Ibid., p. 1497. "...which threatened to have catastrophic consequences: air superiority over the Reich's territory was conceded to the Allied Air Force, which was thereafter able to conduct sustained and devastating raids on city after city, soon including daylight bombing. And suddenly, due to the enemy's superiority in detection and weaponry, the submarine campaign came to an abrupt halt." (translated by the author).

241 Specialised RAF bombers equipped with advanced navigation and target-marking technology, including radar, tasked with leading bomber formations to their targets. They improved bombing accuracy by marking targets with flares or incendiary bombs, enhancing the effectiveness of Allied air raids.

242 Lovell 1991, p. xviii; Burns 2013, pp. 379–385; Harris 1995, Appendix A, No. 74, p. 70.

of 27/28 July, which involved extensive area bombing and resulted in firestorms that caused widespread devastation across the city.[243] The bombing of Hamburg led to a catastrophic loss of life and left the city in ruins, starkly exposing the immense suffering and humanitarian consequences of the air war.

British-American cooperation in radar development, initiated in 1940, led to a more effective utilisation of radar systems against German submarines during World War II. In May 1943, despite a renewed increase in the number of German U-boats operating in the North Atlantic, their operational successes remained limited, while they suffered sustained and untenable losses.[244] This was attributable to new radar systems with enhanced resolution and omnidirectional detection, combined with short-wave direction finding and the decryption of the Navy's ENIGMA, all of which significantly facilitated the anti-submarine campaign.[245]

3.4 Debates and Uncertainties

Some historians, including Rohwer, who contributed to the proceedings of an international historical symposium in 1981[246], identify late 1941 as a turning point in World War II, primarily reflecting the global escalation marked by the German invasion of the Soviet Union in June 1941 and the entry of the United States into the war in December 1941. From a strategic and operational perspective, however, 1943 appears more decisive for the Allies, due to technological advances in radar, short-wave direction finding, cryptanalysis and operational research. Nevertheless, Rohwer argues that the war was, in practical terms, already lost for Germany by 1941. Structural deficiencies, including a two-front war, limited resources, industrial inferiority relative to the Allies, and the neglect of modern technologies such as radar and high-frequency equipment, made sustained warfare impossible, even if Germany initially achieved military successes.

243 Lovell 1991, pp. 173–179.

244 Schramm 1963, p. 1439; Dönitz 2018; pp. 397–422, pp. 483–484; Rahn 1990, p. 84.

245 Dönitz 2018, pp. 484–486; Gießler 1971, pp. 80–82.

246 Rohwer und Jäckel 1984.

4 Basics of Radar Technology

4.1 Radar Principle

Radar is a technique that utilises electromagnetic waves, either reflected or automatically returned, to obtain information about distant objects. The term radar is an acronym derived from Radio Detection and Ranging.[247]

In its basic form, a radar system detects targets by transmitting electromagnetic pulses and analysing the signals reflected back from objects within the beam path. In contrast, secondary radar systems function by transmitting an interrogation signal, which prompts an automatic response from a transponder located on the target, typically at a different frequency.

In general usage, the term radar typically refers to primary radar, unless explicitly stated otherwise. Building on this foundation, radar techniques are employed to support spatial orientation beyond the optical line of sight.

In comparison to optical systems, radar offers lower lateral resolution, and its range resolution is likewise limited.[248] However, the primary frequency bands used for radar currently span from 1 GHz to 100 GHz, corresponding to wavelengths between 30 centimetres and 3 millimetres.

The principal advantage of radar lies in its ability to function over significantly greater distances, independent of lighting or weather conditions. Moreover, radar can simultaneously determine both the distance and velocity of detected targets.

Like any active imaging method, radar can be jammed. During World War II, countermeasures such as Window, Chaff, or Düp-

247 Further literature can be found in Skolnik 2008b; Skolnik 2002; Wehner 1987; Barton 1985; Hovanessian 1984; Ulaby et al. 1981. For German-language sources, Göbel 2011; Klausing und Holpp 2000; Huder 1999; Ludloff 1993.

248 One notable exception is Synthetic Aperture Radar (SAR), which will be briefly touched upon later, since it was developed after 1951 and will not be covered in detail in this book.

pel[249] were used for this purpose. These consisted of numerous small metallic strips or fibers that acted as dipole reflectors. When released in large quantities, they formed radar-reflective clouds that confused enemy radar systems. Since these early forms of interference, electronic countermeasures (ECM) have become significantly more advanced. In addition to passive decoys, modern ECM systems use radar jammers that emit targeted radio signals to disrupt enemy tracking. Some systems go even further by intercepting incoming radar pulses, storing them digitally, subtly modifying certain characteristics, and retransmitting them. This creates false targets or misleading distance information on enemy radar displays, making it much harder to track actual targets.

What began as simple passive interference has evolved into a complex system of active electronic deception.

Radar operates on the echo principle. A transmitter emits focused electromagnetic waves into the observation area via an antenna. These waves are partially absorbed and partially reflected by inhomogeneities and targets (Fig. 4).
The signal energy reflected back to the receiving antenna is processed by the receiver, which analyses amplitude, frequency, phase, polarisation, and time of flight to locate and measure the target. In a narrower sense, a target refers to an object of interest for search, tracking, or measurement, characterised by its target parameters.[250]

Radar fundamentally seeks to address the following enquiries regarding the environment under observation:

- Are there any targets? → Target detection
- Where are the targets located? → Target location
- How many targets are there? → Resolution capability
- What characteristics do the targets have? → Velocity, structural properties.

In accordance with these enquiries, radar systems are employed to detect targets and ascertain their positions in terms of both direction and range. The relative spatial distribution of multiple targets

249 GB: Window; USA: Chaff; D: Düppel.

250 Klausing und Holpp 2000, p. 1; Skolnik 2008a, Chapter 1.1-1.3.

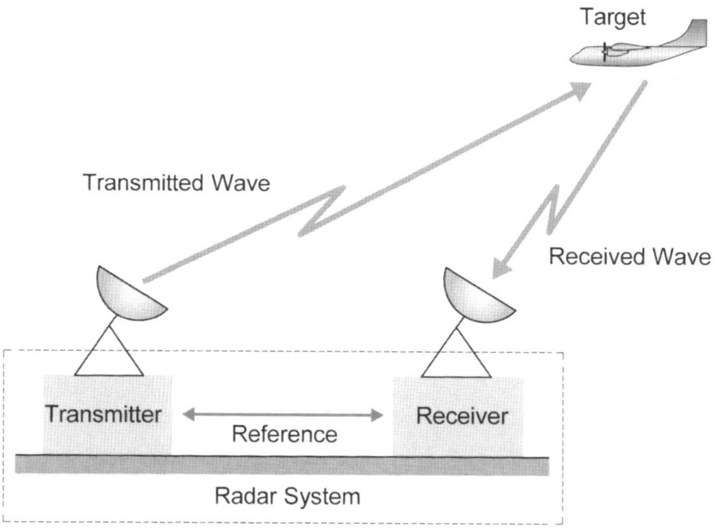

Fig. 4: Radar Principle[251]

can be determined, along with their fundamental attributes and states of motion, including dimensions, approximate velocity vectors, as well as surface texture and reflectivity characteristics.

The accuracy and reliability of the information obtained, as well as the overall measurement performance of radar, are predominantly governed by its operational range, resolution, and measurement precision. These parameters inherently define the scope of feasible applications and the system's limitations.[252]

By analysing the echo signals in the radar receiver, the following information can be obtained for well-defined targets:

- Range
- Angular position:
  - Azimuth (horizontal)
  - Elevation (vertical)
- Height above ground
- Speed (or radial speed)

251 Source: Klausing und Holpp 2000, p. 2.
252 Klausing und Holpp 2000, pp. 1–2.

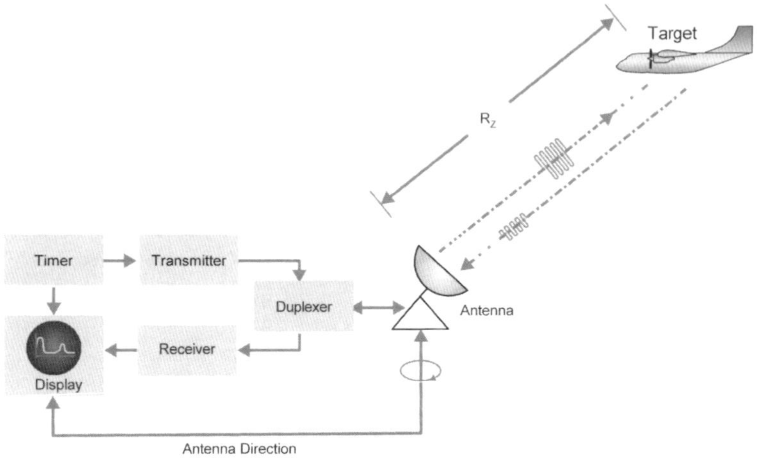

Fig. 5: Pulse Radar[254]

– Size, structure, and material properties.

The range is determined from the time of flight of the radar signal from the transmitter to the target and back to the receiver. The angular position is directly derived from the position of the sharply focused antenna during reception. Radial velocity can be determined from the frequency shift between the transmitted and received signals due to the Doppler effect for moving targets, and the target speed can be calculated from this using additional auxiliary parameters. Furthermore, the intensity, that is the signal amplitude of the backscattered field, provides information about the size and structure of the detected targets.[253]

Pulse radar is the most well-known and widely used radar technique. In classical pulse radar, the two key measured variables, the distance and direction of the target, are determined. Fig. 5 shows a simplified block diagram of a pulse radar system.

During transmission, a high-frequency pulse modulated signal is emitted in a focused manner via the antenna. Pulse modulation is achieved by briefly activating the transmitter using a timer, which defines both the pulse duration and the pulse repetition frequency.

253 Klausing und Holpp 2000, pp. 2–3.

254 Source: Klausing und Holpp 2000, p. 4.

When the same antenna is used for both transmission and reception, a transmit and receive switch (duplexer) is required. This switch connects the antenna to the transmitter during transmission and to the receiver during reception. The transmission pulse is emitted into the observation area via a fixed or rotating antenna. The echo signal reflected from the illuminated target is received by the antenna and routed to the receiver through the transmit and receive switch.

The target distance R_Z is directly derived from the measured time of flight t_L of the radar signal to the target and back.

With the speed of light $c_0 = 3 \cdot 10^8$ m/s, the following applies:

$$2 \cdot R_Z = t_L \cdot c_0.$$

It follows that:

$$R_Z = t_L/2 \cdot c_0.$$

Upon reception, the low-power echo signal undergoes further processing to facilitate visualisation on the display unit. In its most fundamental configuration, the display unit consists of a cathode ray tube (CRT) equipped with a linear time base, which is synchronised by the timer at the precise moment the transmitted pulse is emitted. The received echo signal induces a deflection of the illuminated spot along the horizontal axis, with the magnitude of deflection proportional to the target range. Utilising a calibrated scale, the target distance and corresponding angular position are accurately represented on the screen, referenced against the antenna's azimuth and elevation angles.

The radar operational environment is frequently subject to electromagnetic interference, which may emanate from adjacent radar systems or other proximate radio frequency sources. Furthermore, deliberate electronic countermeasures are employed in military contexts to impair or negate radar detection capabilities. Such measures include the use of jamming transmitters, designed to saturate or confuse radar receivers.[255]

255 Klausing und Holpp 2000, pp. 3–6.

4.2 Types of Radar

As a rule, a single antenna is used for both transmission and reception. Alternatively, two antennas may be employed, provided their separation is small in relation to the distance to the target. A radar system with this configuration is referred to as monostatic radar (Fig. 6). In contrast, bistatic radar systems have separate transmitting and receiving antennas that are located at different sites and are significantly spaced apart (Fig. 7). Multistatic radar involves multiple spatially separated transmitting and receiving antennas.

The radar method discussed so far is referred to as primary radar. In primary radar, the echo signal from a passive target is processed according to the criteria outlined earlier.[256]

Secondary Surveillance Radar (SSR), in contrast to primary radar, does not rely on echoes from passive targets. Instead, it utilises an active response device known as a transponder which is located on board the target, typically an aircraft (Fig. 8). In this system, each interaction with the ground station consists of an interrogation signal which is automatically answered by the transponder.

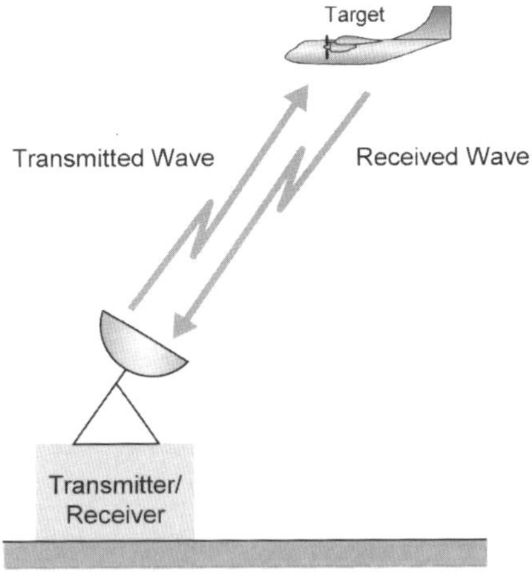

Fig. 6: Monostatic Radar[257]

256 Klausing und Holpp 2000, p. 6.
257 Source: Klausing und Holpp 2000, p. 7.

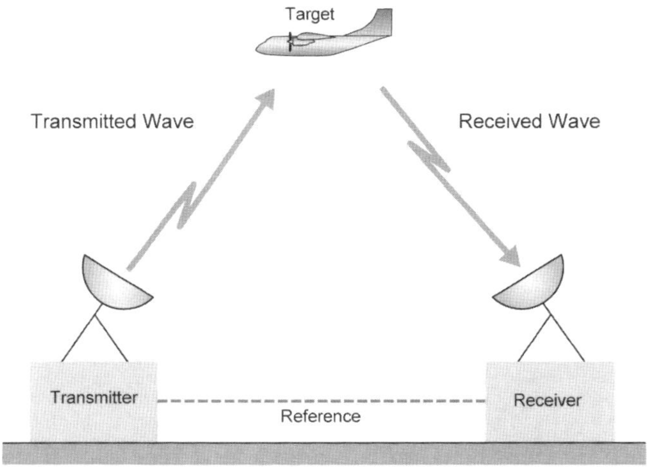

Fig. 7: Bistatic Radar[258]

Secondary radar is primarily employed in civil and military aviation. The response signals transmitted by the transponder provide information such as flight identification number, altitude and, particularly in military applications, friend or foe identification.[259]

4.3 Resolution

A comparison between optical systems and imaging systems operating in the radar frequency range reveals that the image quality achieved by radar falls far short of that attainable with optical methods. Radar waves are up to 100,000 times longer than those visible to the human eye, resulting in significantly coarser angular resolution.

The angular resolution α_w, is primarily determined by the wavelength λ of the transmitted signal, which corresponds to the transmission frequency $f_T = \frac{c_0}{\lambda}$, as well as by the physical dimensions D of the antenna.

258 Source: Klausing und Holpp 2000, p. 7.
259 Klausing und Holpp 2000, pp. 6–8.

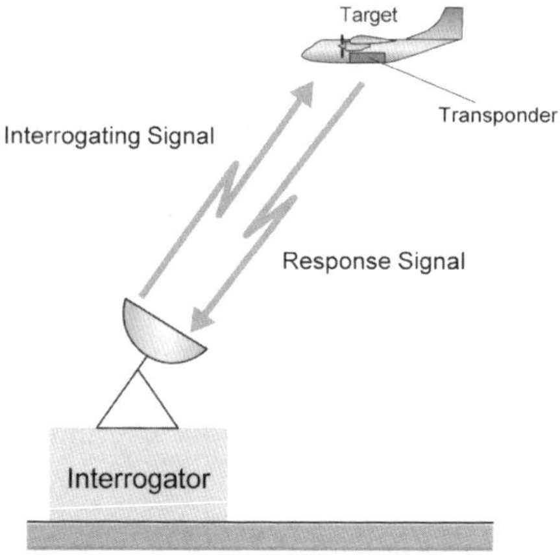

Fig. 8: Secondary Surveillance Radar[260]

The following applies to the angular resolution α_W:

$$\alpha_W \approx \frac{\lambda}{D}.$$

This results in the lateral resolution ΔX at the distance R_Z:

$$\Delta X = \alpha_W \cdot R_z \approx \frac{\lambda}{D} \cdot R_{z.}$$

Fig. 9 shows the relationship between the angle (α_w) and the lateral resolution ΔX.[261]

In order to achieve satisfactory angular or lateral resolution, the transmission frequency f_T can be increased, meaning that the wavelength λ is reduced, or the dimensions of the antenna D can be increased while maintaining the same distance to the target R_Z. However, mechanical limitations are quickly encountered, such as installation constraints on aircraft or, in the case of omnidirectional radar systems, restrictions imposed by the required rotational speed.

The main advantages of shorter wavelengths include:

- Sharper focusing of the antenna beam, resulting in higher resolution

260 Source: Klausing und Holpp 2000, p. 7.

261 Klausing und Holpp 2000, pp. 10–13, pp. 88–90.

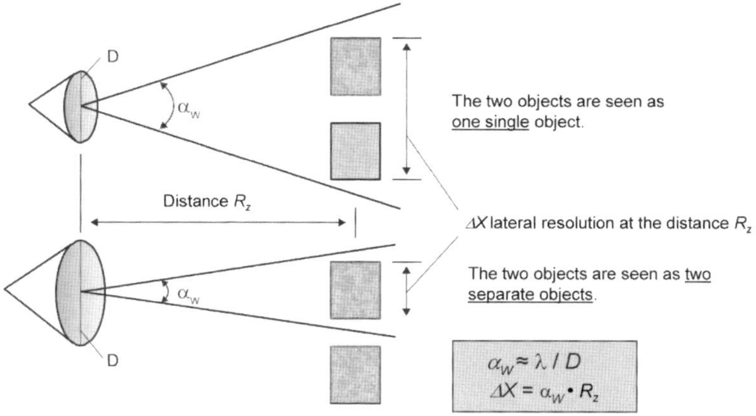

Fig. 9: Relationship between angle and lateral resolution[262]

- Smaller installation dimensions
- The capability to image the Earth's surface
- High immunity to interference.

With shorter wavelengths, that is, higher frequencies, lateral resolution improves while antenna dimensions remain unchanged. However, the shift toward higher frequencies in the centimetre and millimetre wavebands is constrained by increased atmospheric attenuation, primarily due to molecular absorption by oxygen and water vapour, as well as meteorological factors such as precipitation and fog. As frequency rises, either the transmitted power must be increased or the operational range reduced. Since transmission power cannot be scaled proportionally to fully compensate for greater attenuation, the achievable range decreases at progressively higher frequencies. Fig. 10 illustrates the inverse relationship between resolution R and attenuation A as a function of wavelength λ.

For this reason, generating the highest possible transmission power at short wavelengths was crucial and decisive for the deployment of radar equipment on aircraft during World War II. These

262 Source: Klausing und Holpp 2000, p. 10.

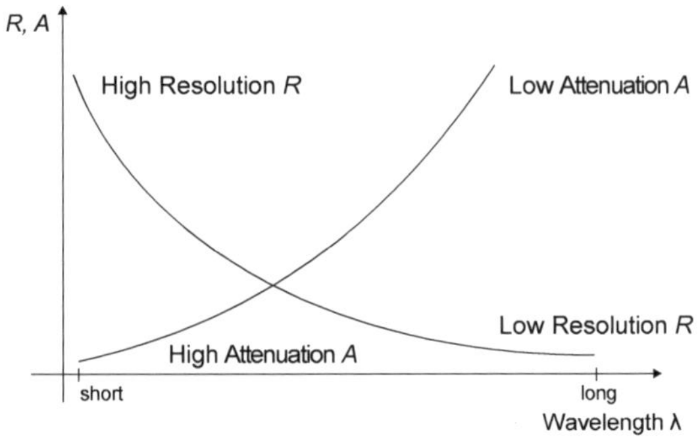

Fig. 10: Relationship between resolution *R* and attenuation *A*[263]

systems were primarily used to map the Earth's surface, akin to creating a chart, in order to accurately guide bombers during attacks on German cities. The use of radar in the centimetre waveband was also vital in the campaign against German submarines. Besides angular and lateral resolution, range resolution played an important role in the early detection and location of targets.

The range resolution ΔR_{Zmin} is achieved by emitting electromagnetic energy in short pulses of duration τ_P, which travel to the target and back at the speed of light. Range resolution is directly related to pulse length: the shorter the pulses, the better the resolution capability. Early advances in pulse technology were made in both Germany and Great Britain. However, for angular, lateral, and range resolution alike, the primary challenge was to generate as much transmitted power as possible at higher frequencies.[264] The following applies for range resolution:

$$\Delta R_{Zmin} = \frac{c_0 \cdot \tau_P}{2} \; .$$

The geometric dimensions of the resolution cell, defined by the lateral and range resolutions, directly impact the achievable imaging accuracy of the radar system.[265]

263 Source: Klausing 1998, p. 16.

264 Klausing und Holpp 2000, pp. 90–92.

265 Klausing und Holpp 2000, pp. 93–95; Klausing 1989, pp. 12–14; Ulaby et al. 1981, pp. 42–47.

5 Origins of Radar Technology

5.1 Early Phase of Radar Development

The development of radio technology, and later radar, illustrates the close interplay between scientific discovery and technological invention. New physical insights gave rise to innovations and applications across a broad range of fields. This process was accompanied by the continual advancement of powerful components and devices, enabling the realisation of emerging technologies. These technological breakthroughs, in turn, opened up new possibilities with far-reaching social and political consequences.[266]

Guglielmo Marconi[267] played a pivotal role in the early development of wireless communication, demonstrating a notable capacity to translate emerging scientific knowledge into commercially viable technology between 1897 and 1900. Until 1908, Marconi maintained a dominant position by leasing his equipment to shipping companies in conjunction with operators trained by his company. These operators were instructed to communicate exclusively with Marconi stations, except in cases of emergency. This monopolistic practice was first challenged by the International Radiotelegraph Convention (IRC) of 1906, which laid the foundations for international regulatory frameworks and facilitated standardised communication between different radio systems.[268]

---

266 Thurn 1913; Wessel 2008, pp. 87–99; Wessel 2004, pp. 173–216.

267 Dragoni 2011; Jäger und Heilbronner 2010, pp. 273–274; for the diverse activities of Marconi, see also Chapter 11.2.

268 Fuchs 2004, p. 93; Fuchs 1998, p. 116; Blumtritt 1988, p. 69; the International Radiotelegraph Convention (IRC), adopted in 1906 and entering into force in 1908, was one of the earliest international agreements aimed at regulating and standardising radiotelegraph communications. It laid the foundation for coordinated global radio communication and addressed key issues such as frequency allocation and interoperability, including the obligation for wireless stations to communicate irrespective of manufacturer or operator.

Around the turn of the century, intense competition between Ferdinand Braun and Adolf Slaby, and their respective companies Siemens & Halske and AEG, gave rise to numerous patent disputes. Siemens & Halske, in collaboration with Braun, primarily served the Army, while AEG, working with Slaby, mainly focussed on naval applications.[269]

Kaiser Wilhelm II recognised the strategic importance of wireless telegraphy for both military communication and the advancement of German industry.

In order to strengthen competitiveness vis-à-vis other nations, particularly Marconi, the Kaiser intervened in the corporate policies of the two companies and accelerated the establishment of a joint venture.[270] On 27 May 1903, the Gesellschaft für drahtlose Telegraphie m.b.H., Berlin, was founded to combine the strengths of both firms, resolve patent disputes, and develop a unified radio system for both the Army and the Navy. Each company contributed equally to the new enterprise.[271]

On 11 November 1903, the Telefunken trademark was registered with the Imperial Patent Office, and the company was thereafter renamed Gesellschaft für drahtlose Telegraphie m.b.H., System Telefunken.[272] Following a shareholders' resolution on 17 April 1923, the company was renamed Telefunken Gesellschaft für drahtlose Telegraphie m.b.H., Berlin.

5.2 The Telemobiloskop

In 1904, Christian Hülsmeyer[273] conducted the first tests with an early precursor to radar, the so-called Telemobiloskop[274], which

269 Fuchs 2004, pp. 83–91.

270 Fuchs 2004, p. 93; Fuchs 1998, p. 119.

271 Thiele 2003, pp. 18–19; Strunk 1999, pp. 32–33; Friedewald 2001, pp. 27–57; Telefunken 1928, pp. 15–18.

272 Thiele 2003, p. 19; Fuchs 2004, pp. 91–95.

273 Jäger und Heilbronner 2010, pp. 205–206; Weiher 1983c, pp. 83–85; Weinreich 2001b, pp. 102–103.

274 Galati 2016, pp. 1–6; Holpp 2014, pp. 179–183; Bauer 2008, pp. 199–224; Bauer 2005b, pp. 1–18; Bauer 2005a, pp. 1–74; Bauer 2004a, pp. 13–57; Rohling 2004,

was primarily intended to prevent collisions at sea. Although his invention was capable of detecting obstacles such as ships using radio waves, and later even determining their distance, it failed to achieve widespread commercial success. This was largely due to the fact that its technological advantages were not fully recognised at the time, and the maritime industry did not yet perceive an urgent need for such a solution.

Another contributing factor was the absence of a clear conceptual distinction between radio and radar technology. The term radar had not yet been properly defined, which led to confusion regarding the actual functioning of Hülsmeyer's device. While it partially applied the principles of radar detection, it was often regarded merely as a type of radio apparatus.

Hülsmeyer also faced significant challenges in further developing his invention, as he lacked access to key wireless communication patents, which were largely controlled by Marconi. In addition, financial constraints ultimately prevented him from continuing his work or bringing the device to market.

As will be discussed later, Telefunken showed no interest in Hülsmeyer's work or patents during his early experiments in 1904. Similarly, nearly three decades later, in late 1933, no collaboration materialised between Telefunken and the NVA in the field of radiolocation. Once again, Telefunken's considerable expertise in radio technology remained unutilised. Moreover, Hülsmeyer's experiments were not widely known at the time.

Christian Hülsmeyer was born on 25 December 1881 in Eydelstedt, near Bremen. After attending the local village school, he enrolled at the teacher training college in Bremen in 1896, where his aptitude for science was recognised at an early stage. He was granted permission to conduct Hertzian mirror experiments in the college's well-equipped physics laboratory, during which he discovered that electromagnetic waves could be reflected from metallic objects and utilised for object detection. Hülsmeyer developed a particular interest in determining the positions of ships. Motivated by a tragic shipwreck on the River Weser near his hometown during

pp. 227–237; Holpp 2004, pp. 175–177; Pritchard 1992, pp. 12–25; Pritchard 1989, pp. 13–29.

conditions of poor visibility, his inventions and ideas came to focus primarily on maritime obstacle warning.[275]

Hülsmeyer left the teacher training college in June 1900 and began an apprenticeship at Siemens-Schuckert in Bremen. In April 1902, he relocated to Düsseldorf to live with his brother Wilhelm, focusing on the practical implementation of his findings, particularly in the optimisation and advancement of transmission and reception technologies. Through a newspaper advertisement, Hülsmeyer came into contact with the investor and entrepreneur Heinrich Mannheim, with whom he founded the Telemobiloskop-Gesellschaft Hülsmeyer & Mannheim on 5 May 1904.[276] The company's name directly referred to his invention, the device known as the Telemobiloskop.

The invention was granted by the Imperial Patent Office in Berlin on 30 April 1904 under the title Verfahren, um entfernte metallische Gegenstände mittels elektrischer Wellen einem Beobachter zu melden[277] (Fig. 11). On 17 May 1904, Hülsmeyer presented his Telemobiloskop for the first time to representatives of major shipping companies and members of the press at the Domhotel in Cologne.[278] The following day, the first official demonstration was carried out beneath the Dom Bridge, now known as the Hohenzollern Bridge, in Cologne.[279] The apparatus was positioned on the riverbank, with its antennas aligned along the course of approaching vessels. It successfully indicated the approach of a tugboat and its tow, located approximately 100 metres away, by activating a bell signal, which ceased only after the vessels had left the detection range.

275 Bauer 2004a, p. 16.

276 Bauer 2008, pp. 215–216; Jäger und Heilbronner 2010, pp. 205–206.

277 Hülsmeyer 1904; Holpp und Klausing 2022, p. 4. "Method for detecting distant metallic objects and reporting them to an observer using electromagnetic waves" (translated by the author). On 10 June 1904, a patent application was filed in Britain under the title "Hertzian-wave Projecting and Receiving Apparatus Adapted to Indicate or Give Warning of the Presence of a Metallic Body, such as a Ship or a Train, in the Line of Projection of such Waves." It was granted on 23 March 1905 (Bauer 2005a, p. 16; Holpp und Klausing 2022, p. 4). Further patent applications followed in Europe and the USA.

278 Bauer 2008, pp. 207–210; Bauer 2005a; pp. 24–25; Kölnische Zeitung; Kölner Tageblatt.

279 Bauer 2004a, pp. 28–29.

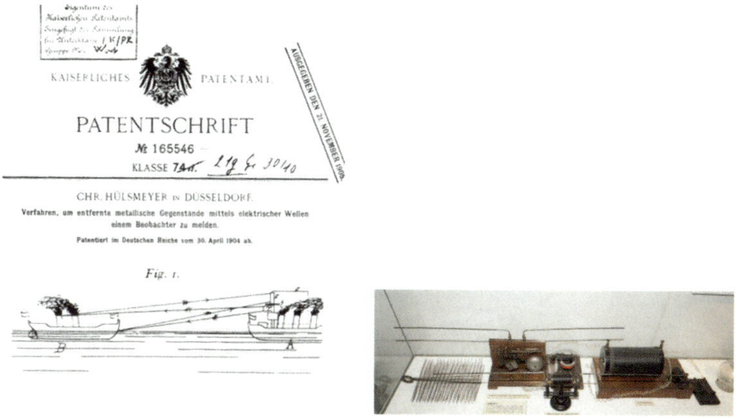

Fig. 11: Telemobiloskop Patent and Experimental Setup[280]

On 9 June 1904, another successful demonstration was held before an international audience on board the tender ship Columbus I of the Holland-America Line (HAL) in the port of Rotterdam, during a shipping conference taking place simultaneously in Scheveningen. A maximum detection range of only a few hundred metres was initially expected.[281] However, direct distance measurement was not yet possible, as the travel-time method using very short pulses was not yet technically feasible.

To address this limitation, Hülsmeyer devised a method for estimating distance by adjusting the angle of inclination of the transmitting antenna and measuring the resulting field strength. Using the known height of the apparatus and the angle of the antenna, the distance could then be calculated based on basic trigonometric principles.[282]

On 9 August 1905, the magazine Prometheus published a detailed report on Hülsmeyer's invention under the title Neue Seessignalapparate.[283] The article provided an in-depth account of the

280 Image credit: Photo left: Stiftung Deutsches Technikmuseum Berlin: Historical Archive; I.2.060-1-23-02; Photo right: Deutsches Museum, München, Archiv, CD_L7071-0026.

281 Bauer 2008, pp. 211–215.

282 Bauer 2008, pp. 216–218; Bauer 2005a, pp. 18–20.

283 Pöthe 1905, pp. 705–709; Kern 1984, pp. 37–40.

demonstrations in Rotterdam. Following a thorough explanation of the invention and its earlier presentation in Cologne, the report went on to describe the equipment's performance in Rotterdam in detail:

> „Für die Versuche in Rotterdam wurde zuerst die gleiche Anordnung mit verstärkten Gebe- und Empfangsdrähten angewandt. Die Apparate waren an Deck des Hafendampfers *Columbus* aufgestellt. Es wurde eine exacte Reflexion bis auf 300 m beobachtet. Mit einem Inductorium von 50 cm Schlagweite und einer Accumulatoren-Batterie von 65 Volt stieg die Reflexion auf 3 km.“[284]

It should be noted that the device demonstrated in Rotterdam differed significantly from the patented design.[285] The results of the tests were also reported:

> „Was die Entfernung bezw. Reichweite anbetrifft, so gelangen die Versuche auf 3 km Entfernung mit einem kleinen an Bambusrohr angeordneten Antennensystem sehr gut. Bei Vereinigung der Auffangdrähte mit der Takelage hofft der Erfinder die Reichweite bis nach Bedarf auf 10 km ausdehnen zu können.“[286]

This demonstrated that the device operated successfully at a range of 3 kilometres. The French and Italian governments also requested a demonstration of the Telemobiloskop. The report concluded with an outlook:

> „Wenn auch die Gefahr der Schiffszusammenstöße wohl niemals ganz aus der Welt geschafft werden kann, so lässt sich doch, wie wir gesehen haben, manches erreichen, um die Sicherheit der Schiffe in grösserem Maasse zu gewährleisten, als dies bis jetzt der Fall ist, wenn nur an den

284 Pöthe 1905, p. 708. “For the experiments in Rotterdam, the same setup was initially used, with reinforced transmission and reception wires. The apparatus was installed on the deck of the harbour steamer Columbus. Clear reflections were observed at distances of up to 300 m. With an Inductorium with a spark length of 50 cm and an accumulator battery providing 65 volts, the reflection range increased to 3 km.” (translated by the author).

285 Kern 1984, pp. 38–39.

286 Pöthe 1905, p. 709. “As for distance or range, the tests conducted at a distance of 3 kilometres using a small antenna system mounted on a bamboo pole were very successful. The inventor expressed confidence that, by connecting the receiving wires to the ship's rigging, the range could be extended as needed and reach up to 10 kilometres.” (translated by the author).

> massgebenden Stellen alle Fortschritte der Technik vollständig nutzbar gemacht werden."[287]

Despite a series of promising demonstrations, neither the naval nor the merchant fleets offered any support for further development. Hülsmeyer's assurances of a substantial increase in range failed to change their stance.

In a letter dated 21 August 1905, the Telefunken company, then still operating under the name Gesellschaft für drahtlose Telegraphie m.b.H., System Telefunken, stated that after reviewing the patent specifications submitted by Hülsmeyer, they found no practical use for the invention.[288] The Navy likewise expressed no interest. Consequently, this significant innovation was all but forgotten for nearly three decades. On 5 October 1905, the company was dissolved by Hülsmeyer and Mannheim.[289]

Moreover, Hülsmeyer's discoveries attracted little attention from either the public or technical authorities at a time when aeroplanes had yet to take to the skies. The Wright brothers had made their first powered, piloted flight on 17 December 1903 in Kitty Hawk, North Carolina, well before technologies such as modern radio, television, transmitter tubes, amplifier components or detectors had come into existence.

As a result, the invention was soon consigned to obscurity. The ongoing debate as to whether the Telemobiloskop should be classified as a radar device in the modern sense requires a more detailed technical assessment. Such an analysis must consider factors including the frequency spectrum used, the absence of certain electronic components at the time, and the methods employed for direct range measurement.

Another likely reason for Hülsmeyer's lack of success was that by 1904 many vessels were already equipped with wireless telegraphy systems. These allowed long distance communication and

287 Ibid., p. 709. "Although the risk of ship collisions may never be entirely eliminated, as we have seen, significant progress can be made to improve vessel safety far beyond what has hitherto been achieved, provided that all relevant technological advances are fully utilised at the appropriate points." (translated by the author).

288 Bauer 2008, pp. 221–222; Bauer 2004a, pp. 51–52.

289 Bauer 2008, p. 222; Bauer 2005a, pp. 59–61.

reasonably accurate position finding using direction finding technology. Furthermore, Hülsmeyer's invention was often presented in the press as based on wireless telegraphy principles. As a result, the distinction between telegraphy and radiolocation was poorly understood, and the potential applications of his device were largely overlooked. Additionally, most shipowners were bound by Marconi's established monopoly on wireless telegraphy equipment, which probably discouraged them from pursuing legal action for breach of contract or from incurring extra costs.[290]

The unfortunate aspect of Hülsmeyer's efforts lies in his failure to convince either military or civilian stakeholders to support his invention on a sustained basis.[291] Moreover, he lacked access to crucial patents held by Braun and Marconi.[292] Despite these obstacles, Hülsmeyer was undoubtedly the first to recognise, demonstrate and patent the fundamental principle of radar technology, namely the use of electromagnetic wave reflection to detect distant objects. It was not until 1933 that radar technology regained prominence in Europe.[293]

Despite Christian Hülsmeyer's early and pioneering work in radar technology, there has been ongoing debate about who should be credited as the true inventor of radar. Over several decades, various engineers and scientists have contributed to the development and refinement of radar systems, leading to differing views and claims in historical and technical literature. Hülsmeyer's contributions were often overlooked or forgotten, which has further complicated the recognition of his role in radar history. For example, the October 2000 Jubilee Issue of the IEEE Aerospace & Electronic Systems Magazine highlights Hülsmeyer's pioneering role and states:

> "Heinrich Hertz, who first demonstrated the basic principle of radar, might be called the first radar scientist. A fellow German, Christian Hulsmeyer, can be called the first radar engineer. In the early 1900s, Hulsmeyer built, patented, demonstrated, and tried to sell to various shipping companies and navies a shipborne radar, based on Hertz's apparatus, for avoiding collisions of ships with one another. He had no suc-

290 Galati 2016, p. 5.

291 Holpp und Klausing 2022, p. 5.

292 Bauer 2008, p. 220.

293 Susskind 1988, pp. 506–517.

cess in selling his radar. His work faded from memory, even in Germany where it had to be reinvented when the need for it finally arose."[294]

5.3 Transitional Period

The so-called transitional period covers the years up to 1933, during which foundational research on electromagnetic wave reflection and the development of vacuum tubes, particularly the triode, which was crucial for early electronic advancements, laid the groundwork for radar technology. This phase was characterised by experimental investigations that gradually revealed the potential for practical and military applications.

Following the development of electron tubes in the early 1900s and their extensive deployment during World War I, it became possible for the first time to transmit undamped continuous waves. Subsequent advances included the construction of direction-finding stations, the introduction of onboard direction-finding equipment, the adoption of short-wave technology, and the establishment of both transmission and reception systems. Notably, Germany's first broadcasting station was officially commissioned on 29 October 1923, employing transmitters supplied by Telefunken.[295]

As early as 1922, Guglielmo Marconi recognised the potential of utilising short-wave electromagnetic radiation to develop a system capable of detecting the reflection of metallic objects and thereby locating ships. Yet, as in the case of Hülsmeyer, his proposal failed to gain meaningful support. On 20 June 1922, he delivered a lecture entitled "Radio Telegraphy"[296] to the American Institute of Electrical Engineers and the Institute of Radio Engineers in New York. After outlining the Marconi Company's efforts in the United Kingdom to develop high power electron tubes and their practical applications, he proceeded to discuss the potential of locating objects using electromagnetic waves:

294 IEEE 2000, p. 27.

295 Strunk 1999, p. 33; AEG 1988, p. 48.

296 Marconi 1922, pp. 215–238; Incidentally, it was during this event on 20 June 1922 that Marconi was awarded the Institute of Radio Engineers' Medal of Honour in recognition of his pioneering contributions to radio telegraphy.

> "Before I conclude I should like to refer to another possible application of these waves which, if successful, would be of great value to navigators.
>
> As was first shown by Hertz, electric waves can be completely reflected by conducting bodies. In some of my tests I have noticed the effects of reflection and deflection of these waves by metallic objects miles away.
>
> It seems to me that it should be possible to design apparatus by means of which a ship could radiate or project a divergent beam of these rays in any desired direction, which rays, if coming across a metallic object, such as another steamer or ship, would be reflected back to a receiver screened from the local transmitter on the sending ship, and thereby immediately reveal the presence and bearing of the other ship in fog or thick weather.
>
> One further great advantage of such an arrangement would be that it would be able to give warning of the presence and bearing of ships, even should these ships be unprovided with any kind of radio…"[297]

In September 1922, before the formal establishment of the Naval Research Laboratory (NRL), two Americans, Albert Taylor[298] and Leo Young[299], conducted pioneering experiments within the U.S. Navy's Radio Division, one of the naval research entities that would later be consolidated into the NRL. They picked up on Marconi's idea and carried out experiments using radio waves with a wavelength of five metres, corresponding to a frequency of approximately 60 MHz.[300] During their work, they observed significant fluctuations in the field strength of the receiver whenever the transmitting antenna was directed towards moving objects. They realised that these variations arose from interference between the emitted waves and those reflected by an object. This breakthrough enabled them, for the first time in the autumn of 1922, to reliably detect the wooden steamer Dorchester on the Potomac River.

Although such phenomena had been understood theoretically for some time, Taylor recognised their potential for practical mili-

297 Ibid., p. 237.

298 Jäger und Heilbronner 2010, p. 427; Watson 2009, p. 114.

299 Watson 2009, p. 115.

300 Brown 1999, p. 42; Swords 2008, p. 52.

tary applications.[301] They shared their findings with the Bureau of Engineering, the U.S. Navy department responsible for the design and procurement of naval technology. The Bureau played a key role in recognising and supporting Taylor and Young's early radar research, facilitating its further development towards practical military use.[302] Following their groundbreaking experiments, Taylor formally communicated their findings to the Bureau of Engineering in a memorandum dated 27 September 1922. This document not only summarised their observations but also highlighted the potential of radio wave reflection as a tool for naval defence, laying the groundwork for further research and development in radar technology.[303] Taylor wrote:

> "If it is possible to detect, with stations one half mile apart, the passage of a wooden vessel, it is believed that with suitable parabolic reflectors at transmitter and receiver, using a concentrated instead of a diffused beam, the passage of vessels, particularly of steel vessels (warships) could be noted at much greater distances. Possibly an arrangement could be worked out whereby destroyers located on a line a number of miles apart could be immediately aware of the passage of an enemy vessel between any two destroyers in the line, irrespective of fog, darkness or smoke screen. It is impossible to say whether this idea is a practical one at the present stage of the work, but it seems worthy of investigation."[304]

There is no documented official response from the Bureau of Engineering. It was obvious that there was no interest.[305]

In July 1923, the Naval Research Laboratory was officially established in Bellevue, Washington, bringing together several existing naval research entities, including the Aviation Radio Laboratory at Anacostia, where Taylor and Young's work continued. Taylor was appointed Superintendent of the Radio Division, with Young as his assistant.

---

301 Hyland 1988, pp. 29–34; Taylor et al. 1934; Brown 1999, p. 42; Swords 2008, p. 52; McKinney 2006, pp. xi – xii; Galati 2016, pp. 60–61; Reuter 1971, pp. 39–40.

302 Swords 2008, p. 52; Galati 2016, p. 60; Reuter 1971, pp. 39–40.

303 Swords 2008, p. 52; Reuter 1971, pp. 39–40.

304 Swords 2008, p. 52; Galati 2016, p. 60; Burns 1988a, pp. 14–15.

305 Swords 2008, p. 52; Kern 1984, pp. 76–77; Allison 1981, p. 55.

The Americans Gregory Breit[306] and Merle Tuve[307] conducted pioneering experiments in 1925 and 1926 to determine the altitude of the ionised layers of the ionosphere using broadband radio pulses. They employed wavelengths of approximately 70 metres, which were particularly suitable for reflection by these layers. Their work provided early empirical confirmation of the ionised layer theory proposed by Arthur E. Kennelly and Oliver Heaviside. Although their focus was atmospheric physics, the technique of emitting and analysing pulsed radio echoes bears fundamental similarities to methods later used in radar.[308]

In Great Britain, Edward Appleton[309] and Miles Barnett[310] were already conducting investigations into the physics of the ionosphere. In 1924, they carried out foundational experiments on radio wave propagation. Appleton's work significantly advanced the understanding of the ionosphere, for which he was awarded the Nobel Prize in Physics in 1947.[311]

Early ideas related to electromagnetic echo detection were also explored by Austrian physicist Heinrich Löwy.[312] In 1923, he filed a patent entitled Einrichtung zum Nachweis und zur Messung des Abstandes elektrisch leitfähiger Massen[313], in which he outlined a method for using reflected electromagnetic waves to detect conductive objects. The concept included blocking the receiver during transmission in order to detect only the reflected signals, an idea that anticipates the pulse echo technique later fundamental to radar.[314] While some retrospective accounts suggest Löwy may have considered similar principles as early as 1911, direct documentary evidence for such an early formulation remains limited. In any

306 Hull 1998, pp. 26–56.

307 Abelson 1996, pp. 407–422.

308 Breit and Tuve 1926; p. 554; Bauer 2005b, pp. 4–5; Swords 2008, pp. 53–59; Brown 1999, pp. 42–43; Galati 2016, p. 58; Holpp 2014, p. 183.

309 Ratcliffe and Hutchins 2012; Clark 1971, pp. 36–48; Watson 2009, pp. 40–41; Weinreich 2001c, pp. 104–105.

310 Clark 1971, pp. 36–37; Swords 2008, p. 53.

311 Clark 1971, pp. 145–146.

312 Kern 1984, pp. 41–50; Swords 2008, pp. 59–60.

313 Reuter 1971, p. 15; Apparatus for the Detection and Measurement of the Distance of Electrically Conductive Masses. (translated by the author).

314 Kern 1984, p. 49.

case, his work did not lead to practical applications at the time and received little contemporary attention.

The technical issue of receiver saturation by the transmitted signal, particularly when the receiver was within the transmitter's radiation field, reappeared in March 1934 during reflection experiments conducted at the NVA in Kiel.

In June 1930, Leo Young and Lawrence Hyland at the U.S. Naval Research Laboratory (NRL) accidentally observed the presence of aircraft using a bistatic continuous-wave radar system operating at approximately 33 MHz.[315] Although these systems were able to detect movement, they could not determine the range or precise location of targets.

5.4 Expansion Phase

The so-called expansion phase began in 1933, when the military potential of radar was recognised, sparking rapid technological progress and international development efforts. By then, the use of reflected electromagnetic waves for target detection was already being actively explored in Europe, particularly in Germany, Great Britain and France. Intensive research followed in these countries as well as in the United States, laying the groundwork for operational radar systems deployed during World War II. From the mid-1930s onwards, other countries including Italy, Japan, the Netherlands and the Soviet Union also accelerated their radar development efforts.

The development of radar in Germany and Great Britain prior to and during World War II is discussed in greater detail in Chapter 7.

After World War II, radar technology advanced rapidly, driven primarily by military demands. Innovations developed during the war firmly established radar's vital role in air and sea surveillance, missile defence, and, subsequently, a wide range of civilian applications. In Germany, radar research and development were ini-

---

315 Hyland 1988, pp. 29–34; Taylor et al. 1934; Swords 2008, p. 52, pp. 101–103; McKinney 2006, pp. xiii – xiv; Galati 2016, p. 61.

tially prohibited after World War II, especially for military purposes. These restrictions were gradually lifted in the late 1940s, and by around 1950, German companies were permitted to resume work on radar technology for civilian applications. This included the licensed production of British radar systems used in air traffic control, maritime navigation along coastlines, and navigation on inland waterways.[316]

Beginning in the early 1950s, radar technology advanced significantly, driven by continuous progress in semiconductor and microprocessor technologies. This trend continues today with ongoing improvements in electronic components.

Today, radar constitutes a key technology with widespread applications in both civilian and military domains. It is used in fields such as aerospace, defence, transport, and environmental monitoring. Radar systems are fundamental to the detection, tracking, and analysis of physical objects and environmental conditions, offering precise measurements of distance, velocity, and position under a wide range of environmental influences.[317]

A major breakthrough came in 1951 with the introduction and advancement of Synthetic Aperture Radar (SAR). This innovation, based on the conventional radar principle, enables the generation of high-resolution images through specialised signal processing techniques.[318] The first description of the fundamental SAR principle, which achieves high angular resolution by analysing the frequency of the received signal using a moving coherent radar system, was provided in June 1951 by Carl Wiley of the Goodyear Aircraft Corporation.[319]

In July 1953, scientists at the University of Illinois reported the first experimental demonstrations of SAR, including the creation of an image of a strip of terrain. By August 1957, a research group at

316 Holpp 2004, p. 187; Gerlitzki 1994, pp. 47–59; for more information on the history of Telefunken radar technology in Ulm from 1954 to 1990, see Gerlitzki 1995.

317 Skolnik 2008a, Chapter 1.9; Ulaby et al. 1981, pp. 23–30; Klausing und Holpp 2000, pp. 361–386.

318 Sullivan 2008, Chapter. 17; Hovanessian 1980, pp. 13–29; Klausing und Holpp 2000, pp. 213–289; Klausing 1989, pp. 14–26; .

319 Wiley 1985, pp. 440–443.

the University of Michigan accomplished the first successful operation of an airborne reconnaissance system with high angular resolution, with images later generated in the laboratory using an optical processor. From that point onwards, the potential of SAR technology was fully recognised, laying the foundation for a revolution in radar technology that is indispensable in both civilian and military applications today.

Another significant application of radar technology is radar satellites, which are widely used today to measure and map the Earth's surface, monitor ocean conditions and natural disasters, analyse the atmosphere, and support emergency services and global communications. Radar satellite technology remains an active and rapidly advancing field of research. These satellites capture breathtaking images of the Earth's surface from very high altitudes, often generating detailed three-dimensional representations that provide valuable insights into terrain features and environmental changes. This capability has revolutionised Earth observation and contributes significantly to fields such as climate science, disaster management, and global monitoring.[320]

320 DLR n.d.

6 Shortest Wave Generation as a Key Technology

Significant progress in the development of radar technology was closely linked to the stable generation and amplification of electromagnetic oscillations. This began with the invention of the electron tube in 1906, which marked a major advancement over earlier spark transmitters and greatly enhanced the capabilities of radio communication. The subsequent refinement of vacuum tube technology in the following decades enabled access to the ultra-shortwave range, with wavelengths of less than 10 metres.[321]

From the early 1930s, the development of high-power magnetrons became a key milestone[322] and an essential component of powerful radar systems.[323] The magnetron had been known in various forms since at least the 1920s. On 29 November 1935, the German inventor Hans E. Hollmann[324] filed a patent for a magnetron with multiple resonators. This patent was subsequently filed in the United States on 27 November 1936 and granted on 12 July 1938.[325] The term magnetron was originally introduced by the American physicist Albert Hull[326] of the General Electric Company, who first described the principle in a publication in September 1921.[327] By 1925, magnetrons capable of producing a high output of 15 kW at

---

321 For readers interested in a more detailed discussion on microwave generation, a related article from DEHs is included in the Appendix.

322 The magnetron is a vacuum tube for generating high-frequency oscillations. It became particularly important in 1940 with the development of a magnetron for the centimetre-wave range, which was used in radar devices to generate the required high pulse power.

323 Leconte 2010, pp. 11–16.

324 Küpfmüller 1972, p. 543; Frühauf 1959, pp. 141–143 (English translation at: URL: https://www.radarworld.org/hollmann1.html).

325 US patent 2,123,728: Magnetron, Telefunken Gesellschaft für Drahtlose Telegraphie m. b. H., Berlin, patented on 12 July 1938. (URL: https://patents.google.com/patent/US2123728A/en).

326 Jäger und Heilbronner 2010, p. 207; Suits and Lafferty 1970, pp. 213–233.

327 Hull 1921, pp. 715–723.

a frequency of 20 kilohertz had been developed, although these early versions were primarily used in control systems for power engineering. For the emerging field of radar, however, magnetrons initially delivered only limited power at the much higher frequencies required. As a result, they were mainly confined to laboratory experiments.

From the early 1930s, companies such as Telefunken began to develop magnetrons specifically designed to generate higher power at higher frequencies.[328] At the time, their intended applications were still seen primarily in radio relay systems.[329]

In 1940, Great Britain achieved a technological breakthrough with the invention of a high-power magnetron operating in the centimetre-wave range. Its compact size and robust design made it ideal for use aboard aircraft, significantly contributing to the Allies' technological superiority in radar technology.

6.1 Germany

6.1.1 Early Developments and Key Challenges

When examining the development of radar technology for air warning, fighter control, and anti-aircraft targeting on the German side up to the middle of World War II, it becomes evident that the wavelengths employed were predominantly in the range of several metres down to 50 centimetres. For instance, the Freya radar, used for air warning and fighter direction, operated at a wavelength of 2.4 metres. Shorter wavelengths were used in certain systems, such as the shipborne Seetakt radar at approximately 80 centimetres, and the Würzburg radar, designed for anti-aircraft fire control, at around 50 centimetres. German airborne radar systems, introduced from 1941 onwards, typically operated at wavelengths of at

328 Runge 1934b, pp. 5–14; Goerth 2010, pp. 17–22; Kern 1984, pp. 101–103; Rukop 1953, pp. 165–176; further contributions on the origin and technical application of the magnetron can be found in Blanchard 2010, pp. 5–10; Blanchard et al. 2013, pp. 244–254. A very good overview of tube development on the British side for the period from 1937 to 1944 can be found in Callick 1990.

329 Runge 1934a, pp. 24–33.

least half a metre.[330] Development of centimetre-wave radar for aircraft did not begin until 1943.

In contrast, by the end of 1942, the British had already equipped 24 aircraft with H_2S radar systems operating at a wavelength of approximately 9 centimetres. Germany's first centimetre-wave radar for airborne use, the Berlin A (FuG 224)[331], was developed by the Telefunken company in 1943. This system was essentially a replica of the British H_2S radar and operated at a similar wavelength of around 9 centimetres.[332]

Interestingly, Germany had a well-established tradition of scientific research and technical development in the field of decimetre and centimetre waves dating back to the 1920s. However, this work was discontinued in November 1942, likely due to resource constraints and a strategic shift towards other technological priorities.

As early as autumn 1933, specific measurements were carried out in Germany at the NVA facility using wavelengths of 13.5 centimetres. However, the transmission power of 100 mW was insufficient to detect reflected echo signals. Later that year, six magnetrons with wavelengths of 50 to 60 centimetres and a continuous wave power of 40 Watts were procured from Philips in Eindhoven.[333] Subsequent tests showed gradual improvements in range.

---

330 Kümmritz 1994, p. 28.

331 Trenkle 1986, p. 145, p. 149; Bauer 2006, pp. 19–26; Oberkommando der Kriegsmarine 1944.

332 Lovell 1991, p. xvii; H_2S (Home Sweet Home) was the codename for a British airborne radar system operating in the centimetre-wave range. From January 1943 onwards, it was used to bomb German cities and urban areas, thanks to its unprecedented surface-imaging accuracy. Pathfinder reconnaissance aircraft marked targets in Germany with flares for the following British bombers. In the latter half of the war, H_2S also played a decisive role in detecting submarines in the Atlantic, contributing significantly to German losses. Two major technological innovations were integrated into H_2S: high-power generation at short wavelengths via the cavity magnetron, and a panoramic overview provided by a rotating antenna (Rowe 2015, pp. 116–117; Reuter 1971, pp. 113–114). Detailed descriptions of H_2S technological developments, including power generation and display equipment, can be found in Callick 1990, pp. 55–76, pp. 107–123.

333 Willisen ca. 1984, p. 3–4; Kroge 2000, pp. 16–17; Kroge 1998, pp. 25–26; for Radar development in the Netherlands before the war see also Staal and Weiller 1988, pp. 235–242.

It should be noted that at these short wavelengths, considerable technical challenges were encountered, including transmitter and receiver instability, as well as signal distortion and multipath interference resulting from specular reflections off smooth surfaces. These phenomena adversely affected signal clarity and complicated the reliable detection of targets, thereby influencing the decision to prioritise research on longer wavelengths.

In January 1936, the NVA, in collaboration with GEMA, initiated a comprehensive development programme to investigate centimetre waves. However, these experiments did not yield practical results due to the limited power output of tube technology at these wavelengths and the assumption that the lack of backscattering was caused by path reflection. Consequently, further research into centimetre waves was considered less important and subsequently discontinued.

In February 1936, a GEMA device operating at a wavelength of 1.80 metres successfully detected aircraft at ranges of up to 28 kilometres. This success paved the way for the development of the Freya radar, which was designed for the long-range detection of incoming aircraft, providing vital information for directing intercepting fighters and coordinating air defence.

In January 1939, the NVA decided to focus on metre and decimetre waves, as existing systems in these ranges were more advanced and reliable. Centimetre waves were considered unsuitable for location purposes due to transmitter and receiver instability and related issues. Consequently, all tests in this area were discontinued.

It is worth noting, however, that advances in magnetron technology in subsequent years, particularly the development of high-power magnetrons capable of generating centimetre waves, would later revolutionise radar systems during and after World War II.

Although fundamental advancements in high-frequency oscillations had been achieved since the late 1920s, the prevailing conviction that centimetre waves were of no practical use persisted until early 1943, with detrimental consequences. In 1928, Hans E. Hollmann completed his doctorate at TH Darmstadt on the generation of very high-frequency electrical oscillations. He subsequently continued his research at the Deutsche Forschungsgemeinschaft from 1928 to 1930, and thereafter at the Heinrich-Hertz-Institut für Schwingungsforschung in Berlin. Together with Gustav Leithäuser[334], the institute's director since 1927, he successfully generated and detected decimetre waves.

In 1933, Hollmann obtained his habilitation at the Technical University of Berlin, and was employed as a scientific researcher at Telefunken, where he secured numerous patents. From mid-1933, he acted as a consultant to the company Tonographie.[335] Following the establishment of GEMA in early 1934, he became a freelance collaborator[336], making significant contributions to the company's initial successes.[337]

Another researcher from the Heinrich-Hertz-Institut, Theodor Schultes[338], was also engaged as a freelance collaborator, and in June 1934 joined GEMA as a permanent employee, becoming a pivotal figure in the development of its devices.

For Paul-Günter Erbslöh and Hans-Karl Freiherr von Willisen, owners of Tonographie and founders and later sole shareholders of GEMA, the collaboration with Hollmann and Schultes proved fortuitous.[339]

In 1936, the second volume of Hollmann's two-volume work "Physik und Technik der ultrakurzen Wellen" was published under the title "Die ultrakurzen Wellen in der Technik".[340] This volume explored the wide range of practical applications, alongside the fundamental phenomena required to master ultrashort waves.[341] Hollmann's work was clearly ahead of its time. As a result, the volumes initially attracted limited attention and circulation. Interest only increased once the significance of ultrashort to decimetre wavelengths for television and emerging radar technologies became evident.

Within the fourth chapter of the second volume, subchapter 3, Hollmann discusses "Sehen mit elektrischen Wellen" and presents

---

334 Jäger und Heilbronner 2010, pp. 260–261; Bruch 1983, pp. 106–108.

335 Tonographie was a German company specialising in acoustic and electronic technologies, particularly in sound recording and analysis.

336 Kroge 2000, p. 16, p. 20; Kroge 1998, p. 25, p. 28.

337 Frühauf 1959, p. 142.

338 Jäger und Heilbronner 2010, pp. 391–392; Weiher 1983b, pp. 171–173; Weinreich 2001a, pp. 106–108.

339 Kroge 2000 pp. 9–14; Kroge 1998, pp. 17–23; Reuter 1971, p. 19.

340 Hollmann 1936.

341 Ibid., foreword.

pioneering findings indicative of the radar principle[342], referring to research conducted at GEMA. However, due to secrecy regulations, his manuscript was submitted to naval censorship, resulting in the excision of numerous passages, contrary to to Hollmann's liberal convictions.

Erbslöh and von Willisen were likewise dismayed to discover that compliance with their client's confidentiality demands was requisite for continued involvement in radar technology. As Hollmann was unwilling to acquiesce to censorship, he severed ties with GEMA. Shortly thereafter, he founded the independent Laboratorium für Hochfrequenztechnik und Elektromedizin in Berlin, where, in collaboration with his brother, he sought to apply physical and technical principles to biological inquiries.[343]

In 1942, Hollmann was appointed scientific director of the newly founded Forschungsgesellschaft für Funk- und Tonfilmtechnik e. V. in Berlin. Alongside his work at the Laboratory for High-Frequency Technology and Electromedicine, he was able to continue his scientific activities.[344] However, he was eventually prohibited by higher state authorities from working on shorter wavelengths. After the war, Hollmann was appointed associate professor of special high-frequency technology at the University of Jena, but soon emigrated to the United States, where he took on new scientific challenges in various roles.[345]

Alongside Hollmann, Abraham Esau[346] contributed to the development of ultrashort waves (VHF) and laid the foundations for later VHF technology through his research on ultrashort wave broadcasting in the mid-1920s. Esau studied mathematics and physics at the Technical University of Berlin and earned his doctorate from the University of Berlin in 1908. In 1912, he joined Telefunken in Berlin, where he became head of the receiver laboratory.

342 Hollmann 1936, pp. 160–161; Kroge 2000, p. 41; Kroge 1998, p. 51; Frühauf 1959, p. 142.

343 Kroge 2000, p. 41; Kroge 1998, p. 51; Frühauf 1959, p. 142.

344 Frühauf 1959, p. 142; Hollmann 1942, pp. 281–293; Hollmann 1943a, pp. 12–21; Hollmann 1943b, pp. 37–44.

345 Frühauf 1959, p. 143.

346 Jäger und Heilbronner 2010, p. 120; Schröter 1959, pp. 640–641; Mauel 1983b, pp. 52–53; Hoffmann und Stutz 2003, pp. 136–179.

In spring 1914, Telefunken sent him to Togo to commission and test the large Kamina radio station. He was captured by the French and held in a remote prisoner-of-war camp in Central Africa before being transferred to France. In 1917, he was moved to an internment camp in Switzerland, where he remained until 1918.[347] After his release, he returned to Berlin and became head of all Telefunken laboratories. In 1925, he accepted a position at the University of Jena, initially as associate professor and director of the Technical Physics Institute. From 1927, he served as a full professor of technical physics, focusing primarily on the development of the ultrashort wave range.[348]

In 1930, Esau and Walter Hahnemann published a report entitled "Report on experiments with electric waves of about 3 metres: their propagation and use".[349] In this report, they discussed experiments on the propagation properties of wavelengths around 3 metres, once sufficient power could be generated. Directional antennas in the form of parabolic reflectors were also successfully employed. Initial experiments on the maximum achievable range were conducted by the Technical-Physical Institute in Jena and later by the Bavarian Post Office from the Herzogstand in the Bavarian Alps. More extensive tests were then carried out by the laboratories of C. Lorenz AG in Berlin in collaboration with the Jena Institute, both from aircraft and from the summit of the Brocken in the Harz Mountains. The results clarified the propagation phenomena of these wavelengths and demonstrated their practical potential for transmitting signals over short distances, particularly for use in fog conditions for shipping and aviation, as well as a possible future method for television signal transmission.[350]

At the end of the report, the following was stated:

> "If it becomes possible to reduce the wavelength somewhat more (down to about 0.5 m), while maintaining as much as possible the transmission energy and the sensitivity of reception obtained for the 3-m wavelength, then still more new applications of the ultra-short

347 Hoffmann und Stutz 2003, pp. 139–142.
348 Hoffmann und Stutz 2003, pp. 142–143; Godt 2003, pp. 234–235.
349 Esau und Hahnemann 1930, pp. 471–489; Kern 1984, pp. 79–80.
350 Esau und Hahnemann 1930, p. 471.

waves will become possible by the fact that for these short wavelengths, reflectors may be used whose dimensions are large as compared with the wavelength. In such cases, however, it will be possible to obtain beams of waves with a directive sharpness similar to that of a beam of rays emitted by a searchlight. The consequences of a development of this kind cannot be foreseen at present."[351]

In addition to his research into VHF technology, Esau also concentrated on radar technology, with the aim of generating wavelengths in the centimetre to millimetre range.[352] He recognised early on that these wavelengths could open up new areas of application.

On 10 December 1937, Esau delivered a lecture at the Deutsche Akademie der Luftfahrtforschung[353] entitled "Elektrische Wellen im Zentimetergebiet"[354], in which he explained:

> „Ich wende mich jetzt dem Wellengebiet zu, das den eigentlichen Inhalt meines heutigen Vortrages bildet, d. h. den Wellen, die kleiner sind als 10 cm. Es handelt sich hierbei um Wellenlängen, die zur Zeit zwar noch nicht für technisch bedeutsame Probleme eingesetzt worden sind, die aber in der allernächsten Zukunft einsatzfähig sein werden, und zwar sowohl für die eigentliche Nachrichtenübertragung als auch für gewisse Probleme, die gerade für die Luftfahrt von ganz besonderer Bedeutung sind und deren Lösung vielleicht mit ihnen erreicht werden kann: die Ortung, Höhenbestimmung von Flugzeugen und andere mehr".[355]

---

351 Ibid., p. 489.

352 In 1938, the International Shortwave Congress was held in Vienna. There, Esau reported on his own experiments at a wavelength of 4 mm. This was a world record and Esau had made a name for himself as a pacesetter in radar technology (Hoffmann und Stutz 2003, p. 143).

353 German Academy of Aviation Research.

354 Esau 1938, pp. 213–221; Electric Waves in the Centimetre Range (translated by the author).

355 Ibid., p. 214. "I now turn to the wavelength range which constitutes the actual subject of my lecture today, that is, waves shorter than 10 centimetres. These are wavelengths which, although not yet applied to technically significant problems at present, will be ready for use in the very near future, both for actual communication transmission and for certain problems which are of particular importance to aviation and whose solution may perhaps be achieved with their help: detection, altitude determination of aircraft, and others." (translated by the author).

He continued:

> „Was die erreichbaren Energien betrifft, so hat sich gezeigt, daß man bei etwa 4 bis 5 cm Wellenlänge maximal etwa 3 Watt herstellen kann, ein Betrag, der zwar gegenüber den erreichten Leistungen im Dezimetergebiet von mehreren 100 Watt sehr klein erscheinen mag, aber im Verhältnis zu den bei anderen Methoden erzielbaren Leistungen schon als außerordentlich hoch angesehen werden kann“.[356]

Later he added:

> „Die bisher erzielten Reichweiten betragen mehr als 30 km bei einer Wellenlänge von etwa 4 bis 4,5 cm, 15 km bei einer Länge von 2,5 bis 3,5 cm und etwa 6 km bei Wellen von 2,3 cm. Wellen von etwa 5 bis 6 cm dürften heute bei vorhandener optischer Sicht Reichweiten von mehr als 100 km ergeben“.[357]

Esau was also aware that research was underway abroad in the field of centimetre wave technology, including for aircraft detection, and that initial tests had already been conducted:

> „Auch im Ausland ist die Lösung dieser größeren Probleme mit Hilfe der Zentimeterwellen ins Auge gefaßt worden, und es sind bereits gewisse Versuche in dieser Richtung vorgenommen worden. Es handelt sich einmal um die Feststellung stationärer Objekte mit Zentimeterwellen unter Ausnutzung des Reflektionseffektes, dann aber auch um bewegte Gegenstände, wie beispielsweise Flugzeuge. Wenn auch zunächst noch die Entfernung, über die solche Versuche durchgeführt werden konnten, als sehr klein angesehen werden muß – es werden etwas mehr als 1 km angegeben – so scheint es doch nur eine Frage

356 Ibid., p. 215. “As for the achievable energies, it has been shown that at a wavelength of about 4 to 5 centimetres, a maximum of around 3 watts can be produced, an amount which, although it may appear very small compared to the several 100 watts reached in the decimetre range, is significant in relation to the outputs attainable by other methods.” (translated by the author).

357 Ibid., p. 219. “The ranges achieved so far amount to more than 30 km at a wavelength of about 4 to 4.5 cm, 15 km at a length of 2.5 to 3.5 cm, and approximately 6 km at waves of 2.3 cm. Waves of about 5 to 6 cm would today, given clear optical visibility, yield ranges of more than 100 km.” (translated by the author).

> der Zeit zu sein, wo man erheblich größere Abstände wird zulassen können“[358]

For Esau, it was therefore only a matter of time before greater ranges could be achieved. He also envisaged other potential applications for this wavelength range, such as beacons for landing and navigation in fog, measuring the ground speed of aircraft, and determining altitude by means of echo methods. He was convinced that this wavelength range was well suited to addressing the aforementioned challenges.

Esau concluded his speech by explaining:

> „Die notwendigen Vorarbeiten sind, wie ich in meinem Vortrag zu zeigen versucht habe, erfolgreich durchgeführt worden. Es beginnt jetzt die zweite Phase! Die Durchbildung der Geräte und ihr Einsatz bei den angeführten Problemen, von der wir hoffen wollen, daß sie sich für die Luftfahrt bald erfolgreich auswirken möge, sowohl in bezug auf den Nachrichtenverkehr als auch für navigatorische Zwecke“.[359]

This lecture was held on 10 December 1937. The second phase did not begin until much later, in February 1943, after parts of a radar operating at a wavelength of around 9 centimetres were discovered in a British bomber shot down near Rotterdam. In 1939, Esau was appointed President of the Physikalisch-Technische Reichsanstalt (PTR), a position he held until 1945. In the same year, he was also appointed full professor of telecommunications technology at the

358 Ibid., p. 221. “The resolution of these major problems using centimetre waves has also been considered abroad, and certain experiments in this direction have already been carried out. These involve, firstly, the detection of stationary objects with centimetre waves by utilising the reflection effect, and secondly, moving objects, such as aircraft. Although the range over which such experiments have initially been conducted must be regarded as very limited – being somewhat more than 1 km – it nevertheless appears to be only a matter of time before considerably greater distances can be achieved.” (translated by the author).

359 Ibid., p. 221. “The necessary preliminary work has, as I have attempted to demonstrate in my lecture, been successfully completed. The second phase now begins: the refinement of the devices and their deployment in the mentioned applications, from which we hope will soon bring successful results for aviation, both in terms of communication and navigational purposes.” (translated by the author).

Technical University of Berlin and was responsible for high-frequency research during World War II, with a particular focus on the development of centimetre wavelengths. At the beginning of 1944, he succeeded Hans Plendl[360] as Bevollmächtigter für die Hochfrequenzforschung (BHF). After the end of the war, he was imprisoned until 1948 due to his role in National Socialism. In 1949, he received an honorary professorship at the Technical University of Aachen for ultrashort wave technology and ultrasound. From 1953 until his death, he headed the newly created Institute for High-Frequency Technology at the Deutsche Versuchsanstalt für Luftfahrt (DVL) in Mülheim an der Ruhr, Germany.[361]

Wilhelm Runge, head of device development at Telefunken, viewed the practical use of centimetre-range wavelengths as crucial, mainly because receiver technology was not yet sufficiently advanced. However, for directional radio links using focused antennas, he believed it was possible to cover distances of up to 100 kilometres with wavelengths of a few decimetres, even at low transmission power. In his lecture "Die praktische Bedeutung der Mikrowellen"[362], delivered alongside Esau on 10 December 1937 at the Deutsche Akademie der Luftfahrtforschung, Runge explained that the technical challenges facing both transmitter and receiver technology could be overcome for wavelengths down to around 20 centimetres. Nevertheless, he emphasised the urgent need for further research and development into shorter wavelengths.

He explained this in his lecture:

> „...Von diesem Gesichtspunkt aus gewertet, kann man heute die Mikrowellen bis etwa 20 cm herunter als technisch völlig erschlossen und einsetzbar bezeichnen. Sendermäßig ist der Bereich bis zu einigen Zentimetern herunter technisch erschlossen, während in diesem Band befriedigende Lösungen des Empfängerproblems erst in der Erschließung begriffen sind. Wie man einmal die Wellen unter einigen Zentimetern betriebsmäßig empfangen wird, weiß, glaube ich, heute noch niemand.

---

360 Seiler 2007, pp. 54–55, pp. 114–117, pp. 133–136; Handel 1999, pp. 57–58; Reuter 1971, pp. 192–193; Watson 2009, pp. 254–255.

361 Hoffmann und Stutz 2003, pp. 170–171.

362 Runge 1938, pp. 222–225; "The Practical Significance of Microwaves" (translated by the author).

> Es ist Sache der freien Forschung, nicht die des wirtschaftlich und termingebunden arbeitenden Ingenieurs, hier den Weg zu finden“.[363]

It should be noted that Telefunken had already been conducting experiments with magnetrons in the decimetre range since the end of 1933, particularly in the field of radio relay technology, and therefore had considerable practical experience. Runge reported on the advantages of using decimetre waves below 1 metre for radio relay connections with bundled antennas in a lecture titled “Die Grundlagen der Dezimeterwellen-Technik”, which he presented to the Vereinigung der Luftfahrtforschung on 21 March 1934.[364] In December, he reported on the technology and operation of various magnetrons in the Telefunken-Zeitung under the title “Schwingungserzeugung mit dem Magnetron”, a topic he had presented on 2 November 1934, to the Vereinigung für Luftfahrtforschung.[365] After most of the remaining research in the field of centimetre waves was discontinued at the beginning of the war due to a lack of personnel, only Telefunken was still working on the research and development of centimetre waves in its laboratory in Berlin under the direction of Waldemar Ilberg.[366] The laboratory had been founded several years before the start of the war but achieved only limited success with the few resources available. Runge also saw the sensitivity of the receivers as a problem in this wavelength range, rather than the power generation by the magnetron.[367]

363 Ibid., p. 225. “...From this perspective, microwaves down to about 20 centimetres can today be regarded as technically fully developed and deployable. On the transmitter side, the range down to a few centimetres is technically mastered, whereas in this band satisfactory solutions to the receiver problem are only beginning to be developed. How operational reception of waves below a few centimetres will be achieved, I believe, is still unknown today. It is a matter for fundamental research, not for the engineer working under economic and time constraints, to find the way forward here.” (translated by the author).

364 Runge 1934a, pp. 24–33; “The Fundamentals of Decimetre Wave Technology.” (translated by the author).

365 Runge 1934b; “Generation of Oscillations with the Magnetron.” (translated by the author).

366 Reuter 1971, p. 197.

367 Ibid., p. 197.

On 30 August 1940, Runge held a lecture titled "Elektromagnetische Rückstrahlung, ihre Technik, Möglichkeiten und Grenzen"[368], alongside the long-established field of communication transmission. He discussed the advantages and disadvantages of backscatter measurements compared to the capabilities of the human eye. After pointing out the lower resolution in the millimetre wave range compared to the human eye, he then discussed the advantages of backscatter technology in terms of distance measurement and depth resolution:

> „Wie weit aber auch die Rückstrahltechnik hinter dem Auge zurückbleiben sollte, zweierlei hat sie vor ihm voraus: das feine Unterscheidungsvermögen für Entfernungen, die Tiefenauflösung und damit die Möglichkeit, Entfernungen einfach und genau zu messen, und die Unabhängigkeit von den Sichtverhältnissen. Mit diesen beiden Möglichkeiten schafft sie in der drahtlosen Technik ein neues Gebiet. Wir erleben in diesen Jahren die ersten Erfolge dieser Technik. Noch viel bleibt zu tun, bis sie als ebenbürtiges Gebiet der drahtlosen Nachrichtentechnik zur Seite stehen wird. Die Möglichkeiten, die die Rückstrahltechnik bietet, sind aber sowohl für den Krieg als auch für den Frieden so bedeutungsvoll, daß sie in wenigen Jahren aus der praktischen Fliegerei nicht mehr wegzudenken sein wird".[369]

A lecture "Ein räumlich peilendes Rückstrahlgerät als Zielgerät"[370] which was given by Runge at the Deutsche Akademie der Luftfahrtforschung on 17 April 1942, pointed out that receiver sensitivity deteriorates drastically at wavelengths of 50 centimetres or less.[371]

---

368 Runge 1941; "Electromagnetic Backscatter, its Technology, Possibilities, and Limitations." (translated by the author).

369 Runge 1941, p. 24. "Although backscatter technology may still fall short of the eye in some respects, it has two advantages over it: its fine ability to distinguish distances, depth resolution, and thus the possibility to measure distances simply and accurately, as well as its independence from visibility conditions. With these two capabilities, it creates a new field in wireless technology. We are witnessing the first successes of this technology in these years. Much remains to be done before it can stand as an equal field alongside wireless communication technology. However, the possibilities offered by backscatter technology are so significant for both war and peace that it will soon become indispensable in practical aviation." (translated by the author).

370 Runge 1942, pp. 4–21; "A direction-finding backscatter device as a targeting instrument." (translated by the author).

371 Ibid., p. 7.

6.1.2 The Turning Point 1942/43

On 22 November 1942, a demonstration of the latest Würzburg radar developments operating at a wavelength of 50 cm was held at the Telefunken plant in Berlin-Zehlendorf. The presentation, organised by Wilhelm Runge and Karl Rottgardt, a member of Telefunken's management board, emphasised once again that centimetre waves tended to be reflected away rather than returned in a useful manner.[372] As such, the 50 cm band was still considered the most effective for radar applications.

Just a few days later, on 27 November 1942, General Erich Fellgiebel[373], Generalbevollmächtigter für technische Nachrichtenmittel (GBN), visited Telefunken together with General Wolfgang Martini[374], head of the Luftnachrichtentruppe and special representative for radar since November 1941, and Admiral Erhard Maertens.

During this visit, Runge reported that transmitters and receivers for centimetre wavelengths developed in Ilberg's laboratory had shown limited promise and required considerable financial resources for further development and production.[375]

In his memoir "Ich und Telefunken", published in 1971, Runge reflects on the events of autumn 1942 against the backdrop of Britain's emerging superiority in centimetre-wave radar technology, which became evident in 1943:

> „...Wir kannten bisher nur die nach unserer Ansicht primitive englische Funkmeßtechnik auf Meterwellen, der wir mit unseren Dezimeterwellen überlegen waren. Jetzt standen wir vor einer uns weit überlegenen Zentimeterwellentechnik.
>
> Unsere Arbeiten mit der Fünfzentimeter-Welle, die schon vor dem Krieg begonnen hatten, hatte ich im Herbst 1942 nach einer Schlußvorführung vor den funktechnisch Verantwortlichen der drei Wehmachtsteile abgebrochen, weil angesichts der dringenden Entwicklungsaufgaben und dem beschränkten Personal solche Zukunftsgebiete schon seit dem Kriegsbeginn nicht mehr intensiv genug bearbeitet

372 Trenkle 1979, p. 43; it should be noted that Prof. Esau of the Physikalisch-Technische Reichsanstalt consistently disputed this view.

373 Klee 2011, p. 148; Wildhagen 1970.

374 Hoffmann 1990, pp. 301–302; Rieger 2018b.

375 Reuter 1971, p. 197.

> werden konnten. Ab nun konzentrierte sich unsere Entwicklungstätigkeit auf Nachbau, Nachentwicklung und Erlernen der englischen Zentimeterwellentechnik, in der dann bis zum Kriegsende auch originelle eigene Weiterentwicklungsarbeit geleistet worden ist".[376]

As a result, Fellgiebel decided to terminate all ongoing centimetre-wave projects. The aim was to avoid further fragmentation of Germany's already limited pool of technical specialists and instead focus on refining the more mature decimetre-wave systems. Consequently, Ilberg's laboratory was dissolved.[377]

Around the same time, in the winter of 1942, a meeting was convened in Wannsee by Hans Plendl in his capacity as BHF. Attended by military and technical experts, the meeting addressed the viability of shorter wavelengths. While it was acknowledged that shorter wavelengths could enhance the effectiveness of signal reflection, it was concluded that this effect diminished at wavelengths below 20 centimetres, rendering them unsuitable. This misjudgement contributed further to the complete cessation of centimetre-wave research in Germany.

A request by General Martini, dated 8 August 1942, for the resumption of centimetre-wave research was formally rejected by

376 Me and Telefunken; Runge 1971, p. 50. "Until then, we had only been familiar with what we regarded as primitive British radar technology operating in the metre-wave range, to which we considered ourselves clearly superior with our decimetre-wave systems. Now, however, we were faced with centimetre-wave technology that was far more advanced than our own. Our work on the five-centimetre wave, which had already begun before the war, was discontinued by me in autumn 1942 following a final demonstration before the radio-technical officials of the three branches of the Wehrmacht. Given the urgent development demands and limited personnel, such forward looking areas could no longer be pursued with sufficient intensity since the beginning of the war. From that point onward, our development efforts focused on replicating, adapting, and studying British centimetre-wave technology, an area in which we also achieved original further developments of our own by the end of the war." (translated by the author).

377 Reuter 1971, p. 197; Godt 2003, p. 239; in Germany, political obstacles were added, such as the Development Stop Order issued on 3 February 1940 by Hermann Göring in his capacity as Commissioner for the Vierjahresplan (Kern 1984, p. 237).

the Technisches Amt of the RLM on 15 January 1943, citing limited chances of success and ongoing personnel shortages.[378]

However, the situation changed abruptly on 9 February 1943, when a British bomber was shot down near Rotterdam. Among the wreckage, German forces discovered a radar device operating at a wavelength of approximately 9 centimetres, most likely an early H_2S system. This unexpected discovery provided clear evidence that Britain had already made substantial progress in the field of centimetre-wave radar, which Germany had only recently discontinued.

The first British bomber raid on Hamburg, conducted on the night of 30 to 31 January 1943, was supported by Pathfinder reconnaissance aircraft equipped with onboard H_2S radar systems operating within the centimetre-wave spectrum. This operation marked the commencement of subsequent devastating raids, including the extensive bombing of Hamburg between July and August 1943, known as Operation Gomorrah.[379]

The deployment of H_2S radar technology proved to be a decisive factor not only in the targeting and destruction of German urban centres and industrial infrastructure but also contributed significantly to the success of the Allied submarine campaign in the Atlantic.

Looking back on developments up to early 1943, Leo Brandt later described what he considered a pivotal moment in the evolution of radar technology, stating:

> „Man könnte den Funkmeßkrieg bis Anfang 1943 vielleicht mit dem Worte *Stellungskrieg* kennzeichnen. Zäh wurde von beiden Seiten in Angriff und Abwehr auf einem technischen Gebiete gerungen, das im wesentlichen durch physikalische Erkenntnisse aus der Vorkriegszeit bestimmt war.
>
> Eine entscheidende technische Wandlung in der gesamten Funkmeßtechnik[380] erfolgte indessen durch den Einsatz der *Zentimeter-Wellen* in England…"[381]

378 Trenkle 1979, p. 43; Godt 2003, p. 239.

379 Lovell 1991, pp. xvii – xviii.

380 Radar technology.

381 Brandt 1962c, p. 69. "Up until early 1943, the radar war might perhaps be described as a *war of positions*. Both sides struggled doggedly in attack and

6.2 Great Britain

6.2.1 Early Developments and Key Challenges

The British efforts in centimetre-wave development were closely tied to the early airborne radar programme, as outlined below:

In both Germany and Great Britain, significant efforts were made to increase transmission power at shorter wavelengths in a stable manner. In the UK, however, the transition to centimetre waves was advocated by Edward Bowen[382] for the use of radar equipment on aircraft even before the war began. He was the first scientist to recognise the necessity and advantages of centimetre waves for radar applications.

In the spring and summer of 1939, Bowen discussed the need for centimetre wave radar with Charles Wright, Director of Scientific Research at the Admiralty and Chairman of the Committee for the Coordination of Valve Development (CVD), during his visits to Bawdsey. The CVD acted on behalf of the Army, Navy, and Air Force. Bowen and Wright agreed that the greatest need for centimetre wave radar was initially on aircraft.[383]

In the summer of 1939, Professor John Cockcroft[384] from the Faculty of Physics at Cambridge University and Professor Mark Oliphant[385] from the University of Birmingham became involved in the top-secret British radar programme. Both became key figures in the development of components that would enable the use of microwave radar.[386]

One problem with long-wave radars, such as those with a wavelength of 1.5 metres and a correspondingly large illumination area, was ground echoes. This occurred because part of the transmitter's energy was radiated directly downwards from the aircraft,

---

defence within a technical domain that was largely shaped by pre-war physical knowledge. A decisive technological shift in the entire field of radar, however, came with the introduction of *centimetre waves* in England…" (translated by the author).

382 Minnett 2015; Bowen 1987; Watson 2009, p. 47.

383 Bowen 1987, p. 143; Callick 1990, p. 61.

384 Allibone 2008.

385 Bleaney 2006.

386 Redhead 2001, p. 321; Burns 1988b, pp. 259–283.

and the strong backscatter signals from the ground masked the weak backscatter signals from an aircraft. As a result, the detection of enemy aircraft at distances greater than the aircraft's altitude, i.e., greater than the distance between the aircraft and the ground, was completely suppressed. Consequently, the AI radar[387] installed in night fighters was unable to detect an enemy aircraft if it was at a distance greater than the fighter's altitude above the ground.[388]

Bowen identified the solution to this problem in directing a narrow antenna beam forward and estimated a beamwidth of 10 degrees for the antenna. The maximum antenna size in the nose of a fighter aircraft led to an operating wavelength of 10 centimetres.[389]

In the summer of 1939, Oliphant strongly recommended the development of an oscillator tube capable of providing a peak power of around 1 kW at a wavelength of 10 centimetres. The advantages were not only the improved resolution of the radar images but also the reduced size of the radar equipment and smaller antenna dimensions. Additionally, the major issue of interfering ground echoes could be mitigated by strongly focusing the antenna beam.[390] This was followed by targeted and coordinated research and development on power generation in the decimetre and centimetre wavelength ranges. In 1939, the CVD at the British Admiralty, which coordinated research and development for all three branches of the armed forces in the UK, tasked various institutions with this work, including Oliphant at the University of Birmingham, Brebis Bleaney at the Clarendon Laboratory at Oxford University, and the University of Bristol, to develop transmitter and receiver tubes for the 10-centimetres wavelength range.[391] At the same time, the Air Ministry was also active and contacted the GEC[392] Research Laboratory in Wembley with the aim of obtaining a development contract for powerful tubes for a radar in the centimetre wave range for use on board aircraft. The exact wavelength

387 Airborne Interception (AI) Radar.
388 Rowe 2015, pp. 76–77.
389 Bowen 1987, p. 143.
390 Redhead 2001, p. 321.
391 Bowen 1987, p. 143.
392 GEC: General Electric Company Ltd.

was not specified but was assumed to lie between 20 and 30 centimetres. This was done against the background that GEC was able to greatly improve the performance of conventional tubes. For example, a tube with a wavelength of 50 centimetres could already generate several kilowatts of power. It was hoped that the wavelength could be reduced to 20 centimetres, which seemed useful for many airborne applications.[393]

Henry Tizard visited the GEC Research Laboratories in Wembley on 6 November 1939 to pursue the matter further.[394] After Robert Watson Watt[395] also contacted GEC, the company was approached on 29 December 1939 regarding a development contract for a short-wave AI radar. The EMI Company in Hayes was also approached to participate in the radar development work. A series of joint meetings took place from 16 January 1940. Edward Bowen was appointed to coordinate the work of GEC and EMI.[396]

6.2.2 Breakthrough

At that time, physicists John Randall[397] and Henry Boot[398] were conducting research at Professor Oliphant's institute to improve the performance of klystrons. They benefited from their detailed investigation of how the klystron functioned and adapted the cavity resonator, originally designed for the klystron, for use in the magnetron.[399] This they did without being overly influenced by existing literature on the magnetron.

They referred to this latter point as follows:

---

393 Bowen 1987, pp. 143–144.

394 Ibid., p. 144.

395 Watt 2017; Jäger und Heilbronner 2010, pp. 452–453; Ratcliffe 1975, pp. 548–568; Burns 2013, pp. 130–131; in 1942 he was knighted in recognition of his significant contributions to the development of radar, after which he was styled Sir Robert Watson-Watt (Zimmerman 2013, p. 305).

396 Bowen 1987, p. 144.

397 Wilkins 2011; Wilkins 1987, pp. 491–535.

398 Nicholls 2011.

399 Burcham and Shearman 2018, pp. 45–49; Kern 1984, pp. 240–241.

> "... Fortunately we did not have the time to survey all the published papers on magnetrons or we would have become completely confused by the multiplicity of theories of operation. ..."[400]

They continued:

> "In November 1939, at the risk of incurring some unpopularity from our fellow workers, we concentrated our thoughts on how we could combine the advantages of the klystron with what we believed to be the more favorable geometry of a magnetron. We had to confine the radio frequency fields in resonators, as did the klystron, in order to get hight circuit efficiency, and it must be made of solid copper to dissipate the power we felt could be put into a magnetron-like geometry ..."[401]

In a working session in November 1939, Randall and Boot finalised the basic design of the cavity magnetron. Due to various internal obstacles, the first model was not completed until 21 February 1940, but it exhibited exceptional performance. The magnetron generated an output of about 400 watts at a wavelength of 9.8 centimetres.[402]

In the period that followed, the team in Birmingham continued efforts to increase the magnetron's performance. By September 1940, successful tests had been conducted at wavelengths of 5 centimetres, 3 centimetres, and 2 centimetres.

In April 1940, the GEC research laboratories in Wembley were commissioned by the CVD to design and manufacture sealed-off versions of Randall and Boot's magnetron. By June 1940, two improved magnetrons had been successfully produced. Magnetron No. 1 was water-cooled, operated at a wavelength of 9.8 centimetres, and delivered a pulse power of about 500 watts.[403] An electromagnet was also used. However, both designs were poorly suited for installation on aircraft. As a result, all subsequent magnetrons were air-cooled.

E.C.S. Megaw of the GEC laboratories made two major technological improvements to the original Randall and Boot mag-

400 Boot and Randall 1976, p. 724.

401 Ibid., p. 724.

402 Klausing and Judkins 2023, p. 12; Randall 1946, p. 249; Redhead 2001, p. 323; Lovell 2004b, p. 284; Kern 1984, p. 241.

403 Redhead 2001, p. 324.

netron design, which led to enhanced performance and a significant reduction in the device's weight. The first improvement was the introduction of pulsed operation at very high voltage, which allowed the magnetron to deliver much higher peak power than continuous wave operation. The second was increasing the diameter of the cathode and implementing an oxide-coated cathode, which improved electron emission and efficiency.

Initially, the magnetron operated in continuous wave mode, but following the development of a high-voltage pulse modulator by Alan Blumlein[404] at EMI[405], the magnetron could generate pulses with power well over one kilowatt at a wavelength of 10 centimetres. According to Bowen, this represented a major breakthrough in radar technology.[406]

The improved magnetron model, designated E1188 and based on the earlier work of Randall and Boot, was smaller and mechanically more robust. It was the first version to be produced in large quantities and quickly reached a peak power output of 10 kilowatts in pulsed operation.[407] These technological advances significantly enhanced the capabilities of airborne radar systems.

Building on these developments, Alan Blumlein, an engineer at EMI, and Bernard Lovell[408], a physicist at the Telecommunications Research Establishment (TRE), made complementary contributions to the advancement of airborne radar technology during the war. While working at different institutions, Blumlein developed key components such as high-voltage pulse modulators, and Lovell contributed to the operational deployment and scientific understanding of radar systems. Together, their efforts were instrumental in the development and successful use of the H_2S radar sys-

404 Alan Dower Blumlein was perhaps the greatest British electronics engineer of the twentieth century. His 128 patents, most of them filed between 1933 and 1940, cover telephony, electrical measurement, sound recording and reproduction, high-definition television, radar, and electronics in general. Blumlein died on 7 June 1942 at the age of just 38 in a plane crash while testing an H_2S radar device (Burns 2014; Alexander 2013; Watson 2009, p. 87).

405 EMI: Electric and Musical Industries.

406 Bowen 1987, p. 147.

407 Klausing and Judkins 2023, p. 13.

408 Murdin 2016; Davies et al. 2016, pp. 323–344.

tem, which provided critical ground-mapping capabilities for Allied bombers.

In France, Maurice Ponte and Henri Gutton[409] had been working since the 1930s for the company Société Française Radioélectrique (SFR), focusing on pulse technology and a monostatic radar configuration at wavelengths of 80 centimetres and 16 centimetres.[410]

In mid-1935, an early obstacle-detection radar developed by Maurice Ponte and Henri Gutton was installed on the liner Normandie. The system operated on a continuous wave (CW) scheme at a wavelength of 16 centimetres and was tested during a transatlantic voyage from late August to early September 1935, with Gutton himself on board to oversee its performance. The radar reportedly achieved detection ranges of 3 to 5 kilometres for ships and up to 10 kilometres for coastlines (Fig. 12, with H. Gutton on the left). Gutton's pioneering work was seminal in demonstrating the feasibility of short-wavelength radar for maritime navigation, paving the way for further developments in both civilian and military radar technology.[411]

In late 1939, they had already achieved a peak power of 1 kilowatt at a wavelength of 16 centimetres with their M-16 magnetron.[412]

By the end of April 1940, as the war situation in France became increasingly critical, Ponte authorised by the French government personally crossed the Channel with two samples of the latest M-16 magnetron developed by SFR in Paris and presented them to the GEC laboratories in Wembley. Just days later in May 1940 the German army broke through the front lines and swiftly advanced towards Paris.[413]

The M-16 magnetron was ideal for pulsed operation. This device was incorporated into the E-1188, which subsequently led to the development of the E-1189. The E-1189 also featured cooling

409 Molyneux-Berry 1988, pp. 45–52.

410 Blanchard 2010, pp. 5–10.

411 Blanchard 2010, p. 8; Blanchard 2004, pp. 83–86, Klausing und Holpp 2022, pp. 22–23.

412 Blanchard 2010, p. 5.

413 Blanchard 2010, p. 10; Klausing und Holpp 2022, p. 23; Redhead 2001, p. 324.

Fig. 12: The SS Normandie's radar in New York harbour[414]

fins for heat dissipation. Air cooling made it suitable for use on aircraft. In addition, the number of resonators in the magnetron was increased from six to eight to improve efficiency, and a permanent magnet replaced the earlier electromagnet, further enhancing its suitability for airborne applications.[415] By July 1940, the further developed E-1189 magnetron, with serial numbers 1 and 2, had already achieved a peak power output of 12 kW at a wavelength of approximately 9.5 centimetres.[416] The model with serial number 12 was sent to North America in September 1940 as part of the Tizard Mission.[417] By September 1940, magnetrons produced by GEC had already reached a peak pulse power of 100 kW at a wavelength of 10 centimetres.[418] By the end of 1941, 2,000 magnetrons had been manufactured by the British Thomson-Houston Company, in collaboration with GEC.[419] Between 1942 and 1943, Western Electric developed a magnetron operating at a wavelength of 3 centimetres,

414 Image credit: J.-M. Gutton Collection.

415 Willshaw 1985, p. 86; Blanchard et al. 2013, p. 251.

416 Callick 1990, pp. 63–64; due to the numerous stages of technological development of magnetrons, the technical data found in the literature often differ from one source to another.

417 Redhead 2001, p. 324.

418 Ibid., p. 325.

419 Redhead 2001, p. 325; Callick 1990, p. 65.

and between 1943 and 1944, the MIT Radiation Laboratory further developed the H_2S radar into the H_2X radar, using this American-built magnetron.[420] Between 1941 and 1944, the cavity magnetron became the centrepiece of more than 150 new radar systems of various types, providing the Allies with a decisive technological advantage during the course of the war.[421]

6.2.3 From Defence to Offence

With the cavity magnetron now firmly established as a core component of airborne radar systems, its strategic implications became increasingly evident. Nowhere was this more apparent than in the transition from Britain's defensive posture to an offensive bombing campaign. The development of the cavity magnetron provided the crucial technological basis for the H_2S radar, marking an important advancement in British aerial warfare by significantly enhancing the capability to conduct precise night-time bombing raids. While the H_2S radar was not the sole factor enabling the shift from a defensive to an offensive strategy this innovation played a crucial role in increasing the effectiveness of the Royal Air Force's bomber fleets.

Combined with other developments such as the growing number of bombers improved tactics and the broader context of the war from 1943 onwards Britain's offensive against Germany intensified markedly. The H_2S radar thus represented a key component in the complex process that ultimately brought about the strategic transition from defence to offence.

This technological innovation was instrumental in enabling the Royal Air Force to conduct its strategic bombing campaigns with greater precision and impact. Using the example of Hamburg J. G. Crowther and R. Whiddington illustrate in their 1948 work "Science at War" how the H_2S radar from 1943 onwards contributed significantly to the success of large-scale bombing raids which, however, also resulted in substantial civilian casualties:

---

420 Klausing and Judkins 2023, pp. 21–22; Lovell 1991, pp. 183–184; Breitkopf 2018, p. 256.

421 Blanchard et al. 2013, p. 251.

"In three great raids, on July 24th, 27th and 29th, 1943, led by Pathfinders using H_2S, Hamburg was laid in ruins. On December 3rd, 1943, Leipzig was found through thick fog and bombed... In November, 1943, the bombers with an improved H_2S at last began to recognize details of Berlin. A series of successful heavy raids followed... Between October 1943 and March 1944, 93 per cent of our strategic bombing effort depended on navigation and bombing with the aid of H_2S."[422]

6.3 Timeline

Fig. 13 shows the development of centimetre-wave technology in Germany, while Fig. 14 illustrates its development in Great Britain.

422 Crowther and Whiddington 1948, p. 69.

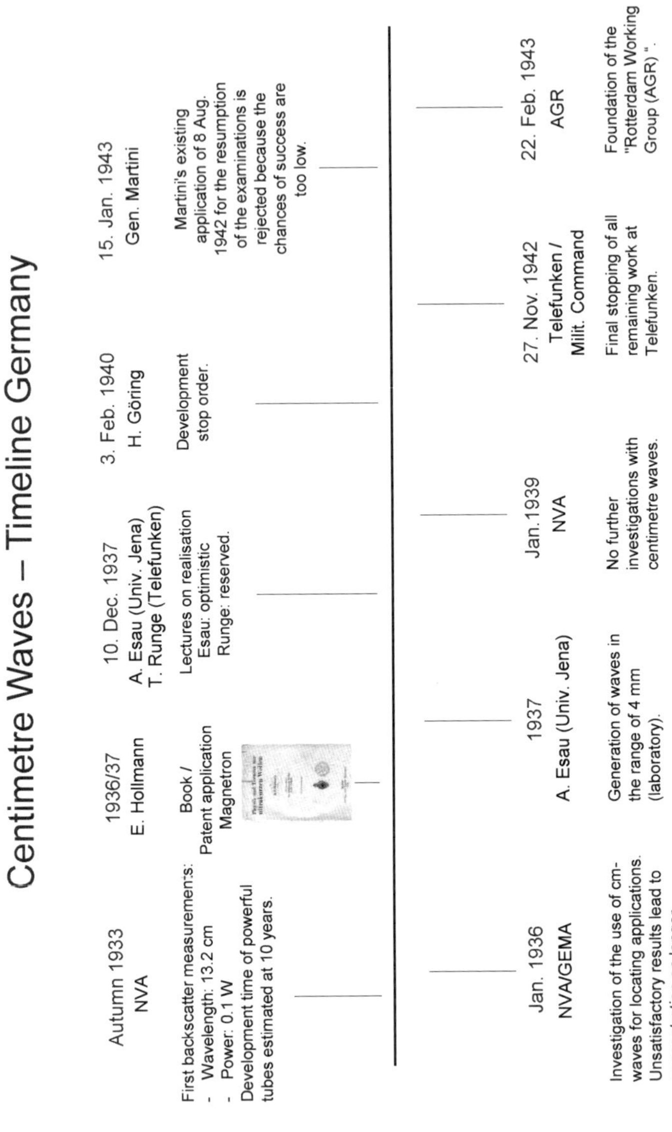

Fig. 13: Centimetre wave technology in Germany

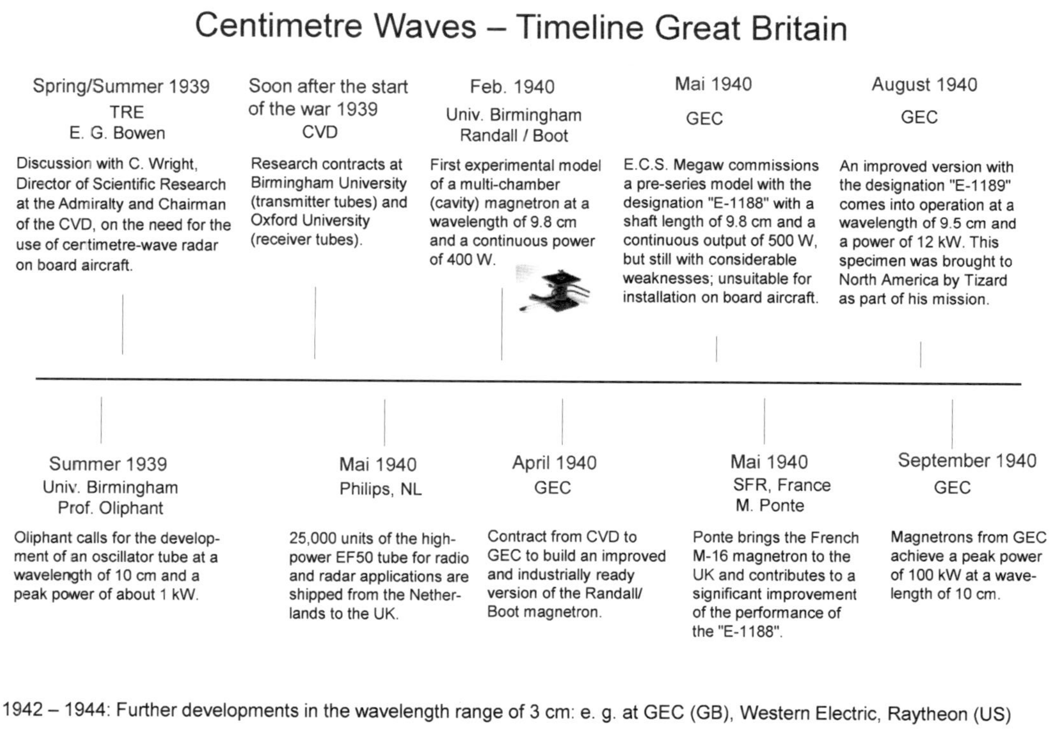

Fig. 14: Centimetre wave technology in Great Britain

6.4 Research in Other Countries

Before World War II, in addition to development work in Germany, Great Britain, and the United States, intensive research on high-power tubes was also carried out in several other countries, including France, the Netherlands, Japan, and the former Czechoslovakia[423] Between 1936 and 1937, the Soviet Union and Poland developed magnetrons in the decimetre and centimetre wave ranges. In the Netherlands, Philips in Eindhoven became Europe's largest tube manufacturer and established the Natuurkundig Laboratorium (Natlab) in 1914 for research purposes. Philips developed a magnetron with a power output of 10 watts at 1.0 GHz and, by autumn 1933, supplied samples with an increased transmission power of 40 watts at a wavelength of approximately 50 cm to the NVA in Germany. These deliveries enabled the NVA to overcome the limitations of low transmission power and conduct the first successful ship measurements. Another significant contribution from Philips was the development and production of the EF50 tube for radio and radar applications. After the German occupation of the Netherlands in May 1940, a British destroyer was sent to The Hague to evacuate around 25,000 EF50 tubes from Eindhoven and transport them to Great Britain.[424]

---

423 Blanchard et al. 2013, pp. 244–254; Borisova 2010.

424 Klausing und Holpp 2022, p. 25.

7 Development of Radar Technology in Germany and Great Britain

7.1 Up to the Outbreak of World War II

7.1.1 Germany

7.1.1.1 First Steps

First attempts to use the reflection of electromagnetic waves for target detection were conducted in Germany in the autumn of 1933 at Kiel harbour by the Nachrichtenmittel-Versuchsanstalt der Marine (NVA).[425] The NVA[426] had been established in 1923 in Kiel to carry out research and development related to communications equipment for warships, with a branch office located on the Bay of Lübeck in Pelzerhaken, near Neustadt/Holstein.[427]

On behalf of the Torpedo-Versuchsanstalt (TVA)[428], the NVA conducted experiments in underwater sound detection, aiming to

---

425 The Nachrichtenmittel-Versuchsanstalt (NVA) was a German naval research institution responsible for the testing and development of communications and signalling technologies. From the early 1930s, with documented experiments around 1933 under the direction of Rudolf Kühnhold, the NVA conducted pioneering trials with the reflection of electromagnetic waves. These marked some of the earliest German steps towards the practical use of radar. While initially tasked with improving naval communications and related equipment, the institute increasingly shifted its focus towards detection and warning systems.

426 Renamed in 1938 to Nachrichtenmittel-Versuchskommando der Kriegsmarine (NVK); Willisen ca. 1984, p. 24.

427 Kroge 2000, p. 7; Kroge 1998, p. 15.

428 In parallel with developments in signalling and radar, the German Navy (Kriegsmarine) also invested in advanced torpedo technologies. The Torpedo-Versuchsanstalt (TVA) was established to research, develop, and test torpedoes for the Kriegsmarine. The TVA played a key role in improving torpedo performance, guidance, and reliability, contributing directly to the operational effectiveness of German naval forces during World War II. Its work included experimental designs, propulsion systems, and control mechanisms aimed at enhancing accuracy and range in combat situations.

improve the accuracy of direction-finding for underwater noises such as those generated by ship propellers. Until then, it had only been possible to distinguish between sounds of large and small ships based on intensity and direction of incidence, but direct distance measurement was not feasible.

In 1928, the TVA commissioned the NVA to develop a precise range and bearing locator for the targeted firing of torpedoes at night and in fog, capable of operating at distances of up to several thousand metres.[429] Since 1929, the NVA had been conducting tests with an underwater horizontal sounder operating with sound pulses.[430]

In 1931, Rudolf Kühnhold[431] of the NVA applied for a patent for a device based on the sum-difference method, which was capable of indicating both direction and distance simultaneously.[432] This method, known today as sonar[433], provided valuable insights for later developments in underwater sound-location and radio technology. It enabled submarines to detect larger surface vessels at distances of 5 to 10 kilometres while submerged, and to locate submarines from moving destroyers at ranges of 3 to 5 kilometres.[434]

However, factors such as noise from the swell, bottom echoes, shallow water depth, and even interference from their own equipment significantly reduced achievable ranges. Additionally, inter-

429 Willisen ca. 1984, p. 2; Watts and Klausing 2025, p. 25; Reuter 1971, p. 16; Kroge 2000, p. 7; Kroge 1998, p. 15.

430 Willisen ca. 1984, pp. 2–3; Watts and Klausing 2025, p. 25; Reuter 1971, p. 16; Swords 2008, p. 92.

431 Kühnhold began working for the NVA after studying physics and earning his doctorate at the University of Göttingen. In April 1933, he became scientific director of the NVA and emerged as a pioneer of radar technology in Germany (Surhone 2010, pp. 1–2; Weinreich 2001g, pp. 108–109; Buderi 1997, p. 202; Reuter 1971, p. 18).

432 Kroge 2000, p. 8; Kroge 1998, p. 16; Willisen ca. 1984, p. 3; Watts and Klausing 2025, p. 25; Gießler 1958, p. 108.

433 Sonar (Sound Navigation and Ranging) is an underwater detection system that uses the propagation of sound in water, typically at a speed of around 1500 m/s depending on temperature, salinity and pressure, to detect, locate and classify vessels such as submarines while submerged. (Grabau 1989, pp. 155–181).

434 Gießler 1971, p. 108.

ference noise increased rapidly as the speed of surface vessels rose.[435]

Another major challenge for naval anti-aircraft artillery in naval ports and for ship artillery was defending against enemy air attacks. It was recognised early on that the acoustic listening devices available in harbours would soon no longer be sufficient for early warning, as aircraft construction was developing rapidly and dynamically, allowing for higher flight altitudes and ever-increasing speeds.[436] Furthermore, the optical rangefinders used by ship artillery to locate ships and aircraft proved inadequate in poor visibility caused by weather conditions such as rain, fog, or cloud cover. Consequently, there was a need for a weather-independent method of detection and location.[437]

In spring 1933, Kühnhold had the idea of utilising Heinrich Hertz's findings on the reflection of electromagnetic waves and using radio waves instead of sound waves to detect and locate targets.[438] He was unaware of Christian Hülsmeyer's experiments some 30 years earlier. Helmuth Gießler[439] reported:

> „Auf Grund dieser Erkenntnisse und angeregt durch Veröffentlichungen in der ausländischen Fachpresse entschloß sich Dr. Kühnhold im Frühjahr 1933, den alten Gedanken von Heinrich Hertz wieder aufzugreifen und sehr kurze elektromagnetische Wellen für eine Echolotung in der Luft anzuwenden…"[440]

In collaboration with Professor Karl Kohl from Erlangen, the company Julius Pintsch AG developed a transmitter and receiver using

435 Ibid., p. 108.

436 Gießler 1958, p. 108; in Great Britain, it was also recognised that acoustic listening devices were becoming increasingly inadequate for detecting aircraft acoustically.

437 Reuter 1971, pp. 16–18.

438 Reuter 1971, p. 18; Kroge 2000, p. 8; Kroge 1998, p. 16; Buderi 1997, p. 202.

439 Helmuth Gießler served as a military adviser to the NVA in Kiel from October 1931 to September 1933; he was promoted to Lieutenant Commander on 1 April 1933 (Gießler 1977, p. 37).

440 Gießler 1958, p. 108. "In spring 1933, based on these insights and encouraged by publications in the foreign technical press, Dr Kühnhold decided to revive Heinrich Hertz's old idea of using very short electromagnetic waves for echo sounding in the air." (translated by the author).

tubes based on the Barkhausen principle, producing a continuous-wave power of 0.1 watts at a wavelength of 13.5 centimetres, corresponding to a frequency of approximately 2.2 GHz.[441] At this high frequency, the antennas achieved good directional sensitivity.

7.1.1.2 Early Experiments Towards Operational Radar

Initial transmission and reception tests carried out in autumn 1933 in Kiel Harbour, between the NVA office building in Düsternbrook and the naval arsenal, over a distance of roughly one nautical mile, showed excellent reception of directly transmitted signals but no reception of reflected echo signals. The reason was that the tube's transmission power was too low to pick up usable echoes.[442]

At that time, tube development in Germany for this wavelength range was still far from achieving higher output power. Professor Kohl from Erlangen estimated that it would take another ten years to develop powerful tubes for the centimetre wave range.[443]

However, as early as 1935, Kühnhold was already convinced of the effectiveness of centimetre waves for accurate target detection and imaging. By then, the NVA had begun initial work in the four-to-ten-centimetre wavelength range, despite the lack of successful results from the Pintsch company regarding the required tube power. The plan was to start practical tests with centimetre waves in January 1936.[444]

Nonetheless, in the following period, the focus shifted to longer wavelengths, that is lower frequencies, and investigations into the centimetre wave range were not resumed until 1943, when the advantages of this wavelength had already been demonstrated by Allied developments and their subsequent successes in large

441 Pintsch's development was already operating at a wavelength roughly equivalent to that of the British H_2S airborne radar, which was used against Germany nearly ten years later.

442 Willisen ca. 1984, p. 3; Watts and Klausing 2025, p. 25; Reuter 1971, pp. 18–19; Kroge 2000, p. 8; Kroge 1998, p. 16; Watson-Watt 1957, p. 105; Watson 2009, pp. 233–234.

443 Willisen ca. 1984, p. 3; Watts and Klausing 2025, p. 25; Kroge 2000, p. 8; Kroge 1998, p. 16.

444 Reuter 1971, p. 22; see also Esau 1938, pp. 212–221.

scale bombing of German cities and anti-submarine warfare with airborne radar.

At the end of 1933, Kühnhold contacted the Telefunken company, which had many years of experience in radio technology. However, a conversation between Kühnhold and Runge, the deputy head of the laboratory at Telefunken, was unsuccessful, and no cooperation materialised. It can be assumed that Kühnhold interpreted Runge's skeptical attitude toward the development of powerful transmitters and corresponding locating devices as an official rejection by Telefunken.[445]

In his memoirs "Ich und Telefunken"[446], Runge writes about the conversation with Kühnhold:

> „Nachträglich erinnere ich mich, dass etwa ein Jahr früher Kühnhold, der technische Chef des NVK, des Nachrichtenversuchskommandos der Marine, mich besucht und aufgefordert hatte, für ihn Rückstrahlarbeiten anzufangen, ein Vorschlag, auf den ich nicht eingegangen war, weil er mir zu abenteuerlich erschien. Diese Ablehnung hat dann Kühnhold als offizielle Stellungnahme der Firma aufgefaßt und die Gründung der Firma Gema veranlaßt, die dann für die Marine die Freya-Entwicklung durchführte."[447]

In the early years of 1933 and 1934, Telefunken's decades of research and development expertise in radio technology could not be

---

445 Kern 1984, pp. 109–110; Reuter 1971, p. 19; Swords 2008, p. 93.

446 Runge 1971; "Me and Telefunken." (translated by the author), unpublished manuscript; Pritchard 1992, pp. 51–101; Pritchard 1989, pp. 57–99.

447 Runge 1971, p. 43. "Looking back, I recall that about a year earlier Kühnhold, the technical head of the Nachrichtenversuchskommando NVK of the Navy, visited me and asked me to start working on radar reflection, a suggestion I had not pursued because it seemed too adventurous to me. Kühnhold then took this rejection as an official statement from the company which led to the founding of the company Gema, which then carried out the Freya development for the Navy." (translated by the author). Reuter also claims that Kühnhold, together with the owners of the company Tonographie, von Willisen and Erblöh, co-founded GEMA (Reuter 1971, p. 19). However, doubts exist as to whether Kühnhold was directly involved in the founding of GEMA (Kroge 2000, pp. 19–20; Kroge 1998, pp. 27–28).

utilised due to a lack of willingness to cooperate with the NVA.[448] Leo Brandt later reflected on this, explaining:

> „Wie auch immer die Gründe waren – Ablehnung aus technischer Fehleinschätzung oder ein theoretisch besserer Geheimhaltungsschutz – man verzichtete mit diesem Schritt auf das Fachwissen und die dreißigjährige Tradition der Funkindustrie und mußte in jeder Hinsicht, vom Physikalischen bis zum Handwerklichen, von der Röhrenentwicklung bis zur Serienproduktion, neu anfangen. Jeder Fachmann wird sich darüber klar sein, daß ein solches Experiment mit größten Schwierigkeiten verbunden sein mußte und nicht den gleichen Erfolg bringen konnte wie eine Gemeinschaftsarbeit auf breiter Basis. Dies kam der deutschen Marine erst spät zu Bewußtsein.“[449]

The scope of Kühnhold's ideas was not yet recognised at the time due to the inadequate performance of valve technology in the centimetre wave range. However, in the summer of 1935, Runge conducted his own reflection tests for the first time.[450]

In contrast to Great Britain, there was no political or military support to consolidate the expertise and decades of experience held by major companies such as Telefunken, AEG, Siemens and Lorenz. Telefunken in particular possessed extensive knowledge in the field

---

448 Reuter 1971, p. 19; Swords believes that Kühnhold probably overestimated Runge's position, as Runge neither had the personnel nor the resources to carry out such research. The head of the device development department was Jewish and understandably reluctant to become involved in the political climate of the time. The board of directors, to whom the request should have been addressed, would most likely have rejected it as well. Swords argues that Runge's refusal ultimately worked in Kühnhold's favour, granting him the freedom to initiate the founding of GEMA. Later, when Runge also became involved in radar experiments, Kühnhold attempted to sue him, claiming that radar was solely his invention (Swords 2008, p. 96).

449 Brandt 1962c, p. 55; Kern 1984, pp. 110–111. "Whatever the reasons – whether rejection due to technical misjudgement or the prospect of theoretically better secrecy protection – this step meant renouncing the expertise and thirty-year tradition of the radio industry and having to start anew in every respect, from physics to craftsmanship, from tube development to mass production. Any expert will understand that such an endeavour was bound to face great difficulties and could not achieve the same success as a broad-based collaborative effort. It was only late that the German Navy became aware of this." (translated by the author).

450 Runge 1971, pp. 42–43; Reuter 1971, p. 32.

of radio technology, especially in radio relay systems.[451] Despite this, the NVA made no further effort to initiate talks between the naval command and the Telefunken board regarding potential co-operation.[452] Instead, the GEMA was founded in Berlin on 16 January 1934.[453]

GEMA was intended to focus on both sound-location and radio technologies. Its co-founders, and from early 1935 the sole owners, were Paul-Günter Erbslöh and Hans-Karl Freiherr von Willisen, proprietors of the company Tonographie, who had already been working with the NVA in a trusting and informal manner. This stood in contrast to the rigid bureaucracy of the NVA, which provided limited resources and repeatedly imposed restrictions on Kühnhold's efforts.[454] As early as 1931, Erbslöh and von Willisen became the owners of Tonographie, a company specialising in the electrical sound recording of speech and music.[455]

Kühnhold found in them two skilled young entrepreneurs, both full of idealism and eager to pursue new technological challenges. In the following years, GEMA made significant contributions to the development and production of radar systems for early warning of aircraft and ships as well as for radio navigation. Alongside Hans E. Hollmann, Theodor Schultes[456] played a major role in GEMA's technical success. GEMA went on to work directly with the NVA in the field of radiolocation using ultra-short electrical waves.

Meanwhile, it became known that the Dutch company Philips in Eindhoven was producing magnetrons with sufficient output power in the wavelength range of 50 to 60 centimetres. Even before GEMA was officially established, six magnetrons with continuous wave output of 40 watts at this wavelength range were purchased from Philips in November 1933, funded by the TVA. The first experimental transmitters using two magnetrons in a push-pull circuit

451 Runge 1953, pp. 182–185.

452 Reuter 1971, p. 19.

453 Kroge 2000, p. 20; Kroge 1998, p. 28; see also Chapter 7.1.1.4.

454 Willisen ca. 1984, pp. 4–5; Watts and Klausing 2025, p. 26; Kroge 2000, p. 16; Kroge 1998, pp. 24–25.

455 Kroge 2000, p. 12; Kroge 1998, p. 21.

456 Jäger und Heilbronner 2010, pp. 391–392; Weiher 1983b, pp. 171–173; Weinreich 2001a, pp. 106–108.

with audio modulation at 1000 Hz and an output of approximately 80 watts were assembled at the Tonographie company in Berlin in December 1933.[457]

At the beginning of January 1934, the newly founded GEMA further developed the device into a stable laboratory model operating at a wavelength of 48 centimetres.[458] On 20 March 1934, the NVA successfully used this device in Kiel to conduct backscatter measurements against the liner Hessen, which was positioned about 500 metres away. The first attempt failed because the receiver was within the direct radiation field of the transmitter, which completely overwhelmed the receiver and rendered it insensitive. Therefore, the initially hoped-for backscatter signal was not detected. However, when the receiver was moved onto a small boat and the boat was positioned outside the direct transmission range, backscatter signals from the ship's hull became apparent at a distance of approximately 20 metres from the ship.[459]

Furthermore, Hans Rindfleisch[460] of the NVA suggested blocking the receiver during the transmission pulses, thus introducing the concept of pulsed operation for the first time.[461] Despite the short range, this represented an important step towards the further development of radar technology. Kühnhold recommended installing transmitters and receivers as close together as possible for shipboard use, enabling operation with a single device.[462] From 7 to 12 May 1934, further measurements were conducted with an improved Pintsch apparatus operating at a wavelength of 13.5 centimetres and a power output of 0.3 watts in Schilksee on the Kiel Fjord. The transmitter and receiver were positioned about 10 metres apart and at a height of 8 metres above the water surface. When measurements were taken against the NVA experimental

457 Willisen ca. 1984, pp. 3–4; Watts and Klausing 2025, p. 26; Reuter 1971, p. 20; Trenkle 1986, p. 24.

458 Willisen ca. 1984, p. 5; Watts and Klausing 2025, p. 26; Trenkle 1986, p. 25; Swords 2008, p. 93.

459 Willisen ca. 1984, p. 5; Watts and Klausing 2025, p. 26; Kroge 2000, p. 23; Kroge 1998, p. 31; Trenkle 1986, p. 25.

460 Weiher 1983a, pp. 144–145.

461 Willisen ca. 1984, pp. 5–6; Watts and Klausing 2025, p. 26; Trenkle 1986, p. 25; Gießler 1971, p. 60.

462 Gießler 1971, p. 60.

vessel Grille, backscatter signals were detectable from distances exceeding 2 kilometres.[463] This development created a competitive situation between the Pintsch company and GEMA.[464]

From 11 October to 2 November 1934, GEMA conducted further experiments at the NVA test site in Pelzerhaken on the Bay of Lübeck, using improved equipment with more stable magnetrons developed for wavelengths of 50 and 60 centimetres, as well as more sensitive receiver tubes.

Initially, satisfactory distance measurements could not be obtained because the transmitter still radiated too strongly into the receiver. This issue was resolved by relocating the receiver to a distance of 200 metres.

Following this adjustment, on 24 October 1934, backscatter signals from the Grille were successfully received at a distance of approximately 12 kilometres.

Additionally, signals from a Junkers W 34 aircraft that happened to fly past were clearly received from a distance of about 700 metres and at an altitude of 200 metres.[465]

As a representative of the TVA was present during the tests, it was possible to secure 70,000 Reichsmarks in funding for further trials.[466] However, strict secrecy was initially imposed, even within the Navy, as it was not yet possible to measure key operational parameters such as range and bearing, which would later prove essential for effective radar technology.[467]

Further developments still seemed necessary before a demonstration for the Naval Command could take place. Key features

463 Willisen ca. 1984, p. 6; Watts and Klausing 2025, pp. 26–27; Kroge 2000, p. 23; Kroge 1998, pp. 31–32; Gießler 1958, p. 109; using these devices, telephony trials were subsequently conducted between the islands of Helgoland and Wangerooge, covering a distance of approximately 43 kilometres and successfully establishing a clear voice connection (Willisen ca. 1984, p. 6; Watts and Klausing 2025, p. 27; Swords 2008, p. 94).

464 Kroge 2000, p. 23; Kroge 1998, p. 32; Swords 2008, p. 94.

465 Willisen ca. 1984, pp. 8–9; Watts and Klausing 2025, p. 28; Kroge 2000, p. 25; Kroge 1998, pp. 33–34; Gießler 1958, p. 109; Trenkle 1986, p. 26.

466 Willisen ca. 1984, pp. 8–9; Kroge 2000, p. 25; Kroge 1998, p. 34.

467 Willisen ca. 1984, p. 9; Gießler 1958, p. 110; GEMA only became aware of the previous inventor Christian Hülsmeyer when patent applications were filed (Kroge 2000, p. 25; Kroge 1998, p. 34).

included the introduction of the pulse method for distance measurement by timing the propagation of very short pulses, with the results displayed on an instrument or the Braun'sche Röhre; blocking of the receiver during transmission pulses; a method for lateral bearing; advancement in high-powered tube technology; and the use of a Tannenbaum radiator.[468] This was intended to fulfil an important requirement for shipboard use, namely that the transmitter and receiver no longer needed to be placed separately but could be integrated into a single device. GEMA was now endeavouring to present the naval command with a method to measure the range and lateral position of naval targets by the summer of 1935 at the latest.[469]

All these developments in 1934 represented a significant step from mere backscatter measurement to radar technology. During the winter, GEMA's development work to meet these requirements progressed well, making the end of 1934 an important milestone on the road to radar technology. Pulse technology and the use of the Braun'sche Röhre to display backscatter signals in relation to distance formed the basis for subsequent radar devices.[470]

By mid-January 1935, an improved receiver operating at a wavelength of around 50 centimetres, or a frequency of approximately 600 MHz, was able to replace the previous model. This was also based on the 1934 experiments, where this wavelength range proved excellent for location purposes. From 29 January to 28 February 1935, GEMA carried out further experiments in Pelzerhaken with the new receiving technology. On 8 February 1935, backscatter signals from a Junkers W 34 seaplane were detected at a distance of approximately 5 to 6 kilometres, and weak backscatter signals could still be received from up to 12 kilometres. The important realisation was that it was also possible to locate aircraft.[471]

At this point, it should be noted that Robert Watson Watt, the British radar pioneer, submitted a memorandum titled "Detection

468 Willisen ca. 1984, p. 9; Watts and Klausing 2025, p. 28; Trenkle 1986, p. 27, p. 33; Gießler 1958, p. 110.

469 Willisen ca. 1984, p. 9; Watts and Klausing 2025, p. 28.

470 Kern 1984, p. 113.

471 Willisen ca. 1984, p. 10; Watts and Klausing 2025, p. 29; Kroge 2000, pp. 27–28; Kroge 1998, p. 37.

and Location of Aircraft by Radio Methods" almost simultaneously, on 27 February 1935. This memorandum contained twenty points with extensive proposals for an effective air defence system using radio location, although not a single essential device existed at that time.[472]

By the beginning of May 1935, GEMA had largely achieved the development goals set during the 1934 trials, and further experiments were conducted in Pelzerhaken using new equipment, such as a newly developed, more stable magnetron operating at a wavelength of 52 centimetres with pulse modulation for distance measurement. The pulse power was 800 watts.[473]

In these tests against the experimental vessel Welle[474], the new equipment achieved a measurement range of approximately 3 kilometres, in contrast to conventional equipment with tone modulation, which still achieved up to 12 kilometres.[475] However, an initial distance measurement was already possible.[476] The entire installation was then re-examined.

This was connected with the realisation that, due to stronger backscattering from the broadside of the ship compared to the narrow side, the amplitudes differed, resulting in higher amplitudes and a greater range for the broadside. It was concluded that the majority of echo signals were likely specular reflections, following optical laws.[477]

As a result of understanding that the strength of the return signal depends directly on the aspect angle of the radar unit to the target and the transmission frequency with higher frequencies and thus shorter wavelengths increasing this dependency, GEMA decided to build a further apparatus operating at a longer wavelength in the two-metre range.[478]

---

472 Watson-Watt 1957, pp. 84–87, pp. 170–174; Austin 1999, pp. 365–372.

473 Willisen ca. 1984, p. 11; Watts and Klausing 2025, p. 29, Trenkle 1986, p. 27.

474 former Grille.

475 Willisen ca. 1984, p. 13; Watts and Klausing 2025, p. 30; Kroge 2000, p. 29; Kroge 1998, pp. 38–39.

476 Willisen ca. 1984, p. 13; Watts and Klausing 2025, p. 30.

477 Willisen ca. 1984, p. 13; Watts and Klausing 2025, p. 30.

478 Willisen ca. 1984, pp. 13–14; Watts and Klausing 2025, p. 30; Trenkle 1986, p. 27; Swords 2008, p. 95.

According to GEMA, this decision was incorrect for locating naval targets, but it laid the foundation for the later development of the Freya radar. Additionally, further development of equipment in the 50 centimetres wavelength range continued, but there was no further focus on shorter wavelengths, which, years later, proved to be a significant mistake. The planned advancement of tube technology at Pintsch progressed only slowly.

In the period that followed, GEMA succeeded in improving the technology of the 50-centimetre system to such an extent that, by July 1935, a measurement range of approximately 8 kilometres could be achieved against the cruiser Königsberg, which was anchored about 4 kilometres off Pelzerhaken in Lübeck Bay. To maintain secrecy, the cruiser was tricked into sailing several courses for the measurements without being told the true reason.[479] In August, further improved equipment enabled a range of approximately 7 kilometres against the Welle, compared to the initial 3 kilometres.

Based on these successes, it was decided to demonstrate the equipment to the Naval Command in the autumn.[480] On 26 September 1935, at Kühnhold's invitation, a demonstration of the 50-centimetre equipment took place at the NVA site in Pelzerhaken. It was attended by senior naval officers, including the Commander-in-Chief, Admiral Raeder, Fleet Commander Admiral Carls, and the Chief of the Naval Weapons Office, Admiral Witzell.[481]

Ranges of approximately 7 kilometres against the Welle and 8.5 kilometres against the larger artillery training vessel Bremse were achieved, with a distance accuracy of ±150 metres and a lateral bearing accuracy of approximately ±0.15 degrees.[482]

The demonstration was a great success, not only for the NVA and Kühnhold, but also for GEMA. The military significance of this new radar technology was finally recognised, and it was agreed that

479 Willisen ca. 1984, p. 14; Watts and Klausing 2025, p. 30; Kroge 1998, pp. 40–41; Kroge 2000, pp. 30–31; Kroge 1998, pp. 40–41; Reuter 1971, pp. 21–22; Trenkle 1986, pp. 27–28.

480 Willisen ca. 1984, p. 14; Watts and Klausing 2025, p. 30.

481 Reuter 1971, p. 23; Gießler 1971, p. 61; Bekker 1980, p. 50.

482 Willisen ca. 1984, p. 15; Watts and Klausing 2025, p. 31; Gießler 1971, p. 61; Kroge 2000, p. 35; Kroge 1998, p. 45.

further funds would be provided by the Naval Command and that trials on naval vessels would be conducted.[483]

This success led to GEMA becoming the Navy's exclusive contractor for radio and sound location technology. However, both the naval authorities involved and GEMA itself were immediately placed under strict secrecy.[484] The secrecy imposed in 1935 allowed GEMA to operate without competition in a field that was still only partially understood by the larger companies.[485] As a result, there was no cooperation with other branches of the Wehrmacht. A brief collaboration did take place with Lorenz AG.[486]

According to Reuter, Lorenz was later able to apply the expertise gained during this collaboration to the development of anti-aircraft fire control systems.[487] In the period that followed, confidentiality regulations were significantly tightened.

As GEMA did not have its own testing site, from autumn 1936 onwards, all ongoing trials were only permitted to take place at the Navy's test facility in Pelzerhaken. Furthermore, GEMA was only allowed to share information with selected representatives of the Navy. Any contact with other branches of the armed forces was strictly prohibited, even though the Air Force had already shown interest in the new technology.[488]

As a result, the Army, Air Force and Navy each demanded and subsequently received completely different equipment for essentially the same purpose.[489] Additionally, the Navy imposed strict secrecy even within GEMA itself in order to safeguard its developmental objectives.

From 1940/41, the term DeTe-Gerät, short for Dezimeter-Telegraphie-Gerät, was used by GEMA as a cover name to conceal the

---

483 Willisen ca. 1984, p. 15; Watts and Klausing 2025, p. 31; Kroge 2000, p. 35; Kroge 1998, p. 45; Reuter 1971, p. 23.

484 Kroge 2000, pp. 35–36, p. 72; Kroge 1998, pp. 45–46, pp. 82–83.

485 Kroge 2000, p. 72; Kroge 1998, pp. 82–83.

486 Kroge 2000, p. 72; Kroge 1998, p. 83.

487 Reuter 1971, p. 34.

488 Kroge 2000, pp. 65–66; Kroge 1998, p. 75.

489 Trenkle 1979, p. 28; this barrier to innovation once again significantly contributed to delays in the progress of radar technology, particularly concerning its performance and potential applications.

true nature of radar technology.[490] From February 1942 onwards, this designation was replaced by the more standardised terms FMG and FuMG. The German Navy generally adopted the designation FuMO from October 1943. Around the same time, the Air Force began using the term FuSE, particularly for systems combining transmitter and receiver functions used in early warning and fire-control roles.

These terms broadly denoted radar systems used by the armed forces, with different branches often employing their own nomenclature for systems of similar function.[491]

To maintain secrecy, various cover names and designations were introduced throughout the war (Fig. 15).

Sound location systems were initially not given specific cover names and were simply referred to as S-Geräte or S-Anlagen, with *S* standing for Sondergerät or Sonderanlage, which will hereinafter be referred to as sonar.[492]

However, funding was allocated based on the Navy's annual budget, which prioritised the construction and enhancement of combat-capable ships, particularly in light of the Anglo-German Naval Agreement of 18 June 1935. This led to limitations in financing other projects.[493] Decisions regarding the order of urgency and the corresponding distribution of funds were made by the Naval Command Office and related institutions, although the final authority lay with Admiral Raeder himself.[494]

Despite the promising results, Admiral Raeder regarded the detection of underwater targets using sound location as the Navy's most pressing task. As a result, he prioritised related work within the NVA, since it was expected that operational devices could become available for frontline use in the foreseeable future. Consequently, the already difficult development efforts at GEMA progressed only slowly, hindered by protracted negotiations within the Naval Command concerning budget allocations and the related contracts.[495]

---

490 Kroge 2000, p. 36; Kroge 1998, p. 46.

491 Trenkle 1986, pp. 190–191; Watts 2025a, pp. 25–26.

492 Kroge 2000, p. 36; Kroge 1998, p. 46.

493 Raeder 1957, p. 46.

494 Raeder 1957, pp. 46–47; Reuter 1971, p. 24.

495 Reuter 1971, pp. 23–24.

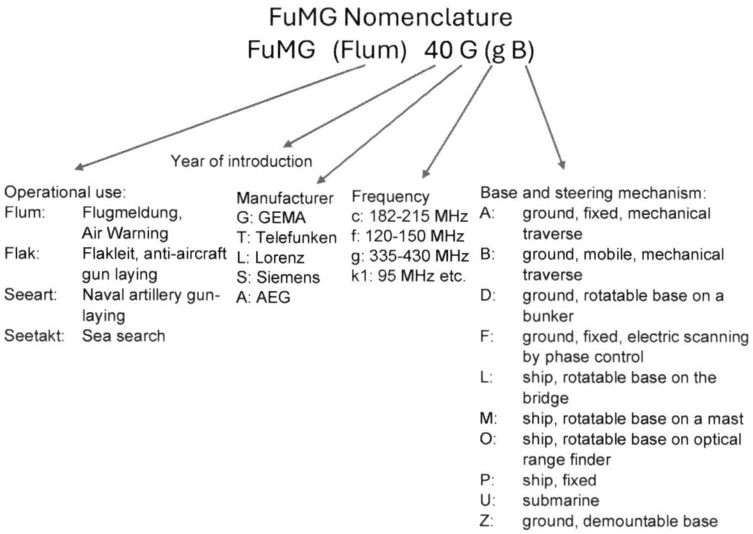

Fig. 15: German radar naming convention[496]

This presented yet another obstacle to innovation, delaying progress in the vital area of radar development. Only after the Marinewaffenamt released the necessary funds was GEMA able to expand its experimental programme in cooperation with NVA scientists, with the aim of largely completing the work by January 1936.[497]

The essential elements of the immediate development programme included a newly developed triode operating at a wavelength of around 50 centimetres using pulse modulation to replace the unstable magnetron; the construction of a pulse-operated device with a wavelength of 10 to 14 centimetres, analogous to earlier experiments at Pintsch, featuring a newly developed magnetron transmitter; the development of an apparatus also operating in pulse mode at a wavelength of approximately 180 to 200 centimetres; improvements in the bandwidth of the intermediate frequency amplifiers; enhanced lateral bearing accuracy for the 50 centimetre

496 Image credit: Watts 2025a, p. 25; adapted from Trenkle.

497 Willisen ca. 1984, p. 17; Watts and Klausing 2025 p. 32.

apparatus; the development of lateral bearing capability at 180 centimetres; and improvements in range measurement accuracy.[498]

Before the equipment was transported back from Pelzerhaken to Berlin, it was experimentally installed on the experimental vessel Welle, where initial measurements against sea and land targets were successfully conducted aboard the vessel. As the vessel was due in Kiel for its scheduled shipyard period in early November 1935, range tests were carried out during this trip against smaller and larger passing ships. This resulted in detection ranges of 6 to 7 kilometres against steamboats and about 2 kilometres against smaller targets such as tugboats. Larger landmarks, including prominent buildings on the island of Fehmarn, were detected at distances of up to 20 kilometres. This was the first demonstration that location was also possible from aboard a ship.[499]

Initially, to improve tube efficiency, the wavelength was changed to 60 centimetres, and the device became the predecessor of the Navy's Seetakt series as a naval tactical radar. In 1938, a wavelength of 81.5 centimetres was ultimately chosen.[500] For GEMA, a strong upward trend in sound and radio location was already apparent at the beginning of 1936.[501]

In February 1936, GEMA developed a new device operating at a wavelength of 1.8 metres, equipped with a newly designed special triode delivering 8 kW pulse power, replacing the magnetron.[502] Early March measurements in the bay of Pelzerhaken against sea targets such as the experimental vessel Welle and larger steamships were unsuccessful, likely due to the antenna being pointed too far upwards.[503]

During subsequent trials, an aircraft passing through the transmission beam at a distance of 8 kilometres was clearly de-

498 Willisen ca. 1984, p. 17; Watts and Klausing 2025, pp. 31–32; Gießler 1971, pp. 62–63.

499 Willisen ca. 1984, pp. 17–18; Watts and Klausing 2025, p. 32; Kroge 2000, pp. 36–37; Kroge 1998, pp. 47–48; Swords 2008, p. 95; Trenkle 1986, p. 28.

500 Trenkle 1986, p. 28, p. 116.

501 Kroge 2000, p. 53; Kroge 1998, p. 60.

502 Willisen ca. 1984, p. 18; Watts and Klausing 2025, p. 32; Swords 2008, p. 97.

503 Willisen ca. 1984, pp. 18–19, Watts and Klausing 2025, p. 32; Kroge 2000, pp. 48–49; Kroge 1998, pp. 57–58; Trenkle 1986, p. 29.

tected, which led to further experiments focused on aircraft detection. A chartered Junkers W 34 was accurately detected at an altitude of 500 metres at distances up to 28 kilometres. Following improvements to the transmission antenna, the detection range increased to over 40 kilometres and soon reached 80 kilometres. It was also found that passing ships could be reliably detected by slightly tilting the antenna downwards.[504] However, altitude measurement was still not possible with this device.[505]

According to GEMA, this marked the birth of the Freya[506] radar, which was used extensively for air warning and fighter control. Von Willisen noted in his records:

> „Dieses war die Geburtsstunde des Luftzielgerätes „Freya", welches im Kriege in großen Stückzahlen eingesetzt wurde. Somit hatte die Nachrichtenmittelversuchsanstalt der Kriegsmarine in Verbindung mit der Firma GEMA die bahnbrechende Tat vollbracht, der die Luftwaffe ihre Rückstrahltechnik zu verdanken hatte."[507]

Most of the subsequent variants of the Freya radar were based on this model.[508] Freya was initially used by the Flugmeldedienst[509] as a mobile air warning radar on land. By 1939, its value for fighter control was also recognised.

By the time of the invasion in 1944, approximately 2,000 GEMA devices of various types were in use with the Navy, both aboard ships and on land. By February 1945, over 6,000 units had been produced for both the Navy and the Air Force.[510]

---

504 Willisen ca. 1984, pp. 18–19; Watts and Klausing 2025, p. 32; Kroge 2000, pp. 48–49; Kroge 1998, pp. 57–58; Trenkle 1986, p. 29; Gießler 1958, p. 111.

505 Reuter 1971, p. 27.

506 Watts 2025a, pp. 1, 3–26, 28; Godt 2003, pp. 15–28; Flugmeldedienst 1943.

507 Willisen ca. 1984, p. 19. "This was the birth of the „Freya" air warning device, which was used in large numbers during the war. In conjunction with the company GEMA, the NVA thus accomplished a groundbreaking feat, to which the Air Force owed its backscatter technology." (translated by the author).

508 Trenkle 1986, p. 72.

509 German military unit responsible for detecting, tracking, and reporting aircraft movements. It provided crucial early warnings to air defence forces during World War II, helping to coordinate defensive actions against enemy air attacks.

510 Kroge 2000, p. 152; Kroge 1998, pp. 187–188.

As a result, transmitter technology was further improved. A Tannenbaum radiator antenna was mounted on a rotating frame, which enabled the detection of aircraft at an altitude of around 3,000 metres during trials in the autumn of 1936. The system achieved a range of approximately 90 kilometres, with a bearing accuracy of ±¼ degree and a range accuracy of ±100 metres.[511]

Following further refinements and a shift in operating wavelength to 2.4 metres, it became possible in the subsequent years to detect targets at distances of up to 150 kilometres, provided they were at a sufficient altitude, with a bearing accuracy of around 2 to 3 degrees. However, altitude measurement with the Freya radar remained impossible.[512]

On 18 April 1936, the Abteilung Technisches Nachrichtenwesen im Oberkommando der Kriegsmarine issued the first military directive requiring that all large battleships and cruisers under construction were to be equipped with radar systems. At the same time, the NVA was tasked with advancing the development of a smaller radar system, deemed urgently necessary, to equip destroyers, torpedo boats and submarines.[513] In parallel, the naval shipyard in Wilhelmshaven, the naval arsenal in Kiel, and the department for warship construction were instructed to coordinate with the NVA regarding the installation conditions.

It was planned to equip each of the new large ships that were to be built with four radar systems: one on each of the rotating domes of the optical rangefinders in the forward and rear positions, and one each on the two forward anti-aircraft gun-laying positions. The technical services recognised the importance of radar technology for smaller warships at an early stage.

Fleet Command, however, showed little initial interest in the development of radar. The Torpedo- und Artillerie-Waffenamt intervened with a demand to improve the accuracy of targeting data for torpedoes and particularly for anti-aircraft artillery. As a result, the armaments departments initially had a greater influence on the development of radar technology than the Naval Command itself,

511 Gießler 1971, p. 63; Gießler 1958, p. 111.

512 Reuter 1971, p. 27.

513 Reuter 1971, pp. 24–25.

which should have been responsible for evaluating the tactical and operational potential of the new systems.[514]

In July 1936, the system, equipped with a newly developed receiver unit, was provisionally installed on board the experimental vessel Welle, and initial sea trials were conducted using a wavelength of approximately 50 centimetres. In August 1936, a demonstration was held for the Chief of Naval Staff and the Inspector of Artillery. During the tests, detection ranges of 9,000 metres were achieved against the battleship Schleswig-Holstein and 5,500 metres against the Grille, proving the practical usefulness of radar on board. To enhance the efficiency of the transmitting and receiving tubes, the system was modified to operate at a wavelength of 60 centimetres.[515] In 1937, the first prototypes were installed on test ships, initially still using this 60-centimetre wavelength.[516]

During a demonstration of the Freya and Seetakt radars in front of Admiral Raeder on 12 May 1937, a range of approximately 10 kilometres was achieved using the 60-centimetre Seetakt radar against the light cruiser Köln, while the Freya radar, operating at a wavelength of approximately 2 metres, detected a seaplane at a distance of about 50 kilometres.[517] Shortly thereafter, experimental vessels were fitted with prototype installations of the 60-centimetre radar equipment.[518].

In 1938, the operating wavelength of the Seetakt radar was standardised at 81.5 centimetres.[519] For the Freya radar, a wavelength of 2.4 metres was ultimately specified.[520] The Seetakt system was assigned the cover designation DeTe I, while the air warning radar Freya, which operated in the metre band and was also referred to as Flum, was designated DeTe II.

The following designations thus applied in 1939/40:

514 Ibid., pp. 25–26.
515 Ibid., p. 27.
516 Kern 1984, p. 126.
517 Kroge 2000, p. 66; Kroge 1998, p. 76; Reuter 1971, p. 27.
518 Reuter 1971, p. 27
519 Trenkle 1986, p. 116, p. 195; Kroge 2000, p. 79; Kroge 1998, p. 93.
520 Trenkle 1986, p. 192; Kroge 2000, p. 79; Kroge 1998, pp. 92–93.

DeTe I: Seetakt, FuMO, FuMG (wavelength 81.5 cm = 368 MHz)

- for Shipboard systems: FuMG 39 G (gO)
- for Ground systems: FuMG 39 G (gB)
- Prototype series: FuMG 38 G (g) (wavelength 60 cm = 500 MHz).[521]

De Te II: Freya, FLUM, FuMO, FuMG (wavelength 2.4 m = 125 MHz)

- for Ground systems: FuMG 40 G (fB).[522]

The technical code name for the radar sets was DeTe-Gerät, while the military code names were Seetakt and Freya. During the course of the war, all radar systems were additionally given technical designations that included continuous type identifiers indicating the year of introduction, manufacturer, frequency band, and antenna mount, alongside the code names.[523] However, this naming convention was discontinued in 1943 to enhance operational security, and subsequent radar designations were deliberately made less descriptive to conceal their type and characteristics. For clarity, the military code names will be used below.

As early as January 1936, the NVA began experiments with centimetre wavelengths. However, further tests produced no significant results. It was assumed that, in addition to transmitter and receiver instability, poor performance was due to a lack of backscattering caused by mirror-like reflections. After unsatisfactory trials using centimetre wavelengths against the experimental vessel Welle, where only a 2,000-metre range was achieved, further testing was halted.[524]

A final decision in favour of longer wavelengths was made in January 1939.[525] At that time it was also assumed that no other country, particularly Great Britain, had developed radar devices for these shorter wavelengths. This misconception persisted and was

521 Trenkle 1986, p. 195.

522 Ibid., p. 192.

523 Trenkle 1986, pp. 190–199; Kroge 2000, pp. 104–105; Kroge 1998, pp. 123–124; for the code names of the German radar and ultra-red developments from 1944 to 1945, see Brandt n.d.

524 Reuter 1971, pp. 29–30; Gießler 1971, p. 63.

525 Reuter 1971, p. 30; Brandt 1962c, pp. 55–56.

even repeated on 22 November 1942 during a presentation on the latest Würzburg developments.[526]

As a result, centimetre-wave development did not resume until February 1943, when a British bomber shot down near Rotterdam on 9 February 1943 was found to contain a radar device operating at approximately 9 centimetres. This discovery led to the painful realisation that halting development in 1936 and later decisions to abandon this frequency range had been serious strategic errors. From February 1943 onward, enormous efforts were made to catch up. Despite notable progress driven by the determined cooperation of all involved, the technological gap could not be fully closed by the end of the war, ultimately preventing effective operational use of centimetre-wave radar.

At the end of 1936, the Freya radar was presented to the head of the Nachrichtenverbindungswesen (NVW)[527] in the RLM, Wolfgang Martini, in Pelzerhaken. Martini then ordered the delivery of 12 units for the Air Reporting Service.[528] The RLM's Technisches Amt later expanded the order to 200 units, so that in 1938 GEMA's order was larger than the navy's.[529] In March 1939, GEMA began manufacturing a test series of around 50 Freya units, after the original order for 200 units had been significantly reduced, as it was likely to exceed GEMA's manufacturing capacity.[530] In October 1939, GEMA began series production of the Freya units. The 12 units ordered for the Air Force were finally delivered in the winter of 1939/40. There was no particular experience in the troops yet.[531]

The year 1937 at GEMA was marked by ongoing improvements in device technology and testing of the Seetakt (DeTe I) and Freya (DeTe II) radars. In the spring of 1937, GEMA tested an im-

526 Trenkle 1986, p. 63.

527 The NVW was the organisational and technical framework employed by the German military during World War II to maintain reliable communications between radar installations, anti-aircraft artillery units, and command headquarters. It was essential for coordinating air defence operations and ensuring effective command and control.

528 Gießler 1971, p. 63; Bekker 1980, pp. 53–54.

529 Gießler 1971, p. 67.

530 Reuter 1971, p. 29; Kroge 2000, p. 81; Kroge 1998, p. 94.

531 Gießler 1958, p. 114.

proved Seetakt device, still with a wavelength of 60 centimetres, as well as a Freya device in Pelzerhaken. Depending on the size of the ship, the Seetakt device could detect targets at a range of 10 to 25 kilometres. The Freya device could detect a Junkers W 34 at a distance of 80 kilometres, and a range measurement was still possible at a distance of up to 60 kilometres.[532]

In September 1937, during Wehrmacht manoeuvres in the Swinemünde area, two early radar devices were installed on board the experimental vessel Strahl.[533] One was a naval radar set operating at a wavelength of 60 centimetres, originally designated DeTe I, which later became part of the Seetakt family. This device reliably detected surface targets at ranges of up to 20 kilometres. The second device, designated DeTe II, functioned as an air reporting unit and marked the first shipboard use of such a device, being installed aboard the Strahl. It later evolved into the early air warning radar known by the military codename Freya, widely used by the Air Force for aircraft detection. With this radar, aircraft types such as the Junkers Ju 52 and Heinkel He 111 could be detected at distances ranging from 40 to 60 kilometres, depending on their altitude.[534]

Between 1937 and 1938, GEMA produced a small pre-series of approximately five of these 60-centimetre naval radar devices and installed them successively on the light cruiser Königsberg, the torpedo boat G10, the armoured cruiser Admiral Graf Spee, and Strahl. Installation on submarines was also tested later in the summer of 1939.[535]

In January 1938, following the evaluation of results from the Swinemünde manoeuvres, a new radar model designated FuMG 38 G[536] was installed aboard the Admiral Graf Spee.[537]

---

532 Kroge 2000, p. 65; Kroge 1998, p. 75.

533 Kroge 2000, p. 68; Kroge 1998, p. 79; Trenkle 1986, p. 116.

534 Kroge 2000, p. 68; Kroge 1998, p. 79; Gießler 1971, p. 65.

535 Trenkle 1986, p. 116.

536 Trenkle 1986, p. 117, p. 195.

537 The Admiral Graf Spee, equipped with the FuMG 38 G naval tactical system, was already in the Atlantic at the outbreak of the war. On 26 September, Captain Langsdorff received official orders to attack Allied merchant ships. On 13 December 1939, the Admiral Graf Spee arrived off the mouth of the Rio de la Plata in Montevideo to repair battle damage and was subsequently

This version still operated on a 60-centimetre wavelength but was housed in a special structure.[538] A radar dome was mounted at the bow, with the antenna placed externally and the transmitter and receiver equipment located inside.[539] The elevated position of the antenna enabled detection ranges of up to 25 kilometres, meeting the requirements of naval artillery. Another advantage of this configuration was that the radar beam and the optical rangefinder remained aligned in the same direction.[540]

In 1938, GEMA produced a pre-series of 31 improved naval radar devices operating at a final wavelength of 81.5 centimetres, corresponding to 368 MHz. These devices were initially designated DeTe I, later DeTe 101, and eventually received the military designations FuMG 39 G (gB) for land-based use and FuMG 39 G (gO) for naval deployment.[541] The 81.5-centimetre wavelength was chosen because it allowed for the use of more powerful transmitter valves and reasonably sized antennas.

From November 1939 onwards, the production version of the FuMG 39 G (gO), later designated FuMO 22, was installed on the bows of major German warships, including the battlecruisers *Gneisenau* and *Scharnhorst*, the heavy cruisers *Admiral Hipper* and *Blücher*, as well as the armoured ships *Admiral Scheer* and

---

encircled by a heavy enemy convoy. After the laytime had expired and with the realisation that a breakthrough to Buenos Aires on the south side of the Río de la Plata seemed hopeless, the ship left the port of Montevideo on 17 December 1939. After traveling about 3 nautical miles, Captain Langsdorff had the ship scuttled (Bekker 1991, pp. 52–53). The ship sank in the shallow waters of the Río de la Plata, only a few metres deep, and burned for three days. The antenna mounted above the forecastle remained visible in the shallow water even after the sinking, allowing the Allies to recognise a ship equipped with radar for the first time. At that time, no enemy ship had such an installation (Kroge 2000, p. 76; Kroge 1998, p. 89; Gießler 1977, p. 41; Kern 1984, pp. 126–127). Although the information about the German radar installation was conclusive, the existence of such radar was debated for a long time by the British (Jones 2009, p. 93; Price 2017, pp. 63–64; Trenkle 1979, p. 169).

538 Reuter 1971, pp. 28–29; Trenkle 1986, p. 117; Holpp 2014, pp. 190–191; this installation took place long before the British and American.

539 Trenkle 1986, p. 117.

540 Trenkle 1986, p. 117; Kern 1984, p. 127.

541 Trenkle 1986, pp. 117–118.

Deutschland. This radar offered a range of 25 kilometres and a bearing accuracy of ±5 degrees.[542]

In the spring of 1941, newly built ships such as the heavy cruiser Prinz Eugen and the battleships Bismarck and Tirpitz were also equipped with this system.[543]

As early as July 1938, Hitler, Göring, Raeder and other senior military leaders visited the naval command. On this occasion, they were shown the Freya radar, now operating at a wavelength of 2.4 metres, on the grounds of the TVA near Eckernförde-Borby. Aircraft such as the Ju 52 and He 111 could still be detected at distances of up to 90 kilometres.[544]

Göring expressed considerable irritation that General Martini, in his capacity as Chief of the NVW, had not been invited to the demonstration. The underlying reason was that the Navy sought to prevent the Air Force from becoming a competitor in the procurement of GEMA equipment.[545] Despite the Navy's assumptions, Göring did respond after the presentation at the TVA.

In view of the advanced development status of GEMA's radar systems, the Technisches Amt of the Air Force had to contact Telefunken and Lorenz to assess the progress of their work on radar systems for anti-aircraft artillery.[546]

By the end of 1939, only eight Freya units and four Seetakt units had been delivered.[547] In 1940, a further 49 Freya units and 27 Seetakt units followed. This delay was partly due to the fact that in July, Raeder had received assurances from Hitler that war was not imminent, and thus the production of radar equipment would not be accelerated.[548]

---

542 Trenkle 1986, pp. 116–118; Reuter 1971, p. 44.

543 Reuter 1971, p. 44; as early as 1936, Telefunken supplied the radio equipment for the Admiral Graf Spee, and later, as part of the fleet construction programme, large radio systems followed for the Gneisenau, Scharnhorst, Tirpitz and Bismarck. Telefunken was also commissioned to equip submarines with radio, detection, and direction-finding equipment, as well as fixed service bases at the shipyards on the North and Baltic Seas, and later also on the Atlantic coast (Thiele 2003, p. 30).

544 Kroge 2000, pp. 75–76; Kroge 1998, p. 88; Gießler 1971; p. 65.

545 Kroge 2000, p. 76; Kroge 1998, p. 88.

546 Kroge 2000, pp. 76–77; Kroge 1998, p. 89.

547 Brennecke 1994, p. 306; Gießler 1971, p. 66.

548 Gießler 1971, p. 66.

After the outbreak of war, follow-up orders were placed with GEMA for 100 Seetakt units, in various versions. The so-called Calais radar systems, designated FuMG 40 G (gB), were subsequently produced as land-based systems.[549] The order for the production of 100 Freya units was awarded to the AEG company.[550]

By early March 1940, eight Freya systems had been installed along the western border, from Freiburg to Filsum in East Frisia, as well as on the North Sea Island of Wangerooge, under the responsibility of the Air Force. In addition, three naval radar sites were established on the islands of Sylt, Helgoland and Borkum. This meant that six months after the start of the war, a radar shield had been established from the Deutsche Bucht to the extreme south-west of Germany.[551]

Industrial production of the Braun tube, used to display targets on the radar screen, had begun around 1937. However, the Navy initially expressed concerns as to whether the tube could withstand vibrations caused by naval gunfire. Early malfunctions, resulting from installation errors, were immediately seized upon as justification for favouring optical rangefinding equipment. This reaction reflected a broader lack of interest in, and in some cases outright rejection of, this forward-looking technological development.[552]

This incident may not appear to have been a major barrier to innovation but it serves as a clear indicator of the fundamental scepticism towards technological change held by many political and military leaders at the time. As early as 1935, C. Lorenz AG had begun initial development work in the field of radar technol-

549 Trenkle 1986, pp. 119-120, p. 195.

550 Reuter 1971, p. 44.

551 Reuter 1971, pp. 45–46; Gießler 1977, p. 41; it should also be noted that on the islands of Sylt, Helgoland and Borkum, the Air Force was also responsible for the operation and organisation of the Freya equipment in addition to the Navy. The exact division of responsibilities depended on the respective military requirements and the strategic importance of the sites. The Air Intelligence Service of the Air Force and the Naval Intelligence Service sometimes worked in parallel or in coordination with each other.

552 Gießler 1968, p. 182.

ogy using decimetre wavelengths, without any official mandate.[553] The high level of advancement in vacuum tube technology stemming from radio relay applications proved advantageous for the further development of higher-performance radar systems. Pulse-based methods were also under investigation.

At the beginning of 1936, the first laboratory device was used to measure the Berlin Cathedral from the roof of the Tempelhof factory at a distance of 7.4 kilometres, operating in impulse mode with a wavelength of 70 centimetres (430 MHz). Over the course of the year, the device's performance improved significantly, including the introduction of new antenna technology with parabolic mirrors, which increased the range to between 12 and 14 kilometres. This device was then presented to the Navy in 1936 in a transportable version in the Kiel Fjord, achieving a range of 7 kilometres against ships of about 4,000 tonnes and approximately 4 kilometres against Junkers W 34 aircraft.[554]

Both Lorenz and Telefunken chose decimetre waves for subsequent developments based on their experience in directional radio technology and refrained from further work in the centimetre wave range.

Despite not being commissioned due to confidentiality concerns arising from the company's international connections, Lorenz continued to enhance the device's performance and achieved a significant increase in range. After a demonstration for Martini, Lorenz received its first development contract from the Air Force for an air defence radar system, designed to detect incoming aircraft and provide early warning to air defence units. Prior to this, the company had to legally commit to confidentiality with its foreign sister companies.[555]

The Heereswaffenamt (HWA), which was still responsible for the development of anti-aircraft artillery at the time, commissioned the companies Lorenz and Telefunken as early as 1935/36 to inves-

553 Müller 1981, pp. 1–65; Trenkle 1986, pp. 30–31; Kern 1984, pp. 118–119; Swords 2008, p. 98.

554 Trenkle 1986, pp. 30–31; Watson 2009, p. 243.

555 Trenkle 1986, p. 31; Kern 1984, p. 119.

tigate the potential use of radio waves for detecting enemy aircraft, regardless of time of day or weather conditions.[556]

The HWA's plans and developments included both active and passive detection methods. The following discussion focuses exclusively on active systems in the sense of radar, which were initially designated as A devices.[557]

After anti-aircraft artillery was formally transferred to the Air Force, radar equipment was divided into two main categories according to tactical and technical requirements:

- Air warning radar with circular search capability. Reliable detection range of 50 kilometres and as accurate a lateral bearing as possible to guide anti-aircraft target devices. A range accuracy of ±1–2 kilometres is sufficient. Target detection possible even at altitudes down to approximately 2,000 metres.
- Fire-control radar with shorter range. Target detection at 20–25 kilometres or beyond, but with high accuracy in lateral and altitude bearing. Range accuracy of ±10 metres. Target detection possible even at altitudes below 2,000 metres. Capability for identification queries. Expandable for automation of the measurement process.[558]

The HWA assigned type designations A1 device (GEMA), A2 device (Lorenz), and A3 device (Telefunken) to the pre-development models. Subsequent series received different designations.[559] For example, the original DeTe designation was later changed to FuMG and FuMO, though the code names were always retained.[560]

---

556 Reuter 1971, pp. 31–32.

557 Hoffmann-Heyden 1954, p. 19; Trenkle 1986, p. 36.

558 Hoffmann-Heyden 1954, pp. 20–21; Trenkle 1986, pp. 36–37.

559 Hoffmann-Heyden 1954, p. 27.

560 Trenkle 1986, pp. 190–191; for example, the designation FuMG 40 G (gB) means Funkmeßgerät, introduced in 1940 and manufactured by the GEMA company, which is indicated by the capital G. The lowercase g indicates the frequency group, covering 335 to 430 MHz. The letter B signifies a ground-mobile unit, meaning it was a land-based installation with the codename Calais B, the FuMG 40 G (gB), as specified by Trenkle, operated at a frequency of 368 MHz. While this represents its primary operating frequency, the device belonged to the g frequency group, which generally covers a range from approximately 335 to 430 MHz (Trenkle 1986, p. 190).

On 15 November 1938, an advance detachment of Flakartillerieschule III was assembled at the testing grounds of the Army Ordnance Office in Lynow, Mark Brandenburg, for the training of operating crews as well as for conducting trials and tests. From this detachment, the Flakartillerieschule III was formally established, comprising training and instructional units alongside a dedicated testing department. This school was stationed in Berlin-Heiligensee from 1940 until 1945.[561]

By the end of 1938, one A1 device, the Freya system from GEMA, was available in Lynow with a wavelength of 2.4 metres[562], and the FuMG 38 L Kurfürst fire-control radar from Lorenz was available as the A2 device with a wavelength of 62.4 centimetres.[563] The FuMG 39 L Kurpfalz followed as a further development of the Kurfürst, featuring higher transmission power, improved measurement accuracy at longer distances, and a portable control unit.[564] The more powerful successor model, the FuMG 40 L Kurmark, with increased transmission power and thus greater range, was initially intended as an interim solution to the FuMG 41 L. However, the FuMG 41 L, which was already 80 percent complete, was discontinued in 1941 in favour of the Würzburg system from Telefunken. This marked Lorenz's final withdrawal from the development of anti-aircraft fire control radar.[565] By the end of 1940, Lorenz had delivered 20 FuMG 39 L anti-aircraft fire control radar systems. In 1941, the significantly improved FuMG 40 L was delivered, also in a series of 20 units, until all development work was discontinued.[566]

The FuMG 40T Mainz, as an A3 device from Telefunken, was used as an operational prototype in a pre-series of 50 units during 1941/42 in troop trials and deployments, until large-scale production of the FuMG 41T Mannheim finally began in 1942.[567] At AEG,

561 Hoffmann-Heyden 1954, p. 27; Trenkle 1986, p. 37.

562 Hoffmann-Heyden 1954, pp. 28–30; Trenkle 1986, p. 38.

563 Hoffmann-Heyden 1954, pp. 30–33; Trenkle 1986, p. 39.

564 Trenkle 1986, p. 42, p. 197; Hoffmann-Heyden 1954, pp. 40–41; Müller 1992, p. 5; Watson 2009, p. 245.

565 Trenkle 1986, p. 51.

566 Reuter 1971, p. 47.

567 Hoffmann-Heyden 1954, p. 39; Trenkle 1986, pp. 51–54, p. 197.

high-frequency research was largely excluded, as it was mostly conducted by their subsidiary Telefunken.[568]

In 1934, Telefunken began research and development work under the direction of Runge, who had headed the receiver laboratory since 1924, aiming to advance into the short-wave range around 50 centimetres for radio relay connections.[569]

That same year, Telefunken established a research focus to investigate possible applications in this wavelength range with two main goals: radio relay and navigation.[570]

In the winter of 1932/33, Runge had already conducted range tests in the decimetre wave range at 50 centimetres. He set up a small laboratory for this purpose, and the tests were carried out over the Landwehr Canal between Tempelhofer Ufer 9 and Hallesches Ufer 12.[571] In 1935, Telefunken developed radio relay links in the decimetre wave range to transmit radio reports from the Olympic Stadium to the broadcasting centre.

Subsequently, with radio relay technology, distances of 50 to 70 kilometres could reliably be covered under line-of-sight conditions, and in some cases even up to 100 kilometres.[572] General Fellgiebel, in his role as GBN, held the new technology in high regard and urged Telefunken to pursue its further development.[573]

During experiments with radio relay transmission, it was repeatedly observed that moving vehicles, especially those crossing the radio path, produced backscatter signals.[574] This phenomenon prompted Runge to plan and carry out dedicated backscatter experiments, applying his expertise in radio relay technology. In the summer of 1935, he conducted his first trials in aircraft detection.

---

568 Weiss 2002, p 110; a detailed account of AEG's diverse activities in the field of armaments research up to 1945 can also be found in Weiss 2002, pp 109–141.

569 Reuter 1971, p. 32; Runge 1953, p. 183; Runge had already discussed in detail the propagation characteristics, generation and reception of decimetre waves, as well as their possible applications, in his lecture "Die Grundlagen der Dezimeterwellen-Technik" presented to the Vereinigung für Luftfahrtforschung on 21 March 1934 (Runge 1934a, pp. 24–33).

570 Runge 1971, p. 38.

571 Runge 1971, p. 36; Thiele 2003, p. 28.

572 Runge 1971, p. 38.

573 Runge 1971, pp. 38–39; Thiele 2003, p. 28.

574 Runge 1971, p. 42.

With a 1 m^2 antenna pointed vertically upwards, a tone-modulated transmitter with a power of just 0.1 watts operating at a wavelength of around 50 centimetres, and an adjacent receiving antenna coupled to a detector, the goal was to investigate the possibility of receiving backscatter signals from aircraft during overflight. A factory-owned Ju 52, made available for these tests, was successfully detected at an altitude of 5,000 metres.[575]

The expected enthusiasm from Telefunken management failed to materialise, and Runge was only provided with funds for further research into radio relay links.[576] However, Runge continued the work without the knowledge of the Telefunken board. These activities were also kept secret from Telefunken's parent companies, AEG and Siemens, due to concerns that this work might be taken away from Telefunken.[577]

This situation can be seen as a characteristic barrier to innovation on the path to technological progress. Over the following years, it became increasingly clear how important the discovery of aircraft location would prove to be. In addition to the development of more powerful tubes, the demand for pulse operation to measure the range of detected targets, as well as lateral resolution, that is the distinction between adjacent targets, played an important role.

By using high-pulse modulation of the transmission pulses, Telefunken succeeded in significantly increasing the transmission power at a wavelength of around 50 centimetres.[578] These advancements met the demands of the NVA and other military authorities.[579]

During this period, industrial production of the Braun'sche Röhre became a key element in the development of radar technology.

In 1935/36, Telefunken began cooperating with military authorities. At this time, the anti-aircraft artillery was fully aware that

575 Runge 1971, pp. 42–43; Runge 1988, pp. 227–228; Trenkle 1986, p. 28–29; Swords 2008, pp. 96–97.

576 Runge 1971, p. 43; Reuter 1971, p. 32; Trenkle 1986, p. 29.

577 Reuter 1971, p. 32; in 1941, AEG took over the Telefunken shares from Siemens & Halske and continued to run the company as a wholly owned subsidiary.

578 Trenkle 1986, p. 29.

579 Kümmritz 1994, p. 27.

its tactical and technical requirements needed a fundamental reorientation, as previous methods of locating aircraft using listening devices and searchlights would no longer be effective against future air attacks occurring in poor weather conditions and at increasingly higher speeds.[580] As a result, there arose a need for new tracking systems that could locate attacking aircraft, especially bombers, regardless of the time of day or weather, at sufficient range, and report their positions.[581] The HWA, which at that time was still responsible for anti-aircraft artillery and all related development projects, awarded contracts to the companies Lorenz and Telefunken to investigate the possibilities of radio wave location.[582] These devices were intended for fine-tuning in their final form, while GEMA's developments for wide-area target acquisition with a search radar system were already well advanced.[583] However, neither Telefunken nor the HWA took concrete steps to promptly fulfil the order; instead, it was assumed the request was non-binding, especially since even Runge was unaware of any order from the HWA.[584]

Up to this point, Runge had been able to keep his experiments secret from the two parent companies, AEG and Siemens, so there was no danger of the work being taken away. This changed in the spring of 1936, when the HWA and the Technisches Amt of the RLM, as well as the management boards of AEG, Siemens, and Telefunken, were able to attend multiple demonstrations of the equipment starting from March 1936 and were continuously updated in several meetings about the ongoing progress of the work. On 1 July 1937, a demonstration was held for the HWA showcasing the now very satisfactorily functioning device components for distance measurement, as well as lateral and altitude determination.[585]

Runge's work was continued from 1936 by one of his employees, Wilhelm Stepp.[586] In January 1937, they replaced the flat antenna with a parabolic reflector three metres in diameter for both

580 Kern 1984, p. 117.

581 Hoffmann-Heyden 1954, pp. 18–19; Reuter 1971, p. 31.

582 Hoffmann-Heyden 1954, p. 27; Reuter 1971, pp. 31–32.

583 Hoffmann-Heyden 1954, p. 27.

584 Kern 1984, p. 117; Reuter 1971, p. 32.

585 Reuter 1971, p. 32.

586 Stepp 1947.

transmission and reception, combined with a constantly rotating defocused dipole, known in German as the Quirl. The dipole rotated around the focal point of the parabolic reflector. This setup made it possible to obtain precise bearings as well as lateral and altitude measurements. Distance was determined by measuring the time of flight of the pulses. Stepp named this advanced prototype of the fire-control radar, the A3 radar, after his hometown Darmstadt.[587] This prototype laid the foundation for the development of future radar systems.

The Darmstadt radar served as the basis for the later fire-control radars Mainz and Mannheim and was fundamental to the successful development of the Würzburg radar family.[588]

The board of Telefunken approved 10,000 RM for the work, and in 1938, the HWA ordered 10 prototypes of the Darmstadt device for testing by the anti-aircraft artillery, although there was no apparent pressure for rapid delivery.[589] In 1937, Telefunken received a development contract from the RLM for an air reporting system capable of determining distance, lateral, and altitude measurements, with a range of 25 to 35 kilometres, as well as for an anti-aircraft fire-control system. This task was jointly prepared by the Air Force Operations Staff, the Anti-Aircraft Artillery Inspectorate, the NVW Staff, and the Technical Bureau after Martini had taken over responsibility for the air reporting service from the anti-aircraft artillery. The task already included references to the use of centimetre waves for the anti-aircraft fire-control to achieve the required accuracy values.[590]

Leo Brandt later explained:

> „Die Telefunken GmbH nahm sich der auferlegten Arbeit in glücklichster Weise an und erkannte von vornherein alle verfahrenstechnischen Grundsätze. Sie wählte die kürzeste, aus dem damaligen Stand der Röhrenentwicklung sich ergebende Welle von 50 cm. Telefunken hatte seit 1934 einen Forschungsschwerpunkt auf diesem Wellenlängen-

587 Runge 1971, p. 44; Trenkle 1979, p. 26; Reuter 1971, p. 33; Kroge 2000, p. 77; Kroge 1998, p. 90.

588 Trenkle 1986, p. 39.

589 Reuter 1971, p. 33.

590 Brandt 1962c, p. 56.

> gebiet gebildet und Grundlagen für vielseitige Anwendungszwecke – z. B. für Dezimeterstrecken (als Kabelersatz) und zur Kollisionssicherung von Flugzeugen – geschaffen, so daß sich die 50-cm-Welle für die Rückstrahlaufgabe zwanglos anbot…"[591]

In the summer of 1938, the Darmstadt device was demonstrated to the Generalluftzeugmeister (GL), General Ernst Udet.[592] His initial reaction was marked by a deep-seated scepticism towards this groundbreaking development. Udet is said to have responded to Runge with the words, "Wissen Sie, wenn Sie so was einführen, dann macht ja die ganze Fliegerei keinen Spaß mehr!"[593] Then he turned around and left.[594]

Similarly dismissive, Hermann Göring remarked three years later during the demonstration of a radar system for night fighters, "Ich will keinen Kintopp (Kino) im Flugzeug".[595]

For GEMA, which already had almost 800 employees by the end of 1938, the period from 1938 onwards was marked by an increasing demand for radar and sound location systems. In addition

591 Ibid., p. 56. "Telefunken GmbH approached the task in the most successful manner, immediately recognising all fundamental procedural principles. It chose the shortest wavelength of 50 centimetres, which resulted from the state of tube development at the time. Since 1934, Telefunken had concentrated its research on this wavelength range, thereby creating the basis for a variety of applications, for example for decimetre-range radio relay links as a cable substitute and to prevent aircraft collisions, so that the wavelength of 50 centimetres was a natural choice for the backscatter task…" (translated by the author).

592 Udet was appointed Colonel in the Air Force under Göring on 1 June 1935 and, on 10 February 1936, became Inspector of Fighter Aircraft and Dive Bombers. Four months later, Göring made him Head of the Technisches Amt. Promoted to Lieutenant General on 1 November 1938, he also became Generalluftzeugmeister on 1 February 1939 while retaining his position as Head of the Technisches Amt. He committed suicide on 17 November 1941(Hümmelchen 2011b, pp. 258–264).

593 "You know, if you introduce something like that, flying just won't be fun anymore!" (translated by the author); Runge 1971, p. 44; Trenkle 1986, p. 41.

594 Trenkle 1986, p. 41; Runge later reported that he had approached the wrong person in Udet, since in practice, anti-aircraft artillery remained under Army control. While Runge could understand Udet's reaction, he could not accept it (Runge 1971, p. 44).

595 "I don't want any cinema nonsense in the aircraft." (translated by the author); Trenkle 1986, p. 41.

to the Navy, orders from the Air Force were also expected. The situation intensified following the Münchener Abkommen in September 1938 and the subsequent incorporation of the Sudetenland into the German Reich in October. After the Wehrmacht's invasion of Prague on 15 March 1939, demand for radar equipment increased further. The Air Force announced a requirement for 250 Freya radar sets, which led to tensions with the Navy. After considerable effort, GEMA, together with the Navy and Air Force, managed to establish the frequency allocations, and a compromise was eventually reached regarding the final operating wavelength for the Freya radar.

This radar, which initially operated at wavelengths of 1.8 metres and later 2.1 metres, was ultimately assigned a final wavelength of 2.4 metres, allowing for higher transmission power and greater range. For the Navy's Seetakt system, a wavelength of 80 centimetres was initially chosen.[596] These specifications laid the foundation for all subsequent generations of radar systems. However, the wavelength for the Seetakt system was later revised and finally set at 81.5 centimetres.[597] By January 1939, the Air Force had six Freya radar sets.

At GEMA, in collaboration with the Zeiss company in Jena, work was underway on suitable turntables. At the same time, the Navy had seven Seetakt systems available, though still without precise bearing capability, for installation on the new battleships and heavy cruisers.

Following the annexation of the Sudetenland and the resulting tensions, the Air Force installed a Freya radar on the Geisinger Berg near Altenberg in the Erzgebirge and on the Großer Schneeberg near Glatz in the Sudetenland to monitor the airspace over Czechoslovakia.

It was observed that although the devices were positioned at an altitude of around 800 metres, which allowed for long-range detection, the accurate location of aircraft was hindered by numerous ground echoes. As a result, one device was relocated to the Marchebene plain near Vienna, where it enabled the monitoring of Czech airspace without interference.

---

596 Kroge 2000, p. 79; Kroge 1998, pp. 92–93.

597 Trenkle 1986, p. 116, p. 195.

Fig. 16: Freya Radar on Wangerooge[598]

Following the annexation of the Sudetenland, the Freya experimental radar system was initially tested near Vienna. It was then transported to the Luftnachrichten-Versuchsregiment Köthen. From December 1939 onwards, the regiment, led on Wangerooge by Lieutenant Hermann Diehl, operated the Freya experimental radar system on the island alongside another unit (Fig. 16).

At that time, fixed Freya radar laboratory installations were under construction on the islands of Borkum, Sylt, and Norderney, while two mobile systems were being developed in the western region.[599]

On 18 December 1939, an experimental radar device on Wangerooge detected a formation of RAF bombers flying into the Deutsche Bucht towards Wilhelmshaven at a distance of 113 kilometres.

598 Image credit: Stiftung Deutsches Technikmuseum, Berlin, Historical Archive; I.2.060 FS 032-1-16-01

599 Reuter 1971, p. 46; Diehl 1953, p. 62.

In this instance, the radar demonstrated not only its effectiveness as an early warning system bu also its potential for directing fighter aircraft. Using radio signals based on the radar's information, German fighter pilots under the guidance of Lieutenant Diehl of the Luftnachrichten-Versuchsregiment Köthen intercepted the British formation. The ensuing air battle resulted in heavy losses for the RAF.[600]

By the end of 1939, GEMA had delivered a total of eight Freya devices and six Seetakt devices, of which six Freya devices were deployed on the islands of Helgoland, Borkum and Sylt. Two Freya systems were installed for air warning purposes, while one Seetakt device was used specifically for detecting low-flying aircraft.[601]

The NVK was responsible for operating the devices and training the deployed personnel. By the beginning of March 1940, a Freya device had been installed at various locations from East Frisia to the Black Forest along the western border. The installation and operation of the air warning systems in this region were ensured by specialists seconded from GEMA.

Together with the Navy's three Seetakt devices stationed on the islands, a protective screen was established over the entire Deutsche Bucht, extending as far as the south-western corner of Germany. This made it possible to monitor the airspace up to approximately 100 kilometres beyond the western border.[602]

As early as the beginning of 1939, against the backdrop of the Sudetenkrise, GEMA had already begun to increase production of both Seetakt and Freya devices. However, delays in training operating personnel within the Navy and Air Force prevented the effective deployment of the systems. According to GEMA, more than fifty radar sets could have been made operational by the Navy and Air Force at the outbreak of World War II if sufficient operating capacity had been secured at an early stage.

Moreover, neither organisational structures nor clear areas of responsibility had been established.[603] This was also a consequence

600 Diehl 1953, p. 63; Reuter 1971, p. 81; Brandt 1962c, p. 58; Kroge 2000, p. 90; Kroge 1998, p. 106.

601 Reuter 1971, p. 43; Kroge 2000, p. 86; Kroge 1998, p. 100.

602 Reuter 1971, pp. 45–46.

603 Kroge 2000, p. 86; Kroge 1998, pp. 99–100.

of the polycratic structure of the Nazi regime, which in this case clearly emerged as a barrier to innovation.

With Great Britain and France entering the war on 3 September 1939, pressure on GEMA increased, as a larger number of devices now had to be produced for the Navy and Air Force. There was also a significant need to train personnel from both branches in the operation of the supplied devices.[604] GEMA was consequently classified as a defence contractor and became a primary supplier for armed forces orders as part of the radar programme, which considerably enhanced its importance, particularly as a manufacturing company.[605]

At the end of July 1939, shortly before the outbreak of war, the Technisches Amt of the RLM presented the technical status of the entire Air Force armament to Hitler, Göring, Milch, Udet and other leading figures at the Air Force test centre in Rechlin. An example of internal conflicts within the top leadership was that the head of the NVW, Martini, was not invited.[606]

The Darmstadt radar by Telefunken was also demonstrated, and Hitler remarked that, unlike the Freya radar, the Darmstadt radar could also measure altitude.

The Technisches Amt of the RLM initially failed to respond promptly to the need for advanced radar systems until after the outbreak of war. However, the placement of a major development order for radar equipment was primarily the result of the foresight and determination of Senior Staff Engineer Werner Beckmann, head of the LC 4 department within the Technisches Amt, responsible for radio and navigation systems.

The order concerned the serial production of a compact short-range radar, designed to provide measurements of distance, azimuth, and altitude over a range of 25 to 35 km.[607] This decision was not based on a comprehensive strategic plan, but rather on Beckmann's personal initiative and conviction regarding the military importance of radar.[608] His actions directly led to the com-

604 Kroge 2000, p. 85; Kroge 1998, pp. 98–99.
605 Kroge 2000, p. 86; Kroge 1998, p. 100.
606 Brandt 1962c, p. 57.
607 Reuter 1971, p. 33.
608 Reuter 1971, p. 33; Brandt 1968, p. 13.

missioning of the Würzburg radar family, marking the beginning of a major development programme.[609]

As a result, Telefunken was able to successfully test new processes in development, design, and production, implementing them almost simultaneously. However, all this happened without the involvement of Göring, who was actually responsible.[610]

At a meeting held on 23 October 1939 at the Technisches Amt, attended by Beckmann, the head of the NVW, Martini, and the Inspekteur der Flakwaffe[611], General Günther Rüdel, the expected demand for radar equipment in 1941 was assessed and procurement was authorised. A large order was placed, including 3,900 Würzburg devices and 200 Freya units. These were intended for use by anti-aircraft artillery units, the air warning service, and the Navy, which planned to deploy Würzburg radars on submarines and surface vessels as a replacement for the GEMA Seetakt system.[612]

As a result, some essential development projects could not be pursued. This included the suspension of the development of broadband antennas, meaning that radar devices could only operate on a single frequency at a time. In the following years, the enemy recognised this limitation and was able to jam the radars accordingly.[613] The fact that the Army, Air Force and Navy each required different devices for the same purpose placed a considerable burden on development and production capacities. Strict secrecy regulations between the branches of the Wehrmacht posed an additional obstacle. Another issue recognised as early as 1937 was the identification of friendly aircraft and ships at long range. That same year, GEMA began developing a special identification

---

609 Bauer 1992; Oberkommando der Kriegsmarine 1943a; Oberkommando der Kriegsmarine 1943b.

610 Reuter 1971, p. 34.

611 Inspector of Anti-Aircraft Artillery was the official responsible for overseeing the weapons, equipment, and military units used to detect, target, and shoot down enemy aircraft, including anti-aircraft guns, radar systems, and fire-control devices.

612 Reuter 1971, pp. 46–47.

613 This was painfully demonstrated by the use of Window by the RAF in July 1943 during the attack on Hamburg.

system designed to automatically distinguish friendly units from enemy forces, thereby helping to prevent friendly fire incidents.[614]

This work laid the foundation for today's air and maritime surveillance systems. In particular, the so-called secondary radar method has become an essential component of both civil air traffic control and military friend-or-foe identification systems, ensuring the safety of modern airspace operations.[615]

Since this identification method was intended for use not only on ships but also on aircraft, a meeting was held in November 1938 between the Navy and the Technisches Amt of the Air Force at the RLM. At this meeting, the Navy was casually informed that the companies Lorenz and Telefunken had been commissioned to develop radar equipment operating in the wavelength range of 50 to 60 centimetres. While Telefunken's work led to the development of the Würzburg fire-control radar[616] with a wavelength of approximately 53 centimetres, Lorenz's system at 60 centimetres was ultimately not produced.[617]

This example highlights the lack of transparency and cooperation between the branches of the Wehrmacht, which created obstacles to innovation and unnecessarily delayed technical progress. Because the Navy and the Air Force developed radar systems separately and operated in different frequency ranges, no unified solution for friend-or-foe identification was found. Throughout the war, attempts were made to agree on a single identification procedure that would work across all frequency ranges used. However, these negotiations ultimately failed to produce a satisfactory result.[618]

In the autumn of 1939, a fully developed GEMA Erstling identifier for the Freya radar, operating at a wavelength of 2.4 metres, was demonstrated to Martini, after which the Air Force requested a

614 Kroge 2000, p. 78; Kroge 1998, p. 91.

615 Klausing und Holpp 2000, pp. 6–8.

616 The Würzburg radar system, a German World War II radar for target detection and fire-control of anti-aircraft artillery, played a crucial role in enabling antiaircraft guns to accurately engage enemy aircraft.

617 Gießler 1971, p. 67.

618 Reuter 1971, pp. 34–35; Kroge 2000, p. 78; Kroge 1998, p. 91.

production run of 2,000 to 3,000 units.[619] Additionally, an identification method for the Würzburg radar, operating at a wavelength of 53 centimetres, had already been developed and presented by Hans Plendl before the outbreak of the war.[620] However, in early 1941, the Technisches Amt of the Air Force ordered 25,000 units of the Zwilling identification device from Telefunken, which later proved to be completely unusable.[621]

It was not until 10 January 1942 that production of the Zwilling was halted, and on 9 July 1942, the Erstling device was declared ready for use, although it suffered from a very limited frequency range. The development of an identification device called Neuling, intended to cover all frequency bands used by radar systems for friend-or-foe identification, began at Telefunken in the spring of 1944. However, it was not deployed before the end of the war.[622]

Even before the outbreak of the war, the Air Force and the Navy agreed to pursue common development goals within the framework of ongoing exchanges. One of the first outcomes was that the Navy accepted the Air Force's request for a radar device to support anti-aircraft artillery. Martini consistently aimed for constructive cooperation with the Navy; however, the Technisches Amt within the RLM followed its own agenda, resulting in a lack of genuine collaboration in research, development, and production well into the war years. A prime example of this disconnect is the problematic development of identification devices.[623]

7.1.1.3 LZ 130 Graf Zeppelin Reconnaissance Flight

In early August 1939, four weeks before the outbreak of World War II, General Martini ordered the Zeppelin LZ 130 Graf Zeppelin to carry out a reconnaissance mission along the British east coast up to the Shetland Islands.[624]

619 Reuter 1971, p. 35; Brandt 1962c, p. 59; Trenkle 1986, pp. 174–175; Kroge 2000, p. 109; Kroge 1998, p. 131.

620 Reuter 1971, p. 35; Brandt 1962c, pp. 59–60.

621 Trenkle 1986, pp. 172–174; Reuter 1971, p. 35; Diehl 1953, p. 66.

622 Reuter 1971, pp. 35–36; Trenkle 1986, pp. 179–181.

623 Reuter 1971, p. 36.

624 Breuning 1989, pp. 156–170; Latham and Stobbs 1999, pp. 51–52; Latham and Stobbs 1997, pp. 2–3; Dekker 2014, p. 15.

The objective was to gather intelligence on the British Chain Home radar system, believed to be similar to the German Freya radar and deployed for coastal defence.

The Zeppelin was selected for its ability to carry a large number of measuring instruments and to hold a nearly stationary position, which made it suitable for prolonged signal monitoring in areas where Chain Home towers were visible.

The most important flight began on the evening of 2 August and ended on the evening of 4 August, covering approximately 4,200 kilometres.[625] During this time, the airship was tracked by British radar stations.

On board, specialised radio equipment searched for emissions similar to those of the Freya system, focusing on transmission frequency, antenna rotation, and high pulse repetition rates. The search did not detect any such signals, which led German authorities to conclude, incorrectly, that Britain had no effective air defence radar.

Several factors contributed to this error. The crew focused on the 100 MHz range, while Chain Home operated between 20 and 30 MHz.[626] In addition, Chain Home used fixed antennas that directed signals in a narrow beam, unlike the rotating antennas used in the Freya system.

The German team also misinterpreted the low-frequency radar signals. The 25 Hz pulse repetition was attributed to faults in the British power grid or to unrelated radio systems. In reality, the signals were generated deliberately using the national grid to synchronise radar transmissions.

The complex metal structure of the Zeppelin may have further interfered with signal reception. At the same time, ionospheric experiments were being conducted under Martini's direction. These tests involved transmitting pulses at various frequencies to study atmospheric reflections from the Heaviside layer and to determine optimal conditions for radio communication.

As a result, the receivers on board the Zeppelin experienced strong interference, especially in the 10 and 12-metre bands, which

625 Breuning 1989, p. 167; Price 2017, pp. 19–20, p. 81.

626 Johnson 2004, pp. 62–64; Johnson 1983, pp. 63–65.

prevented successful detection and analysis of Chain Home signals.[627]

Due to the lack of confirmed radar activity, the German command saw no reason to prioritise attacks on British radar sites. Hitler also remained convinced that Britain and France would not enter the war and believed that integrated air defence using radar played only a minor role. It was assumed that British defences still relied primarily on wireless direction finding, as they had in World War I.[628]

7.1.1.4 The Rapid Growth of GEMA

The GEMA[629], Gesellschaft für elektroakustische und mechanische Apparate m.b.H., was founded in Berlin on 16 January 1934 and registered in the commercial register.[630] The company was established as a complement to Tonographie, which had been operating since 1931 and specialised in the electrical recording of speech and music using devices developed specifically for this purpose.[631] According to the commercial register entry, GEMA's business purpose was the development, manufacture and distribution of devices and systems in the fields of mechanical, electrical, acoustic, photo-technical, photo-chemical and related technologies.

The initial shareholders were Paul-Günter Erbslöh, Hans-Karl Freiherr von Willisen, Georg Neumann and Erich Rickmann. However, Neumann and Rickmann left the company at the beginning of 1935, leaving Erbslöh and von Willisen as the sole shareholders after acquiring their shares. The company's initial aim was to develop and produce prototypes and small series.[632]

GEMA began with two employees, while Hans E. Hollmann and Theodor Schultes initially worked as freelancers. They later be-

627 Breuning 1989, p. 169; Judkins and Bauer 2014, p. 17.

628 Judkins and Bauer 2014, p. 18.

629 A comprehensive account of the foundation and activities of GEMA can be found in Kroge 2000; Kroge 1998 and Willisen ca. 1984.

630 Kroge 2000, p. 20; Kroge 1998, p. 7; Willisen ca. 1984, p. 4; Godt 2003, pp. 221–232; Kern 1984, p. 110; Reuter 1971, pp. 19–20.

631 Kroge 2000, p. 12; Kroge 1998, p. 21.

632 Kroge 2000, pp. 19–20; Kroge 1998, pp. 26–28.

came crucial to GEMA's further business development and success in the field of radar technology.[633] Schultes went on to become head of the high-frequency laboratory, while Walter Brandt, another employee, became head of the low-frequency laboratory.[634]

To better understand GEMA's foundation and early orientation, it is worth looking back at the experiences of its founders.

Shortly after the end of World War I, the Potsdam students Erbslöh and von Willisen discovered their shared interest in wireless telegraphy and telephony.[635] Starting with the construction of microphones for radio stations, they went on to design electrical recording devices for vinyl records, an achievement that had previously only been possible using purely acoustic methods. In the years that followed, Erbslöh and von Willisen also worked for the Tonographie sound studio, for which they built complete recording facilities. In 1930, Erbslöh travelled to Java for several months to record Malay music.[636]

In the meantime, both had established contacts with high-frequency experts such as Hollmann from the Heinrich-Hertz-Institut, as well as with the Reichspostzentralamt in Berlin and the rector of the Musikhochschule in Charlottenburg, which helped raise public awareness of their development work. In 1931, Erbslöh and von Willisen acquired the company Tonographie, which, due to positive business developments, expanded from merely designing and manufacturing recording devices into a firm that also sourced equipment from contracted specialist companies. This enabled the company not only to work for well-known record labels but also to record concerts, political speeches and similar events. Thanks to a well-qualified team, Tonographie was able to continue operating independently, allowing Erbslöh and von Willisen the freedom to focus on new challenges in the field of radio location. At Hollmann's suggestion, von Willisen continued to pursue the development of radio-based technologies.[637]

633 Kroge 2000, p. 20; Kroge 1998, p. 28.
634 Kroge 2000, p. vii.; Kroge 1998, p. 6.
635 Kroge 2000, p. 9; Kroge 1998, p. 17; Godt 2003, p. 222.
636 Kroge 2000, p. 12; Kroge 1998, pp. 20–21.
637 Kroge 2000, p. 12; Kroge 1998, p. 21.

In 1932, a connection was established with the HWA, which required recordings of muzzle blasts and bullet sounds for training purposes. These recordings were produced to the client's complete satisfaction.[638] At the end of 1932, contact was made with the NVA. The Navy faced the challenge of recording sounds from passing watercraft, especially submarines, for use in auditory training courses. Erbslöh and von Willisen approached these challenges with great personal dedication and achieved a very high quality of recordings, much to the delight of the NVA.

Rudolf Kühnhold, a scientific associate and head of the acoustics group at the NVA, was occasionally present during the recordings and became acquainted with Erbslöh and von Willisen at that time. Kühnhold was seeking a high-power acoustic pulse generator operating in the frequency range of 10 to 20 kHz to advance his research on underwater acoustic detection of ships, particularly submarines. The objective was to detect underwater targets followed by an attack using torpedoes. After consulting with friends and confidants, von Willisen agreed to accept an order from Kühnhold to develop and manufacture an acoustic pulse generator through the company Tonographie. By then, Kühnhold had already been appointed scientific director of the NVA.

The first demonstration of the acoustic pulse generator took place in mid-1933, fully meeting the NVA's requirements and was immediately accepted, allowing Kühnhold to continue his experiments until a specification for a standard naval detection system was drawn up.[639]

The period following late 1933 was characterised by close and trusting cooperation between Kühnhold and his team and Erbslöh and von Willisen, despite no formal business contract yet being established with Tonographie. At the time, the NVA had limited development and procurement budgets, subject to price controls. Nevertheless, Tonographie's revenues were strong enough that Erbslöh and von Willisen provided services in advance, despite potentially long payment delays from the NVA.

---

638 Kroge 2000, p. 13; Kroge 1998, p. 22; Godt 2003, p. 222.

639 Kroge 2000, pp. 13–14; Kroge 1998, pp. 22–23.

The subsequent founding of GEMA in January 1934 helped ensure Tonographie avoided financial difficulties linked to its involvement with the NVA. Although GEMA initially faced challenges in its establishment and organisational and technical obstacles, it experienced rapid growth until its official dissolution on 31 May 1945.

Employee numbers rose from 200 in autumn 1937 to 500 in the summer of 1938 to 800 by the end of that year, 1,643 by the end of 1939, over 2,500 by the end of 1940, 3,000 in 1941, and nearly 6,000 by 1944.[640]

From 1941 onwards, forced labourers from Italy, France, the Netherlands, Belgium, and Croatia were employed; from mid-1942, they were joined by Soviet forced labourers.[641]

7.1.2 Great Britain

In Great Britain, targeted radar development[642] did not begin until the mid-1930s but gathered speed due to the threat of an emerging military conflict with Germany and finally during the war.[643]

The consideration of radar development must not be limited to scientific and technological progress alone, but the interaction between politics, the military, science and industry must be analysed comprehensively.[644] In the case of radar development in par-

640 Kroge 2000, pp. 73, 75, 79, 89, 101, 113, 151; Kroge 1998, pp. 82, 87, 92, 105, 120, 137, 186.

641 Fransecky 2003, p. 117; Noack 1998, pp. 53–64; Kroge 2000, p. 130; Kroge 1998, p. 158.

642 The term radar originated in the USA, probably around 1940. Before that, the British used the term Radio Direction Finding (RDF), which was not originally a cover term for radar, since the word radar did not yet exist. RDF referred to the technique of radio direction finding, a method distinct from radar.

643 Detailed accounts of the development of British radar throughout World War II can be found, for example, in Judkins 2007, Watson-Watt 1957, Bowen 1987, Bragg 1988, Brown 1999, Buderi 1997, Rowe 2015, Latham and Stobbs 1999, Swords 2008, and White 2007b, as well as in German in Godt 2003, Kern 1984 and Dekker 2014. For the development of transmitter and receiver technologies, see Callick 1990.

644 Klausing 2022, pp. 31–42.

ticular, the holistic approach involving all those responsible and later the use of radar in conjunction with other technological developments or operational measures such as Signals Intelligence, radio direction finding and operational research ultimately proved to be a decisive factor in the success of the war.[645]

As already explained, British radar development was characterised by a strong awareness among the political and military leadership of the threat to the country from German air raids, against the background of the experiences from World War I.

Guided by this threat and the realisation of the country's own vulnerability due to its relatively small geographical size and high population density, the government's defence policy was sharply attacked by opposition members, particularly in the person of Winston Churchill.[646] He particularly criticised the government's appeasement policy, especially the politicians Stanley Baldwin and Neville Chamberlain.

The sentence by British Prime Minister Baldwin on 10 November 1932:

> „...the bomber will always get through..."

had an impact on politicians, ministries, the military and broad sections of the population.[647] This situation prompted the government and military leadership to seek air defence options which, in contrast to radar development in Germany, initially focused solely on establishing an early warning system to protect the coastline against German air attacks. From the outset, all work related to the development of an air defence system was centralised, thereby largely avoiding duplication and internal competition, as was the

645 In this context, Kaiser's article offers an excellent analysis of both the development of British radar technology and appeasement policy, highlighting the significance of the history of technology for understanding political history (Kaiser 1996a; Kaiser 1996b).

646 Churchill, who held no government post during the 1930s, returned to office as First Lord of the Admiralty at the outbreak of World War II in September 1939. He was appointed Prime Minister and Minister of Defence in May 1940, following the resignation of Neville Chamberlain.

647 See also Chapter 3.1.

case in Germany.[648] The serious threat posed by German air attacks created pressure that led to the rapid coordination of joint efforts among various actors, including representatives from politics, the Air Ministry, Defence Committees, the RAF, science, and industry.

1934 was an eventful year for air defence.[649] Following domestic and foreign policy shifts in Germany, and amidst rising tensions in Asia, Germany announced its withdrawal from the League of Nations on 14 October 1933.

Germany was increasingly recognised as a potential aggressor. In 1933, the Committee of Imperial Defence (C.I.D.) established the Defence Requirements Sub-Committee (DRC) on 14 November.[650] The C.I.D., composed of senior political and military leaders, was the supreme body for the strategic defence of the British Empire, while the DRC was tasked with investigating significant deficiencies in the British defence apparatus. It produced studies on rearmament and defence policy. After heated debate, the committee recommended substantial funding to expand the armed forces, including an increase in the RAF's operational strength.[651]

Exercises in the summer of 1934, aimed at intercepting enemy bombers with defending fighters, highlighted the need for more precise information on the attackers' flight movements, something the existing sound location methods could not provide. During these summer exercises, mock attacks on the Air Ministry, Parliament, and other key facilities revealed the vulnerabilities of British defences.[652]

As Air Force officers, Air Ministry officials, and politicians, including Winston Churchill, realised the gravity of the situation, the pressure grew to take swift action.[653]

648 Kern 1984, p. 160.

649 Judkins 2007, pp. 99–104.

650 Neilson 2003, pp. 651–684; Zimmerman 2013, p. 60; Judkins 2007, pp. 99–100.

651 Zimmerman 2013, p. 60.

652 Latham and Stobbs 1999, p. 8; Clark 1965, p. 106; Buderi 1997, p. 53.

653 Swords 2008, p. 82.

Ronald Clark[654] described this situation, encompassing the political decisions, the work of the radar pioneers, the developments surrounding the Tizard Committee, and the role of the Air Force, as

„The Political Birth of Radar."[655]

Churchill was at the forefront of the general call for rearmament and railed against the destructive potential of air warfare, referring to it as "the cursed, hellish invention and development of war from the air".[656]

His close friend Frederick Lindemann[657], later Lord Cherwell, publicly questioned the government's position on air defence.

7.1.2.1 First Steps

In June 1934, Albert Rowe began an unofficial analysis of existing detection systems for air defence purposes at the British Air Ministry. Rowe was convinced that early warning against attacking aircraft using conventional technology, such as acoustic listening devices, would no longer be sufficient. Airspace surveillance with constantly flying patrols would also have far exceeded the available resources of aircraft and personnel.[658] At that time, Rowe was the only employee at the Directorate of Scientific Research of the Air Ministry working on defence-related issues, even though the Directorate had been explicitly told that defence was not within its remit.[659] Rowe was a personal assistant to Harry Wimperis[660], the Director of Scientific Research at the Air Ministry.[661] Additionally, research into wireless communication was explicitly not Wimperis' responsibility, as this task lay with the Royal Aircraft Establishment

654 Biographer of Henry T. Tizard.
655 Clark 1965, pp. 105–127.
656 Buderi 1997, p. 53.
657 Blake 2008; Birkenhead 1961; Burns 2013.
658 Watson-Watt 1957, p. 80; Kern 1984, p. 162.
659 Fisher 1989, p. 25; Rowe 2015, p. 1.
660 Serby and Bradley 2020.
661 Rowe 2015, p. 1.

(RAE) in Farnborough, which already had established connections with the radio industry.[662]

Rowe searched through all available files on air defence problems. Among the very large number of scientific proposals, he found only 53 that dealt with this topic, but not a single one was useful.[663]

As a consequence, in response to Baldwin's claim that the bombers would always get through, the only solution, according to Trenchard's strategy[664], would be to build more bombers. However, this was not a viable solution.

Rowe reported on the results of his investigations:

> "In June 1934, then, against a background of frustration in armament work and gloom over the War Office work on early warning of air attack, I undertook an informal survey of the problem of air defence and to this end collected every available file on the subject; there were fifty-three of them. It was clear that the Air Staff had given conscientious thought and effort to the design of fighter aircraft, to methods of using them without early warning and to balloon defences. It was also clear however that little or no effort had been made to call on science to find a way out. I therefore wrote a memorandum summarizing the unhappy position and proposing that the Director of Scientific Research should tell the Secretary of State for Air of the dangers ahead. The memorandum said that unless science evolved some new method of aiding air defence, we were likely to lose the next war if it started within ten years. Unfortunately, I was not clever enough to think of a new method."[665]

7.1.2.2 Important Defence Committees

The situation changed in November 1934 when Wimperis realised that without scientific support, likely to come from outside the Air Ministry, no effective solution to the pressing air defence challenge was possible. At that time, there was little connection between scientific and military thinking.[666]

---

662 Judkins 2007, p. 101.

663 Fisher 1989, p. 25; Swords 2008, p. 82; Kirby 2003, p. 66.

664 See also Chapter 8.2.

665 Rowe 2015, pp. 4–5; Jones 2017, p. 14; Latham and Stobbs 1999, pp. 8–9.

666 Rowe 2015, p. 5.

On 12 November 1934, in a memorandum addressed to the Secretary of State for Air, Charles Londonderry[667], as well as other senior figures from the Air Ministry and military circles, including Air Marshal Hugh Dowding[668], Wimperis proposed the establishment of a committee to carry out a scientific survey of air defence methods.[669] The first sentence was one that certainly provoked thought:

> "When one looks back to the stupendous technical advances of the last 50 years one cannot but wonder what equally striking advances can possibly lie ahead of us in the next equal period of years ..."[670]

This led to the formation of the Committee for the Scientific Survey of Air Defence (CSSAD), chaired by Henry Tizard[671], who was then Rector of the Imperial College of Science and Technology in London. The CSSAD, commonly known as the Tizard Committee[672], was officially formed in November 1934 with the task of investigating how scientific and technological advances could strengthen current air defence methods.[673]

Subsequently, the CSSAD was incorporated under the umbrella of the newly established Air Defence Research Sub-Committee (ADRC), which was founded on 10 April 1935 as part of the C.I.D.

The ADRC, often referred to as the Swinton Committee[674] after its chairman Philip Cunliffe-Lister, Viscount Swinton[675], was

---

667 Jackson 2008; Lord Londonderry was Secretary of State for Air from 1931 to 1935. A detailed account of his ambivalent relationship with Germany and his closeness to the Nazi regime can be found in Kershaw 2005; Kershaw 2004.

668 Orange 2008.

669 Swords 2008, p. 83; White 2007b, p. 4; Bragg 1988, p. 20; Buderi 1997, p. 54.

670 Bragg 1988, p. 20.

671 Clark 1965; Piller 2008; Farren 1961, pp. 313–348; Watson 2009, pp. 42–43.

672 Guerlac 1987, pp. 127–135; Phelps 2010, pp. 17–28; Bragg 1988, p. 22.

673 Rowe 2015, p. 5; Air Ministry 1963, pp. 3–4.

674 Kaiser 1996a, p. 34; Kaiser 1996b, p. 149; Zimmerman 2013, pp. 104–112.

675 Robbins 2008; Philip Cunliffe-Lister served as Secretary of State for Air from 1935 to 1938 and as Minister for Civil Aviation from 1944 to 1945. In November 1935, he was elevated to the hereditary peerage as Viscount Swinton, and in May 1955, he was created Earl of Swinton.

charged with accelerating technological development and allocating financial resources for air defence.

Both Tizard and Dowding were members of the ADRC alongside Winston Churchill. The ADRC had a remit very similar to that of the CSSAD. To avoid duplication and streamline work, the ADRC conducted its scientific research through the CSSAD, effectively making the CSSAD a subcommittee within the C.I.D. structure.[676]

Fig. 17 illustrates the relationships between the major defence committees chaired or influenced by Tizard between 1935 and 1940. It is somewhat confusing that the Tizard Committee, initiated by the Air Ministry in November 1934, was subsequently incorporated into the ADRC, officially established on 10 April 1935.

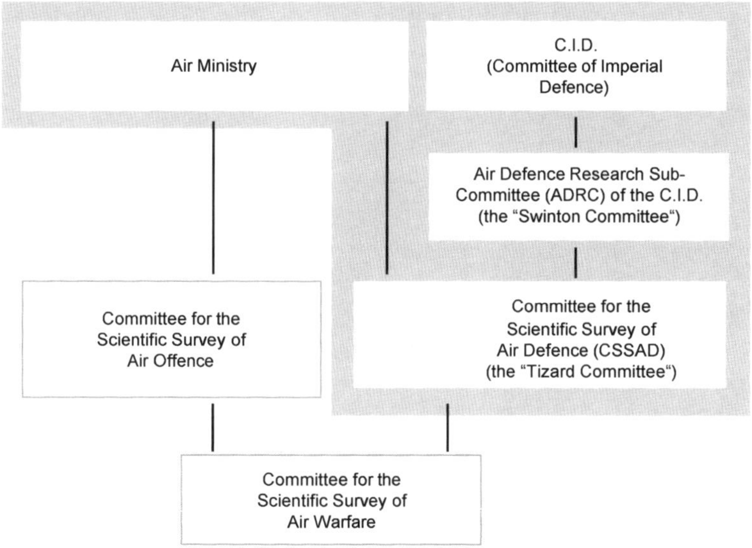

Fig. 17: Important Defence Committees[677]

In early 1937, a second committee was established alongside the CSSAD, known as the Committee for the Scientific Survey of Air

676 Judkins 2007, pp. 108–109.
677 According to Clark 1965, p. 128.

Offence (CSSAO), which dealt with the subject of air attack and was also headed by Tizard.[678] In October 1939, the two committees were merged to form the Committee for the Scientific Survey of Air Warfare (CSSAW).[679] This committee existed until June 1940. While both committees played important roles, the CSSAD was particularly influential in the development of radar technology, which is the primary focus here.

The first meeting of the CSSAD took place on 28 January 1935.[680] The appointed members present were Tizard as chairman, Professor Patrick Blackett[681] from Birkbeck College, Professor Archibald Hill[682] from University College London, and Wimperis of the Air Ministry.[683] Rowe was appointed as secretary of the committee.[684]

Blackett was a prominent student of Rutherford and later won the Nobel Prize in Physics in 1948. Hill was one of the world's leading physiologists and had already been awarded the Nobel Prize in Medicine in 1922. Beyond their scientific achievements, Blackett's and Hill's military experience likely played a significant role in their selection for the committee. Blackett had served as an active naval officer until he began his studies after World War I. Hill led a research group focused on ballistics for anti-aircraft defence during World War I and authored a seminal work on anti-aircraft artillery. This military connection was a key factor in the committee's success, as it enabled scientific findings to be effectively communicated

---

678 Clark 1965, pp. 161–163.

679 Clark 1965, p. 128, p. 196.

680 Watson-Watt 1957, p. 108; White 2007b, p. 4; Swords 2008, p. 83.

681 Nye 2011; Swords 2008, p. 84; Blackett was a key driver of Operational Research.

682 Katz 2009; Swords 2008, pp. 83–84.

683 Phil Judkins reports that the founding of the CSSAD by Wimperis was less straightforward than most sources suggest, occurring in a highly complex environment. The appointments of Tizard and Blackett to the CSSAD are said to have been no coincidence. As Tizard and Blackett were already members of the Aeronautical Research Committee of the C.I.D., and Hill was carefully integrated into the CSSAD due to his reputation as a physiologist, these selections helped to avoid sensitive political issues (Judkins 2007, pp. 101–102).

684 Rowe 2015, p. 5; Judkins 2007, p. 102, p. 169.

to military leaders from the outset and developed under real operational conditions until ready for deployment.[685]

Tizard and Lindemann also brought experience from World War I, having served as pilots in the Royal Flying Corps. Wimperis suggested that the committee's terms of reference should be broad enough to encompass all possible developments:

> "To consider how far recent advances in scientific and technical knowledge can be used to strengthen the present methods of defence against hostile aircraft."[686]

Tizard was born in 1885 in Gillingham, Kent, the son of a naval officer, but was unable to join the Navy due to an eye condition. He was educated at Westminster School and the University of Oxford, where he earned a degree in chemistry in 1908. He then spent a year in Berlin studying under Professor Walther Nernst[687], who later won the Nobel Prize in Chemistry in 1920. It was in Berlin in 1908 that he met his future rival Lindemann, Nernst's favourite student.[688]

In the autumn of 1909, after a brief period as a science master at Eton College, Tizard became a freelance researcher at the Davy-Faraday Laboratory of the Royal Institution, where he established his reputation with two significant papers on the colour changes of indicators. In 1911, he also translated the sixth edition of Nernst's Theoretical Chemistry. That same year, Tizard returned to Oxford as a tutorial fellow at Oriel College and worked as a demonstrator in the electrical laboratory, leading to several co-authored papers on the motion of ions in gases.

During World War I, Tizard distinguished himself through his contributions to air force research. He played a key role in developing air defence technologies and was involved in testing aviation equipment. Additionally, he was involved in flying tests for the Royal Flying Corps.

685 Snow 2013, pp. 22–23; Snow 1961b, pp. 26–27.

686 Burns 2013, p. 98; Bragg 1988, p. 21.

687 Jäger und Heilbronner 2010, p. 303; Mendelssohn 1976; Mendelssohn 1973; Bartel 2014, pp. 49–53.

688 Mendelssohn 1976, p. 93.

In 1920, Tizard became Assistant Secretary, and in 1927, Secretary of the Department of Scientific and Industrial Research (DSIR), which had been established in 1915 to promote scientific research in Britain during World War I.[689] In this capacity, he was directly responsible to a minister.[690] From 1929, Tizard held a senior position at Imperial College of Science and Technology in London, while maintaining close ties to the British civil service and continuing his leadership role within the DSIR.[691] Tizard's ability to act as both an administrator and a scientist, combined with his scientific and military research experience, made him particularly well suited to serve as chairman of the CSSAD, which had to address both technical and political challenges.[692]

Lindemann was born on 5 April 1886 in Baden-Baden to wealthy parents. He studied in Berlin and earned his doctorate in physical chemistry under Professor Walther Nernst, where he also met Tizard. At the start of World War I, he became associated with the Royal Flying Corps, like Tizard, and became known for his work on preventing aircraft spins. In 1919, he took up the Chair of Physics at Oxford and became Director of the Clarendon Laboratory.[693]

Lindemann's life was significantly shaped by his close friendship with Winston Churchill, whom he met in 1921 and for whom he became the closest scientific adviser. This friendship lasted until Lindemann's death in 1957. Lindemann's influence on Churchill often led to conflicts in politics and within defence committees. One of the most notable disputes was with his former friend Tizard in the CSSAD, which contributed to the temporary dissolution and reorganisation of the committee.

Even before the CSSAD convened for its first meeting on 28 January 1935 at the Air Ministry, there had been an unofficial meeting on 18 January between Wimperis and Watson Watt, who was

689 Zimmerman 2013, p. 80.

690 Snow 2013, p. 18; Snow 1961b, p. 21.

691 Clark 1965, pp. 79–80; Zimmerman 2013, p. 81; Snow 2013, p. 18; Snow 1961b, p. 22.

692 Zimmerman 2013, p. 81; Rowe 2015, p. 5; Tizard 1946, pp. 338–339.

693 Blake 2008, p. 2.

then head of the Radio Department at the National Physical Laboratory (NPL) in Teddington.

Watson Watt was born on 13 April 1892 in Brechin, Scotland. He studied at University College Dundee, then part of the University of St Andrews, and graduated with honours in Applied Mathematics and Electrical Engineering in 1912.

With the outbreak of World War I, he volunteered for military service and, in September 1915, took up a post as a meteorologist at the Meteorological Office, located at the Royal Aircraft Factory in Farnborough. There, he applied his expertise in detecting thunderstorms and the atmospheric disturbances they caused in order to warn airmen.

By 1917, Watson Watt's work on the detection of thunderstorms using radio techniques had moved to the Aldershot Wireless Station, where he began to supervise activities in this emerging field.[694] His work increasingly came under the coordination of the newly established DSIR, which supported research of national importance.

In 1924, his work was transferred from Aldershot to Slough, where the Radio Research Station (RRS) was established under the DSIR. In 1927, the RRS became administratively part of the National Physical Laboratory (NPL), though it remained physically separate. Watson Watt was appointed its Superintendent. The RRS then operated as an outstation of the NPL. Following a further reorganisation, he became head of the newly established Radio Department at the NPL headquarters in Teddington in 1933.[695]

Although working in different areas, both Watson Watt and Tizard were closely associated with the DSIR, which provided the institutional foundation for much of Britain's scientific innovation in the interwar years.

As early as 1933, Kühnhold of the German Navy's NVA recognised that conventional sound-location methods were no longer adequate for effective early warning of air raids, a view later shared by Watson Watt:

694 Watson 2009, p. 39.

695 Rowe 2015, p. 12; Bowen 1987, p. 7; Ratcliffe 1975, pp. 551–552; Watson 2009, pp. 37–39.

> "Up to the beginning of 1935 no alternative to the acoustic method was available or in sight, and warning schemes using these large 200-foot acoustic mirrors had been planned for the United Kingdom and for some overseas stations and territories."[696]

Early efforts were made to detect engine noise along coastal regions using large concave acoustic mirrors fitted with microphones. However, this method of location soon proved inadequate for several reasons. For example, the acoustic signatures of ships could easily be mistaken for those of aircraft, and the perceived intensity of a sound source varied depending on its distance from the mirror. A further major limitation of sound location for early warning and interception of incoming bombers whose speed and operational altitude were steadily increasing was the time delay involved. In many cases, aircraft had already passed over the detection site before an accurate location could be established.[697] Additionally,

---

696 Watson-Watt 1957, p. 80.

697 Zimmerman 2013, pp. 29–50; Zimmerman 1997, pp. 73–99; Rogers 2018, pp. 13–15; Galati 2016, pp. 80–84. During World War I, acoustic listening devices (acoustical mirrors, sometimes known as sound mirrors) were used in Britain for early warning against attacking Zeppelins and later against aircraft. The period following the war was characterised by the need to erect increasingly larger acoustic mirrors along the east and south-east coasts in order to extend detection range and provide earlier warning of potential air attacks. This was a direct consequence of the continuous increase in aircraft speed, driven by major advances in aviation technology. Moreover, there was a growing need to differentiate between various types of aircraft noises and to filter out environmental sounds, such as those from ships or land vehicles (Judkins and Bauer 2014, p. 11). The leadership of the Royal Engineers requested the assistance of physicist William Tucker (Zimmerman 2013, pp. 26–28), who had been involved in acoustic early warning research since the end of World War I. After the formation of the Air Defence Experimental Establishment (ADEE) in 1924, Tucker played a leading role in its acoustics research. The ADEE, which had originated from the Searchlight Experimental Establishment (SLEE) during World War I, focused on the development and practical testing of air defence technologies, including early radar and acoustic systems. In October 1929, the acoustics section was relocated to Biggin Hill to continue field experiments under more realistic conditions (Zimmerman 2013, pp. 30–31). In the summer of 1934, the results of Tucker's twelve years of work in the hills of Kent attracted the attention of senior officials from the Air Ministry, including Robert Watson Watt, who showed great interest in the technical methods for defending against air attacks. Despite concerns about the sensitivity of sound mirrors to environmental noise

sound location was also unable to distinguish between friendly and hostile aircraft and was vulnerable to interference from ambient noise.[698]

Wimperis represented the Air Ministry on the Radio Research Board (RRB) of the DSIR and, through this role, became acquainted with some of Watson Watt's work.[699] As a result, he invited Watson Watt to an informal meeting at the Air Ministry, as already mentioned, on 18 January. The main topic of discussion was the so-called "Death Ray" theory, a speculative idea that high-energy electromagnetic waves might be used to disrupt aircraft systems, detonate ammunition or incapacitate aircrew.[700]

Watson Watt later recalled:

> "He then asked for my opinion on the possible utility of some kind of damaging radiation as an aid in defence against air attack. This I translated mentally into the low-brow "Death Ray", and provisionally rejected as very unpromising. I promised, however, to do some think-

---

and their limited warning time, the Air Ministry decided by the end of January 1935 to proceed with the installation of a comprehensive acoustic early warning system (Galati 2016, pp. 81–82). Tucker was tasked with planning the necessary communication infrastructure, which primarily consisted of telephone lines and switching systems. By June 1935, more than 500 personnel from the Air Ministry were already involved in work on acoustic detection devices (Galati 2016, p. 82). However, in August 1935, the Air Ministry decided to suspend all work on the acoustic mirror system in the Thames Estuary until the end of March 1936, in light of emerging alternative detection technologies (Swords 2008, pp. 179–180; Zimmerman 2025, pp. 36–41). This marked the beginning of efforts to detect aircraft using radio waves, the dawn of the radar era in Britain, which truly gained momentum in the second half of the 1930s. In 1939, the ADEE moved to Christchurch from its original site at Biggin Hill. In the same year, the Ministry of Supply took over responsibility for the establishment from the War Office. The ADEE was later renamed the Air Defence Research and Development Establishment (ADRDE) in 1941, reflecting its expanded role in research and development during the wartime effort. According to modern British historians, the communication infrastructure built for the acoustic detection system later served as a valuable foundation for linking the Chain Home radar stations. Meanwhile, Germany pursued a similar path in acoustic detection. Advanced versions of acoustic direction finders, such as the Ringtrichter-Richtungshörer, continued to be used well into World War II (Müller 2002, p. 47, p. 101).

698 Judkins 2007, p. 101.

699 Watson-Watt 1957, p. 81.

700 Watson-Watt 1957, p. 81; Burns 2013, p. 127; Jones 2017, p. 16.

> ing and some arithmetic and to give an early but considered answer in writing."[701]

The theory of "Death Rays" was published in numerous sources of varying credibility and was promoted by politicians and the military as a potential wonder weapon. On 11 July 1934, the renowned physicist Nikola Tesla[702] published an article in The New York Times in which he claimed that such "Death Rays" could be realised with tremendous destructive power.[703] Watson Watt was subsequently tasked with calculating the energy that would need to be generated and transmitted in order to achieve the proposed effects. Arnold Wilkins[704], his assistant at the National Physical Laboratory (NPL), demonstrated through straightforward calculations that generating such "Death Rays" was not technically feasible. For example, the frequently cited goal of incapacitating a pilot by increasing his body temperature by just 2 degrees Celsius would have required a transmission power far beyond existing technology.[705] However, Wilkins pointed out that the energy reflected from an aircraft might be sufficient to detect it at a useful distance. He recalled that postal engineers in 1931 had observed interference with radio reception whenever aircraft flew near their receivers and suggested that this phenomenon could be harnessed to detect enemy aircraft.[706]

7.1.2.3 Watson Watt Memorandum and the Daventry Experiment

At the first meeting of the CSSAD on 28 January 1935, an initial memorandum from Watson Watt was already available. It expressed a negative assessment of "Death Rays" (Radio Destruction) but also contained the first indications of a detection method based on the reflection of electromagnetic waves (Radio Detection).[707]

---

701 Watson-Watt 1957, p. 81.

702 Jäger und Heilbronner 2010, p. 429; Neumann 2021, pp. 50–53; Tragner 2004, pp. 217–245.

703 Watson 2009, p. 43; Tesla Memorial Society of New York.

704 Latham and Stobbs 2011, pp. 1–4.

705 Zimmerman 2013, pp. 84–85; White 2007b, p. 4.

706 Watson Watt 1957, pp. 93–94; Judkins 2007, p. 105; Jones 2017, p. 16; Watson 2009, p. 44.

707 Judkins 2007, p. 105; Brown 1999, pp. 50–51.

Watson Watt was commissioned to further develop and submit his calculations and proposals. In his letter of 12 February 1935 to Rowe at the Air Ministry, he enclosed a memorandum entitled "Detection of Aircraft by Radio Methods".[708] This was followed on 27 February 1935 by a second memorandum, "Detection and Location of Aircraft by Radio Methods", which outlined in twenty points how effective air defence could be achieved through radio detection.[709] Remarkably, radar technology was already being considered a viable solution for air defence at a time when no actual device yet existed, and development was expected to take three to four years.

Watson Watt later described this second memorandum as the basis for the invention of radar:

> "...I regard this Memorandum on the "Detection and Location of Aircraft by Radio Methods" as marking the birth of radar and as being in fact the invention of radar."[710]

At the beginning of the memorandum, Watson Watt explained that the detection and location of an enemy aircraft using its own emissions, such as light, radio waves, infrared radiation or propeller noise, was hardly feasible. Light and infrared radiation from the engines would not be suitable for location due to atmospheric attenuation. Compared to electromagnetic waves, active sound detection methods are unsuitable for locating aircraft because of the low propagation speed of sound waves. Watson Watt concluded that only the evaluation of reflected electromagnetic waves was suitable for detecting and localising enemy aircraft.[711] The aircraft would have to fly through a wall of radio waves, with the metallic wings of the aircraft acting as a reflective antenna. The best reflection intensity would be achieved when the emitted wavelength matched the target size. In his calculations, Watson Watt assumed the target was a typical night bomber with metal wings and a wingspan of around 25 metres. The wing structure could, to a first approximation, be

708 Watson-Watt 1957, p. 83.

709 Watson-Watt 1957, pp. 83–87, pp. 470–474; Austin 1999, pp. 365–372; Swords 2008, pp. 278–285; Reuter 1971, pp. 37–38.

710 Watson-Watt 1957, p. 83.

711 Ibid., p. 470, pt. 2.

considered a linear oscillator with a fundamental wavelength of 50 metres and low ohmic resistance.[712] Watson Watt favoured operation at a wavelength of 50 metres, as there was already experience with pulse operation, which made it possible to measure the range of enemy aircraft. The goal was to transition to shorter wavelengths, which would improve the resolution capability. Watson Watt also pointed out that the direction of an echo signal could be visualised for a wavelength of 50 metres using the cathode ray direction-finding technique developed in Slough.[713]

Shortly before the memorandum was submitted to CSSAD on 27 February 1935, Watson Watt and his assistant Wilkins conducted an initial field test near the BBC transmitter in Daventry on 26 February 1935. The aim was to prove that an illuminated aircraft would reflect enough energy to be detected.[714] This test was prompted by a request from Air Marshal Dowding for a practical demonstration of the idea of using radio waves. The BBC had erected its most powerful radio transmitter in England, the Empire Radio Station near Daventry, with a transmission power of 10 kW at a wavelength of 49 metres. The beam of the transmitting antenna had an azimuth of 30 degrees and an elevation of 10 degrees. A Handley Page Heyford bomber was ordered to fly through the transmission beam at altitudes between 1,000 feet and 6,000 feet. Watson Watt and his team positioned an old lorry with electronic equipment, including a receiver coupled to a cathode-ray tube, near the transmitter. The cathode-ray tube displayed both the signal emitted directly from the radio transmitter and echo signals reflected from the aircraft, with a detection range of up to 8 miles.[715]

Watson Watt was so impressed by this result that he turned to Rowe and said:

712 Ibid., p. 470, pt. 4.

713 Ibid., p. 473, pt. 12.

714 Watson-Watt 1957, pp. 110–112; Latham and Stobbs 2011, pp. 13–20; Griffiths and Austin 2016, p. 3–8; Zimmerman 2013, pp. 85–91; Swords 2008, pp. 180–186; Austin 1999, pp. 370–371; Judkins 2007, p. 107, p. 171.

715 Latham and Stobbs 1999, pp. 9–10.

"Great Britain has become an island again."[716]

Watson Watt was aware that this test result did not yet constitute a fully realised aircraft detection system, but merely demonstrated that an aircraft would, in principle, reflect enough energy to form the basis of effective air defence. Rowe later explained:

> "Overnight one of Watson-Watt's assistants, A. F. Wilkins, had erected equipment in a van near Daventry. All that was hoped of this equipment was that it would show that an aircraft, when in the Daventry beam, would reflect enough of the beam's energy for its presence somewhere in the vicinity to be inferred from observations in the van. This is just what happened on 26 February 1935. So far from the demonstration being witnessed by distinguished officers from the Services, the sole representative from the Service departments was one humble civilian scientific worker – myself. Watson-Watt and I were pleased with the demonstration, since reflections from the aircraft were obtained when it was estimated to be about eight miles away, but we knew that we had not seen the location of an aircraft by radio."[717]

These initial, elementary measurement results were sufficient to secure the necessary funding of £10,000 from the Air Ministry's budget, at a time when both politicians and military leaders recognised the urgent need for an early warning system.[718] This is another impressive example of how innovation was strongly supported by political and military leaders from the very beginning. Watson Watt later wrote:

> "… Even £10,000 was "real money" in 1935, and, fortunately or no, we did not then know that we were spending the first instalment of a multi-million-pound radar programme… "[719]

The months following the experiment at Daventry were characterised by focused further development of key radar components,

716 Watson-Watt 1957, p. 117; Latham and Stobbs 1999, p. 10.

717 Rowe 2015, p. 8.

718 Watson Watt 1957, p. 113; Bragg 1988, p. 33; Judkins 2007, p. 107; Kaiser 1996a, p. 35; Kaiser 1996b, p. 150.

719 Watson-Watt 1957, p. 113; Watson-Watt 1946, p. 14.

such as transmitters and receivers, alongside resolving existing antenna issues.[720]

7.1.2.4 Radar Systems and Air Defence

Watson Watt was authorised to form a small team of specialists from the NPL staff, with Wilkins and Bowen, the latter joining shortly afterwards, as senior assistants. They decided that their experiments would need to be carried out over the North Sea, as enemy aircraft were expected to approach from that direction. On 1 March 1935, Watson Watt visited Orfordness in Suffolk and decided to establish an independent laboratory there for the development of new detection methods, using the cover name Ionospheric Research Station, under his direction.[721]

On 13 May, the team relocated from the NPL in Slough to Orfordness, situated about 140 kilometres northeast of London on the North Sea coast. The group initially comprised Bowen, Wilkins, and two other NPL employees.[722] They were all fully aware of the importance of maintaining secrecy around their research.

During its first year, the team never exceeded six members. It was formally led by Watson Watt, who remained based in Slough and visited Orfordness at weekends to supervise progress. Although all staff were organisationally part of the NPL, they were funded by the Air Ministry. In practice, since Watson Watt was rarely on site, the day-to-day leadership of the team fell to Wilkins.[723]

7.1.2.4.1 Chain Home

Work on developing a radar system along the coast to provide early warning of enemy aircraft progressed rapidly. On 5 June, three pairs of wooden masts with aerials were erected, each 75 feet high, with one pair used for transmitting and two pairs for receiving. The

720 Swords 2008, p. 182.

721 Watson-Watt 1957, p. 126; Kern 1984, pp. 172–173; Zimmerman 2013, p. 91.

722 Bowen 1987, p. 11; Rowe 2015, pp. 13–14; Latham and Stobbs 1999, p. 10; Latham and Stobbs 2011, pp. 23–26.

723 Swords 2008, p. 186; Bowen 1987, p. 14.

project was referred to as Radio Direction Finding (RDF) for reasons of secrecy, as this terminology was already in use and unlikely to attract public attention.[724]

During a visit by the CSSAD on 15 June 1935, backscatter measurements were taken from a Vickers Valentia aircraft at a distance of approximately 17 miles, under very difficult weather conditions.[725] Watson Watt reported that the poor weather had forced the aircraft to alter its flight path, resulting in only intermittent echo signals being received.

Further attempts on the following day remained entirely unsatisfactory. The Officer Commanding the Air Ministry Wireless Station informed us that, at the time of the trials, weather conditions had not been so unfavourable for the propagation of short waves in the previous 15 years.

However, the team had gained insights into how performance could be improved.[726] These included the use of taller masts, specifically 200 feet in height, and shorter wavelengths, where technically feasible.

Subsequent demonstrations using the higher antenna masts, 200 feet tall, were conducted by 16 July. A Bristol Type 120 aircraft was detected at a distance of approximately 38 miles. By this time, the wavelength had been reduced to 26 metres, enabling aircraft to be tracked at ranges between 5 and 38 miles, depending on conditions.[727]

A further milestone in radar development was reached on 24 July 1935, when echo signals from three other aircraft of the Hawker Hart type were displayed on the cathode ray tube during measurement tests with a Westland Wallace research aircraft at a distance of around 34 miles. This was the first radar observation of a formation of aeroplanes with an associated determination of their number.[728] Watson Watt summarised:

> "In sum, then, a known aircraft had been followed to 42 miles, a known aircraft detected at 39 miles, an unknown aircraft detected at 33 miles,

---

724 Latham and Stobbs 1999, p. 10; Bowen 1987, pp. 11–14, Swords 2008, p. 190.

725 Rowe 2015, p. 14.

726 Watson-Watt 1957, pp. 127–128.

727 Swords 2008, p. 191; Kern 1984, p. 173.

728 Rowe 2015, p. 16; Swords 2008, p. 191.

> a known aircraft followed out to 2–3° above the horizon. These were our interim jewels, but the most shining one for me was that of 24 July. On that day we were using a *Wallace* as target aircraft, and had decided to photograph typical stages of the picture on our cathode-ray tube. The *Wallace* was followed out to 34½ miles. Then as we awaited its return, an unexpected echo intruded, beyond 20 miles in range. By its fluctuations we inferred (as promised in the Memorandum) that it was not from a single aircraft but from a formation of three. Then our *Wallace* reappeared beyond 34 miles, to be photographed at 32½ miles."

and continued:

> "This was the very first radar observation of a formation, the first counting of numbers in a formation, the first diagnosis of a change in formation, the first recount. By great good fortune, and by the quality of the *Wallace* pilot, we had from him the full confirmation afforded by his own visual detection of the *Harts* which comprised the formation."[729]

In 1935, there was still no reliable method for identifying friend or foe, and this challenge was not resolved until 1939.

In a review dated 9 September 1935, Watson Watt informed the CSSAD that aircraft could be detected at ranges of approximately up to 58 miles. If the arrival of aircraft was anticipated within a 20-minute window, detection was possible at more than 40 miles. Even if aircraft arrived unexpectedly, they could still be detected at over 32 miles, with an accuracy of about half a mile.[730]

Following earlier discussions in August about a coastal early warning system, Watson Watt now submitted a formal proposal. It included the construction of a chain of radar stations along the coastline to be defended, with transmitters placed every 20 miles and receivers at every second station, or every 40 miles.[731]

This review of 9 September 1935 is regarded as a key document in the development of British radar.

Yet in the first half of September, Wilkins measured the flying height of an aircraft flying at 7,000 feet and at a distance of 15 miles,

729 Watson-Watt 1957, pp. 131–132.

730 Watson-Watt 1957, pp. 132–134; Judkins 2007, p. 111.

731 Swords 2008, p. 288; Watson 2009, p. 51; Bragg 1988, p. 43.

with an estimated error in height measurement of 1,200 feet. Trials had been made with six types of land aircraft and five types of marine aircraft at flying heights of 1,000, 5,000, 7,000 and over 10,000 feet, with comparable performance on wavelengths of 50, 28, 27, 26 and 25 metres.[732]

The original wavelength of 50 metres, which had proved suitable for detecting the aircraft types in use at the time, had to be reduced because of interference with commercial radio transmissions. It was initially shortened to 26 metres without any loss in performance, and later further reduced to a range of 10 to 13 metres. This wavelength range was retained until the end of the war and became the standard for all Chain Home radar stations.[733]

7.1.2.4.1.1 Bawdsey Manor

Regardless of developments at the CSSAD, work on the construction of a coastal radar system to provide early warning of enemy aircraft continued apace at Orfordness. At the meeting of the ADRC on 16 September 1935, it was reported that progress on the coastal radar system had far exceeded expectations, achieving in just five months what was initially estimated to take five years, and at a meeting on 24 October, military requirements were closely aligned with technical challenges.

Consequently, the committee approved a proposal to construct a chain of radar stations between the River Tyne and Southampton[734] and concluded that a larger number of stations would be needed to ensure reliable detection and tracking of enemy aircraft at distances ranging from 5 to 65 miles off the coast, a recommendation made all the more remarkable given that no fully operational prototype radar station yet existed.

At the same meeting, it was proposed that Bawdsey Manor, located approximately 14 miles north of Orfordness, be established as

732 Watson-Watt 1957, p. 132.

733 Swords 2008, pp. 188–207, pp. 225–229; Neale 1988, pp. 132–150; Neale 1985, pp. 73–83; with regard to the technical specifications of the transmitters, receivers and antenna systems used, it should be noted that there were numerous variations, particularly between 1936 and 1940.

734 Watson-Watt 1957, p. 134; Kern 1984, p. 173, p. 175.

the centre for research and serve as headquarters for the organisation responsible for implementing the radar chain. The plan envisaged creating an initial operational station there to serve as a model for subsequent installations. Following this proposal, Watson Watt and Wilkins helped identify Bawdsey Manor as an ideal location due to its elevation of about 70 to 80 feet above sea level and sufficient space for the required 360-foot transmitter and 240-foot receiver masts. After securing the owner's agreement, the British government purchased the estate in late 1935 for £23,000. On 19 December, the Treasury approved a plan to establish five radar stations to protect the Thames Estuary, with Bawdsey as the northernmost station and South Foreland as the southernmost.[735]

The meetings on 16 September and 24 October 1935, which closely aligned military requirements with technical challenges, marked a significant innovative step in the early development of radar. On 24 October 1935, the first official conference was held under the leadership of Air Marshal Dowding, then Deputy Chief of the Air Staff, to address for the first time the challenges of military operations involving radar. In the following months, Bawdsey became the focal point for radar research, training, and administration, serving as the operational centre for the emerging radar network.[736]

Following the decisions made at the ADRC meeting on 16 September 1935, Bawdsey Manor became the centre for research, training, and the coordination of radar station installations. The research team relocated to Bawdsey Manor as early as February 1936, and the move was completed by May 1936. The site was then officially designated as Bawdsey Research Station (BRS). Staff remained at Bawdsey Manor until 2 September 1939, just before the outbreak of war.[737]

As the concept of fixed long-range radar stations began to take shape, Bawdsey Manor became the focal point for developing what would become the Chain Home system (Fig. 18). From the outset, the goal was clearly defined: to create an effective early warning network capable of detecting incoming air attacks. This objective

735 White 2007b, p. 6.

736 Watson-Watt 1957, p. 141; Kern 1984, p. 175.

737 Swords 2008, p. 192.

Fig. 18: Bawdsey Manor[738]

was pursued with determination, leading to the installation of radar equipment not only at ground stations, but also on aircraft and naval vessels to strengthen Britain's overall defensive capabilities. On 1 August 1936, the team of researchers who had previously worked at NPL was transferred to the Air Ministry, with Robert Watson Watt appointed as the first Superintendent.[739] During 1936 and 1937, work at Bawdsey Manor focused on the construction of the initial five Chain Home radar stations and the fitting of radar equipment to aircraft.[740]

In March 1936, Watson Watt submitted a list of points to the CSSAD, explicitly outlining areas that still required investigation. These included determining the location of enemy aircraft for fighter control and the installation of onboard radar equipment in aircraft.[741] However, priority was initially given to the construction and improvement of the five Chain Home radar stations along the coast (Fig. 19).

738 Image credit: Obtained from the Bawdsey Radar Trust.

739 Bowen 1987, p. 22; in May 1938, Watson Watt was appointed Director of Communications Development at the Air Ministry, and Rowe succeeded him as Superintendent of the BRS (Lovell 1991, pp. 4–5; Watson 2009, p. 59).

740 Godt 2003, p. 195.

741 Watson-Watt 1957, p. 146; Kern 1984, pp. 175–176.

Fig. 19: Bawdsey Manor with CH radar station[742]

However, there were also technical mishaps in the UK during hasty demonstrations, which also triggered discussions at a political and military level. One example is an air defence exercise involving the demonstration of radar for airspace surveillance, which was originally scheduled for 31 August to 11 September 1936 in Bawdsey.[743] Only the radar station in Bawdsey could be put into operation, as the construction of the other four stations was not yet sufficiently advanced for them to take part. Around one hundred aircraft took part in the exercise over a period of three days. Half of these were bombers, which were to carry out simulated attacks from about 100 miles seaward. The other half consisted of fighter aircraft, which were to intercept them on the instructions of the radar station in Bawdsey.[744]

In addition to the entire CSSAD, Air Marshal Dowding and the Secretary of State for Air, Swinton, had been invited to attend the exercise.[745] A major failure occurred on the first day of the exercise, when no echoes could be received from the approaching

742 Image credit: unknown.

743 Swords 2008, p. 192; Bowen 1987, pp. 24–26.

744 Bowen 1987, p. 24; Watson 2009, pp. 55–56.

745 Hanbury Brown 1991, p. 15.

aircraft at a sufficient distance to intercept them. This was also extremely embarrassing due to the presence of Swinton and Dowding.[746] The aircraft could only be detected by the engine noise as they flew over the receiving area.[747] Bowen was not involved in the exercise at the time but was working on tests for an airborne radar with a transmitter modelled on Orfordness and a miniature receiver and indicator for installation in an aircraft. This enabled him to receive very good echo signals from approaching aircraft from the ground up to a distance of 50 miles. He only found out about the debacle at midday. Incoming aircraft could only be located at a distance of 10 to 15 miles and the fighters could no longer be given instructions. Watson Watt was furious, and Tizard was also very angry about the demonstration. After Bowen had then reported his measurements to Watson Watt, Dowding, with the help of Bowen's equipment, succeeded in presenting impressive echoes up to a distance of 50 miles.[748] Nevertheless, the measurement results immediately afterwards were still inadequate. After powering up the Bawdsey receiver with the Orfordness transmitter the next morning and making further technical improvements, the rest of the exercises went without incident and a high percentage of intercepts were achieved.[749] In conclusion, it can be said that the exercise was scheduled far too early due to the many design changes made during the construction of the radar chain, which were often carried out at the same time with the result that hardly any mature equipment technology was available and without test

---

746 Bowen 1987, p. 24.

747 Judkins 2007, p. 124; Bowen 1987, p. 24; Judkins reconstructs this day as 17 September.

748 Bowen 1987, p. 24.

749 Bowen 1987, pp. 25–26; in his analysis, Judkins provides more detail about the reactions of political and military leaders, in addition to the course of the air defence exercise already described here. Tizard in particular, who often had to assert himself against powerful opponents such as Churchill and Lindemann, came under fire for having supported a very expensive and demonstrably ineffective system. Watson Watt was also viewed as a failure, and Dowding, who in his leading military role had placed great hopes on the new radar system, called Wilkins a charlatan. Fortunately, Dowding was able to see for himself the following morning that the radar system was working (Judkins 2007, p. 124).

equipment.[750] Useful results were only achieved after recalibrating the devices on 22 September.[751]

Watson Watt also reports that the initial failure was later offset by a well-deserved success, after motivation at the highest scientific level had faltered.

New motivation came through a message from Tizard, who wrote to Watson Watt that he had just seen Air Marshal Joubert de la Ferté[752], who told him that:

> "if we could always guarantee results as good as on the last two days of the Exercise, he would have almost as much information as he needed, provided he could get height."[753]

Watson Watt said:

> "...The supreme encouragement, from C.A.S.[754], meant more work for us, and we were glad. It was now decided that the Exercise using Bawdsey, Canewdon and Dover should be flown in the Spring of 1937..."[755]

In March 1937, the first prototype radar station, built at Bawdsey, was handed over to RAF personnel, becoming the world's first fully operational radar station. This prototype for subsequent Chain Home radar stations was designated Air Ministry Experimental Station Type 1 (AMES Type 1).[756] Further stations were commissioned at Dover and Canewdon before the end of 1937, followed by Great Bromley and Dunkirk in 1938.[757]

Home defence exercises took place in August 1938, with all five radar stations fully operational. Civilian and scientific person-

750 Bowen 1987, p. 26.

751 Swords 2008, p. 192.

752 On 1 September 1936, Air Marshal Philip Bennet Joubert de la Ferté assumed command of RAF Coastal Command as Air Officer Commanding-in-Chief.

753 Watson-Watt 1957, p. 173; Clark 1965, pp. 153–154.

754 Chief of the Air Staff; however, Joubert was not C.A.S.; this was Edward Ellington.

755 Watson-Watt 1957, p. 173.

756 Swords 2008, p. 193.

757 Swords 2008, p. 194; Kinsey 1983, p. 15; Dunkirk was completed but not yet calibrated.

nel were involved in operating the equipment at all sites, particularly at Bawdsey. Shortcomings in evaluation procedures were still apparent, and the insufficient radar coverage at low altitudes was confirmed. In September 1938, the five stations at Bawdsey (Suffolk), Great Bromley (Essex), Canewdon (Essex), Dunkirk (Kent), and Dover (Kent) were placed on 24-hour operational status. These were supplemented by mobile units at Drone Hill (covering the Forth-Clyde area), West Beckham (covering the Wash), and Ravenscar (covering the Tyne-Tees area).[758] Together, the fixed stations between Suffolk and southern Kent ensured that the Thames Estuary was protected against surprise air attacks targeting London.

Aircraft flying at 10,000 feet could typically be detected at ranges of up to 80 miles, although this varied with factors such as atmospheric conditions, terrain, and the calibration state of the individual radar station.[759] An identification system to distinguish friend from foe was also in place.[760]

The radar stations covering the Thames Estuary were operational by the time Prime Minister Chamberlain flew to Munich in September 1938, and they remained active throughout the remainder of the war.[761] Watson Watt explained:

> "The "Munich Crisis" brought the abbreviated chain to continuous watch, twenty-four hours of every day, save for brief maintenance periods carefully staggered among adjacent stations. The watch from these five sites was not to be interrupted until after victory was achieved in Europe and assured in the Pacific … "[762]

At the start of the war on 3 September 1939, 18 Chain Home (CH) radar stations were operational and connected to a central evaluation unit known as the Filter Room at Fighter Command Headquarters in Bentley Priory. Two additional stations, Netherbutton and School Hill in Scotland, were also operational but had not yet been connected to the Filter Room.[763]

---

758 Swords 2008, p. 195, p. 296.

759 Watson-Watt 1957, p. 187.

760 IFF: Identification Friend or Foe; Watson-Watt 1953, p. 35, p. 43; Latham and Stobbs 2011, pp. 61–66; Swords 2008, pp. 33–36; Reuter 1971, p. 38.

761 Watson-Watt 1953, p. 35, p. 43.

762 Watson-Watt 1957, p. 187.

763 Swords 2008, p. 195.

Fig. 20: Bawdsey radar towers[764]

Fig. 20 shows the transmitter aerial towers at Bawdsey Chain Home radar station, Suffolk, May 1945.

It quickly became apparent that aircraft approaching from the sea at low altitudes could not be detected effectively using the long wavelengths employed by the original radar sets. As early as 1939, the Chain Home (CH) radar network was supplemented by the development of Chain Home Low (CHL) systems designed to detect enemy aircraft flying at lower altitudes.

While CH radars (AMES Type 1), operating at frequencies of approximately 22 to 50 MHz, were capable of detecting targets flying at around 500 metres altitude at a range of 40 kilometres, and at 4,000 metres altitude up to around 135 kilometres, CHL radars (AMES Type 2) became operational for the first time in November 1939.[765] These used a wavelength of 1.5 metres, corresponding to a frequency of about 200 MHz, and enabled detection of aircraft flying as low as 150 metres within a range of 40 kilometres.

764 Image credit: IWM (CH 15337).

765 Swords 2008, p. 188, pp. 236–242.

By 1943, the development of Chain Home Extra Low (CHEL) radar operating at 3 GHz, corresponding to a wavelength of 10 cm, allowed for the detection of targets flying at 1,800 metres at ranges up to 145 kilometres, and aircraft flying at altitudes as low as 15 metres could still be tracked within 30 kilometres.

With the complete radar system in place, an advance warning time of 15 to 20 minutes before enemy aircraft reached the coastline could be achieved.[766]

7.1.2.4.2 Operational Defence

The foresight demonstrated by Tizard and Dowding in prioritising the development of a comprehensive air defence system in parallel with radar technology was instrumental in the timely and effective defence of Great Britain. In this process, they established clear objectives and operational boundaries. Using their authority and political influence, they ensured the active involvement of scientists, engineers, and operational personnel in the development of the system.[767] Moreover, political and military leadership, along with relevant government ministries, were informed of the programme's goals at an early stage and remained engaged throughout its progression. Consequently, these efforts contributed significantly to the successful repulsion of a potential invasion of Great Britain during the Battle of Britain in 1940.

7.1.2.4.2.1 Biggin Hill Exercises

The first field test conducted by Watson Watt and his assistant Wilkins at Daventry in February 1935, which successfully detected an aircraft using electromagnetic waves, is widely recognised as a milestone in the development of radar.

Further decisive progress was made in 1936 and 1937, following experiments initiated by Tizard in the autumn of 1935 and supported by Dowding. These efforts aimed to develop an integrated air defence system capable of filtering and evaluating all available

766 Reuter 1971, p. 38.
767 Corrigan 2008, pp. 7–8.

information to ensure the optimised deployment of air defence resources.[768]

Based on data concerning the position and direction of incoming enemy aircraft, the system was designed to enable the timely and targeted selection of appropriate squadrons, followed by precise instructions via radio link to guide interceptors to their targets with sufficient accuracy.[769]

The idea of a comprehensive air defence system was conceived even before the first prototype of a radar station in Bawdsey was handed over to the RAF in March 1937, and the next stations were completed in the following years. Flight trials were to begin as soon as operational radar stations were available. Work on this coordinated air defence system was carried out at Biggin Hill in Kent.[770] On 13 July 1936, Tizard outlined his proposals for what became known as the Biggin Hill Experiments to a number of senior RAF officers at the Air Ministry.[771]

On 27 July 1936, he proposed conducting a series of experiments to intercept attacking aircraft based on a defined warning time that radar would provide. It was here that Tizard's strength in combining scientific expertise and its translation into operational action came to the fore. Tizard explained that personnel on the ground should be forewarned from the moment the attacking aircraft were seventy-five miles away. He proposed a series of experiments to explore two key points. One dealt with the location of attacking aircraft and the associated continuous improvement of position determination by measurement at certain time intervals. The second point involved investigating the possibility of using radiotelephony to guide an interceptor towards an attacking bomber if its position and flight path could be determined from

---

768 Zimmerman 2013, pp. 156–166; Latham and Stobbs 1999, pp. 12–14; Bowen 1987, p. 26; Clark 1965, pp. 149–163.

769 Judkins 2007, pp. 125–126.

770 Clark 1965, pp. 149–163; Latham and Stobbs 1999, pp. 12–13; Biggin Hill airfield is located 17 miles (27 kilometres) southeast of London and was founded in 1917. It played a significant role in both world wars and served as a key RAF base during World War II. The airfield was attacked several times by the German Luftwaffe in 1940 and 1941.

771 Clark 1965, p. 150; Tizard 1946, p. 339.

the ground. The experiments were to begin on 4 August 1936 and last two months.[772]

It was thanks to the foresight of Tizard and Dowding that the experiments went ahead despite the reservations of some senior officers.[773] Tizard recognised that while radar could provide valuable information on direction, range, altitude, and approximate speed of incoming aircraft, entirely new interception techniques had to be developed. This would be achieved through an integrated warning and tracking system that could map the overall operational situation.

As acoustic listening devices were gradually replaced by radar, standing patrols could become redundant, resulting in significant savings in personnel and aircraft deployment.[774] Tizard and Dowding agreed that radar development must be finalised urgently, but meanwhile they conducted their own tests using radio direction finding. For this purpose, a ground-based system known as Pipsqueak was employed.[775] Pipsqueak was a British radio-location system used during World War II. Each fighter aircraft carried a radio set that automatically transmitted a constant 1 kHz tone for 14 seconds at fixed intervals every minute. Ground direction-finding stations picked up these signals and, by triangulation, determined the aircraft's position. This allowed RAF controllers to track and direct their squadrons effectively, particularly during the Battle of Britain.

The preparation and refinement of interception procedures during these trials proved invaluable in accelerating the operational use of radar data once available. Tizard also understood that all these trials needed to be conducted in peacetime.[776]

The outcome was an air defence system composed of a network of radar stations to warn of incoming aircraft, combined with a hierarchical filter system to collate and evaluate data, and an

772 Clark 1965, p. 150.
773 Ibid., p. 150.
774 Ibid., pp. 149–150.
775 Latham and Stobbs 1999, p. 13.
776 Ibid., p. 12.

efficient communications network to alert interceptors and guide them to their targets.[777]

At Dowding's instigation, a group from the RAF led by Squadron Leader Raymund Hart was sent to Bawdsey to coordinate work on the radar chain, which became a joint project between scientists and the RAF.[778] In February 1937, Hart opened a training school for RAF personnel at Bawdsey.[779] In the autumn of 1937, another exercise involving all five stations around the estuary and the filter system was held, achieving resounding success.[780] This laid the foundations for the forthcoming Battle of Britain. In October 1938, the Filter Room was opened at Bentley Priory near Stanmore, Middlesex, the headquarters of Fighter Command.[781]

7.1.2.4.2.2 Dowding System

By May and June 1940, the Air Defence System, later known as the Dowding System[782], had largely matured and was fully operational.

Although improvements in communication and coordination continued, it already functioned as the world's first radar-based air interception network, covering airspace from northern Scotland to England's southern coast. The coastline was divided into sectors; each integrated into a comprehensive command and communication structure.[783] The system featured well-defined communication channels, including an extensive telephone network, with some lines dating back to World War I, alongside integrated radio and telex systems. This infrastructure enabled the efficient collection and filtering of incoming air raid data, creating a real-time operational picture from which precise orders could be issued to fighter squadrons and anti-aircraft guns.

777 Bowen 1987, p. 26; in contrast to Germany, British pilots accepted command and control from the ground, partly because Tizard promoted the integration of ground control into RAF doctrine.

778 Ibid., p. 26.

779 Swords 2008, p. 193.

780 Bowen 1987, p. 26

781 Swords 2008, pp. 195; Price 2009, pp. 5–9; Price 1978, pp. 16–19.

782 The name Dowding System was only coined later.

783 Air Ministry 1941, pp. 6–10.

Information on enemy attacks passed through a hierarchical reporting chain to the Operations Room at Fighter Command Headquarters (HQ) in Bentley Priory, the highest command authority. One of Dowding's key aims was to streamline the flow of information so that only essential details reached pilots quickly and accurately.

Upstream of the Operations Room was the Filter Room, also at Bentley Priory, which received raw data from CH radar stations. The Filter Room verified and consolidated this radar information into a coherent air picture before forwarding it to the Operations Room, where data were plotted on large map tables with enemy and friendly aircraft marked in different colours to aid tactical decisions.

In parallel, the Royal Observer Corps (ROC), a volunteer organisation, played a vital role in Britain's air defence by visually identifying and tracking enemy aircraft from a nationwide network of observation posts. The individual posts reported to their respective Group Operations Rooms, providing information on the position, estimated altitude and direction of the approaching force, along with an assessment of the number of aircraft and, in most cases, their types. The Group Operations Rooms then checked, collated, and processed the reports before forwarding the consolidated data to the ROC Control Centre, which produced a complete operational picture. This picture was subsequently forwarded to the Group and Sector Operations Rooms, as well as to the Central Filter Room at Bentley Priory, where it was incorporated into the overall Fighter Command picture and used to coordinate air defence measures. Sector Operations Rooms and individual squadrons received their information via their respective Group Operations Rooms, and in some cases also directly from the ROC Control Centre, particularly to supplement radar coverage for low-flying aircraft. The Filter Room at Bentley Priory primarily received raw radar information from the Chain Home stations. These reports were processed to remove duplicate plots and false returns, producing a reliable air picture. In addition, the Filter Room also made use of certain reports from the ROC Control Centre to verify and supplement radar information. Once the information had been consolidated and checked, it was passed to the Group Operations Rooms, where it formed the basis for operational decisions and the

Fig. 21: Barrage Balloons[784]

direction of fighter aircraft.[785] Alongside fighter squadrons, anti-aircraft guns[786] and searchlights[787], and barrage balloons[788], the ROC significantly contributed to Britain's layered air defence. Barrage balloons were used across Great Britain during World War II as a passive air defence measure against low-flying enemy aircraft. They protected cities, industrial sites, ports, airfields, and military installations by forcing aircraft to fly higher or risk striking steel cables.

Barrage balloons played a key role during the Blitz and were deployed in large numbers during Operation Overlord to defend invasion bases and supply routes. Similar balloons were also deployed by Allied forces in other theatres of war.

784 Image credit: IWM (A 23093).
785 Corrigan 2008, pp. 3–6.
786 Price 2009, pp. 9 – 12.
787 Ibid., pp. 9–12.
788 Rogers 2018, pp. 199–205; Price 2009, pp. 12–14; Price 1978, pp. 22–23, p. 28.

Fig. 22: Royal Observer Corps members at their post[789]

The deployment of anti-aircraft guns and searchlights was coordinated by Anti-Aircraft Command (AAC)[790], the British Army organisation responsible for Great Britain's ground-based air defences, while barrage balloons were operated separately under RAF Balloon Command.

Fig. 21 shows barrage balloons deployed as a defensive measure, with Fig. 22 showing members of the Royal Observer Corps (ROC) monitoring aircraft from an observation post.

The Operations Room in Bentley Priory formed the central command and control core of Britain's air defence. Its efficient coordination enabled rapid and accurate responses to air threats, proving crucial to the system's success.

Mission orders were typically issued from Group or Sector Operations Rooms, usually housed in purpose-built control centres away from the operational airfields themselves.

During the Battle of Britain, the interceptor force was organised into four Groups, each covering a large geographical area and equipped with its own Group Operations Room:

789 Image credit: IWM (CH 8215).

790 Price 2009, pp. 9–10.

- No. 10 Group (South-west England and Wales)
- No. 11 Group (South-east England, including London)
- No. 12 Group (Midlands, East Anglia and parts of northern England)
- No. 13 Group (Rest of northern England and southern Scotland).

The central Operations Room at Bentley Priory, the headquarters of Fighter Command, displayed squadron readiness at the Group level and issued directives to the Group Operations Rooms. Each Group Operations Room coordinated air operations across its area, supervised several Sector Operations Rooms, and allocated resources according to operational priorities. Sector Operations Rooms, typically located at operational airfields, implemented these orders locally, exercising direct tactical control over fighter squadrons, including scrambling and interception vectoring. They maintained continuous radio contact during missions, coordinated with anti-aircraft guns and searchlight units, and provided updates on flight movements at intervals of approximately two to five minutes, ensuring comprehensive responses to enemy air attacks.

The command hierarchy ran as follows for Chain Home Radar Stations and Royal Observer Groups (Fig. 23).

This layered command structure ensured that intelligence was processed efficiently, and fighter resources were deployed where they were most needed, contributing decisively to Britain's air defence during the Battle of Britain.

An example is the No. 10 Group Operations Room at RAF Box near Colerne, Wiltshire (Fig. 24).

This centre directed fighter squadrons, coordinated with ground defences, and maintained real-time tactical oversight during missions. Although operational during the Battle of Britain, its Sector Operations Room only fully developed after the battle, reflecting the growing strategic importance of decentralised command.

While the organisational framework was well established, the system's effectiveness increasingly relied on emerging technologies. The Pipsqueak allowed ground stations to locate RAF fighters, complementing radar tracking and aiding tactical coordination. However, ground-controlled Interception was not yet central dur-

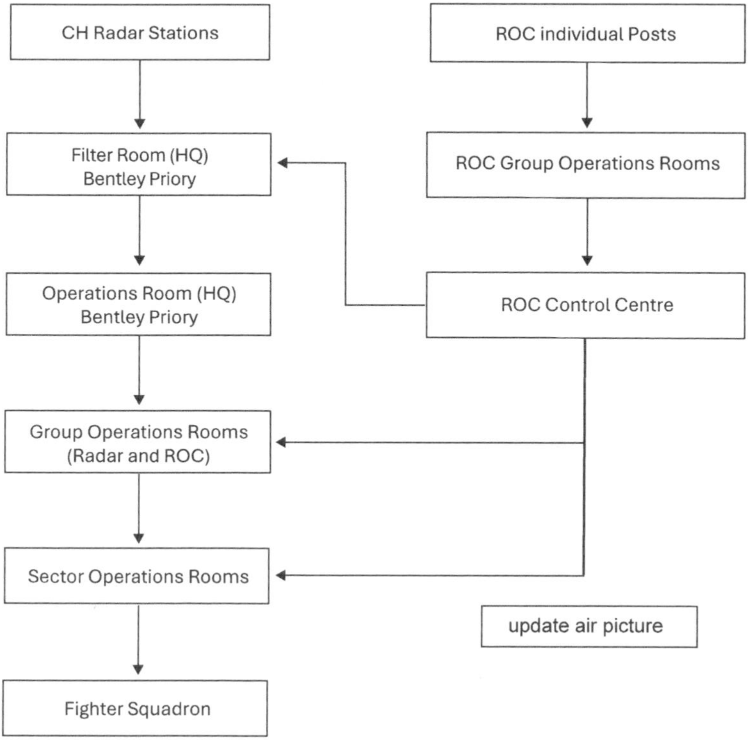

Fig. 23: Command Hierarchy of the Dowding System

ing the Battle of Britain and only became widespread later, notably enhancing night-fighter operations.

Anti-aircraft guns and barrage balloons formed essential parts of the multi-layered defence network. These ground-based assets were coordinated through Sector Operations Rooms, providing protective coverage over key sites and supporting fighter aircraft in countering enemy incursions.

Fig. 24: No. 10 Group Operations Room, RAF Box[791]

7.1.2.4.2.2.1 Key Components and Information Flows in the Dowding System

This section presents summary tables that complement the detailed description of the Dowding System provided in the main text. The tables offer a clear and structured overview of the key air defence facilities, their functions, and the hierarchical communication and command flows during the Battle of Britain.[792]

Table 3 lists the key facilities of the Dowding System and summarises their primary roles in detecting, processing, and coordinating responses to enemy aircraft.

791 Image credit: IWM (CH 13680).

792 Air Ministry 1941, p. 9; Corrigan 2008, p. 4.

Facility / Station	Function / Description
Headquarters (Bentley Priory)	Central command headquarters of RAF Fighter Command, directing overall air defence.
Filter Room (Bentley Priory)	Processed raw radar and selected observer data, filtered out duplicates and noise.
Group Operations Rooms	Responsible for geographical RAF areas, coordinating missions and communications.
Sector Operations Rooms	Local command posts at airfields, directly controlling fighter squadrons.
Radar Stations (Chain Home)	Early warning system detecting incoming aircraft via radar reflections.
Royal Observer Corps (ROC)	Visual aircraft spotting supplementing radar, especially for low-flying targets.
Ground-Controlled Interception (GCI) Stations	Later-developed radar stations guiding fighters by radio to enemy aircraft (post-1940).
Operations Plotting Rooms	Displayed aircraft movements on maps for situational awareness and tactical decisions.

Table 3: Key facilities and functions in the Dowding System

Table 4 shows the flow of information from radar stations and the Royal Observer Corps to the relevant command posts

Source / Input	Initial Receiver	Further Distribution
Radar Stations	Filter Room	Group Operations Rooms → Sector Operations Rooms.
ROC Observations	ROC Control Centre	Filter Room /Group Operations Rooms /Sector Operations Rooms.

Table 4: Information flow from radar and ROC in the Dowding System

Table 5 illustrates how orders and operational information were passed from Fighter Command Headquarters down to Group and Sector Operations Rooms, and ultimately to fighter squadrons, anti-aircraft units, and barrage balloon commands.

Command Source	Controller	Controlled Unit / Function
Fighter Command Headquarters	Group Operations Rooms	Strategic direction and distribution of filtered air situation to Groups.
Group Operations Room	Sector Operations Room	Coordination of operations within the Group and tactical control delegated to Sectors.
Sector Operations Room	Fighter squadrons	Tactical control of local fighter squadrons and start orders, target assignment, in-flight instructions.
Anti-Aircraft Command	Anti-Aircraft Units	Coordination and control of anti-aircraft defences (guns and batteries).
Barrage Balloon Command	Barrage Balloon Units	Deployment, positioning, and management of barrage balloons.

Table 5: Command and order flow in the Dowding System

Pilots initially had to adapt to receiving instructions from ground control, which is why experienced pilots were often assigned as controllers. These controllers were required to visit the workstations of ground personnel to gain a thorough understanding of their roles and perspectives. It was largely due to the persistence of Tizard and Blackett that pilots were eventually convinced to rely on accurate information and strict instructions rather than instinct alone.[793]

793 Corrigan 2008, p. 5; Like Lindemann, Tizard had achieved significant success working closely with aviation and air power during World War I. All of them,

Even before the war, practical exercises were conducted to improve interception rates, efforts that paid off during the Battle of Britain, often resulting in a tactical advantage.[794] Care was taken to distribute the burden of combat missions as evenly as possible among squadrons. New formations and tactics were practiced between missions in phased training. Furthermore, no squadron was deployed without ensuring full combat readiness; they were properly prepared, and operations were typically led by experienced squadron leaders.[795]

7.1.2.4.3 Airborne Radar

After the move to Bawdsey was completed in May 1936, Bowen and a small group of researchers began in September to investigate the feasibility of installing radar equipment in aircraft for use in air combat. This was prompted by the limitations of the Chain Home radar system, which provided inadequate tracking of enemy aircraft at night and in poor visibility. During daylight, friendly fighters could visually detect the enemy at ranges of 5 to 10 miles, but at night, detection distances reduced to less than 1,000 feet, depending on sky clarity and moonlight.

By this time, the technical requirements set by Watson Watt and the Tizard Committee for CH radar to determine the range, direction, and altitude of incoming aircraft had been fully met. Tizard was convinced that the nearly 100-mile detection range already achieved, combined with interception procedures and the filtering of radar data as tested in the Biggin Hill experiments, offered strong capabilities for daytime air defence. However, he also believed that the German Air Force would likely respond by shifting to night bombing raids. Tizard expressed this concern in a letter to Air Marshal Dowding.

At this time, Lindemann, as a member of the CSSAD, continuously proposed new ideas for air defence, such as air mines sus-

especially Tizard, could easily identify with the pilots and aircrew of the RAF, which fostered a climate of acceptance. Göring and Udet, who had not yet come to terms with their experiences from World War I, were highly sceptical of radar.

794 Air Ministry 1941, p. 8.

795 Ibid., p. 10.

pended from balloons on wires and the use of infrared (IR) technology to detect enemy aircraft.[796] While Lindemann did not fundamentally oppose radar development, he criticised the slow progress and often delayed the Tizard Committee's work by promoting numerous alternative suggestions. Despite Lindemann's proposals, two fundamental requirements remained essential for effective air defence, both day and night:

1. The need to locate and track enemy raiders in good time to effect an interception by day or by night (with greater accuracy being required in the case of the latter).
2. The ability to destroy the enemy once he had been found, in daylight or in darkness.[797]

Ultimately, two technological options were under consideration: one was the development of a radar system compact and lightweight enough to be installed in an aircraft; the other was the detection of enemy aircraft via their exhaust emissions using infrared sensors. Coincidentally, Lindemann's former doctoral student, Reginald Jones[798], was conducting research in infrared spectroscopy. He built a prototype detector but soon realised that it would take years before such a device could be installed in an aircraft and achieve the required performance.[799]

By early 1936, with the success of Chain Home (CH) against German bombers by day assured, the much more difficult issue of intercepting night bombers arose again.[800] With efforts at Bawdsey focused on the construction of CH radars, Watson Watt decided to divide the personnel. The CH radar installations remained under the responsibility of Wilkins, while the problem of night bomber defence was assigned to Bowen.[801]

796 White 2007b, p. 7; the idea of illuminating the night sky with powerful searchlights was abandoned early on due to the enormous energy requirements and the vast number of searchlights needed (White 2007b, p. 7).

797 White 2007b, p. 7.

798 Martin 2011; Cook 1999, pp. 239–254.

799 White 2007b, p. 7; Jones 2017, pp. 19–20.

800 Bowen 1987, p. 30.

801 White 2007b, p. 7.

After discussions with two well-known engineers[802] from RAF Martlesham Heath[803], Bowen began talks with the Air Ministry and Fighter Command. At that time, however, neither the Air Ministry nor Fighter Command had a department specifically responsible for night defence.[804] Nonetheless, there were many opinions and assumptions about what characteristics a night-capable fighter aircraft should possess. From these discussions, Bowen developed a list of necessary characteristics for an airborne radar system. These included considerations such as weight, installation dimensions, maximum power consumption, antenna size, and whether the system could be operated by the pilot or an onboard observer.[805]

Bowen was fortunate to receive a tuned high-frequency receiver developed by the EMI Company for the BBC, operating at a wavelength of 6.7 metres and offering a sensitivity not previously achieved in the UK.[806] This receiver, combined with the display of received echo signals on a cathode-ray tube, formed the basis of the entire airborne radar test programme over the following two years.[807]

After Bowen had set up his laboratory in the attic rooms of the Red Tower and White Tower at Bawdsey Manor and recruited additional staff, the first tests were conducted against aircraft using a wavelength of 6.7 metres. The transmitter was installed on the roof of the Red Tower and the receiver on the White Tower. Ranges of between 40 and 50 miles were soon achieved. However, it quickly became apparent that a transmitter operating at this wavelength was far too large to be installed in an aircraft.[808]

As a result, the initial plans envisaged situating a powerful transmitter on the ground at Bawdsey, with a receiver and target display carried aboard the aircraft. In the autumn of 1936, the first

802 Fred Rowarth and N. E. Rowe had an excellent reputation as men who were very familiar with the mechanical processes involved in aviation and knew exactly what was and what was not possible (Bowen 1987, p. 32).

803 RAF Martlesham Heath was a former RAF station located southwest of Woodbridge, Suffolk. It was operated between 1917 and 1963 and played an important role in the development of Airborne Interception (AI) radar.

804 Bowen 1987, p. 32.

805 Bowen 1987, p. 32; White 2007b, pp. 7–8. For a brief overview of some the highlights from the early days of airborne radar see Simon Watts 2025b.

806 Bowen 1987, pp. 32–33.

807 Bowen 1987, pp. 32–33; White 2007b, p. 8.

808 Bowen 1987, pp. 33–35; White 2007b, p. 8.

flight trials were undertaken using this arrangement, employing a Handley Page Heyford bomber from Martlesham Heath.[809] The receiving aerial consisted of a half-wave dipole strung between the aircraft's wheels, whilst power was supplied by a combination of accumulators and dry-cell batteries positioned on the aircraft's floor.

Bowen, serving as an observer aboard, reported that on the very first flight echoes were received from aircraft in the vicinity of Martlesham Heath. The flight was conducted at an altitude of between 2,000 and 3,000 feet above the Bawdsey transmitter. The initial range was 8 to 10 miles, with a maximum of approximately 12 miles achieved on subsequent sorties.[810]

The Air Ministry assigned the designation RDF1 to the ground-based CH early warning radar and RDF2 to the airborne radar. Bowen's team referred to the combined system with the transmitter on the ground and the receiver onboard the aircraft as RDF1.5, while the Air Ministry officially designated the ground-based component RDF1R.[811]

RDF1R had an effective range of approximately 20 miles at low altitude and did not require a complex ground control infrastructure.[812]

However, the system still faced several challenges. These included the necessity to synchronise the airborne receiver with the ground transmitter and the requirement that the attacking aircraft, that is the enemy bomber, had to be illuminated by the transmitter. In situations where the attacking aircraft was positioned behind the ground transmitter the receiver aircraft could not detect its presence because the radar beam no longer illuminated the target. Distance measurements were accurate only when the attacking aircraft was located directly between the ground transmitter and the receiver aircraft. In all other cases the measured distance was underestimated owing to the indirect angle of the radar signal path which did not correspond to a direct line of sight return.[813]

---

809 Bowen 1987, pp. 36–37; Smith et al. 1985, p. 359.

810 Bowen 1987, pp. 36–37; White 2007b, p. 9.

811 Bowen 1987, p. 37; according to White, the designations RDF1 and RDF2 were given by the Air Ministry, which ultimately chose the designation RDF1R for Bowen's RDF1.5 (White 2007b, p. 13, pt. 32).

812 Bowen 1987, p. 37; White 2007, p. 9.

813 Bowen 1987, p. 37; White 2007b, p. 9.

Watson Watt remained unconvinced of RDF1.5's merits and the trials were subsequently cancelled. Bowen regarded this decision as a grave mistake seeing it as a missed opportunity to gain early experience in introducing and testing sophisticated radar technology for night use in aircraft. He expressed his disappointment thus:

> "With hindsight, it is now clear that this was a grave mistake. If RDF1.5 had been given proper Service trials in the RAF, there would have been several important consequences. In the first place, it would have given them an interim device on which test interceptions could have been carried out at night, two whole years before the outbreak of war. This would have provided pilots and observers with training in the techniques of night interception, something which they did not actually get until war was declared. Lastly, it would have given us experience of the immense problems, still to be faced, of introducing new and sophisticated equipment into Service use."[814]

It became clear that devices needed to be developed for shorter wavelengths, that is to say, higher frequencies, in order to reduce their size and ultimately allow both the transmitter and receiver to be installed within an aircraft. In March 1937, a compact transmitter operating at a wavelength of 6.7 metres was installed in a Handley Page Heyford bomber, utilising vacuum tubes from the American company Western Electric. During the trials, the new transmitter proved to be powerful, and clear echo signals were received from Harwich docks and from ships approaching the harbour from the sea. However, the range to the ships could not be determined with certainty as the aircraft was not authorised to fly over the sea. The maximum range achieved was 3 to 4 miles.[815] Following these successful trials, Bowen managed to secure a twin-engined Avro Anson, which was used by Coastal Command as a reconnaissance aircraft and was therefore suitable for operations over the sea, to serve as a test platform. The Royal Air Force eventually provided two Avro Anson aircraft, along with five pilots and supporting ground crew.[816]

814 Bowen 1987, p. 37.
815 Bowen 1987, pp. 37–38; White 2007b, p. 10.
816 Bowen 1987, pp. 38–39; White 2007b, p. 10.

Bowen later remarked:

> "These proved to be the foundation on which the whole of British airborne radar was subsequently based and must rank as one of the best investments in the future that the Royal Air Force ever made…"[817]

The configuration of the radar unit was subsequently modified, and a wavelength of 1.25 metres was eventually adopted. However, this wavelength represented a practical limit, as any further reduction would have resulted in a significant decrease in transmission power.

In August 1937, an improved version of the transmitter, receiver and display unit was installed in one Anson, while the other served as a target aircraft. On 17 August 1937, an Avro Anson took off from Martlesham Heath for its first flight equipped with the new system. A range of 2 to 3 miles was achieved against ships off the Suffolk coast.[818]

With the aid of reference measurements using test equipment installed on the White Tower at Bawdsey Manor, it became possible to determine the optimum wavelength. It was found that a slight increase to 1.5 metres resulted in a significant improvement in sensitivity.

The results of airborne and simultaneous ground-based measurements demonstrated that the wavelength range around 1.5 metres was particularly effective. This range subsequently became the standard for radar systems in aircraft, ships and on land, and land-based systems.[819] It was not until the introduction of centimetric radar technology in 1942, with the H_2S radar system, that this wavelength range was superseded for airborne use.

A few days after the successful trials, Bowen received a call from Watson Watt, asking whether he would be interested in taking part in a naval exercise scheduled for September. This was a joint operation involving ships from both the Royal Navy and Coastal Command. The exercise was set to begin on 4 September and was to last for 48 hours. Coastal Command participated with 48 aircraft,

---

817 Bowen 1987, pp. 38–39.

818 Bowen 1987, pp. 41–42; White 2007b, pp. 10–11.

819 Bowen 1987, pp. 41–42; White 2007b, p. 11.

roughly about half were Avro Ansons. Their task was to intercept and report on the fleet as it moved through the English Channel into the North Sea.

Bowen later remarked:

> "This was an opportunity too good to be missed."[820]

During several flights over the fleet, Bowen was impressed by the strong echo signals returned from the large ships. As the Avro Anson, with Bowen on board, continued its approach to the fleet, it came under fire with practice ammunition and was intercepted by aircraft launched from the aircraft carrier HMS Courageous. Bowen later reported:

> "We were told later that she was carrying 15 Swordfish at the ready and all 15 took off to intercept us. For the very first time we saw aircraft echoes between ourselves and the sea at a range of about one mile."[821]

This was the first clear evidence that radar pulses from one aircraft could be reflected by another aircraft, providing the final proof that a radar-based defence system against both air and sea targets was feasible.

The results of the measurements attracted considerable interest from both the Army RDF group at Bawdsey and the RAF. After demonstration flights had already been carried out with representatives from both services, a flight with Henry Tizard took place on 18 October. This was Tizard's first flight in a radar-equipped aircraft. From Bowen's perspective, this flight was one of the most rewarding events of the entire development programme.

By the end of 1937, alongside the development of the Chain Home (CH) radar stations and the planning of a radar network for the defence against enemy air attacks, significant progress had also been made in the development of airborne radar capable of detecting both air and sea targets. Bowen and his team had succeeded in reducing the size and weight of the transmitters and receivers to such an extent that they could now be installed in aircraft.

---

820 Bowen 1987, p. 42.

821 Ibid., p. 43.

As a result, two main applications for airborne radar had emerged: AI (Airborne Interception) radar for fighter aircraft, and ASV (Air to Surface Vessel) radar for maritime reconnaissance aircraft tasked with locating ships.[822] In the years that followed, particularly 1938 and 1939, the focus shifted toward fitting radar equipment into aircraft of both the Royal Air Force and the Royal Navy.

Bowen later recalled:

> "By the end of 1937 our two projects, ship detection and air interception, had been graced with the inevitable acronyms – *ASV* for Air to Surface Vessel and *AI* for Air Interception. As ships were a much easier target than aircraft, we concentrated our efforts on ship detection, knowing that any improvement in the detection of ships would automatically benefit the detection of aircraft. In fact, 1938 became the *ASV* year during which we carried out the greater part of the development of ship detection equipment."[823]

The performance of the radar systems continued to improve throughout 1938. The ability to detect sea targets was significantly enhanced by the introduction of side-facing antenna arrays. On 9 May 1938, a radar image of the well-known aircraft carrier HMS Courageous was successfully obtained at a distance of 12.8 kilometres using such an array. At that time, the maximum detection range for a 10,000-ton ship was between 24 and 32 kilometres, and up to 48 kilometres for coastal targets. By the end of 1938, development of the ASV radar operating at a wavelength of 1.5 metres was largely complete, and the focus of further work shifted toward air defence using the AI radar system.[824] Bowen noted that, during this period, they saw far less of Watson Watt than in earlier years, as he had been promoted to Director of Communications Development at the Air Ministry.

His sage advice and the in-depth technical discussions that had once characterised his involvement at Bawdsey were greatly missed.[825]

822 Air Ministry 1947, Chapter 1–2; Reuter 1971, pp. 38–39.

823 Bowen 1987, p. 47.

824 Bowen 1988a, pp. 180–181.

825 Bowen 1987, p. 49.

In May 1939, a prototype of a complete Airborne Interception (AI) radar system operating at a wavelength of 1.5 metres was installed in a Fairey Battle aircraft, and the first test flight took place on 1 June.[826] Bowen recognised that the minimum range at which a pilot could detect and identify an attacking aircraft at night was a critical factor for operational success.[827]

On 20 June 1939, Winston Churchill visited Bawdsey Manor, the centre of radar development, to see the coastal defence radar systems in operation. On his return journey to London, Churchill stopped at Martlesham Heath to observe how a Fairey Battle had been equipped with AI radar.[828] Although Churchill did not have time for a flight demonstration, he witnessed a ground demonstration of the system's capabilities.[829] According to Bowen, Churchill concluded the visit with an impromptu 15-minute speech. While Bowen could not recall Churchill's exact words, he believed the famous phrase, "...we will fight them on the beaches, we will fight them in the streets..." came close to his memories. According to Bowen, there was not one person present who was not inspired and moved by his words.[830]

A few weeks later, in late June or early July, a further demonstration of the air defence system was held for Air Chief Marshal Hugh Dowding, head of RAF Fighter Command. An important outcome was that the Fairey Battle was deemed unsuitable as a night fighter and was subsequently replaced by the Bristol Blenheim.[831] On 10 July 1939, likely influenced by Churchill's visit, Professor Frederick Lindemann, Churchill's scientific adviser, attended Martlesham Heath for his own demonstration of the radar system and requested a flight in the Battle, which was arranged on that day. By the end of July 1939, Bowen was satisfied that the AI radar installation in the Bristol Blenheim was performing well. Bowen and his team looked forward to orders being placed for pro-

826 Bowen 1987, p. 67; Bowen 1988a, p. 181.

827 Bowen 1987, pp. 67–68.

828 Ibid., p. 72.

829 Ibid., p. 72.

830 Ibid., p. 73; Churchill officially spoke these words on 4 June 1940 in a speech to the House of Commons.

831 Bowen 1988a, p. 181.

totypes of both AI and ASV radars. Up to that point, six units were planned.[832]

Then, with virtually no notice, Bowen was informed in early August that 30 transmitters and power supplies had been ordered from Metro-Vickers, along with 30 receivers from Pye Radio. These components not only had to be manufactured but also installed as AI radar systems in 30 Blenheim aircraft within just 30 days, by 1 September.[833] This urgent demand from the Air Ministry came shortly after Dowding's visit. However, it was clear that such a small team could not realistically equip 30 aircraft within that timeframe. Ultimately, six AI Mk. I[834] radar units were installed in Blenheim aircraft and delivered to RAF Northolt by August, far fewer than originally planned.[835]

This initial version of the AI Mk. I radar still had fundamental limitations in terms of range and technology and remained largely experimental.[836]

7.1.2.4.4 Army and Royal Navy

At this point, we will now briefly discuss radar development in the Army and the Royal Navy. The official history of radar development in the Army was published in 1950 by Brigadier Arthur Sayer.[837] Detailed accounts of the Army's radar progress can also be found in works by Tomlin[838], while the Royal Navy's radar history is covered by Coales and Rawlinson[839], Howse[840], and Kingsley.[841]

---

832 Bowen 1987, p. 78.
833 Bowen 1988a, p. 181; Bowen 1987, p. 78; Lovell 1991, p. 12.
834 AI Mk. I: Airborne Interception Radar Mark I, the first British airborne radar system for night and poor-visibility interception, introduced in 1939–1940.
835 Lovell 1991, p. 13.
836 Ibid., p. 13.
837 Sayer 2022, pp. 237–263.
838 Tomlin 1988, pp. 284–295.
839 Coales and Rawlinson 1988, pp. 53–96.
840 Howse 1993.
841 Kingsley 1995.

National defence responsibilities were divided among the Army, the Royal Air Force, and the Royal Navy. The Army handled defence on land and in coastal waters; air defence was the remit of the Royal Air Force; and sea defence fell to the Royal Navy. Thus, the Army had a dual role: overseeing both air defence and coastal defence, including countering enemy ships approaching the coast.[842]

Initial radar experiments by Watson Watt and his team took place at Daventry and Orfordness, and later at Bawdsey on behalf of the Air Ministry. These were closely monitored by the War Office. Notably, the Army was the first to consider RDF[843] for detection purposes when W. A. S. Butement and P. E. Pollard submitted a patent in 1931 demonstrating that signals with a wavelength of 50 centimetres could reflect from boats or sheets of metal.[844] However, the Army showed little initial interest until 1936, when Watson Watt's work at Bawdsey became known. Consequently, a liaison team was established at Bawdsey in October 1936, officially named the Military Applications Department but commonly called the Army Cell.[845]Their most urgent task was to improve the accuracy of the air defence system, leading to the development of the first British Army radars: Gun Laying No. I (GL I) and Coast Defence No. I (CD I).[846]

In contrast, the Royal Navy sent only one officer to Bawdsey for six weeks in late summer 1936 to explore radiolocation possibilities for the fleet.[847] However, they refused to allow their radar research to be conducted at Bawdsey, preferring to use their own research laboratory at HM Signal School in Portsmouth. This annoyed Watson Watt, who wished for all radar research to be centralised. HM Signal School was no stranger to high-frequency technology, being a world leader in equipping ships with high-frequency direction finders.[848]

---

842 Hutchinson 2022, p. 6.

843 The code name for RDF in the army was Cuckoo.

844 Tomlin 1988, p. 284.

845 Sayer 2022, pp. 249–254; Brown 1999, pp. 58–62; Watson 2009, p. 68.

846 Hutchinson 2022, p. 6; Swords 2008, pp. 85–87; Watson 2009, p. 69.

847 Hutchinson 2022, p. 6; Coales and Rawlinson 1988, p. 55.

848 Brown 1999, p. 62; Swords 2008, p. 87.

In September 1935, the Navy set preliminary requirements for radar equipment to be used against aircraft and ships:

- Aircraft: warning on approach at 60 miles; precise location at 10 miles
- Ships: warning on approach at 10 miles; precise location at 5 miles.[849]

On 30 September 1935, HM Signal School was tasked with developing devices to meet these requirements.[850]

In the years that followed, radar development and trials continued across various frequency ranges aboard different capital ships, including cruisers and battleships, well into the war years.[851]

The Royal Navy made a key contribution to the success of the British CH air defence system by developing and supplying extremely powerful electron tubes. Originally designed for continuous-wave radio communication, these tubes also proved suitable for pulse operation and were crucial for aircraft detection. Under Bowen at Bawdsey, despite official instructions not to interfere, the operating voltage of the tubes was doubled, enabling the generation of sufficient power to detect aircraft at long range. This breakthrough, along with subsequent refinements, significantly enhanced the effectiveness of the CH system.[852]

In 1940, recognising the need for more powerful technology at shorter wavelengths and with higher transmission power in more compact radar systems to improve anti-submarine warfare, the Royal Navy awarded a contract to the University of Birmingham to develop a solution.

By the end of 1941, the Royal Navy had installed fifty of the newly developed Type 271 radar sets on escort vessels.[853] Operating in the centimetre-wave range (3,000 MHz), these systems used small antennas that could be easily installed on warships of

849 Swords 2008, p. 87; Coales und Rawlinson 1988, pp. 55–57.

850 Coales and Rawlinson 1988, p. 57.

851 Coales and Rawlinson 1988, pp. 90–93; Swords 2008, pp. 87–88; as early as 1939, the Germans also equipped some warships with operational radar equipment.

852 Hutchinson 2022, p. 6.

853 Coales and Rawlinson 1988, p. 69; Klausing and Judkins 2023, pp. 18–20.

any size.[854] This radar represented a significant advance, greatly enhancing the Royal Navy's effectiveness against German submarines.[855] The Army also adapted this design to meet some of its own requirements.[856]

7.1.2.5 Dispute between Tizard and Lindemann

After Baldwin became Prime Minister again on 7 June 1935, and Cunliffe-Lister was appointed Secretary of State for Air, Baldwin invited one of the most prominent critics of British defence policy, Winston Churchill, to join the ADRC. At that time, the establishment of a new scientific subcommittee on air defence within the CID had already been recommended. This proposal had been submitted three weeks earlier to Prime Minister Ramsay MacDonald by Frederick Lindemann through Churchill.[857] Although Lindemann played an influential role in promoting the creation of the committee, he was not formally appointed as a member of the ADRC. Churchill and Henry Tizard subsequently worked closely together within the ADRC. Churchill agreed to join the subcommittee only on the condition that he retained the freedom to criticise the actions of the Air Ministry in particular, and that Lindemann, his scientific adviser, be included in a technical sub-committee.[858] Lindemann later became a member of the CSSAD, which was soon marked by considerable disputes between him and other committee members.[859]

Owing to his close friendship with Churchill, Lindemann exercised significant personal influence not only over him but also within senior political and social circles. Through this relationship, Churchill gained early insights into emerging technological developments and operational planning well before he took office

854 Coales and Rawlinson 1988, p. 69.
855 Ibid., pp. 69–70.
856 Hutchinson 2022, p. 6.
857 Judkins 2007, p. 106.
858 Churchill 1954, p. 187; Churchill 2005, p. 135; Brown 1999, p. 55; Snow 2013, p. 27; Snow 1961b, pp. 30–31.
859 Burns 2013, pp. 529–541; Zimmerman 2013, pp. 135–155; Snow 2013, pp. 27–30; Snow 1961b, pp. 31–33.

as Prime Minister in May 1940, insights he consistently acquired through close consultation with Lindemann.[860]

Winston Churchill addressed the House of Commons on 7 June 1935:

> "The point [I said] is limited, and largely scientific in its character. It is concerned with the methods which can be invented or adopted or discovered to enable the Earth to control the Air, to enable defence from the ground to exercise control – indeed domination – upon aeroplanes high above its surface ... My experience is that in these matters, when the need is fully explained by military and political authorities, science is always able to provide something..."[861]

Churchill, who was First Lord of the Admiralty at the outbreak of the war and became Prime Minister and Minister of Defence on 10 May 1940, later reflected on these developments in his memoirs of World War II:

> "The highest priority was assigned to all forms of Radar study and application. Scientists and technicians were engaged and organised on a very large scale. Labour and material was made available to the fullest extent. Other methods of striking down the hostile bomber were sought tirelessly, and for many months to come these efforts were spurred by repeated, costly, and bloody raids upon our ports and cities..."[862]

Lindemann's appointment soon became a serious burden for the CSSAD. Fierce disputes arose between him and Tizard, which unsettled the entire committee and significantly hampered its work.[863] He regarded defence against night attacks as the main problem, rather than attacks during the day. Accordingly, Lindemann demanded that radar development be postponed in favour of infrared detection systems[864], as well as the development of parachute bombs and parachute mines intended to be dropped in the path of enemy aircraft.

860 Churchill 2005, p. 135; Churchill 1954, p. 187.
861 Churchill 2005, p. 134; Churchill 1954, p. 185.
862 Churchill 1985, p. 346.
863 Jones 2017, pp. 13–20.
864 Jones 2009, pp. 34–44; Zimmerman 2013, pp. 197–198.

His participation in the CSSAD was marked by a deep personal dislike of Tizard, which severely hindered the committee's progress. Lindemann launched sharp attacks against Tizard, claiming there was a fundamental disagreement over the urgency with which a solution to air attacks should be pursued. In his view, no substantial progress had been made since the CSSAD was established. By June 1936, relations between Tizard and Lindemann had deteriorated to such an extent that Tizard wrote the following letter on 18 June:

> "*Dear Lindemann,*
>
> No doubt you already know that as a result of your personal criticism to Winston Churchill he made a written attack on the Research Committee without taking trouble to ascertain my views first. I was obliged to answer this categorically, whereupon he followed up in Committee with other wild criticisms presumable based on information from you. Needless to say I have no objection to your discussing with him the work of our Committee – on the contrary everyone concerned would welcome this if the object were to produce fresh ideas and constructive suggestions. But if the only result is to produce ill-founded criticisms then I am bound to say that however good your ultimate motives are, the only effect of your actions is to retard progress. I would really enjoy working with you if you were ready to work as a member of a team, but if you are playing another game I don't think it is possible for us to go on collaborating without continual friction. I have told Swinton this – so you ought to know.
>
> I wish we could have settled such differences of opinion that exist in a friendly manner in our own Committee, but you have made things very difficult, if not impossible.
>
> I am writing a general statement about the policy underlying the priority attached to different items of our work. I will let you have this next week. If agreed by members it will be circulated to the C.I.D. Committee.
>
> Yours sincerely,
>
> H. T. Tizard"[865]

According to Birkenhead, after consultation with Churchill, Lindemann's reply was promptly given on 23 June:

865 Birkenhead 1961, pp. 186–187; Clark 1965, p. 139.

"*Dear Tizard,*

I thank you for your letter from the 18th June.

It is not worth while for me to attempt to discuss the relative wildness of the various statements which may or may not have been made in my absence at the meeting of the C.I.D. Committee, but I am sure that anything deriving from me was strictly accurate. If you would send me a copy of the statement of which you complain and of your "categorial" reply I would be glad to deal with the matter.

The point on which we seem to be in complete disagreement is the different urgency which we attach to our endeavours to find some method to deal with air attack. Your procedure would no doubt be excellent if we had 10 or 15 years' time. I believe that the period available is to be measured in month.

Apart from Watson-Watt's work you will scarcely claim that any appreciable advance towards a solution of the problem has been made since the Committee has been in being. You appear to be perfectly satisfied with the rate of progress. I am not.

In view of the immense importance of the question and holding the views I do as to its urgency, you will not be surprised that I have used every means at my disposal to accelerate progress and that I am determined to continue so. I am very sorry if this offends you, but the matter is too vital to justify one in refraining from action in order to salve anybody's *amour propre.*

Yours sincerely,

F. A. Lindemann"[866]

As a result of the increasingly acrimonious correspondence the tone of the debate grew more hostile and Rowe was drawn into the dispute.[867] Following these exchanges with Tizard Lindemann once again sought confrontation within the CSSAD particularly at a meeting on 15 July 1936. The central issue was whether absolute priority should be given to investigating aerial mines either suspended from balloons or dropped by parachute and infrared detection systems. The majority of the committee except Lindemann believed these experimental approaches should not be prioritised until their

866 Birkenhead 1961, pp. 187–188; Clark 1965, pp. 139–140.
867 Clark 1965, pp. 140–142.

practical effectiveness such as the correct placement of minefields was established.[868]

Although Lindemann did not fundamentally oppose radar development he continued to promote alternative air defence methods which he considered more promising. Up to July 1936 he remained sceptical about radar's operational reliability especially its limitations for night defence. Lindemann categorically rejected all mediation attempts and was so confrontational that Blackett and Hill resigned from the committee in protest on 15 July, no longer willing to tolerate his provocations and personal attacks on Tizard.[869] Subsequently the CSSAD was reorganised. Tizard remained chairman and Blackett and Hill were reinstated and Lindemann was replaced by Edward Appleton[870] a physicist noted for his work on electromagnetic wave propagation.

In 1924 Appleton experimentally confirmed the existence of an ionised upper atmosphere layer that reflects radio waves as predicted by Kennelly and Heaviside in 1902. Known as the Kennelly-Heaviside or Appleton layer it forms part of the ionosphere.[871] Appleton's work greatly advanced the understanding of long-distance radio transmission and strongly influenced radar theory. He was awarded the Nobel Prize in Physics in 1947 in recognition of these contributions.

With this change in personnel the CSSAD continued its work without further disruption and with renewed focus. Although no longer on the committee Lindemann remained influential. When Churchill became Prime Minister in 1940 Lindemann was appointed his chief scientific adviser acting as a key link between the scientific community and military leadership. In 1941 on Churchill's recommendation Lindemann was elevated to the peerage as Lord Cherwell.

868 Zimmerman 2013, pp. 198–205; Judkins 2007, p. 122.

869 Judkins 2007, pp. 122–123; Jones 2009, pp. 35–36; Snow 2013, p. 30; Snow 1961b, pp. 33–34.

870 Dobinson 2010, p. 139; Clark 1971, p. 102; Zimmerman 2013, p. 145.

871 Clark 1971, pp. 60–61.

7.1.2.5.1 Perspectives on the Dispute

C. P. Snow[872] describes the consistently controversial relationship between Tizard and Lindemann from various social and socio-political perspectives, beginning with their time studying together under Professor Nernst in Berlin and continuing through the ongoing disputes between Lindemann and members of the CSSAD, particularly Tizard and Hill. The close friendship between Churchill and Lindemann played a significant role in these conflicts.

In May 1962, Birkenhead's biography of Lindemann provoked heated discussions in England.[873] Birkenhead made it clear that he was deeply upset by Snow's harsh criticism of Lindemann's behaviour and presented a far more positive portrayal of him. David Edgerton also criticises Lindemann in his book "Warfare State. Britain, 1920–1970" and likewise strongly challenges Snow's depiction. A balanced obituary reflecting Lindemann's multi-layered personality can be found in the writings of R. V. Jones, Lindemann's former doctoral student.[874]

In an effort to objectively assess the major advances and challenges that led to the development of an integrated air defence system in the UK, both the political and technical processes must be analysed in greater depth and, in some cases, reassessed. This also applies to the well-known dispute between Tizard and Lindemann within the CSSAD. The profound conflict between two former close friends, which escalated into open hostility, was examined in the 1960 Godkin Lectures at Harvard by C. P. Snow. He published these lectures prominently in his 1961 book "Science and Government", thereby significantly shaping prevailing opinion on the conflict between Tizard and Lindemann.

Snow argued that Lindemann had little understanding of radar development and was primarily focused on promoting his ideas of

872 Weintraub 2011; Snow delivered his influential Two Cultures lecture in 1959, arguing that the humanities and literature had become culturally separated from science and technology. In 1961, he published Science and Government, based on his 1960 Godkin Lectures at Harvard, examining the relationship between science and political leadership through the example of Britain's radar development from 1934 to the early years of World War II.

873 Birkenhead 1961.

874 Jones 1962, p. 1107.

air minefields, parachute bombs, and the development of infrared detectors for target detection. This portrayal was widely adopted and cited, including in German literature. However, following the publication of Snow's work, his interpretation was criticised by several sources, including Lord Birkenhead, Lindemann's biographer, and R. V. Jones, who was both a colleague and friend of Lindemann as well as a friend of Tizard.[875] According to Jones, his own experiences with the two rarely matched Snow's descriptions.

In 1962, Snow published "A Postscript to Science and Government".[876] In this work, responding to Birkenhead's biography, the official history of the strategic bombing offensive 1939–1945, and other sources, he acknowledged that a reassessment was necessary in light of new information. Nonetheless, Snow made it clear that he essentially stood by his original statements. He noted that only the tone and presentation of the conflict as a kind of parable had been softened, while his conclusions remained unchanged.[877]

In the epilogue titled "Lindemann, Snow, and the Godkin Lectures", Russell Burns provides a detailed analysis in his book of Snow's partly misleading statements in the Godkin Lectures, considering them against the backdrop of the relationships between Lindemann, Tizard, and Churchill. Burns begins his remarks with the following statement:

> "In 1961 Dr C P Snow, physicist turned novelist, was invited by the University of Harvard to deliver the Godkin Lectures, a set of three lectures. Snow chose for his theme 'Science and Government' and the influence of certain influential scientists in the governmental decision making processes.[878] Tizard and Lindemann were selected to illustrate Snow's thesis. For his purpose Snow mentions in his Preface that his main written sources for his lectures were the Tizard papers; namely, some autobiographical notes and Tizard's diaries. Curiously, Snow does not refer to any of the very extensive collection of papers left by Lord Cherwell to Nuffield College, University of Oxford. In addition Snow states that he had talks with many of the people who

875 The biography of Tizard by Ronald W. Clark did not appear until 1965 (Clark 1965).

876 Snow 1962.

877 Ibid., pp. 1–3.

878 Snow 1961a.

> were involved in the events he describes, but omits their names, other than those of Dr A V Hill, Professor P M S Blackett, and A P Rowe, who were all supporters of Tizard. Prima facie, Snow sets out on his task by showing partiality to Tizard."[879]

Burns continues with the statement:

> "Much of Snow's set of lectures is a travesty..."[880]

Summary:

In this epilogue, Russell Burns provides a critical analysis of C. P. Snow's portrayal of the conflict between Frederick Lindemann and Henry Tizard in his 1960 Godkin Lectures. Snow used their dispute as an example of the tensions between science, politics, and military leadership, portraying Lindemann as lacking expertise in radar development and stubbornly fixated on his own ideas.

However, Burns shows that Snow's account is somewhat misleading and biased, relying mainly on Tizard's papers while overlooking important documents and perspectives from Lindemann's archives.

Burns emphasises that the relationship between Lindemann and Tizard was more complex, characterised by personal friendship, political power struggles, and differing visions for Britain's air defence. Lindemann's influence, especially through his close relationship with Winston Churchill, is acknowledged as significant, although he could be uncompromising at times.

The epilogue calls for a more nuanced view of the historical events and warns against accepting Snow's lectures as definitive truth, instead presenting them as one perspective that requires further research and critical reflection.

879 Burns 2013, p. 530.
880 Ibid., p. 530.

7.2 Up to the End of the War

7.2.1 Germany

7.2.1.1 Technical Progress and Misjudgements

When the war broke out in September 1939, the Wehrmacht was only partly prepared. In particular, the Navy took no urgent steps to equip its units with radar technology or to ensure adequate personnel training.

In July 1939, Hitler had assured the Commander-in-Chief of the Navy, Admiral Raeder, that war was not expected imminently. As a result, no measures were taken to expedite radar development and production.[881] At the war's outset, Germany possessed technically advanced radar sets, and subsequent work focused on their continuous improvement and increased output. However, these devices still fell short of requirements in terms of design, technological maturity and operational reliability. Developing electron tubes capable of transmitting at the highest possible frequencies with sufficient power, together with the corresponding receiver technology, proved to be a major challenge.[882]

At the time, Germany had no knowledge of comparable Allied developments, so there was little incentive to continue research into other wavebands, such as the 2.4 m, 80 cm, and 50 cm ranges, especially as existing systems already achieved very good performance.[883] Research into centimetric waves was likewise not pursued. Moreover, no efforts were made to develop a concept for integrating radar systems into a coordinated defence network, as was done in Britain.

Whereas Germany initially focused on offensive applications, Britain prioritised radar as part of an integrated air defence system. As the war progressed, this situation was gradually reversed, significantly shaping how each side's radar technology evolved.

Winston Churchill later remarked:

881 Gießler 1977, p. 41.
882 Reuter 1971, p. 43.
883 Gießler 1958, p. 114.

> "... The Germans would not have been surprised to hear our Radar pulses, for they had developed a technically efficient Radar system which was in some respects ahead of our own. What would have surprised them however was the extent to which we had turned our discoveries to practical effect, and woven all into our general air defence system. In this we led the world, and it was operational efficiency rather than novelty of equipment that was the British achievement."[884]

The fact that technically advanced radar equipment was already being developed and produced in Germany at the start of the war was largely due to the early research and development efforts of the NVA and the GEMA company. However, coordination between the relevant departments of the Wehrmacht and industry was limited. Research institutes were involved only in the context of specific, short-term tasks.

There was also little attention given to the development of defensive measures against enemy air attacks, as it was widely assumed that the war would be short. In contrast to Britain, where structured cooperation existed between principal organisations such as the TRE, no comparable collaborative framework was established in Germany.

Although the Air Force and Navy agreed shortly before the outbreak of war to define joint development objectives through an exchange of ideas, there was neither open nor efficient coordination in research, development, and production well into the war.[885] Germany's initial advantage in radar technology at the start of the conflict was based on systems developed during the 1930s, operating at wavelengths from several metres down to 50 centimetres.

Research into centimetric wavelengths, which had begun in the 1930s, was largely discontinued in 1942. On 27 November 1942, following the presentation of the Würzburg device, which operated at a wavelength of 50 centimetres, military authorities decided to halt all further research and development in this field. The remaining research into centimetre waves at Telefunken was also discontinued. The laboratory responsible, led by Waldemar Ilberg, was closed, and the staff were reassigned to ongoing projects.[886] This

884 Churchill 2005, p. 140; Churchill 1954, p. 195.

885 Reuter 1971, p. 36.

886 Reuter 1971, p. 197; Trenkle 1986, p. 63; Runge 1971, p. 50.

occurred despite the number of engineers and technicians involved in development at Telefunken increasing from 250 in 1933 to 1,500 by 1943. The personnel released were deployed to prepare production documentation for Wehrmacht equipment. Furthermore, the company remained convinced that its own decimetre-wave technology was far superior to the British metre-wave systems.[887]

Other German radar developments were largely shaped by the Telefunken company, particularly with the Würzburg radar family.[888] As early as 1937, Telefunken had developed a small radar, the Würzburg (FLAK) FuMG 39 T, for close-range location and identification purposes on behalf of the RLM. In March 1939, it was demonstrated for the first time to the Luftnachrichtentruppe and the Fliegertruppe, and in late autumn also to the FLAK.[889]

Mass production of the device was prepared so that initial deliveries could begin in summer 1940. The first version, later designated with the index A, was designed for one-man operation in line with its original intended use (Fig. 25).[890]

This radar operated simultaneously at a frequency of around 560 MHz with a pulse power of 8 kilowatts. The rotatable parabolic reflector had a diameter of 3 metres and was equipped with a permanently installed dipole and reflector. The detection range was between 20 and 30 kilometres, with a distance measurement accuracy of ±100 metres and an angular accuracy of approximately ±1.4 to ±1.7 degrees in both azimuth and elevation.[891]

On 9 April 1940, a first-series Würzburg radar device was demonstrated for the first time at the Air Force test centre in Rechlin. Although it was originally intended as a fire-control and target-tracking device, it was first deployed with an anti-aircraft battery near Berlin, in Falkensee. After its relocation to Frintrop, near Essen, it achieved the first confirmed night-time shootdown of an

887 Thiele 2003, pp. 35–36.

888 Godt 2003, pp. 28–36; a very good overview of the development and structure of the Würzburg radar family can be found in Bauer 1992.

889 Trenkle 1986, p. 43.

890 Trenkle 1979, p. 28, p. 200; Trenkle 1986, p. 43, p. 197; Kroge 2000, pp. 115–116; Kroge 1998, p. 139; Bauer 1992, pp. 8–11.

891 Trenkle 1979, p. 28, p. 200; Trenkle 1986, p. 43, p. 197; Holpp 2014, p. 186 (values given in the literature often differ from each other).

Fig. 25: Würzburg A[892]

enemy bomber in September 1940.[893] This marked the beginning of its use by anti-aircraft artillery in night operations and poor visibility conditions.[894]

With the help of preliminary developments from the Darmstadt device, short-term adaptations were made to make it suitable for anti-aircraft artillery, achieving satisfactory accuracy in range, lateral and height measurements. In addition to improving accuracy, urgent work was underway to develop a system that would automatically transmit the Würzburg's precise location data to the air defence command system, as until then, location results had to be relayed via telephone, causing significant delays.

After these necessary improvements were made, the Würzburg radar, along with its subsequent developments, became the stan-

892 Image credit: Stiftung Deutsches Technikmuseum Berlin, Historical Archive; I.2.060 FS 032-1-13-01.

893 Reuter 1971, p. 34.

894 Trenkle 1986, p. 44.

dard radar system for anti-aircraft artillery during the war. The Würzburg C model was based on technical developments that had begun as early as 1939 and was formally introduced towards the end of 1940. It offered improved accuracy and range and marked a significant advance over earlier versions. In the summer of 1941, the Würzburg D model followed, incorporating further enhancements in range and operational capability. Together, the Würzburg C and D models became the standard radar systems used by German anti-aircraft artillery throughout the war.[895]

In 1940, an improved version of the Würzburg A, designated Würzburg B, was also developed. It included an additional coupled infrared device intended to improve target tracking at night. However, this version did not enter production.[896]

The use of an eccentrically rotating dipole antenna, the so-called Quirl, first introduced with the Würzburg C, resulted in a significant improvement in bearing accuracy.[897] For the Würzburg C, at a distance of 10 kilometres, the error in lateral and height measurements was between ±80 and 100 metres, and the error in range was between ±80 and 120 metres. By the summer of 1941, development of the Würzburg D device had already been completed.[898] With the addition of the Emil device, the Würzburg D was able to significantly reduce the error rate in range measurements from 1942 onwards. Bearing accuracy was improved to ±0.4 or ±0.5 degrees for lateral and height measurements. At a distance of 10 kilometres, the error in lateral and height was between 60 and 80 metres, and the error in range between 25 and 50 metres.[899] The operating frequency, pulse power, and range of 20 to 30 kilometres remained unchanged.[900] Emil made it possible to feed range data directly into the transmission system of the anti-aircraft battery,

895 Trenkle 1979, p. 30; Reuter 1971, p. 34; Godt 2003, pp. 30–31.

896 Trenkle 1979, p. 30.

897 Trenkle 1979, p. 30, p. 200; Trenkle 1986, p. 44, p. 197; Holpp 2014, p. 186 (values given in the literature often differ from one another).

898 Hoffmann-Heyden 1954, pp. 48–50; Reuter 1971, p. 246.

899 Sitzungsprotokolle der Arbeitsgemeinschaft Rotterdam (AGR) n.d., AGR supplementary protocol, 1 September 1944, p. 6.

900 Trenkle 1986, p. 45, p. 197.

which reduced reaction time and improved target tracking accuracy.[901]

This technical improvement contributed significantly to the effectiveness of anti-aircraft batteries, especially during night operations or in poor visibility. The achievable range of the Würzburg D device was sufficient for use as a fire-control radar but was not suitable as a target tracking radar. To complement the Freya radar in fighter guidance, the Würzburg D was further developed into the Würzburg-Riese.[902] This upgraded version extended its range to 40 to 60 kilometres, providing enhanced target tracking capabilities.[903]

Together with the Freya radar, which was primarily used for long-range detection and tracking of incoming enemy aircraft at ranges typically up to 120 kilometres and, under optimal conditions, up to 160 kilometres, the Würzburg-Riese formed a complementary radar system. Freya provided the initial radar data used to direct fighter aircraft and coordinate anti-aircraft defences, supplying early warning information to fighter control units.

This was achieved by the Würzburg-Riese enlarging the parabolic reflector to a diameter of 7.4 metres (Fig. 26).[904] The error in range was still between 25 and 50 metres. At a distance of 10 kilometres, the error in lateral and height measurements was between ±30 and ±50 metres.[905]

A total of approximately 4,000 Würzburg radar systems of various models were produced during the war.[906]

The Würzburg radar sets A and C, as well as the Würzburg-Riese, operated at a frequency of around 560 MHz.[907] In 1942, the Mannheim radar, originally designated FuMG (Flak) 41T and soon

901 Trenkle 1986, pp. 45–46.

902 Würzburg Giant.

903 Bauer 1992, pp. 24–32; Trenkle 1986, pp. 46–49, p. 198; Müller 1992, pp. 16–22; Reuter 1971, pp. 246–247 (values given in the literature often differ from each other).

904 Trenkle 1986, pp. 46–49, p. 198.

905 Sitzungsprotokolle der Arbeitsgemeinschaft Rotterdam (AGR) n.d.; AGR Supplementary Protocol, 1 September 1944, p. 6.

906 Brandt 1968, p. 13; Holpp 2014, p. 187.

907 Kümmritz 1994, pp. 35–36.

Fig. 26: Würzburg-Riese[908]

renamed FuSE 64, was introduced, operating at approximately the same frequency.[909]

Also developed by Telefunken, the Mannheim radar was the first German radar system to feature automatic target tracking and real-time transmission of targeting data to command centres or fire control units. In addition, the system was considered for other applications, such as the remote control of missiles, following a potential modification of the high-frequency components. However, development was not completed by the end of the war.[910]

Further developments followed, such as the Knickebein[911] guidance system, a long-range radio navigation aid developed around 1940 to 1941 to assist German bombers in locating their tar-

908 Image credit: Stiftung Deutsches Technikmuseum Berlin, Historical Archive; I.2.060 FS 032-1-26-02.

909 Hoffmann-Heyden 1954, pp. 96–123; Trenkle 1986, pp. 51–54, p. 197; Müller 1992, pp. 24–27.

910 Hoffmann-Heyden 1954, p. 97; Trenkle 1986, pp. 52–53; Müller 1992, p. 33.

911 Crooked Leg.

gets. It had a range of up to 300 kilometres and was developed by the Telefunken company, based on the Lorenz blind landing system.[912]

The Freya LZ (FuMG (FLUM) 42 G (fZ)), later designated FuMG 401, was one of the most successful developments by GEMA and represented a significant advancement in air warning technology during World War II. The first series of this newly developed radar system was ceremonially handed over in November 1942.[913] With a range of up to 150 kilometres and improved accuracy, the system significantly enhanced early warning capabilities.

It was equipped with a rotating platform mount developed by the Luftschiffbau Zeppelin GmbH. However, it was designed more for stationary installation and not intended for mobile deployment.[914]

The further development of FuMG 401 led to the improved FuMG 401 C, which featured a manual Identification Friend or Foe capability.[915] This function enabled the radar to distinguish between friendly and enemy aircraft by detecting response signals from cooperative aircraft. However, the identification process was not fully automated and required manual verification by the operator. While this system marked a notable improvement, it remained less efficient than the fully automated systems introduced later in the war. Despite these limitations, the FuMG 401 C became a central component of Germany's air defence network, playing a key role in early detection and aircraft identification. The introduction of the FuMG 401 C significantly contributed to the German Air Force's integrated air defence system, improving early warning capabilities and enabling more effective coordination of interceptor responses during World War II.

Fig. 27 shows the Freya LZ/C (FuMG 401 C) by GEMA from 1942.[916]

---

912 Jones 2017, pp. 92–105; Johnson 2004, pp. 11–61; Johnson 1983, pp. 11–62; Trenkle 1986, pp. 98–99; Holpp 2004, p. 181.

913 Kroge 2000, p. 127; Kroge 1998, p. 156.

914 Kroge 2000, pp. 127–128, p. 190; Kroge 1998, pp. 156–157, p. 220; Trenkle 1986, p. 83, p. 192.

915 Trenkle 1986, p. 83

916 Ibid., p. 83, p. 192.

Fig. 27: Freya LZ/C[917]

A comparison between the Freya LZ and the British Chain Home radar shows that, while both were designed with similar fundamental objectives, they differed in how they were integrated into military structures. In particular, the Freya early warning radar, the Freya LZ (Luftziel), was designed specifically for detecting air targets, with a range of up to 150 kilometres. Unlike the British Chain Home system, which was fully embedded in a centralised command-and-control structure, the Freya LZ initially operated in a more fragmented way within the Air Force's air-defence network.

Early on, there was a demand for so-called panoramic or overview devices capable of simultaneously monitoring the entire horizon or sky dome. Initial concepts were documented by GEMA in a patent application filed in 1936/37. However, due to numerous pressing development priorities, work on the panoramic radar

917 Image credit: Stiftung Deutsches Technikmuseum Berlin, Historical Archive; I.4.048 F 050-007-02.

only began in 1939 following a direct request from military authorities.[918]

Development started with preliminary trials conducted by GEMA on a brick tower near Tremmen, west of Berlin. Between 1940 and late 1941, the first panoramic test facility, capable of 360-degree surveillance, had been installed.[919] The range of this so-called Tremmen device was initially well below 100 kilometres and reached up to 120 kilometres by 1943, thanks to technical advancements and improved components.[920]

Following the completion of broadband developments, the broadband equipment was incorporated into the panoramic radar system and categorised into three wavelength bands:

- 1.2 m – 1.9 m
- 1.9 m – 2.5 m
- 2.4 m – 4.0 m.[921]

After the initial trials had been carried out by GEMA on the tower in Tremmen, further development of the device into a series-production model was undertaken from autumn 1942 onwards by Siemens & Halske in Siemensstadt. This led to the creation of the Jagdschloß[922] radar system, which operated in two wavelength ranges:

- 1.2 m – 1.9 m
- 1.9 m – 2.5 m.[923]

The first production model of the Jagdschloß unit (FuMG 404) was delivered at the end of 1943.[924]

By April 1945, a total of 80 systems had been delivered, divided as follows:

- 18 units operating in the 1.2 m – 1.9 m range

918 Schultes 1953, pp 56–57.
919 Trenkle 1986, p. 104–105; Kroge 2000, p. 117; Kroge 1998, p. 141.
920 Trenkle 1986, p. 194.
921 Schultes 1953, p. 57.
922 Godt 2003, pp. 42–49; Trenkle 1986, pp. 104–109; Reuter 1971, pp. 245–246, p. 269.
923 Schultes 1953, p. 58.
924 Schultes 1953, p. 58; Reuter 1971, pp. 245–246; Müller 1992, p. 47.

– 62 units operating in the 1.9 m – 2.5 m range.[925]

It remains unclear how many of these were actually deployed at the front by the end of the war.

The Jagdschloß FuMG 404 had a range varying between approximately 80 and 200 kilometres, depending on conditions (Fig. 28).[926]

Fig. 28: Jagdschloß[927]

The Jagdschloß system met the requirements for wide-area airspace surveillance even when subjected to countermeasures such as Window and jamming. Although it was not completely immune to these forms of radar deception, it proved significantly more resilient than earlier radar systems thanks to its rotating 360-degree scanning and technical filtering methods.

A key feature of the Jagdschloß system was the image transmission, codenamed Landbriefträger.[928] This allowed the radar dis-

925 Schultes 1953, p. 58; Diehl 1953, p. 73; here, the sources in the literature sometimes contradict each other significantly.

926 Trenkle 1986, pp. 106–109, p. 194.

927 Image credit: Deutsches Museum, München, Archiv, BN16420

928 Schultes 1953, pp. 77–78; Trenkle 1986, pp. 107–108.

play to be transmitted in real time via cable or VHF radio relay to remote evaluation centres, enabling faster situational awareness and more efficient coordination within the air defence network.

Among the many different device developments, the Mammut and Wassermann systems, both derived from the Freya radar and used as long range early warning systems, should also be mentioned.[929]

The Mammut I (FuMG 41 G (fF))[930] was developed by GEMA in close cooperation with the Nachrichtenmittel-Versuchskommando (NVK), which guided the technical requirements and operational deployment. It consisted of a fixed array of eight interconnected Freya antennas arranged in a large rectangular configuration approximately 10 metres high and 25 metres wide. Lateral beam steering of up to ±50 degrees was achieved electronically using phase-shifting techniques, allowing the system to scan wide sectors without mechanical rotation. The Mammut I had an operational detection range of approximately 120 to 200 kilometres depending on altitude and environmental conditions.[931]

The Mammut II (FuMG 42 G (fF))[932], introduced in 1942, was an improved successor system featuring multiple antenna arrays and enhanced electronics. It had a redesigned antenna array measuring 11.1 metres in height and 28.5 metres in width, powered by 100 kW transmitters.[933] This configuration enabled detection of aerial targets flying at altitudes up to 8,000 metres at ranges between 200 and 300 kilometres. The directional beam could be electronically steered laterally by up to ±50 degrees. The system provided a range accuracy of approximately ±300 metres and a bearing accuracy of ±0.5 degrees. Unlike Mammut I, Mammut II was equipped with antenna arrays at both the front and rear, enabling the simultaneous tracking of both approaching and receding aircraft.

The Mammut systems produced for the Air Force were primarily installed along the western coast of occupied Europe. The

929 Reuter 1971, p. 91; Godt 2003, pp. 36–42; Trenkle 1986, pp. 91–97; Müller 1992, pp. 39–43.

930 Trenkle 1986, p. 93.

931 Trenkle 1986, p. 193.

932 Trenkle 1986, pp. 93–94, p. 193.

933 Trenkle 1986, p. 93; Müller 1992, p. 42.

Navy also operated similar systems, although in a different configuration. Air Force versions of the Mammut I were mounted on four steel lattice masts, while naval variants typically used only three. In 1941, GEMA developed several prototype systems for the Kriegsmarine for long-range air and sea surveillance on the 80-centimetre wavelength. One of these, the Mammut-Gustav (FuMG 41 G (gA))[934], had a range of 40 to 60 kilometres, comparable to that of the Würzburg-Riese. Later models, such as the Mammut-Cäsar (FuMG 41 G (cF))[935], featured antenna arrays similar to those used by the Air Force and were capable of detecting aircraft at ranges between 200 and 300 kilometres.[936] The Mammut-Cäsar was produced only in a small series for the Navy and the systems were deployed along the coast from northern Norway to southern France.[937]

Soon after the outbreak of war, GEMA was commissioned to develop a radar system that would greatly exceed the range capabilities of the existing Freya devices.[938] This led to the creation of the Wassermann L (FuMG 41)[939], L standing for leicht meaning light, and the Wassermann S (FuMG 42)[940], S standing for schwer meaning heavy.[941]

Initially, GEMA mounted four Freya antennas vertically on a rotating lattice mast approximately 36 metres high. This Wassermann L system offered an effective detection range of up to 200 kilometres with a lateral bearing accuracy of ±4 degrees. The Wassermann S system featured twice as many antenna arrays and was installed on a considerably taller mast reaching 60 metres in height. It boasted an extended effective range of up to 300 kilometres and a markedly improved lateral bearing accuracy of ±0.3 degrees.[942] A total of ten Wassermann S units were manufactured.

934 Trenkle 1986, pp. 134–135; p. 196.
935 Ibid., p. 136, p. 196.
936 Ibid., p. 193.
937 Ibid., p. 136.
938 Trenkle 1986, pp. 91–92; Müller 1992, p. 39.
939 Trenkle 1986, p. 193.
940 Ibid., p. 193.
941 Trenkle 1986, pp. 91–92; Müller 1992, p. 39.
942 Trenkle 1986, p. 92; Müller 1992, p. 39.

In October 1942, an agreement was reached between GEMA and Siemens for Siemens to take over the Wassermann project.[943] Improved versions were subsequently developed there, beginning with the first Wassermann M I (FuMG 43)[944], which was mounted on a 36 m mast, and continuing up to the final designs, Wassermann M IV and Wassermann M V, which were mounted on masts 51 m and 60 m high, respectively.[945] For the Wassermann M IV the bearing accuracy was ±0.25 degrees to the side and ± 0.75 degrees in height.[946] The system was designed for a measuring range of 300 kilometres, although even greater ranges of up to 380 kilometres could be achieved over the sea.[947]

Systems such as Jagdschloß, Mammut and Wassermann represented the culmination of German radar development within the metre and decimetre wavelength bands. By 1943, however, it became increasingly apparent that centimetre-wave technology offered significant advantages, especially for airborne and fire-control applications. In response, substantial progress was made in this field. Notably, in 1944, the Kulmbach (FuMG 74)[948] circular air warning radar and the Marbach (FuMG 76)[949] fire-control radar, both operating at approximately 3.3 GHz with a wavelength of around 9 centimetres, were developed.

Kulmbach functioned as a radar system for circular search and target designation, providing coverage of up to around 50 kilometres and detecting airborne targets at distances of 25 to 35 kilometres, with more powerful versions reaching up to 40 to 60 kilometres.[950] These targets were then passed on to the Marbach radar, which had a range of 30 to 50 kilometres and a distance-measurement accuracy of approximately ±25 – ±35 metres.[951] Combined with the Bayern[952] command vehicle, which integrated the control and display elements

943 Kroge 2000, p. 123, p. 124; Kroge 1998, p. 149, p. 152.
944 Trenkle 1986, p. 193.
945 Trenkle 1986, pp. 95–97, p. 193.
946 Trenkle 1986, p. 97, p. 193; Müller 1992, p. 39.
947 Trenkle 1986, pp. 95–97, p. 193; Müller 1992, p. 39.
948 Hoffmann-Heyden 1954, pp. 248–264; Trenkle 1986, p. 198.
949 Hoffmann-Heyden 1954, pp. 264–269; Trenkle 1986, p. 198.
950 Trenkle 1986, pp. 67–69, p. 198.
951 Trenkle 1986, p. 66, p. 198; Müller 1992, p. 32.
952 Hoffmann-Heyden 1954, p. 269; Trenkle 1986, pp. 69–70.

of both systems, they formed the Egerland system. This was an advanced and fully integrated radar-based solution for Flak fire-control.[953] Despite its technical sophistication and modular structure, only two Egerland systems were put into operation by the end of the war, which significantly limited their overall impact.[954]

Following the successful use of the Freya radar on 18 December 1939 to detect and repel British aircraft over the Deutsche Bucht, the Royal Air Force significantly reduced its daytime bomber operations and increasingly shifted its focus to night-time attacks.[955] This marked the beginning of a new field of application for German radar detection, namely in night fighter operations.[956] Further developments of the Freya and Würzburg radar systems led to initial improvements in location accuracy. At first, these ground-based radar systems were sufficient for tracking and directing fighters against small formations or individual aircraft. However, as British bombing raids grew in size and complexity, it became necessary to equip German night fighters with their own onboard radar systems. Unlike Britain, Germany had no airborne radar available for maritime reconnaissance or night fighters at the beginning of the war.

The development of airborne radar systems by German authorities only began in earnest after the outbreak of war, around December 1939. In that year, Telefunken produced an early radar unit operating at a wavelength of approximately 60 centimetres, which could have been installed in aircraft. However, in the absence of defined tactical requirements, the system was initially used only in experimental altimetric applications.[957] A few days after the outbreak of the war, a working prototype was presented to the RLM. The Technisches Amt showed little interest, arguing that existing altimeters were sufficient and that aircraft should not be burdened with additional radio equipment. This response reflected a broader lack of foresight regarding the possible applications of radar in air-

953 Hoffmann-Heyden 1954, pp. 246–248; Trenkle 1986, pp. 66–70, p. 198; Reuter 1971, p. 148; Müller 1992, p. 32.

954 Reuter 1971, p. 148.

955 Reuter 1971, pp. 82–83; Webster and Frankland 2006, p. 397.

956 Reuter 1971, p. 83.

957 Trenkle 1979, p. 84; Bekker 1980, p. 105.

craft. No consideration was given to its potential use for terrain display, or to what might have been achieved by shifting the radar beam from vertical to horizontal in order to observe the airspace ahead of the aircraft.[958]

In contrast to Germany's initial reluctance to explore airborne radar applications, developments in Great Britain progressed rapidly. Beginning in 1936, Edward Bowen and his team initiated experiments to adapt the large ground-based radar systems of the time into aircraft-compatible airborne radar. In June 1937, the first test flights took place aboard a Heyford bomber, using a radar operating in the wavelength range of 6–8 metres, with a detection range of 8–10 miles.[959] With the help of more powerful transmitter tubes and higher frequencies, radar systems with a wavelength of 1.5 metres were successfully developed, including the ASV (Air-to-Surface Vessel) and AI (Airborne Interception) radars, which would later become standard. By January 1940, tested radar sets, such as the ASV Mk. I, were available for installation on Hudson and Sunderland aircraft.[960]

The systematic development of airborne radar equipment for detecting sea and air targets began in Germany in February 1940, following several inquiries by the RLM to various companies and research institutes. Building on previously developed altimeter technology, initial test devices were created at multiple locations. However, their further development and deployment were significantly delayed due to technical and regulatory constraints. The heavy test models were difficult to install, and the Technisches Amt insisted that all antennas be housed within the aircraft fuselage. This requirement was only relaxed in early summer 1941, permitting the use of external antennas. Consequently, the first devices could not be tested until autumn that year.[961] In autumn 1941, Telefunken introduced the Lichtenstein BC radar, the first operational airborne radar for night fighters.[962]

958 Bekker 1980, p. 105.

959 Swords 2008, pp. 243–244.

960 Watts 2018a, p. 2-1; Swords 2008, pp. 245–246; Smith et al. 1985, p. 365.

961 Trenkle 1979, p. 84.

962 Holpp 2004, p. 181; Trenkle 1979, pp. 104; a good overview of Telefunken's Lichtenstein radar sets can be found in Trenkle 1979, pp. 104–116.

Although the existing laboratory model was not intended for series production, procurement was ordered through a Führer decree issued in spring 1941, and manufacturing began immediately. The first prototype was tested at Rechlin during the summer of 1941.[963] The radar, designated FuG 202[964], was ready for operational use by autumn. It operated at a frequency of 490 MHz with a peak output power of 1.5 kW. The antenna array consisted of four quadrants, each comprising four dipoles with reflectors. The system had a range of approximately 0,2–4 kilometres, with a range measurement accuracy of ±100 metres and a angular accuracy of ±2.5 degrees for both azimuth and elevation[965] Fig. 29 shows the Lichtenstein BC radar, with its characteristic antenna array installed on the nose of a Ju 88.

Fig. 29: Lichtenstein BC[966]

963 Reuter 1971, p. 88.
964 Trenkle 1986, p. 196; Bauer 2006, pp. 2–7.
965 Trenkle 1979, pp. 104–106, p. 201; Holpp 2004, p. 181.
966 Image credit: Stiftung Deutsches Technikmuseum Berlin, Historical Archive; I.2.060 FS 032-1-24-05.

Another variant, the Lichtenstein SN 2 (FuG 220)[967] radar, operated at multiple frequencies of 73, 81, and 91 MHz with a peak transmission power of 2 kW. It featured the ability to switch between these frequencies, which was crucial in countering Allied jamming techniques such as Window. This improved resistance to jamming ensured an effective detection range of approximately 5 kilometres.[968] By 1 April 1945, over 2,000 units of the Lichtenstein SN 2 had been delivered to night fighter units.[969]

In spring 1943, the Flugfunk-Forschungsinstitut Oberpfaffenhofen (FFO), in collaboration with Siemens, commenced preliminary studies for a lightweight radar system designed for single-seat night fighters. By autumn of the same year, the first prototype of this system, the Neptun V (FuG 216 V), had been completed and installed in a Focke-Wulf Fw 190.[970] This was subsequently followed by the series-produced FuG 218 V[971], which operated within the frequency range of 162 to 187 MHz and possessed a peak transmission power of 2 kW. It offered an effective detection range of approximately 5 kilometres, with a distance measurement accuracy of ±200 metres and an angular accuracy of ±2.0 degrees in both azimuth and elevation. In total, over 150 units were manufactured.[972]

---

967 Trenkle 1979, p. 201; Bauer 2006, pp. 7–18.

968 Trenkle 1979, p. 110, p. 201; Reuter 1971, p. 247; Bauer points out that there were different versions of the Lichtenstein radar set produced by Telefunken (Bauer 2006, pp. 2–18). He further explains that the British had assumed a frequency of 125 MHz based on analysis of camera images, believing that this wavelength was adequately covered by existing jammers designed against the Freya radar, which they assumed operated at the same frequency. However, this assumption was misleading as it did not take into account the presence of extension coils, meaning the radar remained undisturbed for almost nine months, effectively an eternity during the war. This misjudgement came to light following a navigational error by a German pilot, who accidentally landed his Ju 88 in Britain on 13 July 1944. As a result, the British were able to develop effective countermeasures within ten days (Bauer 2006, p. 18).

969 Trenkle 1979, p. 201; Holpp 2004, p. 181.

970 Trenkle 1979, p. 94, p. 201; Holpp 2014, p. 189.

971 The designation FuG 218 V is widely accepted as the official number for the Neptun V radar. Some sources, such as Trenkle, also mention FuG 216 V, likely referring to an early version or alternative numbering. However, FuG 218 V remains the most commonly used designation in the literature.

972 Trenkle 1979, p. 102, p. 201.

The first airborne radar system to operate in the centimetre-wave band was developed by Telefunken in 1943 and designated Berlin N1A (FuG 240/1).[973] This radar operated at a frequency of 3.3 GHz with a transmission peak power of 15 kilowatts, enabling it to achieve a detection range of approximately 9 kilometres. The system exhibited an angular accuracy of ±1.0 degree in both azimuth and elevation. Its antenna consisted of a parabolic reflector with a diameter of 70 centimetres coupled with a rotating feed dipole.[974] The entire assembly was housed within a wooden radome. In total, 25 units of the Berlin radar were manufactured.[975]

The archival file RL 2 V / 172, dated 30 July 1944 and titled Funkmessgeräte der Luftwaffe, contains a list of all radar systems used by the Air Force, including technical specifications and images. This includes both ground-based and airborne radar systems.

Following the end of World War II, all radar development in Germany was brought to a complete halt. The country was prohibited from conducting any research or development in this field until about 1950. Thereafter, German firms were permitted to manufacture British radar systems under licence, primarily for civilian purposes such as air traffic control and maritime surveillance.

7.2.1.2 Overview of German Radar Developments

The following table provides an overview of the development of German radar systems during World War II. They are organised according to their primary function: ground-based air warning, fire-control, tracking, airborne, and ship-based radar. Each entry lists the year of development or introduction, the company responsible, the radar designation, and key performance data such as wavelength, detection range, and antenna characteristics.

The overview includes both operational systems that were widely deployed and experimental projects that remained in prototype form by the end of the war. Together, these developments illustrate the rapid technological progress achieved in Germany during

973 Ibid., p. 118, p. 201.

974 Trenkle 1979, p. 201; Reuter 1971, p. 148; Holpp 2014, pp. 189–190.

975 Holpp 2004, p. 182.

World War II, ranging from early long-wave detection systems to advanced centimetric radars that anticipated post-war designs.

Ground-Based Air Warning Radar

Year	Company	Radar Unit	Performance data
1939	GEMA	Freya	➢ λ=2,4 m ➢ Range: up to 120 kilometres
1940	GEMA / Siemens	Wassermann	➢ λ=2,4 m ➢ Range: up to 300 kilometres
1941	GEMA / Siemens / NVA	Mammut	➢ λ=2,4 m ➢ Range: up to 300 kilometres
1942	GEMA	Seetakt[1]	➢ λ=82 cm[1] ➢ Range: up to 40 kilometres[2]
1943	Siemens	Jagdschloß	➢ λ=1.2 to 2.4 m ➢ Range: 80 to 200 kilometres
1944/45	Lorenz	Jagdhaus[2]	➢ λ=1.4 to 1.8 m ➢ Range: approx. 150 kilometres
1945	Telefunken	Forsthaus Z[3] (not operational)	➢ λ=9 cm ➢ Range: ?
1945	Siemens	Jagschloß Z[3] (not operational)	➢ λ=9 cm ➢ Range: ?

1 Seetakt was originally developed as a ship-based radar but and was first deployed on the North Sea islands in 1939 for short-range air and sea target detection. From 1942, it was also installed in coastal bunkers as a ground-based air-warning radar, although it was not primarily intended as a long-range air-warning system.

2 Jagdhaus was an experimental long-range air-warning radar developed by Lorenz. Only a single prototype was built, and it never entered operational service. Sources differ on its expected range: some estimate about 150 kilometres, while others suggest it might have reached 200–300 kilometres under ideal conditions.

3 These centimetre-wave systems were developed and tested but never entered regular operational service before the end of the war.

Fire-Control Radar

Year	Company	Radar unit	Performance data
1940	Lorenz	Kurfürst	➢ λ=62.4 cm ➢ Range: 8 to 12 kilometres
1940	Telefunken	Darmstadt	➢ λ=50 cm ➢ Range: 5 to 8 kilometres
1940	Lorenz	Kurpfalz	➢ λ=62.5 cm ➢ Range: 10 to 25 kilometres
1940	Lorenz	Kurmark	➢ λ=53.5 cm ➢ Range: 25 to 40 kilometres
1940	Telefunken	Würzburg A	➢ λ=53 cm ➢ Range: up to 30 kilometres ➢ Parabolic antenna: Ø 3 metres
1941	Telefunken	Mainz	➢ λ=53.5 cm ➢ Range: 25 to 35 kilometres
1941	Telefunken	Mannheim	➢ λ=53.5 cm ➢ Range: 25 to 35 kilometres
1944/45	Telefunken	Kulmbach	➢ λ=9 cm ➢ Range: 25 to 40 kilometres ➢ Parabolic antenna: Ø 3.0 metres
1945	Telefunken	Marbach	➢ λ=9 cm ➢ Range: 30 to 50 kilometres ➢ Parabolic antenna: Ø 4.5 metres

Tracking Radar

Year	Company	Radar unit	Performance data
1941	Telefunken	Würzburg-Riese	➢ λ=53 cm ➢ Range: up to 70 kilometres ➢ Parabolic antenna: Ø 7,4 metres

Airborne Radar

Year	**Company**	**Radar unit**	**Performance data**
1942	Telefunken	Lichtenstein BC	➢ λ=63 cm ➢ Range: 0.2 to 4 kilometres
1943	Lorenz	Hohentwiel (maritime reconnais-sance)	➢ λ=54.4 cm ➢ Range: 80 kilometres against ships
1943	Telefunken	Lichtenstein SN2	➢ λ=3,3 m ➢ Range: 0.5 to 5 kilometres
1944/45	Siemens	FuG 218 V; part of the Neptun radar series	➢ λ=1,8 m ➢ Range: up to 5 kilometres
1945	Telefunken	Berlin N 1A	➢ λ=approx. 9 cm ➢ Range: 0.5 to 9 kilometres

Ship-Based Radar

Year	**Company**	**Radar unit**	**Performance data**
1939	GEMA	Seetakt	➢ λ=82 cm ➢ Range: up to 20 kilometres

Table 6: Overview of German radar developments[976]

976 See also Kümmritz 1988, pp. 210–226; Kümmritz 1994, pp. 25–46; reported values vary across different sources. Years indicate the start of operational service; prototypes or experimental systems are marked separately. Wavelength (λ) denotes the operating wavelength of the radar system, with frequencies ranging approximately from 125 MHz (2.4 m) to 3.3 GHz (9 cm). Range values indicate the maximum detection distance under ideal conditions (e.g. high-altitude targets, clear line of sight at sea level). Actual performance may vary due to terrain, weather, target characteristics, and antenna design. Special notes that apply solely to individual radar units are provided directly alongside those entries.

7.2.2 Great Britain

7.2.2.1 From Bawdsey to Malvern

Following the outbreak of war in Europe on 1 September 1939 and the subsequent British declaration of war on 3 September, the Air Ministry quickly realised that the geographical location of Bawdsey Manor posed a significant risk in the event of air raids because of its proximity to the European mainland. As a result, Robert Watson Watt contacted the rector of his alma mater, University College Dundee, Scotland, to request space for the relocation of the radar research team.

At that time, Rowe's group at Bawdsey consisted of approximately 250 personnel, including around 100 scientists.[977] On the evening of 2 September 1939, one day after Germany's invasion of Poland and just before Britain entered the war, the evacuation of Bawdsey Manor was ordered. A special train was arranged at a rural station near Bawdsey. That night, in complete darkness, the train was loaded with packed equipment and much of the heavy laboratory apparatus, and it departed the next day for Dundee.[978]

Upon arrival, however, the team discovered that the university was not adequately prepared to accommodate them. This was probably due to a misunderstanding with the university authorities. Furthermore, the autumn term had already begun, which made the lack of space for the researchers even more problematic.[979]

Following the move to Dundee, the Bawdsey Research Station (BRS) was renamed the Air Ministry Research Establishment (AMRE).[980]

As Albert Rowe later wrote:

---

977 Buderi 1997, pp. 77–78.

978 Judkins 2007, p. 209; Watson 2009, p. 78.

979 Judkins 2007, p. 209; Bowen 1987, p. 87; Buderi 1997, p. 78; Watson 2009, pp. 81–82.

980 Meanwhile, another key research establishment, the Air Defence Experimental Establishment (ADEE), was relocated to Christchurch near Bournemouth to reduce the risk of enemy bombing. It later became the Air Defence Research and Development Establishment (ADRDE), focusing on innovations in radar and air defence systems. In late 1940, John Cockcroft was appointed Chief Superintendent of the ADRDE at Christchurch, where he led the establishment's vital contributions to technologies essential for the British war effort.

> "We could hardly continue to be called Bawdsey Research Station when working in the middle of Dundee. We might, of course, have become known as the Dundee Research Station, but perhaps some prophetic sense told Headquarters that Dundee was to be but an incident and that our travels were only beginning; at any rate the new title chosen for us was the Air Ministry Research Establishment (A.M.R.E.) which effectively disguised our concern with radar. After writing of the pioneer work at Orfordness and of the great days of Bawdsey, I confess I find it difficult to write seriously of our stay in Dundee. We started badly enough..."[981]

Due to severe space shortages at the University College, Dundee, the teams working on Airborne Interception (AI) and Air to Surface Vessel (ASV) radar under the leadership of Bowen were relocated to RAF Scone, a civilian airfield near Perth about 20 miles west of Dundee.[982] The facilities at Scone proved inadequate and working conditions were entirely unsuitable for the complex radar research being conducted, particularly for AI. As a result, priority was given to ASV development.[983]

Before this relocation, the team had used Martlesham Heath, an airfield well equipped for radar work and conveniently located near Bawdsey. In contrast, Scone had originally been intended for civilian use and basic pilot training for the RAF. Bowen's team was assigned one of only two small hangars, offering minimal space for laboratory activities. It soon became clear that the airfield operator would require the premises back, further complicating the research effort.[984]

Due to the unsuitable conditions at Scone, a new location had to be found.[985] In mid-October 1939, Bowen and the airborne-radar installation group were informed that they would be relocated to St Athan in South Wales. However, this site also proved to be only partially suitable. Originally established for airframe and engine maintenance, St Athan had now been given the additional responsibility of specialising in the installation of various types of airborne radar.

981 Rowe 2015, p. 54; Godt 2003, p. 196.

982 Bowen's unit comprised about two dozen employees (Judkins 2007, p. 355; Buderi 1997, p. 78).

983 Judkins 2007, p. 355.

984 Bowen 1987, pp. 87–88; White 2007b, p. 30; Lovell 1991, pp. 12–13.

985 Bowen 1987, pp. 87–91.

The task assigned to Bowen and his team was to instruct the local personnel in the installation and testing of onboard radar equipment so that they could eventually return to their core responsibility: research and development. At St Athan, the Special Duties Flight (SDF) was formed to carry out flight trials.[986]

It was never intended that Bowen and his team remain at St Athan for any extended period. However, this was not always understood by the newly recruited scientific staff from the universities, many of whom had expected to engage in actual research and development work, something that, at least for the time being, was not possible.[987]

It also became clear that maintaining contact with headquarters in Dundee, hundreds of miles away, was extremely difficult due to long communication and postal routes. Bowen later complained about the harsh working conditions during the winter of 1939 to 1940, and he placed the blame for the situation squarely on Watson Watt and Rowe:

> "We might have got by with those canvas partitions if it were summer, but it was already November and for several months there was no source of heat whatsoever. For a large part of the time our people wore greatcoats and gloves inside this improvised laboratory space… The conditions were simply appalling – and any member of the airborne group will say that this was by far the most unpleasant and least productive part of their lives. Here was one of the most sophisticated of defence developments being introduced to the Royal Air Force, and it was being done under conditions which would have produced a riot in a prison farm. I often wonder how much better the night defence of Great Britain would have been in the latter half of 1940 if we had been provided with better facilities and greater manpower during those critical days at Perth and St Athan. It is easy to be wise after the event, but I blame it all on that precipitate flight from Bawdsey; responsibility for this must be placed firmly on the shoulders of Rowe and Watson Watt. They had the task of deciding what to do and where to go – and they so decided. As was already abundantly clear, Bawdsey and Martlesham had not been singled out for special treatment by the enemy and, once

986 Williams 2018, p. 7.
987 Bowen 1987, p. 92.

> this was obvious, I will never understand why we were not allowed to go back."[988]

By early 1940, it had become clear that Dundee was not a suitable long-term location for the dispersed radar groups. Consequently, the core operational and research teams, including Bowen's airborne-radar team were relocated to Worth Matravers in Dorset on the south coast of England, with Swanage as the nearest town.[989] This location offered a good vantage point across the English Channel.[990] The move took place in early May 1940, involving around 200 people, the majority coming directly from Dundee with a smaller contingent from St Athan. This relocation brought together the previously dispersed groups into a single site, providing suitable facilities for further radar development and flight testing.[991] Rowe later gave a rather bleak assessment of their time in Dundee:

> "It is perhaps worth while noting why the Dundee era was a failure. First, the team built up by Watson-Watt, and extended during my period of Superintendentship at Bawdsey, had been scattered, and although some brilliant men had joined us after the outbreak of war, the task of mixing them with what was left of the pre-war team was not easy. Secondly, the facilities for our work were inadequate and there was little chance of improving morale by achieving successes. To make matters worse I had risked offending my old colleagues by appointing Lewis as my deputy, though he had been with us but a few months. Living conditions in Dundee did nothing to raise our spirits and what was surely the blackest black-out in Great Britain was not made more bearable by the barricades of sandbags on the pavements.
>
> Dundee is an example of a conviction I have regarding the conduct of an applied research Establishment. If it is in the doldrums, do something big; disband it, move it elsewhere or bring in fresh blood at the top. We moved elsewhere."[992]

Simultaneously with the move, the working group's name was changed to the Ministry of Aircraft Production Research Establish-

---

988 Bowen 1987, pp. 93–94; White 2007b, p. 33.

989 Rowe 2015, pp. 56–57; pp. 59–74.

990 Rowe 2015, p. 56; Watson 2009 pp. 94–95.

991 Rowe 2015, p. 57; the figures differ greatly in individual reports.

992 Rowe 2015, pp. 57–58.

ment (MAPRE), establishing it as the RAF's central research group for radar applications. The team working on AI and ASV radars under Bowen's direction also relocated from St Athan to Worth Matravers, joining MAPRE.[993] In November 1940, the organisation was renamed again, becoming the Telecommunications Research Establishment (TRE).[994]

Within eight months, Bowen and his team had transferred from Bawdsey to Christchurch near Worth Matravers, via Dundee, Perth (Scone), and St Athan. Except for Martlesham Heath near Bawdsey, all previous airfields had proven unsuitable for their work.[995] The airfield at Christchurch was the most suitable option, although it was about 20 miles from Worth Matravers.[996] Bowen's team was tasked with installing AI radars on aircraft, including wiring and fitting instruments, training operators and maintenance personnel, and exploring new applications. However, confusion between Coastal Command and the Navy over prioritising AI and ASV radar installations doubled the team's workload. This led to delays in fitting AI radars on fighter aircraft, setbacks in developing ground control procedures, and insufficient work on tactical applications of AI radar. For instance, minimum range requirements were poorly defined.[997]

Moreover, tensions arose between Bowen and Rowe, who had succeeded Watson-Watt as Superintendent at Bawdsey in May 1938. According to Bowen, Rowe did not get along well with him or with Wilkins, both pioneers of radar development. These personal tensions were symptomatic of broader difficulties. Bowen's frustration reflects the wider challenges faced by the radar research teams during the early months of the war. Although their work laid the foundation for later successes in airborne radar, their temporary displacement and the lack of adequate facilities significantly hindered progress at a critical time.

---

993 Lovell 1991, p. xii.

994 Rowe 2015, p. 82; Watson 2009, p. 94; the TRE was renamed the Radar Research Establishment in 1953 and again in 1957 as the Royal Radar Establishment.

995 Hanbury Brown 1991, pp. 45–46; Judkins 2007, p. 353.

996 Rowe 2015, p. 56.

997 Judkins 2007, p. 353.

The delays in the radar research programme caused by the various relocations, particularly those affecting Bowen's group, posed a significant obstacle to innovation at a time when the RAF was in urgent need of AI and ASV radar systems. Nevertheless, valuable, if not yet fully satisfactory, results were achieved, especially in receiver technology during flight testing.

Rowe later recalled that by the spring of 1942, there was no indication of any planned departure from Worth Matravers. The poor working conditions previously experienced in Dundee had faded into the background, and the working facilities were steadily improving. After the small airfield at Christchurch was abandoned in favour of a nearby, better-equipped airfield, and with overall conditions favourable for continued progress, sudden news arrived: German paratroopers were reportedly preparing to attack TRE from across the Channel, prompting the Prime Minister to order an immediate evacuation. Military measures to protect staff and infrastructure were swiftly implemented.

Operation Biting[998], carried out in late February 1942, likely heightened fears of a similar German raid against Worth Matravers. Concerns intensified in April when it was reported that a German paratrooper unit had taken up positions near Cherbourg, although their intentions remained unclear. In this tense atmosphere, the Prime Minister took the threat seriously and ordered an evacuation from the vulnerable south coast.[999]

998 Operation Biting, also known as the Bruneval Raid, was a British Army commando operation during World War II. From 27 to 28 February 1942, a German position near Bruneval in France, equipped with a Würzburg A radar device, was raided, and important parts of the radar equipment were brought to England (Goodchild 2013, pp. 263–290; Jones 2009, pp. 233–249; Price 2017, pp. 68–77).

999 Jones 2017, p. 247; Hutchinson 2022, p. 12; Rowe 2015, p. 128; Rowe reports that they were informed the Prime Minister had ordered a relocation away from the south coast before the next full moon. Although Rowe was uncertain whether German paratroopers were actually poised for deployment across the Channel, he was aware that the Prime Minister took the evacuation very seriously (Rowe 2015, p. 129; Hutchinson 2022, p. 12). The ADRDE, which was also based in Christchurch, was relocated to Malvern during this period (Hutchinson 2022, p. 12, p. 137). The ADRDE's primary focus was the development of radar systems for the British Army. Following the move, the

The move to Malvern in Worcestershire began on 25 May 1942.[1000] From then on, Malvern became the principal site for the development of RAF radar systems, both ground-based and airborne, and in particular for aircraft radar used by the Fleet Air Arm (FAA) of the Royal Navy.[1001]

7.2.2.1.1 Malvern and TRE

Malvern was known for its healing springs and its long-standing tradition as a spa town. However, from 1939 onwards, it began to lose its appeal. The small amount of local industry declined, and tourists stopped visiting after excursions in the Malvern Hills. The town increasingly came to be seen as 'a sleepy place, home to schools, colleges, and old ladies'.[1002]

Against this backdrop, the TRE team was surprised by the initial resistance to their arrival from the local population. The relocation of the Telecommunications Research Establishment to Malvern took place in 1942, primarily for security reasons, as the original site at Worth Matravers on the south coast was considered vulnerable to enemy action. Therefore, in addition to the well-known advances in radar development, it is important to highlight the remarkable logistical achievements involved in the move from Worth Matravers to Malvern, as well as the complex mix of local resistance and support they encountered.

Rowe, who served as Superintendent of the TRE until 1941 and remained in senior leadership until the end of the war[1003], reported:

ADRDE was renamed the Radar Research and Development Establishment (RRDE) in 1944 (Hutchinson 2022, p. 137).

1000 Rowe 2015, p. 134, Jones 2017, p. 247; Rowe reports that a week after the move to Malvern, Swanage was bombed (Rowe 2015, p. 134). Worth Matravers was spared in this attack, as German air raids typically targeted more obvious objectives such as towns and harbours. During World War II, Swanage suffered heavily from six major air raids, which were the result of rapid coastal attacks by German aircraft approaching across the Channel.

1001 Putley 2009, p. 4.

1002 Hutchinson 2022, p. 12.

1003 Rowe served as Superintendent and thus the highest-ranking official at the TRE until 1941. In this role, he played a central part in the organisational development and coordination of the establishment during the early war

"There is no doubt that the arrival of T.R.E. was a shock to many who lived at Malvern, for the war had made little impact upon the town. A stray bomb was alleged to have fallen a few miles away but no one quite knew where. There had been no influx of industry, making heavy demands on accommodation. In our innocence we thought that there would be no great difficulty in increasing the population by about seven per cent at a time of great emergency, but we were soon disillusioned. The first problem was to find somewhere for 1,000 men and women to eat and sleep on their arrival. To this end the attractive Winter Gardens at Malvern were taken over as a temporary measure and modified to serve as a canteen... To make matters more difficult from a living point of view, the Air Defence Experimental Establishment, which was the radar Establishment serving the Army, also moved to Malvern at this time ...

The process of settling down to live in Malvern was influenced by two diametrically opposed attitudes to T.R.E. The average citizen of Malvern clearly did not want us. On the now somewhat crowded buses, the once undisturbed citizens complained that all had been well until we came. Billeting is at best a miserable business but more than usual difficulty was experienced at Malvern. Potential billetors fell ill with alarming regularity and the number of destitute aunts who were being given permanent homes in a few days' time passed all bounds of reason. Some gave shelter on the understanding that billetees were in by ten o'clock at night while others gave it on the understanding that they stayed out, somewhere, until the same hour. There were, of course, many who gladly did everything they could to provide homes for men and women who had been sent to Malvern without any choice in the matter, but it is idle to deny that feeling ran high in the beginning.

Fortunately, there were other influences at work. Never in our travels had we received aid from local authorities, but at Malvern, from the beginning to the end of our stay, the Malvern Council could not do too much for us. The friendly attitude adopted by the Council on our arrival was strengthened later by their meeting with Sir Stafford Cripps, the Minister for Aircraft Production, who, as far as security considerations allowed, told them something of our achievements and of our future tasks. Other local bodies, such as the Women's Voluntary

years. In 1941, John Cockcroft assumed the scientific leadership of the TRE, marking a stronger focus on research and innovation, while Rowe continued to hold significant management responsibilities and acted as an important link between research teams and operational commands.

> Service, gave their help in full measure at a time when we most needed it.
>
> It is pleasant to record that, as the years went by, we became an integral part of Malvern … But what we most valued, at the end of the war, were the sincere efforts of the Malvern Council to secure T.R.E. as a permanent Establishment in their town."[1004]

In addition to initial difficulties with accommodation, there were further reservations among the local population regarding the TRE staff. One issue was their public demeanour. Many of the scientific personnel were young, fit, and highly capable men and women. They wore no uniform and did not appear to be engaged in active service or contributing directly to the war effort. Consequently, they were often viewed with suspicion or even disdain by local residents whose own young people had enlisted and were risking their lives or were otherwise occupied with war-related duties. Although the scientists made a vital contribution to the war effort, their work was highly classified and could not be disclosed.[1005]

By the end of the war, around 2,000 people were employed at the TRE. One of Rowe's greatest achievements was to integrate this large and diverse group, comprising individuals with varied skills and temperaments, into a coherent and effective organisation.[1006]

Following the relocation of the AMRE from Dundee to Worth Matravers in May 1940, and the integration of the radar group working on airborne systems from St Athan, Bowen gradually withdrew from day-to-day development activities to assume broader strategic responsibilities.[1007]

Bowen played a major role in late August 1940 as part of the Tizard Mission to the United States.[1008] He helped to initiate and advance research on centimetre-wave radar at the Massachusetts Institute of Technology (MIT), based on the British cavity magnetron,

---

1004 Rowe 2015, pp. 135–136.

1005 Malvern Museum, (URL: https://malvernmuseum.co.uk/tre/); Putley 2009.

1006 Lovell 2004a, p. 1; the number of employees differs considerably in some of the sources. For example, Bill Penley goes even further with the number of employees and mentions 3,500 employees plus many service employees at the end of the war (Penley 2011).

1007 Hanbury Brown et al. 1992, p. 51.

1008 Minnett 2015, p. 1.

and fostered close cooperation between the United States and Great Britain. His efforts were crucial to the mission's success.[1009] In late May 1943, Bowen travelled to Australia, where he joined the Radiophysics Laboratory of the Council for Scientific and Industrial Research (CSIR). He became head of the laboratory in 1946, a post he held until 1971.

In 1941, the TRE adapted the AI radar to meet the needs of Bomber and Coastal Commands as well as the Fleet Air Arm. Bowen's team in Christchurch was responsible for installing AI radars in aircraft. Other projects included the Identification Friend or Foe (IFF) system and navigation beacons. These activities required an increasing number of test aircraft. The Special Duties Flight (SDF), based at St Athan, was transferred to Christchurch in April 1940. Following the establishment of the Telecommunications Flying Unit (TFU) at RAF Hurn airfield in August 1941, the SDF became the research section of the TFU. The creation of the TFU and the commissioning of RAF Hurn were closely linked, as the airfield served as the base for this specialised British unit from the outset.[1010] In addition to its military role, RAF Hurn was also a centre for the development and testing of radar technologies. Hurn was closely associated with the nearby research base at Christchurch and the work of the TRE, which made important advances in radar technology during this period.[1011]

7.2.2.1.2 Defford Airfield

Construction work on Defford Airfield[1012] began in 1940, and the airfield became operational in the autumn of 1941. The project involved an enormous effort, both in the building work and in creating temporary accommodation for RAF personnel due to arrive shortly, as well as for the large number of workers involved in the construction.[1013]

1009 Godt 2003, p. 196.
1010 From 1942 it was also used by the United States Army Air Forces (USAAF).
1011 Williams 2018, p. 8.
1012 Ibid., pp. 1–53.
1013 Hutchinson 2022, p. 15.

Defford Airfield was designed for military purposes from the outset and became a central hub for the development and testing of radar technologies during World War II. In May 1942, the TFU relocated from RAF Hurn to Defford, coinciding with the move of the TRE from Worth Matravers to Malvern. This transfer enabled closer cooperation with the TRE and marked a critical phase in the advancement of airborne radar systems.

Personnel numbers soon swelled to 2,200, including several hundred female members of the Women's Auxiliary Air Force (WAAF) and the Women's Royal Naval Service (WRNS), all part of an organisation far more complex than that of a typical RAF station. Members of the WAAF and WRNS undertook a variety of tasks ranging from administrative and support duties to specialised technical roles[1014]

The operation and maintenance of aircraft were managed by the RAF, but the installation of test equipment was overseen by civilian scientists and engineers, who often flew on the trials themselves. These boffins worked for TRE Defford, a part of the TRE organisation, which was under the control of the Ministry of Aircraft Production. In addition, many civilians from the surrounding region worked in support roles for the RAF and TRE at Defford, bringing the total number of staff at the station to nearly 3,000.[1015]

A closer examination of the role played by WAAF and WRNS personnel at RAF Defford, in coordination with the TRE in nearby Malvern, highlights the contributions women made to the highly technical and operationally complex work of radar development during the war. The significant technical and operational work at RAF Defford relied heavily on the skills and dedication of WAAF and WRNS personnel, whose efforts merit close attention. These women were part of a much larger force, with similar technical, operational, and administrative duties carried out across Great Britain and beyond.

1014 Williams 2018, p. 10.
1015 Ibid., p. 10.

7.2.2.1.3 WAAF and WRNS

During World War II, women played an important role in the armed forces of both the United Kingdom and Germany.[1016]

In Great Britain, the WAAF and the WRNS were independent military organisations with clearly defined ranks and extensive responsibilities within the Air Force and Navy.[1017] Members of these services undertook a wide range of duties, including operating radar and communications equipment, performing technical maintenance, providing administrative support, and assisting with experimental flight trials. Their contributions were vital in advancing radar and communication technologies, which significantly supported the Allied war effort.

The WAAF was established in 1939 to recruit women into supporting and technical roles, thereby freeing men for front-line combat duties. Key responsibilities included operating and monitoring radar and communications equipment, air traffic control, administrative and logistical support, and technical maintenance of equipment and machinery.

Dame Katherine Jane Trefusis Forbes[1018], who served as the first Director of the Women's Auxiliary Air Force from 1939 to 1943 with the rank of Air Chief Commandant, married Robert Watson-Watt in 1966 and was thereafter known as Dame Katherine Jane Watson-Watt.

Under her leadership, the service grew rapidly and became an indispensable component of the Royal Air Force's operational efficiency, particularly in air defence and technological development.

1016 Escott 1989; Senger und Etterlin von 2021, p. 184.

1017 WAAF and WRNS personnel were active throughout Great Britain during World War II, with around 250,000 women serving in a wide range of technical, operational, and administrative roles. Many WAAF members were stationed at key intelligence centres such as Bletchley Park, where they operated code-breaking machinery and supported vital cryptanalysis efforts. At RAF Defford, these women contributed directly to the highly technical and operational work of radar development in coordination with the TRE at nearby Malvern. Their efforts were part of the broader contribution of women across Great Britain, which was crucial to the success of the Allied war effort.

1018 Stone 2008.

The WRNS served as the women's branch of the Royal Navy. Reactivated for World War II, members undertook roles such as radio and communications operations, administrative and clerical work, technical support and maintenance, and monitoring radar and detection systems. Although women in both services were generally not deployed in direct combat, their technical and organisational expertise was essential to military operations.

RAF Defford became a key centre for the development and testing of radar technologies. Several hundred women from both services served there, operating radar systems during flight trials, providing administrative and organisational support, and assisting with technical tasks that supported scientific research. Their work formed part of a sophisticated and technically demanding organisation and was indispensable to radar advancements that had a significant impact on the outcome of the war.

Beyond radar development at sites such as Defford and Malvern, women contributed to other highly classified operations. Thousands served at Bletchley Park as cryptanalysts, radio operators, clerical staff, and machine operators, playing a decisive role in breaking enemy codes and supporting Allied intelligence.

In Germany, there was no directly comparable counterpart with independent military status. Women served as auxiliaries in the Air Force, the Wehrmacht, or the Navy, primarily in supportive roles such as anti-aircraft artillery, signals, or medical services. They were mostly stationed in rear areas or on the home front and were not organised into independent units like their British counterparts.[1019] This contrast highlights the differences in organisation and recognition of female auxiliaries in the two countries while underlining the critical contribution of women to military, technological, and intelligence operations during the war.

In several other countries during World War II, women also served in various military and supporting roles.[1020]

1019 Senger und Etterlin von 2021, pp. 4–165.
1020 Ibid., pp. 166–219.

7.2.2.2 Airborne Radar

At the outset of World War II, Great Britain had already made significant progress in radar technology, particularly with ground-based stations for air defence. The next challenge was to adapt this technology for aircraft, enabling the detection and interception of enemy aircraft during night operations and adverse weather.[1021] Maritime surveillance was equally vital, safeguarding supply routes and countering enemy submarines, particularly during the Battle of the Atlantic.[1022]

Radar technology advanced rapidly during the war, notably with the development of autonomous airborne systems and the transition from metre-wave to centimetre-wave radar. Early airborne radars provided a foundation but were limited in resolution and targeting precision. The invention of the cavity magnetron, producing signals at approximately 9 cm wavelength (around 3.3 GHz), marked a decisive breakthrough. This allowed high-power transmissions at high frequencies, significantly improving target detection, night operations, navigation, and long-range landmark identification.

1021 Bowen 1987, pp. 30–46.

1022 Key sources provide detailed information on ASV and AI radar systems. The "Introductory Survey of Radar, Part II" (Air Ministry, 1947) offers a non-technical overview, with Chapters I (Aircraft Interception) and II (H_2S and ASV) being most relevant. Simon Watts' important book, "Airborne Maritime Surveillance Radar: Volume 1, British ASV radars in World War II 1939–1945", gives a thorough account of ASV development; he also authored several other notable publications on ASV, and his second book briefly covers post-war developments; R. A. Smith et al.: "ASV: the detection of surface vessels by airborne radar", provides a valuable contribution to the understanding of ASV radar systems. The Air Ministry volumes on AI (Volume III, Aircraft Radio, 1956) and ASV (Volume VI, Radio in Maritime Warfare, 1954) document these systems in detail. Ian White's book, "The History of Air Intercept (AI) Radar and the British Night-Fighter 1935–1959", presents AI evolution and operational variants. The September 2016 e-Newsletter of the Defence Electronics History Society (DEHS), entitled "Introduction: H_2S and ASV: Air Publication 1093D", critically reviews the content of Vol. I, Chapter 2 of Air Publication 1093D. Further literature on H_2S can be found in the same e-newsletter published by Thompson and Air Publication 1093C. Collectively, these references illustrate the many radar variants and technical improvements introduced during World War II, which explains why the literature sometimes presents differing accounts and emphases.

The Telecommunications Research Establishment (TRE) was the central British research centre for radar, developing airborne systems for both the RAF and the FAA. The RAF conducted air combat, bombing, and reconnaissance, with RAF Coastal Command specialising in maritime patrol and anti-submarine warfare. The Fleet Air Arm (FAA) operated carrier-based and shore-based aircraft. Key developments included the Airborne Interception (AI) radar for night fighters and the Air to Surface Vessel (ASV) radar for detecting ships and submarines, both critical for Britain's air defence and the protection of Allied sea routes.[1023]

The first centimetric airborne radar, the AI Mk. VII, was introduced in November 1941 for interceptors targeting low-flying enemy aircraft.[1024] Test flights revealed strong echoes from ground objects, despite the system not being intended for ground mapping. This led to the concept of a radar specifically designed to scan the ground beneath the aircraft, resulting in the H_2S Mk. I radar. Such a system allowed Bomber Command to identify targets when operating beyond the limited range of existing radio navigation aids, which was about 350 km. The H_2S Mk. I entered service in early 1943, followed shortly by the improved H_2S Mk. II.[1025]

Centimetric radar principles were applied to maritime operations in the ASV Mk. III, deployed from January 1943 on Coastal Command Vickers Wellingtons. The nose-mounted scanner greatly enhanced detection of surface vessels and submarines.[1026] Subsequent variants included the ASV Mk. VI, introduced in January 1944, the ASV Mk. VIA with additional improvements, and the ASV Mk. VIB, intended for bulk production but entering service in May 1945; all were optimised for anti-submarine warfare. From 1943 the FAA introduced ASV Mks. XI and XIII, which were smaller radars operating in X-band.

The development of centimetric radar represented a decisive technological advance, improving targeting accuracy, operational range, and the overall effectiveness of both RAF and FAA opera-

1023 Smith et al. 1985, pp. 359–384.
1024 Air Ministry 1947; Chapter 1, para. 7, para. 52–54.
1025 Ibid., Chapter 2, para. 2.
1026 Lovell 1991, pp. 159–160.

tions. It marked a key step in the evolution of modern radar systems and played a vital role in the Allied war effort.

The general advances in radar technology provided a foundation for more specialised applications. In particular, attention soon focused on maritime operations, where Air to Surface Vessel (ASV) radar enabled aircraft to detect and engage enemy ships and submarines with greater precision.[1027]

7.2.2.2.1 Air to Surface Vessel (ASV) Radar

As early as 1938, the British Admiralty recognised the strategic potential of ASV radar for maritime reconnaissance and anti-submarine operations.[1028] Preliminary evaluations indicated that radar could provide a significant advantage in detecting surface vessels and surfaced submarines, particularly under conditions of reduced visibility.

By August 1939, as the likelihood of war became increasingly apparent, the urgency of operationalising ASV radar systems intensified. It was evident that such equipment would be essential for equipping aircraft assigned to maritime patrol and convoy escort duties. The initial deployment concept focused on using ASV radar to monitor maritime traffic and provide early warning of enemy naval movements, thereby enhancing the protection of Allied convoys.[1029]

RAF Coastal Command, RAF Fighter Command, and the Fleet Air Arm (FAA) all expressed interest in the new technology. Nevertheless, Coastal Command lagged behind Fighter Command in implementing radar equipment on its aircraft, partly due to logistical and technical challenges.[1030]

The development and limited production of the ASV Mk. I[1031] began in late August 1939. By December of that year, the first oper-

1027 Air Ministry 1947; this document provides a comprehensive overview of the various versions of ASV and AI radar systems, documenting the technical developments during the early phase of World War II.

1028 Watts 2018a; Watts 2018b.

1029 Watts 2019, p. 50.

1030 Bowen 1988a, p. 186.

1031 A thorough account of the early development and operational use of the ASV Mk. I is provided by Simon Watts, who, despite the limited coverage in ational squadrons had been equipped.[1032] As of mid-January 1940, ASV Mk. I sets had been installed on 12 Lockheed Hudsons and 12 Short Sunderlands under Coastal Command.[1033] However, the ASV Mk. I was still at an early stage of development and faced numerous operational shortcomings. These included an absence of spare parts, inadequate ground testing equipment, a lack of formalised training and documentation, and significant reliability issues. Equipment failures were frequent, reflecting the experimental nature of the system.

Nonetheless, the deployment of the ASV Mk. I provided invaluable operational experience and technical feedback. This informed the development of a significantly improved successor system, the ASV Mk. II, which entered service in late 1940.[1034]

Although both the ASV Mk. I and the AI Mk. I operated at a wavelength of approximately 1.5 metres, the two systems differed substantially in terms of antenna design, orientation, and operational purpose. While AI radar was optimised for airborne target detection and engagement, ASV was tailored to the detection of surface targets, such as ships and surfaced submarines.

The ASV Mk. II radar was taken to the United States as part of the Tizard Mission and became the first operational airborne radar in the US when installed on a US Navy PBY aircraft.[1035] It was also directly copied, with versions produced in the United States, Canada and Australia. Over the course of World War II, the ASV Mk. II became one of the most widely deployed airborne radar systems.[1036] Its enhanced performance and reliability made a significant contribution to Allied anti-submarine efforts, particularly during the Battle of the Atlantic, enabling long-range patrol aircraft

existing literature, offers a well-supported analysis of this pioneering radar system based on official records and archival sources (Watts 2019, pp. 50–61).

1032 Bowen 1988a, pp. 186–187.

1033 Watts 2019, p. 50; Bowen 1988a, p. 186; Swords 2008, p. 246.

1034 Watts 2019, p. 50; Swords 2008, p. 246; Air Ministry 1947, Chapter 2; para. 55–61.

1035 Bowen 1988b, pp. 298–300.

1036 According to Swords, more than 10,000 complete ASV Mk. II radar sets were manufactured during the war, making it one of the most widely used radar system of the conflict (Swords 2008, p. 246).

to detect and engage German U-boats before they could threaten merchant convoys.

In March 1942, trials were conducted at Ballykelly, Northern Ireland. A high-power 1.5-metre variant of the ASV Mk. II was installed in a long-range LRASV Mk. II Wellington, equipped with an experimental 100 kW transmitter. The receiver was a standard ASV Mk. II model. Reliable maximum detection ranges against a fully surfaced submarine were 14 miles at 1,500 feet, 11 miles at 1,000 feet, and 7 miles at 500 feet. Against a surfaced submarine trimmed for underwater running, the ranges were 7 miles at 1,500 feet, 6 miles at 1,000 feet, and 4 miles at 500 feet. Minimum detection ranges varied between 1 and 3 miles.[1037]

A prototype installation of ASV Mk. III[1038] in a Wellington, referred to as the approximately 10 cm ASV variant, produced significantly more impressive results. Under average weather conditions, the most effective operational altitude ranged from 500 to 800 feet, extending up to 2,000 feet in calm seas. Large convoys could be detected from as far as 40 miles away at an altitude of 500 feet.[1039]

ASV radar was of great importance for the FAA, which operated from aircraft carriers. It was essential for anti-submarine warfare, convoy protection, and anti-ship attacks using torpedo bombers. Additionally, the FAA relied on ASV for high precision navigation to enable aircraft to return safely to their carriers.

Certain aircraft, such as the Fairey Swordfish, a torpedo bomber and reconnaissance biplane, were equipped with the ASV Mk. II radar. This version featured a more compact transmitter suitable for small carrier-based aircraft.

However, only the forward facing ASV Mk. II could be installed on these aircraft, limiting its effectiveness for long-range wide-area surveillance. This limitation was subsequently addressed through the development of the Long-Range Air to Surface Vessel (LRASV) configuration, which incorporated additional broadside antenna arrays. These significantly increased the effective detection range, particularly against surfaced submarines, and for the first time enabled wide area maritime surveillance from the air.

1037 Watts 2018a, p. 7-1.

1038 Air Ministry 1947, Chapter 2, para. 3.

1039 Watts 2018a, pp. 7-1, 7-2; Swords 2008, p. 246.

It was quickly realised that a radar device operating at a much shorter wavelength would be more satisfactory. When the H_2S radar was developed for Bomber Command and also planned for Coastal Command aircraft, it was adapted and introduced as the ASV Mk. III, operating at a wavelength of 9.1 centimetres, also known as S-band.

Recognising the limitations of the metric wavelength ASV Mk. II and the bulky centimetric radars such as the ASV Mk. III for carrier operations, developers sought a lightweight radar operating at even higher frequencies around 10 GHz, i.e., 3-centimetres wavelength, which corresponds to the X-band. Such a system promised higher resolution and significantly smaller antenna and installation sizes.

In 1941, the TRE began developing an ASV radar in this centimetric wavelength band specifically for the FAA. The Fairey Barracuda, a light torpedo bomber and reconnaissance aircraft intended to replace the Swordfish, was initially selected as the platform for this new system.

Between 1941 and 1943, German U-boat attacks inflicted severe losses on Allied convoys beyond the reach of land-based air cover. Escort ships were insufficiently fast and too few in number to provide adequate protection, creating an urgent demand for airborne radars capable of detecting surfaced submarines at ranges of at least seven miles with full 360-degree azimuth coverage.

Since the Barracuda had not yet entered service, it was decided to fit the ASV Mk. XI[1040] (known experimentally as ASV.X) radar into the Fairey Swordfish for anti-submarine reconnaissance.[1041] Modifications began in July 1942, but progress was slow, and preproduction equipment was not installed until February 1943. After trials, the first operational Swordfish squadron equipped with ASV Mk. XI was ready by November 1943.

Simultaneously, six experimental X-band H_2S radar sets were installed in Lancaster bombers and used during raids on Berlin and Leipzig in late 1943.[1042] From 1944 onwards, improved versions of the H_2S Mk. II at 9 centimetres and H_2S Mk. III at 3 centimetres

---

1040 Watts 2024, pp. 3–4.

1041 Ibid., p. 3.

1042 Lovell 1991, p. xix.

(both X-band) were introduced into the Pathfinder Force and Main Force. The H_2S Mk. III eventually became the standard centimetric radar for the remainder of the war.

While ASV radar addressed the demands of maritime reconnaissance and anti-submarine warfare, parallel developments concentrated on air defence. Airborne Interception (AI) radar was designed to equip night fighters with the capability to locate and intercept enemy aircraft.

7.2.2.2.2 Airborne Interception (AI) Radar

Airborne Interception (AI) radar, developed in the late 1930s, enabled British night fighters to locate enemy aircraft in darkness or poor visibility, forming a key element of the nation's night air defence.[1043]

The AI Mk. I radar unit available at the start of the war still had the status of a prototype. In November 1939, testing began on an improved version with the designation Mk. II. Even the improved version, which was introduced at the beginning of 1940, still had considerable shortcomings in terms of production and testing quality and was ultimately only suitable for research and training purposes.[1044]

On the other hand, there was still a demand for a reduction in the minimum range for recognising enemy aircraft. At this time, RAF Fighter Command had already requested a reduction from one thousand feet to around six hundred feet, as one thousand feet was completely inadequate for visual identification at night.[1045] Despite these shortcomings, the radars showed potential for use in interceptors and encouraged the RAF, particularly Dowding, to continue the research programme.[1046]

The RAF's Coastal Command and the FAA were very interested in a radar to monitor shipping and support aircraft used to escort convoys.[1047] In addition, the detection of ships on the water

1043 Air Ministry 1956; Chapter 1–4; White 2007b; White 2007a.
1044 White 2007b, p. 38.
1045 Lovell 1991, p. 15.
1046 White 2007b, p. 38.
1047 Watts 2019, p. 50.

surface was initially easier than the use of radar in interceptors. The reason for this was that the ASV radar only presented a two-dimensional challenge, while air combat took place in three-dimensional space and ship targets tended to be stationary.[1048]

After the first versions AI Mk. I, AI Mk. II, and AI Mk. III, which still had considerable shortcomings, the AI Mk. IV, developed at RAE Farnborough with an EMI modulator, became the first series-produced radar to achieve substantial improvements. It performed exceptionally well in tests, with a maximum range of 20,000 ft against a Blenheim target and a minimum range of 500 ft and was first installed on Blenheims and shortly thereafter on Beaufighters for RAF night fighter operations in August 1940.[1049]

The fact that Rowe's deputy, Lewis, had signed a contract with EMI to investigate the minimum range problem led to great progress. The group at EMI included electronics specialists Alan Blumlein and E.L.C White, who developed an excellent transmitter modulator which, together with modifications to the receiver, not only reduced the minimum range of the AI Mk. III but also allowed a modest increase in maximum range.[1050]

However, the AI Mk. IV was initially of little use in the heavy German night raids of autumn 1940, due to lack of adequate ground control, which was still being developed.[1051] However, it soon became standard equipment for RAF night fighters and played a major role in the defence of Great Britain during the German Air Force's night bombing campaign in late 1940 and early 1941, which eventually contributed to the end of the raids. Several more advanced versions of the AI Mk IV were also produced, providing head-up displays for the pilot and options for use in single seat aircraft. However, these developments were overtaken by the rapid improvements in microwave systems, and both the Mk. V and Mk. VI were only produced and used on a limited basis.[1052] The Mk. IV,

1048 Bowen 1988a, p. 186.

1049 Hanbury Brown 1991, pp. 61–62.

1050 White 2007b, p. 59.

1051 Hanbury Brown 1991, p. 62; White 2007b, p. 95.

1052 White 2007b, pp. 91–108.

Mk. V and Mk. VI still operated at a wavelength of approximately 1.5 metres.[1053]

The Mk. VII was the first radar to operate at a frequency of 3.3 GHz, that is at a wavelength of 9.1 centimetres.[1054] It was introduced to the RAF in January 1942 for defence against low flying aircraft. Only about fifty aircraft were equipped to bridge the period during which the Mk. VIII was being finalised. The Mk. VIII entered service in December 1942.[1055]

1053 Air Ministry 1947, Chapter 1, para. 4–7.

1054 Burcham 2016, pp. 35–39.

1055 Air Ministry 1947, Chapter 1, para. 7; an American AI equipment, known as SCR.720 or AI Mk. X, was delivered to the UK and flown by the RAF. This device also used a wavelength of 9.1cm.

8 Political Factors Influencing the Development of Radar Technology

An initial comparison of Germany's and Great Britain's defence efforts reveals significant differences in their operational characteristics, particularly in radar technology. In Great Britain, early developments focused on defence as part of a comprehensive air defence strategy, before later shifting towards offensive applications with the installation of powerful radar systems in aircraft.[1056] Germany, by contrast, initially pursued an offensive approach, continuously refining individual radar devices, and only later turned to technologies aimed at countering Allied attacks.

Assessing these processes in both countries requires evaluating technological progress within the context of their respective political situations.[1057] Central to this are the views of the two key leaders, Hitler and Churchill, whose technical knowledge and therefore capacity to judge research and development shaped strategic priorities. Hitler's technological outlook remained rooted in the mindset of World War I. His preparations for the summer of 1939 envisioned a conflict similar to that war, focusing on increasing manpower for the armaments industry. The need to recruit scientists and engineers for critical wartime research went unrecognised. By contrast, at the outbreak of war, Great Britain had already identified 5,000 scientists and engineers from universities and industry as essential personnel.[1058]

1056 A good overview of British radar development up to the beginning of the war, in the context of a combined political-historical and scientific-technical perspective and taking into account British defence and appeasement policy, can be found in Kaiser 1996a, pp. 29–52, and Kaiser 1996b, pp. 145–160.

1057 Klausing 2022, pp. 31–42; Klausing 2025, pp. 48–53; Schmucker 1992, pp. 379–398.

1058 Johnson 2004, p. 9; Johnson 1983, pp. 9–10.

8.1 Germany

A state's armament efforts must be analysed in relation to government and the military, as well as science, technology, industry, and society. In the case of the Nazi state, the often-cited hierarchical structure of the totalitarian regime, with its seemingly clear chains of command-and-control processes in armaments research, requires critical scrutiny. The polycratic system, with overlapping responsibilities and power distribution on the one hand, and Hitler's absolute claim to leadership on the other, must be examined more closely. As a consequence, the frequently asserted lack of scientific concepts should also be questioned.[1059] Various interest groups and their specific tasks and objectives play a key role in this analysis. Even during the Weimar Republic, there was already a lively exchange between basic and application-oriented research and technical development in the field of armaments, far removed from disputes between different circles vying for dominance over science. In the introduction to the book "Rüstungsforschung im Nationalsozialismus", Helmut Maier, the editor, explains:

> „Überwunden werden muß auch die hierarchisierende Sicht einer Stufenfolge Staat / Militär, Industrie und Institutionen der Rüstungsforschung, die der »top-down«-Vorstellung von Steuerungsprozessen in einem totalitären Regime erliegt. Danach würde auch Rüstungsforschung »angeordnet« oder »untersagt«: je weniger an den Bedürfnissen der Wissensproduktion ausgerichtet, desto schlechter die Resultate. Ein ähnlicher Irrtum besteht hinsichtlich der Entstehung innovativer Rüstungstechnologie, wonach ohne theoretische Grundlagenforschung auch keine innovative Zweckforschung und deshalb keine »überlegene« Rüstung möglich wäre. Derlei Prämissen vernachlässigen die spätestens seit Weimar längst eingeübten Kooperationsformen zwischen Militär, Staat, Industrie und Institutionen der Rüstungsforschung, in denen Grundlagen- und Zweckforschung gleichzeitig stattfanden und in Wechselwirkung zueinander standen Die hier relevanten Funktions- und Kompetenzträger standen dabei in Tauschverhält-

1059 Recent studies on the history of arms research in Germany before and during World War II show that general statements about the respective science policy, both in terms of technological and political conditions, must be considered in a more nuanced way and, in some cases, re-evaluated depending on the specific research area and time frame (Maier 2002, pp. 7–29).

nissen, wobei pragmatisch und zielgerichtet jene Kompetenzen abgerufen wurden, die dem Erreichen des Innovationsziels am besten dienlich schienen. Diese Prozesse liefen fernab der Streitigkeiten zwischen jenen Gruppen, die um die politisch-administrative Vorherrschaft über die Wissenschaft kämpften…"[1060]

However, this realisation does little to change the considerable barriers to innovation, which, due to the existence of diverse power structures before and during the war, led to a loss of technological advantage in radar development, especially after the war had begun. A closer look at the power structures of the Nazi state reveals apparent efforts by individual authorities to win Hitler's favour. This objective was sometimes achieved through opposing and contradictory actions with fatal consequences for the promotion of technology and research.[1061] The Reichsministerium für Wissenschaft, Erziehung und Volksbildung (REM), founded on 1 May 1934 and headed by Bernhard Rust[1062], was formally responsible for managing science, education, and cultural policy in National Socialist Germany. Rust, a staunch National Socialist, led the ministry from its foundation until the end of World War II in 1945.

1060 Maier 2002, p. 13. "The hierarchical view of a sequential order of state/military, industry and defence research institutions must also be overcome, which succumbs to the »top-down« notion of control processes in a totalitarian regime. According to this view, armaments research would either be »ordered« or »forbidden«: the less it is geared to the needs of knowledge production, the worse the results. A similar misunderstanding exists with regard to the development of innovative defence technology, whereby it is assumed that there can be no innovative applied research and thus no »superior« defence without theoretical basic research. Such assumptions neglect the long-standing forms of cooperation between the military, the state, industry and research institutions, which were established at the latest during the Weimar period and in which basic research and applied research took place simultaneously and interacted with each other. The relevant functional and competent actors operated within exchange systems and pragmatically and purposefully drew on the skills that seemed most appropriate for achieving the innovation goal. These processes took place far removed from the struggles between the groups that vied with each other for political and administrative dominance over science…" (translated by the author).

1061 Handel 1999, pp. 51–52; on the power structures of the Nazi state and their contradictions see Broszat 2007, pp. 363–402.

1062 Klee 2011, p. 516; Grüttner 2004, p. 143.

Despite its important remit, however, the ministry held a comparatively weak position.[1063]

Plans to establish a Reichsakademie der Forschung[1064] in 1934 failed due to the complex power structures and internal rivalries within the Nazi regime. The Reichsforschungsrat (RFR) which was founded by decree on 16 March 1937 and initially assigned to the REM, was also unable to achieve its goal of coordinating and managing basic and applied research.[1065] General Karl Becker, Dean of the Faculty of Defence Technology at the TH Berlin and designated head of the HWA, was appointed President of the RFR by Rust.[1066] Following Becker's suicide on 8 April 1940[1067], Bernhard Rust decided to assume the office of president on 9 April. However, Rudolf Mentzel[1068], the President of the Deutsche Forschungsgemeinschaft (DFG), took on the leadership role. As a result of Becker's death, power struggles emerged in science policy aimed at realigning existing research organisations, which lasted until 1942.[1069] The REM's area of responsibility did not include aviation research. It was directly subordinate to the RLM under the direction of Hermann Göring, resulting in competitive situations. This circumstance did not lead to a focus on a unified goal, as the RLM was not prepared to relinquish departmental competencies.

A few hours after the sudden accidental death of Fritz Todt[1070] on 8 February 1942, who until then had been head of the Reichsministerium für Bewaffnung und Munition (RMfBuM)[1071] founded on 17 March 1940, Albert Speer[1072] was appointed by Hitler as Todt's successor in all offices.[1073] In the context of the constantly

1063 Frei 2013, p. 133.

1064 Hammerstein 1999, pp. 163–172; Maier 2002, pp. 14.

1065 A comprehensive overview of the first RFR (1937–1942) can be found in Flachowsky 2008, pp. 232–279 and Flachowsky 2005, pp. 207–208.

1066 Flachowsky 2008, p. 233.

1067 Suicide following accusations that he was responsible for shortages in the supply of ammunition to the troops (Flachowsky 2008, p. 267; Klee 2011, p. 35).

1068 Klee 2011, pp. 403–404.

1069 Flachowsky 2008, pp. 267–279; Flachowsky 2005, p. 208.

1070 Klee 2011, p. 627; Müller 1988, pp. 453–466.

1071 Milward 1966b, pp. 40–58; Müller 1988, pp. 466–485.

1072 Klee 2011, p. 590; Speer 2013.

1073 Moll 1997, p. 234, Doc. 143; Müller 1988, pp. 677–689.

changing military situation, Speer took the initiative to reorganise the RFR and suggested to Hitler that Göring be placed at the head of the RFR as a representative figure. On 9 June 1942, Hitler ordered the formation of the second RFR with the aim of combining all existing forces in the interests of the state and concentrating them on the goals to be achieved.[1074]

In the context of the founding decree, Hitler ordered:

> „Führende Männer der Wissenschaft sollen auf ihren Sondergebieten in Gemeinschaftsarbeit in erster Reihe die Forschung für die Kriegsführung fruchtbar gestalten."[1075]

This was intended to harmonise state and university research more closely with the requirements of the war. Göring was appointed to the office of President on the same day.[1076] However, he subsequently showed little interest in this task.[1077] The RFR was a very complex organisation with numerous specialist divisions and authorised representatives.[1078] By decree of 29 June 1943, Göring established the Planungsamt im Reichsforschungsrat[1079] as an independent institution and appointed Professor Werner Osenberg[1080], who had held the Chair of Machine Tools at the Technical University of Hanover since 1 April 1938, as its head.[1081] The primary aim of the office was to exempt the specialised personnel required for armaments research from military service or to bring them back from the front.[1082] From mid-1942, Osenberg campaigned for bet-

---

1074 A comprehensive overview of the so-called second RFR from 1942–1945 can be found in Flachowsky 2008, pp. 280–462; Flachowsky 2005, p. 208.

1075 Studt 2002, p. 187. "Leading men of science are to collaborate in their specialised fields, taking the forefront to make research fruitful for warfare."

1076 Federspiel 2002, p. 80; Studt 2002, p. 187.

1077 Federspiel 2002, pp. 80–81.

1078 Flachowsky 2008, pp. 300–374; Federspiel 2002, p. 103.

1079 Federspiel 2002, p. 86; Zierold 1968, pp. 248–250.

1080 Professor Dr Werner Osenberg studied mechanical engineering at the TH Munich and TH Dresden from 1920 and obtained his doctorate at the TH Dresden in 1927. He had been a member of the NSDAP since April 1933 and joined the SS in June of the same year (Federspiel 2002, p. 82; Grüttner 2004, p. 127; Klee 2011, p. 445).

1081 Federspiel 2002, pp. 73–74.

1082 Flachowsky 2008, pp. 485–487.

ter utilisation of the work capacity of university institutes with development contracts important to the war effort with his first memorandum, which he sent to various state and party offices. In it, he called for the exemption of researchers and scientific specialists from military service and also gave concrete figures.[1083]

Until the beginning of the war, there was parallel and sometimes uncoordinated research and development in the field of high-frequency and radar technology. The further development of radar equipment for the Navy, Army and Air Force was driven by many individual power promoters, such as Göring, Dönitz, Speer and Milch, with a wide variety of interests. For example, there were research activities in research institutes of the RLM, in the Navy and other organisations under the coordination of the RFR, as well as in industry or other research institutes and universities.[1084] Furthermore, a large number of research centres and institutions were founded with a wide variety of focal points and without coordination with industrial research. Rivalries between the individual parts of the Wehrmacht and various secrecy regulations also sometimes made cooperation between the different parts of the Wehrmacht as well as between the research institutes and industry difficult or even impossible or led to parallel research work. In addition, the lack of scientific and technical personnel became noticeable in the following period, as research in the field of new technologies in particular was no longer considered necessary due to the expected rapid end of the war. The call-up of scientific personnel also meant that continuous research work was hindered by changing staff appointments.[1085]

The political and military command in Germany recognised the overall potential of radar technology too late or failed to take advantage of the opportunities it presented. On 3 February 1940, Göring, in his capacity as Beauftragter für den Vierjahresplan, instructed the Reichswirtschaftsminister to postpone all development projects that would not yield results within a year:

1083 Ludwig 1974, p. 243.
1084 Ibid., pp. 234–241.
1085 Federspiel 2002, p. 73.

„Ministerpräsident, Berlin, den 3. Feb. 1940
Beauftragter für den Vierjahresplan
Generalfeldmarschall Göring

An den
Herrn Reichswirtschaftsminister

Der Führer legt entscheidenden Wert darauf, daß die Rüstung im Jahre 1940 zur größtmöglichen Höhe gebracht wird. Es müssen daher mit allen Mitteln alle die Vorhaben gefördert werden, die im Jahre 1940 bzw. bis zum Frühjahr 1941 zur Auswirkung kommen können. Alle anderen Programme, die sich erst später auswirken, müssen, falls es die Belegung der Wirtschaft erfordert, zu Gunsten der obigen Vorhaben zurückgestellt werden.

Größter Förderung bedürfen der Pulver- und Sprengstoffplan, die Erhöhung der Energieerzeugung und die Erhöhung der Aluminiumproduktion. Zu prüfen wäre, inwieweit die Treibstoffherstellung, der Bau der dritten Bunafabrik, das Zellwolleprogramm und das neu aufgetretene Fettprogramm zu Gunsten obiger Vorhaben zurückgestellt werden müssen.

Ich bitte um Vortrag über das Ergebnis der Prüfung.

gez. Göring“[1086]

Whether this Entwicklungsstopp-Befehl was implemented consistently can be doubted.[1087] The BHF appointed by Göring on

1086 Schramm 1961, p. 962; Suchenwirth 2012, pp. 49–54; Handel 1999, pp. 52–53; Kern 1984, pp. 138–139. "Minister President Berlin, 3 February 1940 Commissioner for the Vierjahresplan
Field Marshal Göring To the Reich Minister of Economics. The Führer attaches the greatest importance to achieving the highest possible level of armaments in 1940. Therefore, all projects that come to fruition in 1940 or by spring 1941 must be promoted. All other programmes that will only have an effect at a later date must, if the economic evidence so requires, be postponed in favour of the above-mentioned priorities. The greatest weight must be given to the powder and explosives plan, the expansion of energy production, and the increase in aluminium production. It should be examined to what extent fuel production, the construction of the third Buna factory, the cellulose programme, and the newly introduced fats programme must be postponed in favour of these priorities. I request a report on the results of this examination. Signed, Göring" (translated by the author).

1087 It should be noted that the tense military situation of Germany during World War II led to a reorganisation of the Reichsforschungsrat (RFR) in June 1942. Hermann Göring was tasked by Hitler with transforming the RFR into a

14 November 1942[1088], Plendl, did not mention this order from 1940 in his accountability report to Heinrich Himmler[1089] dated 16 December 1943 and 7 January 1944[1090] on the occasion of his resignation in December 1943. Osenberg did not mention it either. In the summer of 1942, Albert Speer had special commissions of his military district officers check whether, despite the existing ban on conducting peace planning and development, technical specialists were still dealing with objects that did not directly serve war-critical purposes. Here, too, the success remained doubtful, since plant managers tried everything to maintain their potential of engineers as fully as possible. In addition, in 1942, relatively more skilled workers and technicians were recruited to the armed forces than members of the service professions.[1091] From the beginning of the

legally independent institution. As a result of this restructuring, the RFR was no longer under the control of the Reichsministerium für Wissenschaft, Erziehung und Volksbildung (REM), but rather under the Reichsministerium für Bewaffnung und Munition (RMfBuM), headed by Albert Speer. This restructuring reflected the prioritisation of military research and development efforts and aimed to ensure that innovative technologies and weapons could be made available to the armed forces quickly and efficiently. The RFR's new status allowed for greater flexibility and independence in research, which was crucial given the ongoing war effort. Göring himself became president of the RFR on 9 June 1942 and appointed so-called Government Commissioners for critical areas of war-related research. These Government Commissioners were tasked with intensifying specialised research efforts in their respective fields and commissioning the relevant institutes and industrial companies. The goal was to avoid duplication of efforts and ensure a targeted alignment with the requirements of the war. Though there were restrictions on the awarding of contracts to industry, individual improvements were still allowed to be made to developed devices. Stopped developments, such as broadband antennas, later gave the enemy considerable advantages, as individual German radar sets could be effectively jammed (Trenkle 1986, pp. 41–42). Important companies for radar development such as GEMA, Telefunken and Lorenz continued their research and thus laid the foundation for a change in research policy in 1942 and the subsequent period, which led to a high level of development in radar research (Brown 1999, p. 109; Kroge 2000, p. 93; Kroge 1998, p. 108; Seiler 2007, pp. 114–115; Ludwig 1974, p. 244).

1088 Flachowsky 2005, pp. 203–226; Handel 1999, pp. 57–58; Seiler 2007, pp. 114–116; Reuter 1971, p. 192.

1089 Scheffler 1972, pp. 172–175; Klee 2011, p. 256.

1090 BArch, NS 19 / 2057.

1091 Ludwig 1974, p. 297.

war, there was a great shortage of personnel in the scientific and technical research institutes, as the staff there, like many others, were called up for military service. They were not given preferential deployment due to their specific expertise; instead, they were deployed as field soldiers or in production. This lack of personnel mainly affected scientists, engineers, technicians and students in the natural and engineering sciences.[1092] Added to this was the lack of expertise in technical and scientific matters on the part of those responsible in the Nazi state, which had far-reaching consequences.[1093] Immediately before the start of the war, the German defence authorities assumed that a future war would have to be fought using conventional and traditional means. Unlike in Great Britain, there were no lists of essential scientists or engineers. Although the military had planned for an increase in personnel requirements in the armaments industry, this was only with regard to the necessary production capacity.[1094]

Another problem, in addition to the lack of skilled personnel, was the aforementioned fragmentation of research and development capacities as well as the lack of coordination and prioritisation of the necessary development work.[1095] Himmler was also aware of this and recognised that the restrictions of 1940, particularly in the field of high-frequency technology, often served as a

1092 Handel 1999, pp. 53.

1093 Recent research on the self-mobilisation of science in the context of technical universities in the Third Reich analyses the close interlocking of theory and practice in the development of the Wehrstaat (Dinçkal et al. 2010; Ludwig 1974, pp. 241–244). Further studies show that the common views on the internal power struggles among the National Socialists and the failure of cooperation between the Nazi scientific organisations across departmental egoisms must be critically questioned. This also applies to the frequently emphasised assumption that the National Socialist system rejected scientific and technical research. On the contrary, universities with a scientific-technical and engineering orientation enjoyed special attention from the Nazi regime (Dinçkal et al. 2010, p. 11). As a result, the self-mobilisation of science allocated resources precisely for the regime's criminal goals (Flachowsky 2005, pp. 206–207). For the specific role of engineers in National Socialism, see König 2010, pp. 63–81 and for individual examples Hagenauer und Pabst 2014, pp. 50–57.

1094 Ludwig 1974, p. 288.

1095 BArch NS 19 / 2057, Ludwig 1974, p. 259.

convenient alibi. However, the fact that German high-frequency research was lagging behind that of Great Britain was due to other factors.[1096] Industrial companies continued to try to retain all their technical personnel, but in 1942 more technical personnel were called up for military service compared to service occupations.

Hitler's idea of warfare was still in line with the armaments of World War I. His interest was limited to traditional technology, in connection with a reluctance to utilise scientific knowledge. His disinterest in radar technology remained throughout the war. Speer later confirmed this:

> „Hitlers technischer Horizont schloß allerdings, ebenso wie sein Weltbild, seine Kunstauffassung und sein Lebensstil, mit dem Ersten Weltkrieg ab. Seine technischen Interessen waren einseitig auf die traditionellen Waffen des Heeres und der Marine ausgerichtet. Auf diesen Gebieten hatte er sich weitergebildet und seine Kenntnisse ständig vermehrt; hier schlug er des öfteren überzeugende und brauchbare Neuerungen vor. Jedoch hatte er wenig Sinn für Entwicklungen wie beispielsweise das Radarverfahren, die Konstruktion einer Atombombe, für Düsenjäger und Raketen...“[1097]

Göring also remained strongly influenced by his experiences as a successful fighter pilot in World War I and showed a great lack of understanding of the possibilities of radar technology and its applications. Although he was responsible for promoting radar development in his dual role as President of the RFR and as R.d.L., there was no significant improvement in the human and financial resources available for research in the high-frequency field.[1098] Göring largely ignored both the demands of Wolfgang Martini, General of the Communications and Signals Service and Inspec-

1096 Ludwig 1974, p. 258.

1097 Speer 2013, p. 246. "Hitlers technical horizon, however, just like his view of the world, his concept of art and his lifestyle, ended with World War I. His technical interests were narrowly focused on the traditional weapons of the army and navy. He had educated himself in these fields and constantly increased his knowledge; here he often proposed convincing and useful innovations. However, he had little appreciation of developments such as radar, the construction of an atomic bomb, jet fighters and rockets..." (translated by the author).

1098 Handel 1999, pp. 55–56.

tor of the Air Force Signals Corps and special representative for high-frequency technology, and those of industry to recall high-frequency technology specialists from the troops or not to assign them to combat missions. At a meeting with industrialists from the aviation and radar industry in early 1943, he explained:

> „Wir haben nicht zu wenig Arbeiter ..., sondern zu wenig im Gehirnkasten, die Erfindungen zu machen, die notwendig sind... Ich habe mir die Apparate oft angesehen. So überwältigend sieht solch ein Ding doch gar nicht aus; es sind lauter Drähte und noch etwas, und der ganze Apparat ist sowieso merkwürdig primitiv.“[1099]

The planned personnel requirements in the defence industry were focussed on elements of production and, in connection with this, the resulting underestimation of science and technology. Ludwig lists the following:

> „Die militärische Erfassung der Bevölkerung kannte überhaupt nur zwei Bereiche: den der Truppe und den der Produktion ....
>
> Für wissenschaftliche Forschungsinstitute gab es ebenfalls keine Ausnahmeregelung. Ohne ein erkennbares System wurden Forscher eingezogen oder, wie im Bereich der Luftfahrt, zu einem hohen Prozentsatz unmittelbar in den industriellen Produktionsprozess überführt...
>
> Wissenschaftliche Vorstadien der Kriegstechnik im Forschungs- und Entwicklungsbereich blieben jedoch unberücksichtigt, und die hochbedeutsame Frage des technischen Nachwuchses wurde gar nicht erst gestellt.“[1100]

1099 Reuter 1971, p. 196. "We don't have too few workers... but too few with brains to make the necessary inventions... I have often looked at the apparatus. Such a thing doesn't look that overwhelming; it's just wires and something else; and the whole apparatus is strangely primitive anyway."

1100 Ludwig 1974, pp. 288–289. "The military registration of the population was divided into two categories: the military and the production .... Scientific research institutes were no exception. Without any recognisable system, researchers were drafted into the military or, as in the field of aviation, a high percentage were transferred directly into the industrial production process... However, the scientific preliminary stages of war technology in the research and development area were not taken into account, and the highly significant question of the next generation of technicians was not even asked." (translated by the author).

The number of male students enrolled at technical universities fell from 12,045 to 5,984 between the summer semester of 1939 and the first war term in autumn 1939, i.e. by 51 percent. At the universities, the decline was only 33 per cent, but this was favoured by the deferral of medical students.[1101] In electrical engineering, for example, the number of diploma examinations taken by German students fell from 495 in 1936 to 124 in 1940.[1102] A statistical survey by the Nationalsozialistischer Bund Deutscher Technik (NSBDT) in 1943 spoke of a total shortage of 50,000 engineers. However, this shortage initially had little impact on the military replacement authorities or the promotion of young talent.[1103] There had been a shortage of technical and scientific personnel at the beginning of the war, as many specialists were called up for military service or deployed in production. However, this shortage only became apparent at the turn of the year 1941/42, as military research and development had previously played a subordinate role during the Blitzkriege. At the end of 1941, the Technisches Amt of the RLM arranged for the formation of development groups in industry for various branches of aviation technology in order to ensure co-operation between air armament production and research and development.

Leo Brandt, a senior engineer at Telefunken, was appointed head of the Entwicklungsgruppe Funkmeß[1104] in 1941 and in this role was attached to General Erich Fellgiebel, who gave him complete freedom of action. Fellgiebel was appointed in 1938 for the implementation of the Vierjahresplan in the field of radio communications as a whole.[1105] Brandt realised that there were only around 400 engineers, technicians and mechanics working in the German radar industry. In cooperation with the Air Force and Navy Brandt drew up a list of all radar technology tasks that appeared necessary and the engineers, designers, mechanics and support staff required for them, with a total number of 8,000 specialists. As a result of this analysis, he demanded the return of these specialists from the front

1101 Ludwig 1974, p. 276, p. 289.
1102 Ibid., p. 281, p. 290.
1103 Ibid., p. 298.
1104 Reuter 1971, p. 49, p. 194; Flachowsky 2005, p. 211.
1105 Reuter 1971, p. 49.

to industry and succeeded in convincing Martini of this plan.[1106] It seems that this request had little success at first.

Hitler's so-called Schlüsselkraft-Befehl of 19 February 1942 took the needs of the industrial war economy into account by introducing the term key manpower. However, this order applied to war production and industrial control organs, not to research and development tasks.[1107] These were only recognised as decisive in the last phase of the war. It was not until the first half of 1943 that specific recall actions were carried out for high-frequency specialists. In the May issue of the magazine "Hochfrequenztechnik und Elektroakustik"[1108] as well as in the June/July issue of "Funkschau"[1109] and again in more detail in the August/September issue[1110], a call for high-frequency specialists for the Air Force was published:

> „**Hochfrequenz-Fachkräfte für die Luftwaffe**
>
> Im Einvernehmen mit dem R.d.L. und Ob.d.L. geben wir bekannt:
>
> Die Luftwaffe hat laufenden Bedarf an: Physikern, Diplom-Ingenieuren, Ingenieuren, Technikern, Rundfunkmechanikern und Amateuren.
>
> Der Einsatz kann erfolgen als:
>
> a) Angehöriger des Ing.-Korps der Luftwaffe d. B.,
>
> b) Werkmeister d. B.,
>
> c) Reichsangestellter der Luftwaffe oder als Dienstverpflichteter,
>
> d) Soldat der Luftwaffe für technischen Dienst.
>
> Noch nicht ihrem fachlichen Können entsprechend eingesetzte Fachkräfte der HF-Technik sind unter Angabe von Anschrift oder Feldpostnummer, Beruf und Geburtstag zu melden an die Schriftleitung der FUNKSCHAU.
>
> Die Bewerber werden zunächst für ihre Aufgabe gründlichst eingewiesen (falls erforderlich Teilnahme an einem umfassenden Hochfrequenzlehrgang). Sie erfahren somit nicht nur eine große Bereicherung

1106 Reuter 1971, p. 49; Handel 1999, p. 54.
1107 Moll 1997, p. 236–237, Doc. 145; Studt 2002, p. 180.
1108 Hochfrequenztechnik und Elektroakustik 1943, p. 152.
1109 Funkschau 1943a, p. 58.
1110 Funkschau 1943b, p. 73.

ihres technischen Wissens, sondern sie dürfen auch die Genugtuung haben, einen besonderen Beitrag zum Endsieg leisten zu können.“[1111]

As a result of this recall campaign, the first 1,500 high-frequency technology specialists returned from the front in the summer of 1943.[1112] In the August/September 1943 issue of Funkschau, in addition to the technical eligibility criteria and application procedures, the call for applications also emphasised the central importance of high-frequency technology for warfare:

„Im Zeichen des totalen Krieges fragt sich jeder, ob er an der Stelle, an der er steht, die größtmögliche, seinen Fähigkeiten und seinem beruflichen Können entsprechende Leistung zu geben vermag. Der moderne Krieg wird weitgehend durch die Hochfrequenztechnik beeinflußt, und die Forschung in der Hochfrequenztechnik, die Fertigung von HF-Geräten und schließlich als ehrenvollste Aufgabe der soldatische Einsatz am HF-Gerät sind für die Kriegsführung von größter Wichtigkeit. Sie sind damit für jeden selbstverständlich, der diese Leistungen seinem Wissen und seiner fachlichen Ausbildung nach vollbringen kann.“[1113]

1111 “High-Frequency Specialists for the Air Force. In agreement with the R.d.L. and Ob.d.L., we announce: The Air Force has an ongoing need for: physicists, graduate engineers, engineers, technicians, radio mechanics, and amateurs. They can be used as: a) Members of the Air Force d. B. engineering corps b) Foremen d. B. c) Reich employees of the Air Force or as conscripts d) Air Force soldiers for technical service. Specialists in HF technology who have not yet been deployed according to their professional skills should report to the office of this magazine, stating their address or field post number, occupation, and date of birth. The applicants are first thoroughly familiarised with their tasks (if necessary, they take part in a comprehensive high-frequency training course). This not only greatly enriches their technical knowledge but also gives them the satisfaction of being able to make a special contribution to the final victory.” (translated by the author).

1112 Handel 1999, p. 54; here the sources are contradictory. While Reuter (Reuter 1971, pp. 49–50, p. 193) gives earlier dates and higher figures, Ludwig (Ludwig 1974, p. 259, p. 298) reports 1,500 recalled technical specialists and refers to Plendl’s accountability report (BArch, NS 19 / 2057), which also mentions the recall of 1,500 men. In contrast, Brandt reports that 15,000 electrical engineers, designers and mechanics were recalled from the front (Brandt 1962c, p. 69).

1113 Funkschau 1943b, p. 73. “In the context of total war, everyone asks themselves whether they are capable of giving the greatest possible performance in

In the October/December 1943 issue of Funkschau, an article appeared on the subject of "Hochfrequenztechnik im Kriege"[1114] by aviation staff engineer Alfred Six on the many possible uses of high-frequency technology in war.[1115] Directly following the article, with reference to Six's remarks, there was another appeal to all physicists, engineers, technicians and short-wave amateurs who felt called upon to volunteer as high-frequency specialists for the Air Force.

It should be noted that at the beginning of the war, amateur radio was banned, and radios were confiscated. Only a few members of the armed forces loyal to the regime received one of the few war radio permits.

In "Mitteilungen des Deutschen Amateur-Sende- und Empfangsdienstes (DASD)", a bulletin from the president of DASD, SS group leader and retired lieutenant general Ernst Sachs, dated 23 May 1941 and entitled "Schwarzsenden ist Landesverrat: Jedem Schwarzsender droht die Todesstrafe"[1116], Sachs was, however, able to obtain more than a hundred war radio licences for German radio amateurs from the High Command of the German Armed Forces.

This meant that another large potential pool of skilled labour was wasted.

As part of the recall campaign in Funkschau in August/September 1943, shortwave amateurs and DASD members with excellent theoretical and practical knowledge as well as specialised expertise were sought for the first time. In Great Britain, by contrast, radio amateurs were organised into special units, and large numbers of new operators were trained and deployed over the course of the war. A similar development took place in the USA, where since 1925

the position they occupy, in accordance with their abilities and professional skills. Modern warfare is largely influenced by high-frequency technology, and research in high-frequency technology, the production of RF devices, and finally the most honorable task, the soldierly operation of RF equipment, are of the utmost importance for military operations. These tasks are therefore self-evident for anyone who can accomplish them according to their knowledge and professional training." (translated by the author).

1114 High-frequency Technology in War.

1115 Six 1943, p. 89.

1116 "Illegal broadcasting is treason. The death penalty threatens every illegal broadcaster." (translated by the author); Sachs 1941.

radio amateurs had already been promoted on a large scale within the Army and Navy. By the time of the war, thousands were serving in the armed forces, while many more worked in the research and development of aircraft detection devices.

On 16 October 1944, Ernst Sachs, by then SS-Obergruppenführer and General of the Waffen-SS, drew urgent attention to the German situation in a letter to Heinrich Himmler, Reichsführer-SS. He referred to the Allies' elimination of German submarines with the aid of high-frequency technology and attributed the interruption of German air raids on London to British advances in this field. After voicing criticism of Esau in his role as BHF and of industry more generally, Sachs proposed measures to improve Germany's position in high-frequency research. Among other things, he suggested that the Zentralinstitut für Hochfrequenz- und Fernmeldetechnik take over the work, that cooperation be established with the laboratory of German shortwave amateurs, and that systematic interception of conversations among radio amateurs held as prisoners of war be introduced.

In a reply dated 21 October 1944, Himmler rejected the idea of using captured radio amateurs, citing the risk of sabotage. A further request from Sachs, submitted on 22 November 1944, was once again rejected by Himmler in a letter of 14 December 1944.[1117]

On 18 December 1943, the OKW, at the urging of Werner Osenberg, head of the Planungsamt im Reichsforschungsrat since June 1943[1118], initiated the repatriation of 5,000 specialists from the Wehrmacht to work in research, following a Führer order.

However, this met with considerable difficulties. Of the planned 5,000 returns, about 2,000 had taken place by July 1944 and about 3,430 specialists had actually been repatriated by November 1944, comprising scientists, technicians and craftsmen who were transferred from the Wehrmacht to arms research.[1119]

1117 BArch NS 19 / 3912; the file contains a letter from Sachs to Himmler dated 16 October 1944 on the use of high-frequency technology and a letter dated 22 November 1944 on the interception of prisoner-of-war shortwave amateurs.

1118 Federspiel 2002, p. 86.

1119 Grüttner 2001, p. 150; Federspiel 2002, p. 89, pp. 102–104; Ludwig 1974, p. 252; Handel 1999, p. 54.

At an early stage, Osenberg set up a research register of scientists and technicians at universities, institutions and Wehrmacht offices, which formed the basis for the repatriation campaign.

Prior to this, Osenberg had written four memoranda and sent them to responsible persons in science, government, and the Wehrmacht with the aim of promoting science and technology for the war effort.[1120]

In Osenberg's third memorandum of 28 December 1943, entitled "Die Kriegslage als Folge unzureichender Auslastung verfügbarer Kapazitäten der deutschen Forschung"[1121] he once again unequivocally criticised the behaviour of the regime, which did not attach any greater importance to science.[1122]

In this memorandum, under the heading "Beispiele aus dem Frontgeschehen als Auswirkung eines nicht hinreichenden Kriegseinsatzes der deutschen Forschung."[1123] he quotes statements on the subject of radio measurement technology by the head of the Bay of Biscay Destroyer Fleet, Captain at Sea Erdmenger, who reported on a meeting of the Torpedo Main Commission held in Flensburg-Mürwik on 15 December 1943:

> „Bei dem ersten Seegefecht im Kanal am 7.12.1939 verwandte der Feind noch keine Ortungsmittel, so daß dieses Gefecht durch die Versenkung mehrerer Kreuzer für uns erfolgreich verlief. Noch im Oktober 1940 konnte bei einem weiteren Gefecht im englischen Kanal die deutsche Ortungsüberlegenheit beobachtet werden. Im Frühjahr 1941 setzte dann die Überlegenheit des Feindes in der Ortungstechnik ein.
>
> Während der Engländer uns auch bei Nebel und bei Nacht auf große Entfernungen jederzeit funkmeßtechnisch erfassen kann, sind wir vollkommen blind zur See, insbesondere bei Nebel. Wir bekommen infolge der größeren Reichweite der feindlichen Peiltechnik eher Artilleriefeuer, als unsere Funkmeßgeräte den Gegner erfassen.

1120 As shortwave amateurs, for example, were forbidden to carry out any activity until then, these resources could not be utilised.

1121 BArch, R26 III / 120, pp. 1–18. The war situation as a result of insufficient utilisation of available capacities in German research. (translated by the author).

1122 Ludwig 1974, pp. 252–253.

1123 Examples from the front as a result of insufficient war efforts by German research.

> Auch unsere Funkmeßbeobachtungsgeräte, die es ermöglichen sollen, das feindliche Anpeilen festzustellen, können ebenfalls nicht im Bereich feindlicher Artillerie eingesetzt werden. Die Geräte sind außerdem in so geringen Stückzahlen vorhanden, daß von 6 Zerstörern nur einer mit einem derartigen Beobachtungsgerät ausgerüstet ist.
>
> Die Einsatzmöglichkeiten unserer Zerstörer sind infolge der Überlegenheit des Feindes in der Anpeiltechnik stark beschränkt.
>
> Diese Tatsache ist umso schwerwiegender, da die geringe Zahl der uns zur Verfügung stehenden Zerstörer sowohl für den Schutz unserer U-Boote als auch für wirksame Angriffe auf englisch-amerikanische Geleitzüge, z. B. in der Biskaya, nicht ausreicht. Mit der Herrschaft über die Biskaya steht und fällt unser ganzer U-Bootkrieg."[1124]

To support his report on the difficult situation regarding the recruitment of personnel in the research sector and the inadequate repatriation of scientists and specialists, Osenberg maintained extensive correspondence with the authorised representatives of the RFR from the end of 1943 to the beginning of 1944. The responses from those responsible made it clear that demands for the release of experts met with only limited success.[1125]

1124 BArch, R26 III / 120, Absatz III, pp. 4–5. "In the first naval engagement in the Channel on 7 December 1939, the enemy did not yet employ detection equipment, so this engagement was successful for us through the sinking of several cruisers. Even in October 1940, during another engagement in the English Channel, German superiority in detection technology could be observed. In the spring of 1941, however, the enemy began to achieve superiority in detection technology. While the British can detect us at long ranges by radar, even in fog and at night, we are completely blind at sea, particularly in fog. Due to the greater range of the enemy's targeting technology, we are more likely to receive artillery fire before our radar equipment can detect the enemy. Our radar observation devices, intended to allow us to detect enemy targeting, cannot be used within the range of enemy artillery. Furthermore, the devices exist in such small numbers that of six destroyers, only one is equipped with such an observation device. The operational possibilities of our destroyers are therefore severely limited by the enemy's superiority in targeting technology. This situation is all the more serious because the small number of destroyers at our disposal is insufficient both to protect our submarines and to carry out effective attacks on English-American convoys, for example in the Bay of Biscay. Control of the Bay of Biscay is decisive for our entire submarine campaign." (translated by the author).

1125 BArch, R 26 III / 120.

Following a conversation with Albert Speer in August 1944, Osenberg made progress in his efforts to obtain research results urgently needed for wartime decision-making. Additionally, the state research institutes were to be managed in such a way as to maximise their usefulness for the war effort. By order of Göring, a Wehrforschungsgemeinschaft (WFG) was established within the Reichsforschungsrat under Osenberg's leadership.[1126] The significance of the WFG in concentrating on wartime needs, similar to the Planungsamt, is still interpreted differently by historians today.

Werner Osenberg was arrested by the Americans in April 1945, taken to Paris, and subsequently to the United States, where he was interned until November 1947. From 1954, he resumed his post as professor at the Technical University of Hanover, holding the position until 1970.[1127]

As early as 1942, the weaknesses of German air defences became increasingly apparent in the face of British day and night attacks. Up to that point, no coordinated research activity in high-frequency or radar technology existed due to overlapping responsibilities and the distribution of power inherent in the polycratic structure of the Nazi state.[1128] Numerous secrecy regulations further hindered cooperation between individual research projects. Rivalries within the Wehrmacht led to parallel work and prevented focus on areas crucial to all branches of the armed forces.[1129]

Consequently, there was no overall scientific research strategy, and research proceeded uncoordinated across the research institutes of the RLM, universities, other institutes, and industry.[1130] One of the first and most important priorities of the Second RFR was high-frequency research, aimed at coordinating and advancing work in this field. To this end, Plendl was appointed BHF by Göring, in his capacity as President of the RFR, by decree of 14 November 1942. In this role, he reported directly to Göring. Plendl, regarded as an expert in radio navigation due to his academic train-

1126 Federspiel 2002, pp. 102–105; Ludwig 1974, pp. 262–266.

1127 Grüttner 2004, p. 127; Klee 2011, p. 445; Flachowsky 2008, Appendix I (CD), pp. 15–16.

1128 BArch-MA, RL 2 V / 64.

1129 Federspiel 2002, p. 73; Ludwig 1974, pp. 237–238.

1130 Handel 2002, p. 255.

ing and work at the DVL and the Air Force test centre in Rechlin[1131], was tasked with uniting all efforts in technology, intensifying research, and aligning it with the requirements of the war:

> "1. Aus der Fronterfahrung heraus sollen die Forschungsaufgaben gestellt werden. Für die Front sollen die wichtigsten Probleme schnell gelöst werden, damit die Ergebnisse ihr in angemessener Frist zunutze kommen.
>
> 2. Verbindung mit den Chefs des Nachrichtenverbindungswesens der Wehrmachtsteile zu halten und das Forschungsprogramm und seine Dringlichkeit mit ihnen abzustimmen.
>
> 3. Einflußnahme auf alle Hochfrequenzforschungsinstitute des Reiches und der Industrie. Vergebung von Forschungsaufträgen in meinem [Görings] Auftrag.
>
> 4. Verbindung mit den technischen Ämtern der Wehrmachtsteile zu halten, damit die Ergebnisse der Forschung schnellstens zur praktischen Auswertung gelangen".[1132]

Radar research for air and coastal defence was to be of particular importance, taking into account the experience gained at the front. The first task of the BHF was to gather all the research institutes of the Reich and industry in the field of high-frequency technology and to establish central coordination in order to transfer the research results into a practical evaluation as quickly as possible.[1133]

On 16 July 1943, the Reichsstelle für Hochfrequenzforschung (RHF), based in Berlin-Gatow, was created to manage forthcoming work, to be conducted according to principles of private-sector efficiency, against the background of a wide range of technical and

---

1131 Seiler 2007, pp. 54–55.

1132 Reuter 1971, p. 192. "1. Research tasks should be set on the basis of front-line experience. The most important problems for the front line should be solved quickly so that the results can be utilised within a reasonable period of time. 2. To maintain connection with the chiefs of Communications and Signals Service of the armed forces and to coordinate the research programme and its urgency with them. 3. Influencing all high-frequency research institutes of the Reich and industry. Awarding of research contracts on my [Göring's] behalf. 4. To maintain contact with the Technical Offices of the armed forces, so that the research results can be evaluated as quickly as possible in practice." (translated by the author).

1133 Flachowsky 2005, pp. 212–213; Reuter 1971, p. 192.

organisational tasks.[1134] For the scientific management of high-frequency research, Plendl founded 12 working groups on the most important areas of high-frequency technology.[1135] Plendl selected the most technically and organisationally suitable people to head the working groups, which also led to connections with important companies such as Telefunken.[1136] The BHF and its working groups were recognised as the authoritative body for the coordination of high-frequency research and worked together with all the relevant bodies in radar research and development. Between April 1943 and February 1945, the BHF awarded around 540 research contracts to state institutions, but also to industrial research in numerous companies, with the Air Force and Navy benefiting in particular from all three branches of the Wehrmacht.[1137] Due to the war situation and the associated need for a rapid solution, the BHF intensified three research tasks in particular and drew attention to the fact that initial successes had already been achieved:

> "1. Schutz unserer Funkmessgeräte gegen feindliche Störung durch abgeworfene Metallfolien (Düppel).
>
> 2. Erschliessung der cm-Wellen-Technik und Schaffung eines Bord-Rundsuchgerätes (ähnlich dem englischen Rotterdam-Gerät). Dazu Erarbeitung von physikalischen Grundlagen zu den Schutzmassnahmen für unsere Städte gegen Ortung durch das Rotterdam-Gerät. (vgl. Anlage 1 Bericht über die Anfliegbarkeit der Reichshauptstadt mit Rotterdam und die Abwehrmöglichkeiten).
>
> 3. Schutz der U-Boote gegen Funkmess-Ortung (Schwarzes-U-Boot)."[1138]

1134 BArch, NS 19 / 2057; Seiler 2007, pp. 116–117.

1135 BArch, NS 19 / 2057; in addition to the working groups, Plendl founded a further 15 institutes to stimulate scientific research (Handel 2002, p. 256).

1136 Flachowsky 2005, pp. 216–217.

1137 Flachowsky 2005, p. 218; CIOS 1946.

1138 BArch, NS 19 / 2057; Flachowsky 2005, pp. 218–219; "1. Protection of our radar devices against enemy interference by dropped metal foils (chaff/window). 2. Development of centimetre-wave technology and creation of an airborne search radar (similar to the British Rotterdam Device). This includes the establishment of physical principles for protective measures for our cities against detection by the Rotterdam Device (see Appendix 1: Report on the feasibility of approaching the Reich capital with the Rotterdam device and

The main focus of the research work funded by the BHF was in the field of centimetre wave technology.

Hans Plendl had to resign from his position as BHF in December 1943 after Göring demanded his resignation.[1139] The President of the Physikalisch-Technische Reichsanstalt (PTR), Abraham Esau, was appointed as his successor on 2 December.[1140]

In his accountability report of 16 December 1943 and 7 January 1944, Plendl described the situation with regard to the structure and working methods of German high-frequency research when he was appointed to the BHF on 14 November 1942 as follows (extract):

> „Die deutsche Forschungskapazität auf diesem Gebiet war der feindlichen etwa im Verhältnis von 1:10 unterlegen und ausserdem noch auf rd. 100 verschiedene, meist kleine und kleinste Laboratorien aufgeteilt. Die an sich brauchbare Forschungskapazität des Auslandes (Holland, Frankreich) war praktisch nicht ausgenutzt. Infolge dieser Zersplitterung konnten die meisten Institute und Laboratorien die wirklich grossen Probleme der Funktechnik in der Kriegsführung nicht, oder nur unvollkommen bearbeiten, da die Zahl ihrer erfahrenen Mitarbeiter, Messeinrichtungen, Räumlichkeiten usw. naturgemäß unzureichend waren. Ganz allgemein herrschte ein großer Mangel an Fachpersonal und an Hilfsmitteln, was zu einem arbeitsmässigen Brachliegen wertvollster Kräfte der Forschung geführt hatte. Eine einheitliche Ausrichtung der zahlenmässig geringen, wirklich arbeitsfähigen Forschungskräfte auf die kriegswichtigen Schwerpunktaufgaben der Zukunft fehlte weitgehend. Die von den verschiedenen Bedarfsträgern stammenden Forschungsaufträge waren ohne hinreichende Abstimmung verteilt, eine gegenseitige Information der verschiedenen Institute erfolgte kaum, sodass häufig erhebliche Doppelarbeit geleistet wurde…“[1141]

---

the possible countermeasures). 3. Protection of submarines against radar detection (Black Submarine).” (translated by the author).

1139 Seiler 2007, pp. 133–136.

1140 Hoffmann und Stutz 2003, p. 167.

1141 BArch, NS 19 / 2057; Reuter 1971, pp. 192–193. “The German research capacity in this field was inferior to the enemy’s by a ratio of about 1:10 and was also divided among around 100 different, mostly small and very small laboratories. The usable research capacity of foreign countries (Holland, France) was practically unused. As a result of this fragmentation, most institutes and laboratories were unable to address the truly significant problems of radio technology in warfare, or could only do so incompletely, as the number of ex-

Plendl also pointed out the following in his report:

> „Nach einer Aufbauzeit von nunmehr einem dreiviertel Jahr kann ich heute melden, dass die deutsche Hochfrequenzforschung in ihrem wesentlichen Teil führungsmässig erfasst ist und in dem geplanten Sinne voll eingesetzt werden kann.
>
> Da Forschungsarbeit grundsätzlich auf längere Sicht ausgerichtet ist, sind naturgemäß in einem Zeitraum von dreiviertel Jahr unmittelbare grosse Erfolge für die Kriegsführung nicht zu erwarten.“[1142]

Whether the lack of rapid success in research and in the pooling of high-frequency technology was the reason for his replacement remains open to question, as the initial effects of his efforts were already becoming noticeable at that time.[1143] Plendl's challenges in addressing pressing problems in defence and armaments research by mobilising resources and funds must also be seen within the broader state structure, with its complex functions and hierarchies. In the field of high-frequency research, however, Plendl was embedded in the key coordination committees and cooperated with research institutions, industry, government, and military bodies at the level of individual projects. In this way, he was able to address war-related problems in collaboration with various actors.[1144]

---

perienced personnel, measuring equipment, facilities, etc. was naturally insufficient. In general, there was a great shortage of specialised personnel and resources, which led to the idle status of the most valuable research forces. There was a widespread lack of unified focus among the small number of truly capable research staff on the key future tasks critical to the war effort. The research assignments originating from various stakeholders were distributed without adequate coordination, and mutual information exchange between the different institutes was rare, leading to significant duplication of work…" (translated by the author).

1142 BArch, NS 19 / 2057; "After a start-up period of a little over nine months, I am now able to report that German high-frequency research has been largely incorporated into the management structure and is being fully utilised in the intended manner. Since research work is fundamentally geared towards the longer term, major successes for the war effort are not to be expected in the space of nine months." (Translatd from German).

1143 Seiler 2007, p. 134.

1144 Flachowsky 2005, pp. 225–226.

8.2 Great Britain

British radar development was marked by a strong awareness within the political and military leadership of the threat posed by German air attacks. As early as the Weimar Republic, Germany had secretly begun to expand its air force, although this was forbidden under the Treaty of Versailles. In addition to the extensive air rearmament, the expansion of the German navy also aroused great concern in Britain. On 16 March 1935, Hitler introduced universal conscription and repealed the military provisions of the Treaty of Versailles.

On 10 November 1932, the British politician and two-time Prime Minister Stanley Baldwin, in his famous statement "The bomber will always get through," made it clear that Great Britain's island status would no longer offer protection from air raids and that future conflicts would be dominated by massive bombing.[1145]

As early as 1928, British Air Marshal Hugh Trenchard[1146], who was head of the British Air Force from 1919 to 1930, developed the so-called Trenchard Doctrine, which stated that it was strategically more advantageous in war to destroy the enemy's production, transport and communication centres than to seek a direct field battle.[1147] In his "Trenchard Memorandum on the War Object of an Air Force" of May 1928[1148], he went on to state that:

> "...Among military objectives must be included the factories in which war material (including aircraft) is made, the depots in which it is stored, the railway termini and docks at which it is loaded, or troops entrain or embark, and in general the means of communication and transportation of military *personnel* and *material*. Such objectives may be situated in centres of population in which their destruction from the Air will result in casualties also to the neighbouring civilian population, in the same way as the long-range bombardment of a defended coastal town by a naval force results also in the incidental destruction of civilian life and property. The fact that air attack may have that result is no reason for regarding the bombing as illegitimate provided

1145 See also Chapter 3.1.

1146 Orange 2011; Miller 2016.

1147 Böhm 2015, pp. 76–95.

1148 Thin 2008, Appendix 6, pp. 141–144.

all reasonable care is taken to confine the scope of the bombing to the military objective. Otherwise a belligerent would be able to secure complete immunity for his war manufactures and depots merely by locating them in a large city, which would, in effect, become *neutral* territory – a position which the opposing belligerent would never accept…"[1149]

Among other things, Trenchard's doctrine can be seen as the foundation for the later area bombing strategy of the RAF in World War II, effectively legitimising civilian casualties from the outset. Before Trenchard, his Italian contemporary Giulio Douhet[1150] had already written on the mechanisation of the military and in the years after World War I, he argued that high-altitude bombing should play a central role in the use of aircraft. Alongside Trenchard, Douhet is regarded as a pioneer of strategic bombing, and his theories (Douhetism) significantly influenced air war planning in Great Britain and the USA during World War II, ultimately laying the groundwork for area bombing of German cities.

Driven by the threat of air raids and the awareness of the country's vulnerability due to its high population density and relatively small geographical area, the government's defence policy faced sharp criticism from the opposition, particularly Winston Churchill. This criticism focused mainly on the government's appeasement policy, especially the approaches of Baldwin and Chamberlain.

This situation prompted the government and the military command to look for air defence options, which, in contrast to radar development in Germany, initially focused purely on an early warning system against air attacks. From the outset, the political and military commands pooled all the work necessary to develop an air defence system in order to avoid parallel work, long decision-making processes and the loss of valuable time due to competing authorities, as was the case in Germany.

The advantage of British radar development, which became apparent in the course of the war, had its origins in the mid-1930s.

1149 Ibid., Appendix 6, pp. 142–143.

1150 Böhm 2015, pp. 60–64; Eula 1986, pp. 94–99; Douhet and translated by Ferrari 2019; Burns 2013, p. 75.

This advantage was first demonstrated by the successful defence against German air raids by the RAF in 1940, which resulted in the failure of the invasion of Great Britain. Later, the British advanced the development of centimetre-wave radar, which enabled the precise bombing of German cities and facilities and the effective countering of German submarines. Although Germany had a technical advantage over Great Britain in the development of powerful radar equipment at the beginning of the war, this situation changed dramatically as the war progressed

In contrast to Germany, which had long been characterised by the National Socialist command's disinterest in high-frequency and radar technology as well as an unclear science policy, Great Britain managed to achieve a head start in research and development. This was made possible by a clear vision from the top political and military command, forward-looking and comprehensive cooperation with science and industry, and precise targets. Thanks to the consistent overcoming of barriers to innovation and open communication with allied partners, this lead was ultimately secured.

The results of the first elementary measurements in February 1935 were enough for the British Air Ministry to provide rapid funding against the background of the need for an early warning system against air attacks, although there was still no concrete proof of the suitability of the measurement results for locating enemy aircraft. In the period that followed, various committees were formed to utilise technical and scientific findings to defend against enemy attacks and later to develop effective on-board radar equipment for installation in aircraft. These committees were made up of personalities from a wide variety of professional and hierarchical backgrounds, who were selected as equal members from all the circles concerned.

The growing Allied advantage during the war was further accelerated by the combination of the development of powerful radar equipment with key technological inventions, such as the cavity magnetron for generating high-power centimetre waves, which was developed in February 1940 by Randall and Boot at Professor Oliphant's institute at the University of Birmingham. This made it possible to install compact radar equipment in aeroplanes with reduced installation dimensions.

A closer look at British activities in defence against enemy air attacks and in submarine warfare reveals a large number of technical developments, such as radar, as well as successful strategic cooperation with other countries, such as the USA and Canada, as part of the Tizard Mission. These cooperations involved technological developments and expanded production capacities, as is apparent in the example of the magnetron. The basis for this was an open and trusting cooperation between science, industry, politics and the military in the countries concerned.[1151] In addition to radar development for air defence and installation in aircraft, radio direction finding and, above all, the concentrated decoding of German radio traffic at Bletchley Park also played an important role.

Another element unknown to the Germans was Operational Research, which was first applied around 1937 to the statistical analysis of test flights.[1152] Although most studies of radar development in Germany and Great Britain before and during World War II emphasised technological advances, the cooperation factor played a decisive role in the success or failure of both countries.

1151 Phelps 2010.
1152 Kirby 2003; see also Chapter 9.4.

9 Other Technological and Operational Influencing Factors

Before and during World War II, radar technology was one of the most closely guarded secrets of warfare, providing a technological edge that could determine victory or defeat.

In the postwar era, its role shifted from a strictly military asset to a cornerstone of safe air and sea navigation, driving rapid advancements from the 1950s onward, often through international collaboration. This period also allowed former adversaries, such as Robert Watson-Watt, the pioneer of British radar, and Wolfgang Martini, former General of the Luftnachrichtentruppe, to exchange expertise and reflect on the shared history of wartime innovation.

One illustrative moment was a 1953 conference in Frankfurt am Main, which brought together historians and engineers to examine the evolution of German and British radar technology up to 1945.[1153] Publications that followed captured the technical breakthroughs on both sides and assessed their influence on the war's operations. British centimetre-wave radar, enabled by the cavity magnetron, became a decisive tool for navigation, target detection, and anti-submarine warfare, an example of how technological innovation could shape the battlefield.

This chapter expands the focus beyond radar, examining other key technological and operational developments that significantly influenced the course of the war, including:

- Signals Intelligence
- Jamming Methods
- Operational Research
- Intelligence services and enemy reconnaissance.

1153 This conference brought together key players such as Robert Watson-Watt and Reginald Jones on the British side, and Leo Brandt, Wolfgang Martini, Albert Kesselring, Hermann Diehl, Max Dieckmann, Abraham Esau, Hans Rukop, and Christian Hülsmeyer on the German side (Brandt 1953b).

9.1 Signals Intelligence

9.1.1 Radio Reconnaissance

According to Rahn[1154] and Rohwer[1155], radio reconnaissance comprises radio monitoring, traffic analysis, radio direction finding, and decryption. Traffic analysis allows conclusions to be drawn about the structure, subordination, ongoing operations, and spatial distribution of enemy forces, based solely on the characteristics, type, and density of radio messages, even without knowing their content.[1156] Radio direction finding enables the location of transmitters at sea or on land by measuring the bearing angle when a message is sent.[1157] Radio decryption aims to break enemy encrypted messages as quickly as possible and relay the decoded information to higher command centres for operational planning and countermeasures.

In the context of the present work, the focus is on radio decryption (content analysis) and radio direction finding. A central element of Allied warfare, long kept secret, was the decryption of German Wehrmacht radio communications under the cover designation ULTRA[1158] at Bletchley Park[1159], where all British radio reconnaissance efforts were consolidated. In contrast, Germany lacked a comparable organisational hub, and decryption was largely carried out manually. Although German decryption services achieved some successes, the most important Allied codes remained largely inaccessible to them.[1160]

1154 Rahn 2011, pp. 16–17.

1155 Rohwer 1979, pp. 328–329.

1156 Ibid., p. 329.

1157 Rohwer 1979, p. 329; Rohwer 1969a, pp. 324–328, Rohwer 1969b, pp. 360–364; Trenkle 1982, pp. 43–132.

1158 The designation ULTRA was introduced in July 1942, replacing the previously unofficially used designation Special (Pröse 2004, p. 149). It was derived from the realisation that information obtained from intelligence sources was subject to the strictest level of secrecy and had to be kept confidential under all circumstances. ULTRA's level of secrecy was even higher than the highest classification at the time, Most Secret.

1159 Bletchley Park is located in Bletchley, which is now part of Milton Keynes in the county of Buckinghamshire, about 80 km north-west of London.

1160 Hüttenhain 1979, pp. 100–109; Rahn 2011, pp. 36–39.

The combination of these different types of information allowed Allied intelligence to construct an increasingly precise overall picture of the operational situation. Radio decryption, supported by traffic analysis and direction finding, thus formed the cornerstone of signals intelligence and contributed decisively to operational planning and the success of Allied military operations. Radio direction finding is treated separately in section 9.2 due to its operational and historical significance.

9.1.1.1 ULTRA

Until the mid-1970s, it remained top secret that Allied radio reconnaissance[1161], particularly the interception and decryption of enemy signals by British cryptologists under the codename ULTRA, had made a decisive contribution to the Allied victory in World War II. This mainly concerned the fighting of German submarines in the Atlantic, but also included, prior to the Battle of Britain, the provision of intelligence on the upcoming Operation Seelöwe[1162],

---

1161 The meaning of the term Radio Reconnaissance should be expanded to include the fact that ULTRA, a codeword within Signals Intelligence (SIGINT), encompassed more than just traditional military reconnaissance from 1941 onwards. It also enabled forward looking strategic planning by analysing non-military communications, such as diplomatic or economic messages, and other sources. Until 1941, the primary focus was on the interception and basic cryptanalysis of enemy communications (Pröse 2004, p. 6). Until the mid-1970s, it remained top secret that Allied radio reconnaissance, particularly the interception and decryption of enemy signals by British cryptologists under the codename ULTRA, had made a decisive contribution to the Allied victory in World War II. SIGINT consists of Communications Intelligence (COMINT), which involves the analysis of communication signals, and Electronic Intelligence (ELINT), which focuses on non-communicative electronic signals such as radar (Grabau 1986, pp. 19–20; Trenkle 1982, pp. 14–42).

1162 Operation Sea Lion; Churchill 1985, pp. 266–278; Churchill 2007, pp. 388–398; on 16 July 1940, Hitler issued Führer Directive No. 16, directing preparations for a potential invasion of England, known as Operation Sea Lion (Forczyk 2016, p. 6). Already on 17 September 1940, Hitler decided to postpone Operation Sea Lion indefinitely due to the lack of air superiority. Following a presentation by Admiral Raeder, Operation Sea Lion was officially abandoned on 13 February 1942, and all material preparations were halted. Details on the division of tasks between the three services of the Wehrmacht and on which was eventually abandoned. Subsequently, ULTRA provided detailed intelligence on German deployment areas and the operational strength of German units during the Battle of Britain.

The interception and decryption of German radio messages and telexes by ULTRA remained strictly secret even after the war, until the publication of the memoirs of Frederick Winterbotham[1163], former RAF liaison officer with the British Secret Service, in 1974. In his book "The ULTRA Secret"[1164] and its German translation "Aktion ULTRA"[1165] from 1976, he revealed details of the decryption of the German cipher machine ENIGMA.[1166] The book gave rise to numerous speculations about the outcome of World War II and the factors contributing to Germany's defeat, particularly regarding the ongoing decryption of German radio traffic by British cryptologists.[1167]

In 1975, a review in the New York Times Book Review reportedly opened with the words "This book reveals the greatest secret of World War II after the atom bomb", helping it become a bestseller.[1168] Since then, dozens of works, supported by extensive documents from American and British archives, have expanded the story and raised public awareness about codebreaking.[1169] Further publications included "Very Special Intelligence" by Patrick Beesly in 1978.[1170] That same year, an international conference was held in Germany, in Bonn-Bad Godesberg, and Stuttgart, on the role of sig-

---

the specific ports of operation of the German naval forces for Operation Sea Lion can be found in Lohmann und Hildebrand 1956, pp. 1–6.

1163 Nicholls 2004.

1164 Winterbotham 1974; Winterbotham 1999.

1165 Winterbotham 1976.

1166 In the introduction to his book The ULTRA Secret, Winterbotham notes that he had to rely on personal recollections, as he was not granted access to government archives at the time of writing (Winterbotham 1974, p. vii; Winterbotham 1999, p. vii; Winterbotham 1976, p. 9). Subsequently, some of his representations, particularly those concerning the navy, were corrected by scholars such as Patrick Beesly (Beesly 1979, pp. 133–151; Beesly 1976a, pp. 147–164; Beesly 1976b, pp. 368–383; Beesly 1976c, pp. 698–707).

1167 Deutsch 1977, pp. 16–32.

1168 Putney 1987, p. xiv.

1169 Kahn 1996, p. 979.

1170 Beesly 2006; Beesly 1978.

nals intelligence during World War II on both Allied and German sides.[1171]

In 1979, Jürgen Rohwer at the Institut für Zeitgeschichte published a detailed analysis of the current state of research based on available sources, entitled "Der Einfluss der alliierten Funkaufklärung auf den Verlauf des Zweiten Weltkrieges".[1172] It should also be noted that before Winterbotham's work, Władysław Kozaczuk published a book in 1967 describing the decryption of the ENIGMA cipher by Polish cryptologists before World War II.[1173]

Further publications followed in 1979 and 2004, though these were primarily recognised by experts rather than the wider public.[1174] In 1973, the Paris publishing house Plon published Gustave Bertrand's[1175] "Enigma ou la plus grande énigme de la guerre 1939–1945", one of the most important primary sources on ENIGMA decryption.[1176] The book provided the first detailed account of eleven years of Franco-Polish cooperation in decrypting ENIGMA before and during World War II, though it was not reviewed in major journals and sold poorly.[1177]

While in Germany the departments responsible for deciphering enemy radio messages were distributed across various organi-

1171 Both events were held as part of the 45th joint conference of the Arbeitskreis für Wehrforschung, the Clausewitz-Gesellschaft, and the Deutsche Gesellschaft für Wehrtechnik in Bonn and the Historical Colloquium of the Bibliothek für Zeitgeschichte and the Historical Institute of the University of Stuttgart from 15 to 18 November 1978. Cryptologists, telecommunications engineers, key machine designers, intelligence officers, generals, admirals, and historians from the Federal Republic of Germany, Great Britain, the USA, Canada, Poland, France, Italy, Australia, Norway, Sweden, Finland, Switzerland, and Greece took part in these two events for the first time. This marked the first time that insights into the background and context of Allied assessments of important strategic decisions could be significantly expanded (Rohwer und Jäckel 1979).

1172 The Influence of Allied Radio Reconnaissance on the Course of World War II (translated by the author).

1173 Kozaczuk 1967; Rohwer 1979, p. 326; Bauer 1997b, pp. 389–391.

1174 Kozaczuk 1987 (German translation of the original Polish publication Kozaczuk 1979); Kozaczuk and Straszak 2004.

1175 Gustave Bertrand was an employee of the French secret service.

1176 Bertrand 1973; Skillen 2024, pp. 62–147.

1177 Skillen 2024, pp. 62–148.

sations such as the High Command of the Wehrmacht, the Foreign Office, the Security Service, the Research Office, and individual branches of the Army, Air Force and Navy, with no coordination whatsoever, Great Britain was able to draw on its experience in codebreaking from World War I.

Already during World War I, German radio messages were deciphered after Great Britain, as one of its first acts of war in August 1914, cut the German transatlantic cable to the USA. This forced Germany to transmit diplomatic messages to North and South America primarily by radio.[1178]

The British team of experienced specialists responsible for decoding German radio messages was based in the so-called Room 40[1179], under the direction of Reginald Hall.[1180] Room 40 was part of the British Naval Intelligence Division from its inception, and by 1917 it had become more closely coordinated with other intelligence services.[1181] Winston Churchill, as First Lord of the Admiralty in 1914, recognised the strategic potential of intercepting and deciphering enemy radio communications. One of the most significant successes of the British intelligence services during World War I was the decryption of the so-called Zimmermann Telegram in January 1917. Its revelation contributed to a shift in American neutrality and played a major role in the United States' decision to enter the war on 6 April 1917.[1182]

1178 Johnson 2004, p. 306; Johnson 1983, pp. 322–323.

1179 The term Room 40 was the cover designation for the British Admiralty's signals intelligence unit during World War I, tasked with decrypting enemy communications, including German naval and diplomatic messages (Kahn 1996, pp. 266–297; Rahn 2011, p. 17).

1180 O'Halpin 2008; Johnson 2004, p. 306; Johnson 1983, p. 322.

1181 Beesly 2006, pp. 1–2; Beesly 1978, p. 20.

1182 The Zimmermann Telegram was a coded message sent in January 1917 by Arthur Zimmermann, the German Secretary of State for Foreign Affairs, to the German ambassador in Mexico. It was transmitted via diplomatic channels, including a United States transatlantic cable that the German government had been allowed to use thanks to America's then-neutral stance. Unbeknownst to both Germany and the United States, the British had tapped this cable and intercepted the message. The telegram was forwarded to the British cryptographic unit Room 40, which gradually decrypted its contents. The background to the message was Germany's decision to resume unre-

The deciphering of this message by British specialists is considered an outstanding achievement in this field, which had a significant impact on the course of World War I due to its importance. In total, more than 15,000 German radio messages were decoded during the war, thanks to the efforts in Room 40.[1183]

As early as 1914, Churchill ordered the establishment of listening stations to monitor maritime radio communications, which were continuously expanded until World War II.[1184]

After the end of World War I, Room 40 was merged with the deciphering department MI1b of the British Army to form the Government Code and Cypher School (GC&CS). In 1923, it was assigned to the Foreign Office and its Secret Intelligence Service MI6, with its main task being the deciphering of diplomatic messages.[1185]

At the beginning of World War II, the GC&CS was relocated to Bletchley Park under the cover name Station X.[1186] After initial dif-

---

stricted submarine warfare in a bid to defeat Britain by severing its supply lines. German leadership was aware that this move might provoke the United States to enter the war on the side of the Allies. To preempt this threat, Germany proposed a military alliance with Mexico in the event of U.S. entry into the war. In return, Germany promised support for Mexico to reclaim the territories of Texas, New Mexico, and Arizona, which it had lost to the United States in the Mexican-American War of 1846–1848. After deciphering the telegram, the British passed it to U.S. Secretary of State Robert Lansing. When its contents became public, it sparked outrage in the United States and significantly shifted public opinion. President Woodrow Wilson, who had previously maintained a policy of neutrality, requested a declaration of war from Congress. On 6 April, 1917, the United States formally entered the war against Germany. The decryption and exposure of the Zimmermann Telegram was one of the most significant successes of the British intelligence services during World War I. It played a major role in bringing the United States into the conflict and thereby decisively influenced the course and eventual outcome of the war. (Nassua 1992; Singh 2000, pp. 137–145; Kahn 1979, pp. 20–22; Tuchman 1982; Tuchman 2014; Boghardt 2012).

1183 Johnson 2004, p. 306; Johnson 1983, pp. 322–323.

1184 A good overview of the role of radio reconnaissance in naval warfare on the German and Allied side can be found in Heinemann 2011, pp. 15–56 and generally for World War II in Beesly 2006; Beesly 1978; Beesly 1979, pp. 133–151.

1185 Davis 2011, pp. 341–354; Pröse 2004, p. 4.

1186 Calvocoressi 1979, pp. 88–99; McKay 2011, p. 12; Smith 2011; Smith 2010, pp. 26–28; Bletchley Park was easily accessible from Oxford, Cambridge, ficulties with funding and staffing capacity bottlenecks, Churchill, who had been Prime Minister since 1940, ensured the provision of suitable personnel and financial resources.[1187]

In contrast to the German command, Churchill recognised the increasing importance of deciphering radio messages in a modern war, one determined by the movement of operational forces, as early as the period between the world wars. In Germany, the importance of deciphering radio messages was only recognised late. In addition, as already mentioned, the polycratic structure was a major obstacle to the development of its own cipher services, which did not cooperate and also kept their findings secret.[1188] Once again, there was no cooperation between key institutions and experts in Germany. In contrast, both British radar development at Bawdsey Manor and the deciphering of enemy radio messages at Bletchley Park made great progress thanks to the successful cooperation of a large number of experts from all services of the armed forces and civilians alongside scientists from a wide range of disciplines.

Just as many civilians as military personnel worked at Bletchley Park. From 1939 until the end of the war, up to 10,000 people, both civilians and military staff, worked around the clock to decipher German military codes and ciphers.[1189] According to Winterbotham, the design and construction of powerful machines made it possible not only to intercept large volumes of German Wehrmacht radio transmissions but also to decipher them on a scale never seen before, with more than 2,000 enciphered messages being transmit-

and London thanks to its own railway connection, which made travel convenient for scientists from these universities. Additionally, Bletchley Park was able to handle a large volume of messages, as it was situated at a key telecommunications hub. Its remote geographical location also provided a high level of security, making it difficult for unauthorised individuals to intrude.

1187 Pröse 2004, p. 4.

1188 Ibid., pp. 5–6.

1189 Calvocoressi 1979, p. 89; Johnson 2004, p. 305; Johnson 1983, p. 321; from 1939/40 onwards, the recruitment of women became increasingly favoured due to the high demand for personnel. It is estimated that women made up around 75 % of the workforce. They not only held supporting roles but also made significant contributions to decoding.

ted daily at the height of hostilities.[1190] Without the determined support of Churchill and his influence within both military and political circles, the work at Bletchley Park would never have been as successful as it was during World War II.

Once the contribution of cryptanalysts to the Allied victory in World War II had been recognised, the deciphering of intelligence information continued to play an increasingly important role in the post-war period. Secret research by the United States intelligence services, combined with rapid advances in computer and communication technologies, significantly enhanced the ability to gather military, diplomatic, and economic intelligence. The primary focus remained on decrypting radio communications. The Anglo-American intelligence services also continued to study and analyse captured ENIGMA machines, which were in use in other countries under the false belief of security. Many governments, unaware of vulnerabilities in their encryption methods, continued to rely on devices such as ENIGMA and TYPEX. While initially considered secure, these machines were relatively easy to break for technologically advanced countries like the United States, ensuring that Anglo-American intelligence services maintained a strategic advantage.[1191]

The delayed disclosure of ULTRA and the activities at Bletchley Park, made public only through Frederick Winterbotham's 1974 memoir The Ultra Secret, was partly due to continued cryptographic and related intelligence operations after the war. These operations targeted former allies, colonial liberation movements, and governments perceived as potential threats, all in the context of the emerging Cold War and the increasing rivalry between the Western powers and the Soviet Union, including concerns about the Soviet atomic program.[1192]

Reports indicate that some countries continued using these machines for their secret communications until around 1974, by which time Winterbotham's revelations exposed the compromise of messages encrypted with them. Over time, the primary focus of

1190 Winterbotham 1999, p. 1; Winterbotham 1976, p. 13.
1191 Pröse 2004, p. 191.
1192 Ibid., pp. 149–151.

intelligence efforts shifted toward the Soviet Union and its allies, reflecting the broader geopolitical tensions of the Cold War.

9.1.2 Essential Cipher Machines of the German Wehrmacht

The German Wehrmacht essentially used three cipher machines, of which the ENIGMA cipher machine is the best known. However, the Siemens-Geheimschreiber T52 and the Schlüsselzusatz (G-Zusatz) SZ42 from Lorenz played an important role in German radio communications and presented the cryptologists at Bletchley Park with enormous challenges in terms of deciphering them.

9.1.2.1 ENIGMA

The ENIGMA cipher machine was a rotor-based encryption device used to secure German radio communications during World War II.[1193]

In addition to the individual branches of the armed forces, various other Reich authorities and party organisations also used different variants of the ENIGMA as a key communication tool, such as the Reichsbahn from 1940 onward.[1194]

The ENIGMA cipher machine surpassed all known encryption methods of the time in its effectiveness, which is why it was primarily used by the Wehrmacht during the war.[1195]

1193 Judkins 2020, pp. 18–24; Werther 1979, pp. 50–65; Trenkle 1990, pp. 162–167; Johnson 2004, pp. 305–349; Johnson 1983, pp. 321–369; a comprehensive overview of the history and technical details of the ENIGMA family, as well as other cipher machines, can be found on the website of the Crypto Museum in Eindhoven (URL: https://www.cryptomuseum.com/). The Crypto Museum, a virtual museum with a physical collection of historical cipher machines, spy radio sets, and espionage-related artefacts, was founded in 2003 by electronics engineers Marc Simons and Paul Reuvers from Eindhoven, the Netherlands, and is managed by a trust to secure its future.

1194 Pröse 2004, pp. 45–46.

1195 In the present work, the focus is placed on the strategic significance of the ENIGMA machine in terms of its influence on the course of the war, rather than on an analysis of its technical structure or the specific performance of

The technical basis was the rotor principle, which had been known since 1917 and was significantly further developed by Arthur Scherbius[1196], who filed the patent application on 23 February 1918.[1197] Scherbius named his encryption machine ENIGMA after the Greek word *αίνιγμα*, which means riddle.[1198]

He offered his machine to the Navy, which initially gave it a favourable assessment in July 1918. However, procurement was ultimately rejected by the Admiralty Staff, and the project was passed on to the Foreign Office, where it also failed to gain traction.[1199] After World War I, the company Dipl.-Ing. E. Richard Ritter & Co. was initially responsible for utilising the patents, from which Scherbius & Ritter was formed in 1920.[1200] In 1923, Scherbius acquired patents for a cipher machine registered by Hugo Alexander Koch in the Netherlands. Based on both his and Koch's patents, he developed the first cipher machine, the ENIGMA A, in the same year. However, interest in the ENIGMA remained low. Due to a lack of capital, Scherbius sold his patent rights in 1923 to the Gewerkschaft Securitas, which later became Chiffriermaschinen AG.[1201] He and Ritter served together in the company's management.[1202] Chiffriermaschinen AG was founded in Berlin on 9 July 1923. The main purpose of its business activities was the continued development, production, and commercial marketing of the cipher machine based on the patent filed in 1918, along with its successor models, beginning with the first model, ENIGMA A, in 1923, and continuing through models B, C, C (modified), D, E/F/G/H (as test models), and I and II up to 1934.[1203]

---

its individual variants. For technical details, reference is made to foundational works such as Pröse 2004 and Ulbricht 2005 and, in a broader sense, to Bauer 1997a, pp. 31–48.

1196 Jäger und Heilbronner 2010, pp. 377–378.

1197 Scherbius 1918; Pröse 2004, p. 33.

1198 Bauer 1991, pp. 211–214.

1199 Bauer 1991, p. 212; Pröse 2004, p. 33.

1200 Jäger und Heilbronner 2010, p. 377.

1201 Crypto Museum, Eindhoven (https://www.cryptomuseum.com/manuf/chimaag/).

1202 Pröse 2004, p. 33; Kruh and Deavours 2002, p. 1.

1203 Pröse 2004, p. 50.

In 1923, Scherbius described the most recent model of the ENIGMA in the Elektrotechnische Zeitschrift:

> „Es erfüllt vollkommen die Anforderungen, die die Post stellen muß, und ist gleichzeitig für amtliche Mitteilungen, für Handel und Industrie in weitestgehendem Maße verwendbar ...".[1204]

In 1923, the ENIGMA A was showcased at the Congress of the International Postal Union in Bern, Switzerland.[1205] In 1924, another presentation took place at the International Postal Congress in Stockholm, Sweden.[1206]

After the first ENIGMA A was produced, which was still heavy and unwieldy, it was replaced by the modified ENIGMA B and presented extensively at postal exhibitions by Chiffriermaschinen AG.

This attracted the attention of officers from the Reichswehr's Department of Cipher and Communications. As a result, the Reichswehr acquired several copies of the ENIGMA B for testing purposes. The interest of the German officers must also be understood in the context of their experiences with the so-called Zimmermann Telegram from 1917.[1207]

This was followed in 1926 by the ENIGMA C, a significantly improved machine that had the great advantage of being able to encode and decode using the so-called Reversing Roller with the same machine setting. However, the disadvantage of this new process was the weakening of the cipher security.[1208]

The ENIGMA D, featuring new rotor technology[1209], was the first machine to be offered for sale on the commercial market in 1927 and sold in moderate numbers. The ENIGMA D was also purchased by other countries and their government services, where

1204 Scherbius 1923, p. 1035. "It fully meets the requirements that the postal service must impose, while also being widely applicable for official communications as well as for commercial and industrial purposes." (translated by the author).

1205 Kruh and Deavours 2002, p. 1.

1206 Ibid., p. 5.

1207 Pröse 2004, p. 34.

1208 Ibid., p. 35.

1209 ENIGMA D was equipped with interchangeable rotors for the first time, which were now officially referred to as Rollers (Pröse 2004, p. 35).

it was modified for their own purposes. For example, a modified replica of the ENIGMA D was made in Japan.[1210]

The German Navy modified ENIGMA C for shipboard use, deploying it from 1926 until 1934, before later developing more advanced naval versions. From 1934 onwards, the ENIGMA I became the standard model used by all services of the Wehrmacht, with the Army employing its own configuration.[1211]

Arthur Scherbius died on 13 May 1929 following an accident involving a horse-drawn carriage.

Sales in the civilian sector could not ensure the company's continued existence, and as a result, Chiffriermaschinen AG had to close, partly due to the law on the conversion of corporations into partnerships, passed on 5 July 1934. The company was taken over by Rudolf Heimsoeth and Elsbeth Rinke, who already held shares in Chiffriermaschinen AG, and continued under the new name Chiffriermaschinengesellschaft Heimsoeth und Rinke oHG, abbreviated to Heimsoeth und Rinke (H&R), maintaining the same business activities, i.e., the development and production of cipher machines. The company ceased operations shortly before the end of the war.

In the period following the rise of the Nazi regime, the Wehrmacht was rearmed, and the demand for reliable cipher machines increased. Additionally, a significant number of ENIGMA machines were produced for civilian applications. From 1923 to 1945, 25 different versions of the ENIGMA were developed and produced.[1212]

In his study ENIGMA Production at Konski & Krüger[1213], Frode Weierud estimated that approximately 35,616 military ENIGMA machines equipped with a plugboard (Steckerbrett) were produced and delivered to the Wehrmacht.[1214] This figure encompasses all types of military machines with plugboards. Additional

1210 Bauer 1997b, pp. 138–139.

1211 Ibid., p. 50.

1212 Pröse 2004, pp. 50–51.

1213 URL: https://www.cryptocellar.org/enigma/e-prod-history/konski&krueger-production.pdf.

1214 URL: https://www.cryptomuseum.com/crypto/enigma/index.htm#qty.

commercial models were produced separately, but Weierud does not provide a precise total for all ENIGMA machines.

9.1.2.2 Siemens-Geheimschreiber T52

In addition to the ENIGMA, the Germans also developed other cipher machines. Between 1929 and 1932, Siemens & Halske designed the T52 cipher teleprinter, a machine that encrypted messages transmitted via telegraph networks using teleprinter technology.[1215]

It was based on an invention by Siemens & Halske employees August Jipp[1216] and Erhardt Roßberg, entitled "Anordnung zur Nachrichtenübermittlung in Geheimschrift über Telegraphenanlagen".[1217] The main patent was granted in Germany on 18 July 1930, and the corresponding U.S. patent was issued on 6 June 1933.[1218]

The T52 was a cipher machine designed for stationary use due to its unwieldy design, which made it unsuitable for front-line deployment. These cipher machines were primarily used by the German military to transmit highly confidential information to the higher command levels of the Army, Navy, and Air Force, each of which had separate and often competing command structures„ a situation that later proved to be a major disadvantage.[1219]

The T52 was classified by the Germans as more secure than the ENIGMA, which was designed for fast and mobile use in tactical operations. Messages transmitted via the T52 were typically less time-sensitive and required a higher degree of secrecy. For this reason, a highly secure encryption system was essential. The encryption was based on the Baudot-Murray Code, also known as the Murray Code. This teleprinter code was not classified and was

1215 Pröse 2004, pp. 76–80; Mache 1992, pp. 85–94; Mache 1989, pp. 97–117; Trenkle 1990, pp. 168–169.

1216 Jipp 1934.

1217 Beckman 2006, p. 106; System for transmitting messages in cipher via telegraph systems (translated by the author).

1218 USP No. 1,912,983; Mache 1992, p. 85; Mache 1989, pp. 97; Beckh 2005, p. 249.

1219 In the course of the war, the Wehrmacht used various mutually incompatible variants of the T52, including the T52a/b, T52c, T52d, and T52e. The cipher machine was originally developed for the German Navy.

commonly used in international telex and radioteleprinter communications. A key feature of the Murray code was its use of the binary number system, which would later become the foundation of digital computing.[1220] The first teleprinter developed by Siemens & Halske was offered to the Navy and subsequently adopted in 1932, although it still required improvements in radio interference suppression. The Air Force also planned to adopt the T52. However, due to its considerable weight and size, the T52 was primarily designed for wired transmission rather than for radio communication.[1221] The main advantage of the T52 was that it was a teleprinter system with built-in encryption and its own key generation. The operator could enter the text to be transmitted as plain text, the machine encrypted the text, the teletype characters were transmitted in encrypted form, and on receipt, the plain text was decrypted and printed out again by the machine without the operator coming into contact with the encrypted text himself.

Because the operator only had to handle the plaintext, messages could be transmitted more quickly and with less effort.

9.1.2.3 Schlüsselzusatz (G-Zusatz) SZ42

The Schlüsselzusatz (G-Zusatz) SZ40[1222], and later SZ42[1223], was a key component developed by Lorenz AG on behalf of the German Army for the Lorenz Lo15 mechanical teleprinter (Fig. 30). It was part of the Lorenz cipher system, used to generate encryption keys for secure communication.

Lorenz AG had many years of experience in the construction of mechanical teleprinters for the Army, but until then, it had no expertise in the field of cryptography. Teleprinters were capable of transmitting at a much higher data rate but required a fixed connection via cable or an interference-free radio link. The Baudot-

1220 Johnson 2004, pp. 338–339; Johnson 1983, pp. 357–358.

1221 Pröse 2004, p. 77; Mowry 2014, pp. 6–11; Crypto Museum, Eindhoven.

1222 The designation SZ40 is derived from the term Schlüsselzusatz 40, referring to the year of its development in 1940. Subsequent and improved versions were designated SZ42, SZ42a, SZ42b, and SZ42c (Roberts 2018, pp. 65–66).

1223 Pröse 2004, pp. 80–86; Copeland 2010, pp. 39–40; the Lorenz SZ42 was given the code name Tunny by the British codebreakers at Bletchley Park.

Fig. 30: Lorenz SZ42[1224]

Murray code was also used for these transmissions. The aim of developing the SZ40 key supplement and later the SZ42 was to significantly simplify encryption for the operator. With the development of the key supplement for encryption, it became possible to enter plaintext on the sender's side and receive plaintext on the receiver's side, without the operator having to interact with the encrypted text.

In contrast to the Siemens & Halske patent specifications, most of which were openly accessible and largely described the construction and mode of operation of the T52 key teleprinter, the Lorenz-Schlüsselmaschine, of which SZ40 and SZ42 were key components, was a secret development project.[1225] The machine was not very suitable for mobile use due to its weight and size, was technically much more complicated, and was used for the secure transmission of strategically important messages at the highest man-

1224 Image credit: Crypto Museum, Eindhoven.

1225 The German code name for the special radio equipment used with teleprinters was Sägefisch (sawfish), and various related transmitting and receiving systems operating in the short-wave range were also designated under this code name (Weierud 2010, p. 309).

agement level. It was used to transmit messages from the Führer's headquarters to his top military command.[1226] The ENIGMA and the Lorenz cipher machine, the Lorenz-Schlüsselmaschine, were very different cipher systems and had little in common. The latter was much more complex than the ENIGMA.[1227]

9.1.3 Breaking the Cipher Machines

The terms decryption and codebreaking in the following consideration refer specifically to the machine decryption of radio messages. These cryptanalytic machines were instrumental in deciphering the vast number of secret German radio communications. Machine decryption played a central role in the success of ULTRA and gave the Allies a decisive intelligence advantage during World War II.[1228]

9.1.3.1 ENIGMA

After World War I, Germany's opponents realised that only machine decryption could counter machine encryption. The German designers of ENIGMA were convinced that the machine could not be broken using manual methods. They did not consider that a different machine might be able to decrypt their messages.

The history of deciphering ENIGMA began in Poland[1229], a country that was in constant tension with its neighbours, Germany and the Soviet Union.[1230] Polish cryptologists had already gained significant experience in deciphering radio traffic during the Polish-Soviet War of 1919–1920. Their work in monitoring and decoding German radio transmissions laid a crucial foundation for the later successes of the Allies.

During the 1920s, the Polish Cipher Bureau, particularly its German section, BS 4, successfully solved many of the hand-key proce-

---

1226 Pröse 2004, p. 81; Singh 2000, p. 295.

1227 Roberts 2018, pp. 71–78.

1228 Pröse 2004, p. 94.

1229 A good overview of the performance of the Polish decipherment service can be found in Lisicki 1979, pp. 66–86 and McCarthy 2020, pp. 1–17.

1230 Lisicki 1979, pp. 66–67.

dures used by the German Reichswehr.[1231] These procedures involved manually encrypting messages according to fixed substitution tables, rather than using a machine. This allowed Polish cryptologists to decrypt a substantial amount of German radio traffic.

However, this situation changed in February 1926, when the German Navy introduced the ENIGMA C, which could no longer be decrypted using previously successful methods. From July 1928, the German Army began using the ENIGMA G. By this time, Polish cryptologists had already mastered many manual key procedures used by the Reichswehr, but it became clear that machine-based encryption presented a new challenge that required scientific and systematic approaches.

In June 1930, the German Army adopted a modified version of the ENIGMA G, which led to the development of the ENIGMA I. This new machine differed significantly from earlier versions, and decrypting its messages remained extremely difficult.[1232] The Polish Cipher Bureau realised that conventional methods were insufficient, and that progress could only be achieved by applying rigorous scientific techniques to the problem of machine decryption.[1233]

To this end, BS4 specifically sought young people with strong skills in mathematics and the German language, as well as a mindset that transcended conventional thinking.[1234] Following a selection process focused on cryptology skills, Marian Rejewski, Jerzy Różycki, and Henryk Zygalski were recruited by BS4 in the autumn of 1932.[1235] These Polish cryptologists then applied their scientific expertise to carry out valuable and highly successful work from 1932 onwards.[1236] As early as January 1933, they succeeded in breaking ENIGMA C by reconstructing the Navy's radio key.[1237] Their main focus remained on deciphering ENIGMA I, used by the Army, which was more complex than both the civilian and naval versions.

By a lucky coincidence, the Polish intelligence service managed to analyse the technical details of an ENIGMA I machine in-

1231 Rohwer 1979, p. 335.
1232 Pröse 2004, p. 50, p. 96.
1233 Ibid., pp. 96–97.
1234 Lisicki 1979, p. 67; Rohwer 1979, p. 335.
1235 Lisicki 1979, p. 67; Kahn 1996, p. 973; Rohwer 1979, p. 335.
1236 Lisicki 1979, p. 68; Pröse 2004, pp. 96–97.
1237 Lisicki 1979, p. 68; Rohwer 1979, p. 335.

tended for the German embassy in Warsaw, after it was intercepted by customs before delivery.[1238] Although their work was making great progress, there were still major challenges regarding the structure and logic of the ENIGMA. Another coincidence came to their aid when, since November 1931, a German employee of the cipher department in the Reichswehr Ministry, Hans-Thilo Schmidt, known by the code name Asche[1239], had been delivering secret documents and key instructions for ENIGMA to Captain Gustave Bertrand, the French intelligence liaison officer, in exchange for money.[1240] Schmidt handed over a total of 303 secret documents to Bertrand in 19 meetings at various locations in Europe.[1241] Even when Schmidt was transferred to the Air Force Research Office in 1936, he continued to provide information to Bertrand. These documents were passed on to the Poles, among others, and with the help of the key settings supplied, the work could be significantly accelerated. The knowledge gained led to the Polish cryptologists beginning to rebuild ENIGMA machines as early as 1934. By 1939, around 15 machines had been reconstructed. These reconstructions gave the Poles a deep understanding of the encryption principles of ENIGMA. However, one of the greatest challenges remained in deciphering the so-called daily key, i.e., the daily changing settings of the ENIGMA. Eventually, in 1937, mathematician and cryptologist Rejewski and engineer Antoni Palluth from AVA-Werke succeeded in constructing the first machine, camouflaged under the designation Cyclometer. This machine went into operation in 1937 and was able to determine the daily key in approximately 20 minutes.[1242]

On 15 September 1938, the German Army and Air Force introduced a new key system, changing the operating settings of the ENIGMA. The new system eliminated the uniform starting position in the daily key, allowing a free choice of the starting position,

1238 Rohwer 1979, p. 335; Beyrer 1999, p. 115.

1239 Singh 2000, pp. 180–182, p. 191, pp. 194–196, p. 225; Feiveson 2018, pp. 71–72; Dönitz 2018, p. 501.

1240 Bertrand 1973; Rohwer 1979, pp. 335–336; Bauer 1997b, p. 391.

1241 These included user and key instructions and key documents from the years 1932 to 1934 as well as a long ciphertext with the corresponding plaintext (Rohwer 1979, pp. 335–336).

1242 Pröse 2004, p. 98; Rohwer 1969a, p. 336.

which significantly increased the encryption possibilities.[1243] However, the Poles succeeded in building a machine based on a new algorithm developed by Rejewski and a design by Palluth, codenamed Bomba.[1244] By November 1938, six Bomba machines were operational. This enabled them to determine the position of all the rotors and their settings in just under two hours.

The intercepted messages could then be read with the help of a replica ENIGMA.[1245] A short time later, on 15 December 1938, the German side introduced two additional rotors, drastically increasing the number of possible rotor positions from six to sixty. This made the existing Bomba machine design inadequate, as it would have required a significant expansion in the number of machines to cope with the increased complexity. Given their limited resources and specialised personnel, the Poles were unable to scale up the operation effectively.[1246] In early January 1939, the Wehrmacht introduced additional plugboard connections, further complicating the encryption. This made it impossible for the Polish cryptologists to continue their work on their own, prompting them to seek cooperation with their Allied counterparts. In light of the imminent threat of war with Germany, and the strain on their personnel and equipment, the Poles decided to share all their knowledge with the French and British at a secret meeting in Pyry, near Warsaw, in July 1939.[1247] During this meeting, the Poles handed over a replica ENIGMA I, along with other crucial documents and materials. This handover provided the French and British decryption services with the means to continue the work, much to their surprise.[1248]

In August 1939, the Polish material reached London via France and was transferred to GC&CS at Bletchley Park for further processing.[1249]

1243 Rohwer 1979, p. 337.

1244 Pröse 2004, p. 99; Lisicki 1979, pp. 77–78; Bomba is the Polish term for bomb. A detailed description of its technical functioning can be found in Ulbricht 2005, pp. 100–115.

1245 Pröse 2004, p. 100; Lisicki 1979, pp. 78–79.

1246 Pröse 2004, p. 100; Rohwer 1979, p. 337.

1247 Dönitz 2018, p. 501.

1248 Pröse 2004, p. 100; Rohwer 1979, p. 338.

1249 Pröse 2004, p. 100.

From the beginning of the war, experts from a wide range of professions and backgrounds worked at Bletchley Park. The first key figures were Alastair Denniston[1250], head of GC&CS, and Alfred D. Knox[1251], both former employees of Room 40 during World War I.[1252] They were soon joined by mathematicians, linguists, chess masters, and historians. Despite his success with methods of linguistic cryptanalysis, Knox was unable to break ENIGMA I. This led to the realisation that machine-based methods, relying on mathematical approaches, were necessary to effectively tackle the new ciphers used by the ENIGMA machine.[1253] It proved to be a stroke of luck when the mathematician Alan Turing[1254] joined GC&CS at the outbreak of war in 1939 and worked with Knox in his efforts to break ENIGMA I. Turing had already made significant contributions to theoretical computer science with his Turing Machine. With his vast knowledge of logic and mathematical algorithms, Turing was ideally suited for the task of cryptanalysis. He officially began his work at Bletchley Park on 4 September 1939, alongside other prominent scientists, including mathematician Gordon Welchman.[1255] Together, Turing and Welchman would soon prove crucial to the success of Bletchley Park's efforts.[1256] Fig. 31 shows Turings's reconstructed office.

Turing worked on the further development of the Polish Bomba. After finalising his concept for the so-called Turing Bombe[1257] at the beginning of 1940, engineer Harold Keen and his team at the British Tabulating Machine Company built the first prototype, named Victory, in March 1940.[1258] The first electromechanical Bombe went into operation in May. Prior to this, ENIGMA I had been deciphered mechanically in cooperation with

1250 Hinsley 2008.
1251 Batey 2006.
1252 McKay 2011, pp. 12–13.
1253 Pröse 2004, p. 101.
1254 Hodges 2017; Hodges 2012; Kahn 1996, pp. 975–976.
1255 Pröse 2004, p. 102.
1256 McKay 2011, p. 16.
1257 Pröse 2004, pp. 106–109; Rohwer 1979, pp. 339–340.
1258 Singh 2000, p. 218; Bauer 1997b, p. 402; Smith 2010, p. 32.

Fig. 31: Turing's reconstructed office at Bletchley Park[1259]

Polish colleagues working in France; however, this method was far too slow for operational use.[1260] The Turing Bombe was also relatively slow at first, but with significant improvements by Gordon Welchman, its performance could be increased considerably, which is why the name Turing Welchman Bombe is justified. It was deployed in mid-August 1940 and was given the name Agnus.[1261] By the end of the war, 211 Turing Welchman Bombes had been built in various versions[1262]. Once the machines were able to decipher

1259 Image credit: Courtesy of Bletchley Park Trust.
1260 Pröse 2004, p. 107.
1261 Erskine and Smith 2011, p. 48.
1262 Pröse 2004, p. 108.

primarily German Air Force transmissions, Turing's next challenge was the German Navy's ENIGMA M3, which used a more complex key process. The Navy had introduced eight interchangeable rotors, numbered I through VIII, from which three were selected for daily operation.[1263] This dramatically increased the number of possible configurations and posed a far greater cryptographic challenge. There was extreme time pressure to break the Naval ENIGMA M3, and without a breakthrough, it might have taken many months. Additionally, crucial secret documents were missing. As a result, the British undertook targeted military operations against German submarines and naval vessels.[1264] On 12 February 1940, the submarine U-33 was sunk in the Firth of Clyde on the west coast of Scotland. Two ENIGMA rotors, numbers 6 and 7, were recovered from the wreck, and the eighth and final rotor was obtained later, further completing the picture of the naval cipher system.[1265] Although Hut 8[1266] at Bletchley Park now had all eight rotors, it was still extremely difficult to break the Navy's ENIGMA code. Further key cipher documents were captured during Royal Navy raids on German weather observation ships, such as the German weather ship München in early May 1941 and the Lauenburg in June 1941.[1267] On 9 May 1941, the German submarine U-110 was

1263 Pröse 2004, p. 110. (Noskwith 2011, pp. 184–194; Murray 2001, pp. 113–118; Noskwith 2001 pp. 119–122).

1264 Pröse 2004, pp. 110–111.

1265 The action in February involved rotors 6 and 7, referred to as Rollers in the original German sources, whereby whereby the first 5 rotors were already identical to the ENIGMA I; the eighth rotor was captured in August 1940 (Pröse 2004, p. 111; Erskine and Smith 2011, pp. 167–168; Erskine 2002, p. 8).

1266 Hut 8 at Bletchley Park was dedicated to deciphering the Naval ENIGMA, particularly the M3 machine used by the German Navy (Noskwith 2011, pp. 184–194; Noskwith 2001, pp. 119–122; Murray 2001, pp. 113–118). While Gordon Welchman officially led Hut 8, Alan Turing played a pivotal role in breaking the Navy's codes. Hut 6, initially led by John Jeffreys until May 1940, and later by Hugh Alexander until March 1941, focused on decrypting the codes used by the German Army and Air Force. After Alexander moved to Hut 8, Stuart Milner-Barry (Erskine 2009) took over Hut 6 in the autumn of 1943 until the end of the war (Milner-Barry 2001, pp. 89–99; Taunt 2001, pp. 100–112). Although Turing was primarily responsible for the Navy's codes, his groundbreaking work and methodologies influenced the overall decryption efforts at Bletchley Park, including those in Hut 6.

1267 Pröse 2004, p. 110; Erskine 2002, p. 9; Bauer 1997b, p. 379.

badly damaged after an attack by British warships in the North Atlantic and was forced to surface.[1268] The crew abandoned the submarine, and a British boarding party went on board. The British captured the Naval ENIGMA M3 cipher machine and important cipher documents.[1269] These materials were of great value, as breaking the Naval ENIGMA was considered particularly challenging. Commander Fritz-Julius Lemp of U-110 apparently assumed that his damaged submarine would sink quickly, which is why he did not arrange for the complete destruction of the M3 and the cipher documents. However, U-110 remained on the surface long enough for the British to secure the valuable materials. The submarine eventually sank on 11 May 1941 while being towed by British ships.[1270] This capture was a turning point, as it provided the British codebreakers at Bletchley Park, specifically in Hut 8, with critical insights into the German Navy's encryption methods and made their work much easier.

On 1 February 1942, the cryptographers at Bletchley Park suffered a significant setback with the introduction of an improved ENIGMA M4[1271] machine by the German Navy. The ENIGMA M4 was used for secure communication between the U-boat command and the submarines operating in the Atlantic. This machine was more complex than all other ENIGMA variants used by the German Air Force, Army, and intelligence services. In the following months, Alan Turing and his colleagues worked intensively on a solution and made progress, but the existing deciphering machines, known as bombes, remained too slow, and other crucial features for decryption were still missing.[1272] A major breakthrough came on 30 October 1942 when the British Royal Navy captured the U-559 submarine off the coast of Egypt. The crew, convinced that the submarine would sink, abandoned ship without destroying all the codebooks and other cryptographic documents.[1273] British sailors

---

1268 Brennecke 1994, pp. 77–102.

1269 Ibid., p. 99.

1270 Ibid., p. 102.

1271 Pröse 2004, p. 111; Erskine 2002, pp. 9–10; Brennecke 1994, pp. 144–150.

1272 Pröse 2004, p. 112.

1273 The German submariners were not informed that the British had captured the documents.

recovered this vital material, and on 13 December 1942, the first successful decryption of the ENIGMA M4 was achieved.[1274] Following this, Bletchley Park sent a telex to the British Admiralty Operations Centre, providing valuable information about the positions of several German U-boats in the Atlantic. This intelligence contributed significantly to preventing the sinking of numerous ships, thus protecting substantial merchant shipping capacity in December 1942 and January 1943.[1275]

After a visit by Turing to the United States from November 1942 to March 1943, where he worked on developments related to speech encryption at the Bell Laboratories and advised the US Navy on the development of a semi-electronic Bombe to decrypt the ENIGMA M4, significant progress was made.[1276] This device, developed under the leadership of Joseph R. Desch of the National Cash Register Company (NCR) in Dayton, Ohio, was a semi-electronic adaptation of the British Turing Bombe. With Turing's guidance, the US Bombe was refined and entered mass production in April 1943, specifically for decrypting the ENIGMA M4. The US Navy's Cryptanalytic Section then took on much of the responsibility for decrypting the ENIGMA M4 from late 1943 onwards. This decryption breakthrough had a profound impact on the U-Boat War in the Atlantic. Thanks to the intelligence gained from the ULTRA intercepts, the Allies were able to identify the locations of German U-Boats, resulting in the prevention of numerous attacks and the destruction of many submarines. The intelligence obtained through the decryption of the ENIGMA M4 was a key factor in the eventual Allied victory over the German U-Boat threat.[1277]

9.1.3.2 Siemens-Geheimschreiber T52

The fact that the T52 cipher machine was fundamentally designed for cable transmission rather than radio communication made it more difficult for Allied cryptologists to intercept and break. After the German occupation of Norway in 1940, all military commu-

---

1274 Pröse 2004, p. 112; Rohwer 1980, pp. 280–281.

1275 Erskine 2002, pp. 10–11.

1276 Pröse 2004, p. 113, pp. 166–167.

1277 Pröse 2004, pp. 166–167; Erskine 2002, p. 11.

nication between Germany and Norway was conducted via cables leased from neutral Sweden, eliminating the need for radio transmissions.[1278] Fearing that these cables could be tapped, the German military employed the T52a and T52b cipher machines.[1279] However, this precaution did not prevent interception. As early as 15 May 1940, Swedish mathematician Arne Beurling[1280] and his team in Uppsala began analysing intercepted T52a and T52b messages. By July 1940, Beurling had succeeded in manually deciphering the first plaintexts. In the autumn of that year, a decryption device capable of producing plaintext output was constructed at Uppsala University. The remaining challenge was to break the daily changing key. Beurling was aided by operator errors on the German side, such as sending multiple messages with the same key setting.[1281] As a result, neutral Sweden gained access to highly sensitive German information for more than two years. The British naval attaché in Stockholm, Commander Henry Denham, regularly forwarded the contents of the deciphered German telex messages to London.[1282] In Sweden, the military significance of the telex traffic was recognised, and efforts to decrypt Fish, the British codename for all German teleprinter traffic, intensified.[1283] In 1941 alone, Swedish cryptanalysts successfully deciphered more than 41,000 German messages. Additionally, it was discovered that Germany wanted to keep Sweden neutral due to its critical iron ore exports.[1284] On 17 June 1942, the Finnish military attaché warned German authorities that the T52a and T52b machines had likely been compromised. In response, the Germans introduced an improved version, the T52c, on

1278 Pröse 2004, p. 117.

1279 McKay 2010, p. 323.

1280 Beckman 2002; Beckman 2006; Bauer 2003; Arne Beurling's work in breaking the German T52 cipher was groundbreaking, achieved entirely through mathematical analysis of intercepted cable transmissions, rather than access to the physical machine itself. This breakthrough, which provided critical intelligence for Sweden and the Allies, lasted for over two years and greatly benefited British intelligence, facilitated by the naval attaché in Stockholm.

1281 Pröse 2004, p. 117.

1282 Mache 1992, p. 90.

1283 The T52 was given the code name Sturgeon (Stör) by British codebreakers at Bletchley Park; Weierud 2010, pp. 307–327; Weierud 2000, pp. 18–52.

1284 Beckman 2006; Beckman 2002.

21 July 1942, alongside the new SZ42 machine developed by Siemens and Halske. However, the T52c was broken by Swedish cryptanalysts by 13 September 1942, and the SZ42 was also decrypted by 9 April 1943. This allowed Swedish intelligence to continue intercepting and reading messages encrypted with both the T52c and SZ42 until late 1943. In 1944, the German Wehrmacht began using the more secure T52d model along with the SZ42.[1285] The T52d, developed in 1942 and 1943 by modifying the T52a and T52b, no longer exhibited the same cryptographic weaknesses and was not broken by the Swedes. The T52b had also been developed for radio transmission. Bletchley Park found the T52a and T52b machines difficult to decipher, since they were almost exclusively used over cable lines. This meant that Allied cryptologists had limited experience with intercepting relevant radio signals and had very little intercepted material to work with. It was not until the summer of 1941 that teleprinter transmissions from these machines were first identified at Bletchley Park. The significance of the encrypted teleprinter transmissions had already emerged from Swedish intelligence, and the recurring mistakes and habits of German cipher operators could be analysed once again. As a result, the machine was reconstructed quickly. From September 1944 onward, Bletchley Park began referring to the newly developed T52e. However, it remains unclear whether this version was ever fully broken and how well known it was to Allied cryptologists by the end of the war.[1286] According to historian Michael Pröse[1287], cryptanalysts at Bletchley Park focused more on breaking the transmissions encrypted with the SZ42, operated by the army, because they contained valuable strategic insights about the actions of Germany's high command. Consequently, the decryption of T52 transmissions was given lower priority than that of the SZ42.[1288]

1285 Pröse 2004, p. 118.
1286 Pröse 2004, pp. 121–122; Weierud 2000, p. 37.
1287 Pröse 2004, p. 121.
1288 Bauer 1997b, pp. 274–375.

9.1.3.3 Schlüsselzusatz (G-Zusatz) SZ42

Building on the pioneering work of Polish cryptanalysts on the ENIGMA, which laid the crucial foundation for later breakthroughs, in 1941 the British engineer John Tiltman[1289] at Bletchley Park succeeded in deciphering a text encoded using the Lorenz cipher system. This breakthrough occurred after a message was transmitted twice from Athens to Vienna by the Germans during testing, which was strictly forbidden.[1290] Tiltman was able to decode the text by analysing small differences between the two transmitted messages. With the help of Bill Tutte[1291] and his team, the complex workings of the previously unknown Lorenz cipher machine were reconstructed.[1292] In the spring of 1942, the Lorenz cipher machine was successfully broken, although Tutte had never actually seen the machine.[1293]

Historically significant is the fact that for the first time, an electronic computer was developed to decode radio messages, greatly reducing the time required for decryption. The idea to build this computer came from Thomas Flowers[1294], who needed about 1,500 vacuum tubes to construct the first version. Fig. 32 shows a detail of the Colossus machine.

The first functional machine, named Colossus, was completed in December 1943 and deployed at Bletchley Park in January 1944. It began decrypting messages encoded with the SZ42 cipher and quickly became renowned as the first electronic machine designed for automatic decryption. Shortly afterward, an improved version was put into operation. In total, ten Colossus machines were constructed during the war, all based on this design.[1295]

1289 Nicoll 2004.

1290 Bauer 1997b, p. 369; Pröse 2004, pp. 124–125.

1291 Biggs 2008.

1292 Roberts 2018, pp. 79–84; Roberts 2018, pp. 72–73; Pröse 2004, pp. 124–125; Landwehr 2007; the Lorenz SZ42's encrypted messages were given the code name Tunny by the British codebreakers at Bletchley Park.

1293 Roberts 2018, p. 72.

1294 Agar 2008; Copeland 2010, p. 1; Judkins 2023, pp. 1–3.

1295 Bauer 1997b, p. 372.

Fig. 32: Colossus (Detail)[1296]

9.2 Radio Direction Finding

Another technological advance on the Allied side was the development of the High Frequency Direction Finder (HF/DF), also known as Huff-Duff[1297] in American slang, which was used against German submarines. This device was able to determine the position of the enemy's transmitter quite precisely, and as a result, it played a key role in turning the tide in favour of the Allies in the Battle of the Atlantic from 1943 onward. In conjunction with radar, the HF/DF played a central role in the U-boat warfare. A detailed overview of radio direction finding as an Allied weapon against German submarines in World War II can be found in Bauer's work.[1298]

Rohwer provides a general overview of the radio command and control systems used by German submarines.[1299] Rohwer, along with Gießler[1300], offers detailed information on the use of

1296 Image credit: Crypto Museum, Eindhoven.

1297 Bauer 1997a, p. 65; Brennecke 1994, pp. 174–179; Gießler 1971, p. 55.

1298 Bauer 2004b; pp. 1–23.

1299 Rohwer 1969a, pp. 324–328; Rohwer 1969b, pp. 360–364.

1300 Gießler 1971, pp. 55–58.

radio direction finding, which was employed by Allied escort ships against German submarines. Reuter also describes the challenges faced by the U-boat weapon in terms of location.[1301]

Both the German warships and submarines were commanded by radio by the High Command of the Kriegsmarine, respectively BdU[1302], the commander of the submarines, Karl Dönitz, which is where the term radio command originates. Dönitz was commander of the U-boats throughout World War II and was appointed Commander-in-Chief of the Navy in 1943.[1303] The German submarines received their orders for deployment, as well as information about supplies such as fuel, torpedoes, and provisions, via radio in their respective operational areas.[1304]

It was also crucial for the command centre of the BdU to be aware of the U-boats' positions and their operational status in order to guide them toward a convoy to attack. If a submarine had sighted a convoy, it would transmit both its current position and the course it was following via contact reports. The BdU command centre could then direct additional submarines to the area. Command was centralised at the BdU command post, and according to a predetermined schedule, the submarine that had sighted the convoy sent bearing signals to approaching submarines. After analysing and coordinating all the information, the BdU command centre would then send the relevant deployment orders to the submarines by radio.[1305]

The need to transmit information via radio messages carried the risk of the enemy being able to take bearings. The German submarine commanders quickly realised that there was a connection between the surprise detection by Allied escort vessels immediately

1301 Reuter 1971, pp. 149–173.

1302 Befehlshaber der U-Boote.

1303 Bauer 1997a, p. 13; in January 1943 Dönitz achieved the rank of Großadmiral (grand admiral) and replaced Grand Admiral Erich Raeder as Commander-in-Chief of the Navy.

1304 The German submarines deployed until 1944 were primarily designed for surface use, submerging only for protection or when under attack. Their cruising speed was significantly higher above water than when submerged, making surface travel an essential prerequisite for mobility and combat capability (Bauer 1997a, p. 13; Reuter 1971, p. 149).

1305 Bauer 1997a, p. 14; Gießler 1971, pp. 54–55.

after their own radio communications. Effective protection against direction finding by enemy land-based radio stations was therefore achieved by observing radio silence.[1306] However, radio silence had to be broken whenever multiple submarines needed to be coordinated during operations against enemy convoys. As an additional countermeasure, a signal codebook was developed in 1939, which allowed for the transmission of brief signals, containing only a few characters, in 10 to 20 seconds. These signals were encrypted using the ENIGMA cipher machine M, which the Germans believed made them almost impossible to intercept and decode.[1307] The High Command of the Navy was well aware of the dangers of direction finding, but the risks were considered relatively low until 1942/43.[1308] The German U-boats mainly attacked Allied convoys at night and from the surface, using the so-called Wolfsrudeltaktik.[1309] This meant that attacks were no longer carried out by individual submarines, but by coordinated groups. The British command was initially ill-prepared for this new tactic, as their previous efforts had focused mainly on locating submerged submarines using active sonar. With the ASDIC[1310] device, submarines could theoretically be detected at distances of up to 1,500 metres. However, in practice, effective detection was often limited to significantly shorter ranges due to environmental factors and the limitations of the technology.

From the summer of 1940 onwards, the British Navy became aware of the German method of radio command and control and made significant efforts to develop new technologies and procedures to locate submarines, even when they were submerged.[1311] In the period that followed, the development of effective radio direction finding techniques, often in combination with radar, became

1306 But also the avoidance of any unnecessary noise.

1307 Rohwer 1969a, pp. 324–325.

1308 Bauer 1997a, p. 15; Gießler 1971, p. 54.

1309 Dönitz 2018, pp. 22–28.

1310 ASDIC originally stood for the Anti-Submarine Detection Investigation Committee, a British body tasked with developing methods for submarine detection. Later, the name was used for the sonar system developed by this committee; Brennecke 1994, pp. 64–65; Reuter 1971, p. 149

1311 Reuter 1971, p. 149.

a high priority for the British, particularly against the backdrop of dramatic losses of supply ships en route to Great Britain due to German U-boat attacks.[1312] Existing anti-submarine tactics and technologies, some of which dated back to World War I, also had to be further developed and adapted by the Allies.[1313] By the end of 1940, the British Navy had gained a clearer understanding of the operational tactics used by German U-boats. This included recognising the importance of radio messages for the command and control of the submarines, as well as the communications between the submarines themselves.[1314] With the help of land-based direction-finding stations across various countries, conventional direction-finders were able to successfully locate individual submarines or groups of U-boats. However, the main disadvantage of taking bearings from land was the inaccuracy of the submarines' positions due to the long distances involved, and the additional time delay before the position data could be relayed to the combat area.[1315] The convoy, therefore, had no information about the position of enemy submarines relative to its own, and only by chance could an enemy submarine be forced to dive. This threat was particularly evident during night attacks by German U-boats. As a result, the problem could only be solved by developing a direction-finding device aboard escort vessels.[1316]

After hostilities with Allied convoys began, the German side considered radio communications to be harmless, as their positions were already known due to the ongoing combat. Immediately after the Allies obtained information about the German radio system, development of the high-frequency radio direction-finding system HF/DF, the Huff-Duff, began.[1317] Once installation and operational issues had been resolved, the Huff-Duff was fully operational by November 1942 and proved to be a crucial tool for the protection of convoys.

1312 The first radar detection of a submarine took place on 18/19 November 1940 from a Sunderland flying boat of the RAF, which had an ASV Mk. I radar set on board (Rohwer und Hümmelchen 1968, p. 84; Brennecke 1994, p. 47).

1313 Bauer 1997a, p. 63.

1314 Reuter 1971, p. 150.

1315 Bauer 1997a, p. 64.

1316 Rohwer 1969b, pp. 360–362; Bauer 1997a, p. 64, pp. 140–153.

1317 Brennecke 1994; pp. 174–179.

The cooperation between Huff-Duff and radar subsequently proved to be one of the most important elements in the fight against German U-boats.[1318] From the beginning of 1943, at least one escort vessel was equipped with Huff-Duff and was able to detect enemy submarines up to a range of 25 nautical miles. The escort vessel could then use the sounding beam to attack, for example, with depth charges.[1319] The existence of Huff-Duff remained unrecognised on the German side for a long time, as they initially believed that radar was the primary threat.[1320] Even unexplained losses were attributed to the use of enemy radar equipment on airplanes and ships.[1321] Although there had been concrete indications of high-frequency direction-finding systems since the spring of 1943, and by April 1944 even evidence of such equipment not only on land but also aboard ships such as battleships, cruisers and aircraft carriers, the threat was not fully recognised by the German side until June 1944.

However, the threat posed by detection through airborne ASV radar was often overestimated by the German side, while the success of radio direction finding using Huff-Duff was underestimated and generally overshadowed by the perceived dominance of radar technology.[1322]

Although radar detection from aircraft and ships was undoubtedly valuable, particularly for covering wide maritime areas, it complemented rather than replaced the role of Huff-Duff in anti-submarine warfare.[1323] As early as 1941, the Royal Navy began installing Type 271M radar systems aboard escort ships. Operating at a wavelength of approximately 9 centimetres, this radar could detect larger ships at ranges of 8 to 10 nautical miles and surfaced submarines at 4 to 6 nautical miles under normal visibility conditions. However, these ranges were significantly reduced in rough seas or poor weather. By July 1941, 25 ships had been equipped with

---

1318 Reuter 1971, pp. 150–151; Gießler 1971, p. 55.

1319 Gießler 1971, p. 58; Rohwer 1969b, pp. 361–362.

1320 Reuter 1971, pp. 155–156; Bauer 1997a, p. 65; according to Rohwer, Huff–Duff was a technique not inferior to radar in its meaning (Rohwer 1969b, p. 364).

1321 Gießler 1971, p. 57.

1322 Bauer 1997a, p. 65; Reuter 1971, pp. 156–157; Rohwer 1969b, p. 362.

1323 Watson-Watt 1957, pp. 358–366.

the Type 271M radar, and the general rollout to all escort vessels began in September 1941, continuing into 1943.[1324] This technology allowed convoy commanders to maintain an overview of the positions of their own ships at all times and to detect approaching submarines more effectively.

From June 1942 onwards, as the threat from British aircraft to German U-boats increased, particularly while transiting the Bay of Biscay, existing operational tactics were adapted. On 24 June 1942, an order was issued requiring submarines to pass through the bay submerged whenever possible, both by day and night.[1325]

From February 1943, the more advanced ASV Mk. III radar system, operating in the centimetre wavelength range, was installed in British patrol aircraft to improve the detection of German submarines. On 1 March 1943, a Wellington bomber equipped with a powerful searchlight, the Leigh Light, was used for the first time in action. This aircraft could synchronise its searchlight with the radar signal and was deployed over the Bay of Biscay to locate and illuminate surfaced submarines during night operations.[1326] This deficiency on the German side was initially addressed by introducing the Metox anti-radar warning receiver[1327], which successfully alerted submarine crews to approaching Allied aircraft and of which 60 units were initially available for installation.[1328] By August 1942, all outgoing U-boats were equipped with a Metox receiver and a simple dipole antenna mounted on a wooden cross, the so-called Biscay Cross, also referred to as the Biskaya Kreuz in German. This simple, non-technical term described the wooden structure used for the antenna, rather than a coded designation.[1329] This allowed submarines to detect British radar emissions from ASV Mk. II systems. However, this did not stop the continuous and often unexpected attacks on German submarines in operational areas. As a result, transmitting radio messages frequently became

1324 Rohwer 1969a, pp. 325–326.

1325 Reuter 1971, p. 152

1326 Rohwer 1969a, p. 326; Brennecke 1994, pp. 166–169; Reuter 1971, p. 157; Lovell 1991, p. xviii.

1327 Reuter 1971, p. 70; Brennecke 1994, pp. 159–165.

1328 Reuter 1971, p. 152.

1329 Bauer 1997a, pp. 194–203; Reuter 1971, p. 153.

impossible due to the need to dive quickly. The German command attributed the Allied success primarily to radar, prompting Dönitz to demand that all submarines be equipped with radar warning devices, a demand that was not fully met until autumn 1943.[1330] Furthermore, from spring 1943 onward, British aircraft were equipped with the ASV Mk. III radar system, which operated in the centimetric wavelength range and remained undetectable by the Metox receiver.[1331] Despite all warnings, the German submarines achieved their greatest successes in sinking Allied ships in March 1943. In April, however, the successes were fewer, which was attributed to the return of the U-boats to their bases. When the deployment of U-boats in the Atlantic became more favourable again in May, there was hope for further major successes. However, this hope was not realised. May brought very heavy and consistently unacceptable losses.[1332] Subsequently, the Metox device was banned and removed from the U-boats, as it was believed that Metox's radiation revealed the position of the submarine, which was thought to be the cause of the heavy losses.[1333] After it became clear that British aircraft were increasingly equipped with centimetric radar, the focus shifted to radar as the main cause of the losses. Only radar detection was seen as the primary reason for the setbacks in submarine warfare. Even as late as March 1944, there was no mention of the dangers posed by radio direction finding in discussions between technical experts at a conference on the German side.[1334] It was not until June 1944 that the threat posed by Huff-Duff was fully recognised. In December 1943, Karl Küpfmüller[1335] was tasked by Dönitz, Commander-in-Chief of the Kriegsmarine, to establish the WFM.[1336] Küpfmüller

1330 Reuter 1971, p. 154.

1331 The development of a warning receiver against radar detection in the centimetre wave range called Naxos began as early as the first meeting of the Arbeitsgemeinschaft Rotterdam on 23 February 1943 (Bauer 2004c, pp. 1–15; Reuter 1971, p. 115).

1332 Schramm 1961, p. 1439; Reuter 1971, p. 158.

1333 Reuter 1971, pp. 162–163; Gießler 1971, p. 82.

1334 Rohwer 1969b, p. 362.

1335 Jäger und Heilbronner 2010, pp. 250–251; Mielert 1982, p. 230; Kind und Mühe 2006b, pp. 208–212; Hagenauer und Pabst 2014.

1336 Wissenschaftlicher Führungsstab beim Oberkommando der Kriegsmarine (WFM).

began his work as head of the WFM in January 1944.[1337] Dönitz commented on Küpfmüller:

> „...Als vordringlichste Aufgabe sollte er in Zusammenarbeit mit dem Reichsforschungsamt und den hierfür maßgeblichen Persönlichkeiten der Wissenschaft und Industrie neue Ortungs-Empfangsgeräte mit erweiterten Frequenzen für die U-Boote schaffen, so dass sie mehr Ortungsbereiche des Gegners erfassen konnten als bisher und damit gegen überraschende Angriffe sicherer sein würden."[1338]

It is obvious that, in this very critical phase of the submarine war, the German side was clearly focused on the danger of detection by radar. This becomes apparent once again when one considers the remarks made in a circular issued by Dönitz on 11 December 1943:

> „Betr.: Aufstellung eines „Wissenschaftlichen Führungsstabes der Kriegsmarine":
>
> Der U-Bootkrieg ist durch den Gegner seit Monaten praktisch lahmgelegt. Nicht durch überlegene Taktik oder Strategie, sondern durch die Überlegenheit auf dem Gebiete der physikalischen Wissenschaft und Technik, die in dem neuzeitlichen Kampfmittel, der Ortung, ihren Ausdruck findet, hat der Gegner dieses Ziel erreicht. Damit hat er uns die einzige Offensivwaffe gegen die Angelsachsen aus der Hand geschlagen. Es ist von kriegsentscheidender Bedeutung, seinen Vorsprung auf diesem physikalischen Gebiet einzuholen und dem Uboot seine Kampfkraft wiederzugeben. Ich habe deshalb zusätzlich zu den bisher ergriffenen Maßnahmen die Aufstellung eines „Wissenschaftlichen Führungsstabes der Kriegsmarine" mit Sitz in Berlin befohlen. Zu seinem Leiter habe ich den Prof. Küpfmüller ernannt und mir unmittelbar unterstellt. Prof. Küpfmüller ist mit allen zur Durchführung seiner Aufgaben erforderlichen Vollmachten von mir versehen worden. Alle Dienststellen der Kriegsmarine haben den Leiter des Wissenschaftlichen Führungsstabes, seine Mitarbeiter und Beauftragten vorbehaltlos unter vollem Einsatz ihrer Kräfte zu unterstützen.

1337 Reuter 1971, p. 159; Hagenauer und Pabst 2014, p. 18; BArch, R 26 III / 91.

1338 Dönitz 2018, pp. 399–400. "...His most urgent task was to work with the Reichsforschungsamt and key figures in science and industry to create new locating receivers with extended frequencies for the submarines, so that they could detect a wider range of the enemy's positions than before and thus be better protected against surprise attacks." (translated by the author).

Der Oberbefehlshaber der Kriegsmarine, 11. Dez. 1943 (gez. Dönitz)"[1339]

Towards the end of World War II, the German side developed the Kurier short-signal procedure for transmitting radio messages from submarines to protect against direction finding from land-based stations and ships. The technique involved compressing the radio messages to transmit them in less than half a second. However, the Kurier system was never widely implemented and ultimately did not see significant use.[1340]

Rohwer saw the appeal of radar technology as an innovative novelty, in contrast to the long-established technology of radio direction finding, as the reason for the Germans' misjudgement in evaluating radio direction finding. He also identified the existing but inadequate contacts between the operational command, the naval signal stations, the technical departments, and scientific research, along with all the associated organisational and personnel problems, as contributing factors.[1341]

Gießler also pointed out that the combination of radar and the Huff Duff high frequency direction finder, used by the Allies on land and especially aboard ships from November 1942 onwards, marked the beginning of a phase that ultimately led to major losses

1339 BArch: RM 7/1234; "Subject: Establishment of a Scientific Command Staff of the Navy: The submarine war has been practically paralysed by the enemy for months. The opponent has achieved this goal not through superior tactics or strategy, but through his advantage in the field of physical science and technology, which is expressed in the modern means of combat, location. In doing so, he has knocked the only offensive weapon against the Anglo-Saxons out of our hands. It is of decisive importance for the war to catch up with him in this physical area and to restore the submarine's fighting power. In addition to the measures taken so far, I have therefore ordered the establishment of a „Scientific Command Staff of the Navy" based in Berlin. I have appointed Prof Küpfmüller as its head and placed him directly under my command. Prof Küpfmüller has been given all the necessary powers by me to carry out his duties. All departments of the Kriegsmarine must support the head of the Scientific Command Staff, his employees and authorised representatives unconditionally and with full commitment of their forces." The Commander-in-Chief of the Navy, 11 Dec. 1943 (signed Dönitz)" (translated by the author).

1340 Bauer 1997a, pp. 205–225.

1341 Rohwer 1969b, p. 364.

on the German side.[1342] Operational Research was another key element in Allied warfare, something that remained completely unknown to the Germans during the war.[1343]

It should be noted that at the time of the publications by Reuter and Gießler, and partly by Rohwer, the interception and decryption of German radio messages and telexes under the cover name ULTRA, still subject to the strictest secrecy even after the war, were not yet publicly known. It was only with Winterbotham's publications from 1974 onwards that the decisive role of radio intelligence at Bletchley Park, which had made an important contribution to the Allied victory in the Battle of the Atlantic, came to light.

9.3 Jamming

An essential factor in gaining superiority on land, at sea, and in the air is the ability to evade enemy detection through artificial jamming. This requires detailed knowledge of the opponent's radar systems, including key characteristics such as transmission frequency, range, and as many device-specific parameters as possible. A distinction is made between active and passive jamming methods. Following the Western campaign in 1940, Germany was able to closely observe British radar systems. To facilitate this, the NVK established a fixed monitoring station near Boulogne-sur-Mer, France, to intercept and analyse signals emitted by British radar installations, particularly from the Chain Home system. It was not until April 1944 that the RLM decided to establish an Air Force observation station at the site as well.[1344] Between 1940 and early 1942, German forces succeeded in determining the frequency ranges of British radar systems used on land, aboard night fighters, maritime reconnaissance aircraft, and anti-aircraft artillery. Concrete evidence also pointed to the existence of fixed radar installations along the British coast at intervals of approximately 50 to 100 kilometres. These stations were used for air raid warning and fighter control. In addition, the British had already

1342 Gießler 1971, p. 55.
1343 Ibid., p. 89.
1344 Reuter 1971, pp. 102–103.

introduced a system for friend or foe identification (IFF). From late summer 1940, the German Air Force and Navy began using ground-based and airborne jamming transmitters against British air and sea targets along the English Channel.[1345]

Of course, the British also had the opportunity to study German radar equipment. Additionally, there were findings from the investigation of the German armoured cruiser Admiral Graf Spee, which was scuttled at the estuary of the Río de la Plata near Montevideo, Uruguay, on 17 December 1939. The ship carried a GEMA Seetakt radar device that operated at a wavelength of approximately 50 to 60 centimetres. GEMA had initially produced a small batch of five of these devices in 1937/38.[1346.]

In 1938, GEMA was commissioned to build a pre-series of 31 units with a final wavelength of 81.5 centimetres under the general designation DeTe I, which were delivered between October 1939 and December 1940.[1347] The majority of the Seetakt devices were used on land for coastal defence. In March 1941, British radar technicians began to interfere with the German radar sets operating in the 80 centimetres wavelength band along the Channel coast.[1348] This was followed by investigations into the defence against British jamming within the Amtsgruppe Technisches Nachrichtenwesen of the High Command of the Navy, though these efforts were not given high priority.[1349] This changed when a British commando unit parachuted near Bruneval, close to Cap d'Antifer in France, on the night of 27th to 28th February 1942, overrunning a Würzburg A position in a combined operation and capturing the antenna and other high-frequency components.[1350] Three prisoners were also taken to Great Britain. This raid, known as Operation Biting[1351], provided the British with reliable information on the Würzburg radar device.

1345 Ibid., p. 104.

1346 Trenkle 1986, p. 116; Kroge 2000, p. 76; Kroge 1998, pp. 88–89.

1347 Trenkle 1986, p. 116.

1348 Reuter 1971, p. 104.

1349 Ibid., p. 104.

1350 Bauer n.d..

1351 Also known in English as the Bruneval Raid, this term refers directly to the raid itself, while Operation Biting was the overarching codename for the entire military action, which included the execution of the Bruneval Raid (Jones 2017, pp. 233–249; Johnson 2004, pp. 105–110; Johnson 1983, pp. 108–

With the help of the captured components, the exact frequency range could now be determined, and there was also certainty that no built-in jammers were present.[1352]

As early as 1 March 1942, a meeting was held on the German side with General Martini, along with representatives from the Amtsgruppe Technisches Nachrichtenwesen of the High Command of the Navy, the Technical Office of the RLM, and representatives from Telefunken and GEMA, to discuss countermeasures against the frequency interference now expected in all Würzburg radar devices.[1353] This was mainly due to the fact that the German radar sets operated on fixed frequency bands and lacked alternative frequency options to counteract interference.[1354] Various types of interference, such as jamming transmitters on airplanes, parachutes, or dropped aluminium foil strips, were considered potential methods of interference.[1355] These aluminium foil strips were known as Düppel in Germany, Window in the UK, and Chaff in the USA (Fig. 33).[1356]

As a countermeasure, Leo Brandt, who oversaw equipment development at Telefunken and had just been appointed to lead the Entwicklungsgruppe Funkmeß, immediately suggested converting the operational radar sets to other frequency ranges, although this proved technically difficult. The attempt to extend the wavelength ranges with additional equipment, 2.4 metres for the Freya and 0.5 metres for the Würzburg radar sets, was given the camouflage designation Wismar. In this context, German air defence faced a significant threat. Consequently, GEMA and Telefunken were commissioned to deliver radar sets with different frequency bands, but initially only small, fixed additional frequency ranges could be

---

113; Brown 1999, pp. 229–231; Reuter 1971, p. 105; Millar 1974). After Operation Biting, the TRE was relocated to Malvern in Worcestershire to protect it from potential German counterattacks.

1352 Jones 2017, p. 245; Reuter 1971, p. 105.

1353 Reuter 1971, p. 105.

1354 Hoffmann-Heyden 1954, p. 161.

1355 Reuter 1971, pp. 105–106.

1356 Leo Brandt coined the name Düppel after the test site of an anti-aircraft battery near the Düppel estate near Berlin-Zehlendorf (Reuter 1971, pp. 106–107; Brown 1999, p. 296); however, there are other interpretations of the origin of this name (e.g. Galati 2016, p. 109).

Fig. 33: Production of aluminium foil strips in Great Britain[1357]

achieved. The next step was to equip as many radar sets as possible with a large number of alternative frequencies. This new development was given the camouflage designation Vollwismar[1358]

After the Bruneval Raid, there was significant concern on the German side that Britain would implement active jamming measures, which could have devastating consequences for German radar technology at the time.[1359] To counter this, jamming monitoring centres were established to detect and analyse enemy jam-

1357 Image credit: IWM (COL 46).

1358 Reuter 1971, p. 106.

1359 Hoffmann-Heyden 1954, p. 161.

ming tactics. As a result, great efforts were made to research and develop countermeasures.[1360] Since no effective defences existed against Window at that time, Göring prohibited any mention of the term Düppel, the German name for this countermeasure, to avoid drawing the enemy's attention. This method of disrupting radar systems was recognised by both Germany and Britain around the same time[1361], but its use remained highly classified in both countries, which delayed its deployment.[1362] This delay in action, particularly in Germany, hampered protective efforts.[1363]

The reason for the delay was that self-protection and deception methods would naturally be detected immediately upon first use. In addition to various jamming techniques, such as transmitters on aircraft or parachutes, the use of metal strips tuned to the transmission frequency of enemy radar became a crucial countermeasure in the following period. The principle of using metal objects to reflect electromagnetic radiation had long been known in radar technology. In May 1942, initial tests were conducted at the RLM test centre in Rechlin, using $\lambda/2$ dipoles made from thin aluminium sheets and later with aluminium foil glued to stiff glassine strips. These tests showed that if the enemy deployed such reflectors, it would result in a disaster for German radar systems.[1364] In March 1943, further secret trials confirmed the effectiveness of this countermeasure during a large-scale test over the Baltic Sea, specifically against the Würzburg radar systems.[1365]

As early as 1937, Lindemann had suggested to Churchill in Great Britain that dropping strips of aluminium foil could disrupt the display of an attacking bomber on enemy radar screens.[1366] The radar would then no longer be able to distinguish whether the re-

1360 Detailed descriptions of developed methods for active interference suppression and turbidity reduction methods for passive interference can be found in Hoffmann-Heyden 1954, pp. 160–173 and pp. 176–236, as well as in Reuter 1971, pp. 105–113.

1361 A little later in the USA.

1362 Brandt 1962c, p. 65; there were also heated discussions about the use of Window in Britain (Johnson 2004, pp. 116–117; Johnson 1983, pp. 121–122).

1363 Reuter 1971, p. 110; Trenkle 1986, pp. 60–61.

1364 Hoffmann-Heyden 1954, pp. 177-178.

1365 Reuter 1971, p. 110.

1366 Watson-Watt 1957, p. 416.

flections were coming from the aluminium foil strips or from an aircraft. If a whole bundle of aluminium foil strips were dropped, a large part of the German radar network, especially the Würzburg devices operating in the wavelength range of about 50 centimetres, could be neutralised. However, the H_2S radar devices with their wavelengths of about 10 centimetres remained largely unaffected.[1367] This idea was not pursued further by the experts at the time and was only taken up again in early 1942 on Lindemann's initiative in top-secret experiments.

After it turned out that light metal foils, such as those used in chocolate packaging, which can be easily dropped in large quantities from an aircraft, met all the necessary requirements, operational investigations into such an application began. The challenge now was whether the short time available would be sufficient to deceive enemy anti-aircraft artillery and prevent defending fighter planes from intercepting the bombers, as the bundles of foil would be carried away by the wind while the attacking aircraft were moving at much higher speeds.[1368]

Although Bomber Command called for the immediate deployment of Window, Fighter Command insisted on secrecy until a defence mechanism had been developed. The reasoning behind this was the same as in Germany: the enemy would likely adopt this form of disruption quickly. On 22 June 1943, during a conference led by Churchill, the British Bomber and Fighter Command staffs decided to deploy Window, based on the premise that the Germans could also use this type of jamming and deception, as it was believed to significantly reduce bomber losses.[1370] The most significant bombing raids involving Window took place during Operation Gomorrah, the RAF air raids on Hamburg, which were carried out between 24 July and 3 August 1943.[1371] During these raids, Pathfinder units equipped with H_2S radar marked targets and

1367 Johnson 2004, p. 116; Johnson 1983, pp. 120–121.

1368 Churchill 1950, pp. 336–337.

1369 Image credit: IWM (C 5635).

1370 Watson-Watt 1957, p. 418.

1371 Operation Gomorrah; Harris 1995, pp. 18–20; Jones 2017, pp. 300–305; Operation Gomorrah involved five night raids by the RAF and two day raids by the United States Army Air Forces (USAAF).

Fig. 34: RAF Avro Lancaster bomber over Essen[1369]

guided the bombers towards them.[1372] The use of Window proved highly successful.[1373]

The raid on the night of 24 to 25 July marked the first large-scale operational use of Window, with an estimated 40 tons, approximately 92 million strips of Window aluminium foil, being dropped. These strips caused numerous false echo signals to appear on the air defence radar screens, effectively disrupting German radar systems. The most devastating raid, which caused the highest number of casualties, took place on the night of 27 to 28 July 1943, when a firestorm engulfed large parts of Hamburg. An example of the use of Window can be seen in Fig. 34, where a Royal Air Force Avro Lancaster bomber drops Window to interfere with FLAK batteries during a large-scale raid on Essen.[1374]

1372 The bombing raid on Hamburg on the night of 30 to 31 January 1943 marked the first operational use of onboard centimetric radar by British Pathfinder aircraft (Harris 1995, Appendix A, p. 70, pt. 74; Lovell 1991, p. xvii).

1373 Reuter 1971, pp. 123–124; Judkins and Bauer 2014, pp. 62; Brandt 1962c, p. 65.

1374 Johnson 2004, p. 117; Johnson 1983, p. 122; Trenkle 1986, p. 60.

The primary radar systems disrupted were those operating in the 50 centimetres wavelength range, notably the Würzburg ground-based systems and the Lichtenstein onboard radar sets of German night fighters. These were heavily affected by onboard jammers and by Window (chaff) released by the attacking bombers, whereas the Freya radar was not significantly impacted. However, the Freya devices could be affected by electronic jamming transmitters on board the bombers.

The disruption of the Würzburg systems made it impossible to guide German night fighters to the incoming bombers, resulting in very few successful interceptions.[1375] The British were surprised by the remarkable effectiveness of Window, which led to unusually low aircraft losses.[1376]

The bombing raid on Mannheim and Ludwigshafen during the night of 17 to 18 November 1943 marked the first operational use of H_2S centimetric radar by the RAF for a blind attack. H_2S enabled bomber crews to accurately identify and strike targets regardless of visibility. It was the first radar system to allow real-time imaging of the ground and specific targets, offering a crucial advantage during night missions or in poor weather.[1377]

The German Air Force intended to use this highly effective jamming device in attacks against British radar stations. Such operations would have required clear air superiority for a surprise attack, as a counter-attack was expected soon afterwards. This condition had not been met since 1942, so after careful consideration, it was decided to postpone the use of Düppel until a later date.[1378]

9.4 Operational Research

The development of Operational Research in Great Britain marked a new method for employing scientific expertise in military strategy and operations. Until then, science had primarily been seen as a source of ideas for new weapons and a means to improve conven-

1375 Reuter 1971, pp. 123–124.

1376 Hoffmann-Heyden 1954, pp. 176–236; Schulze 1988, pp. 405–409.

1377 Harris 1995, Appendix A, p. 71, pt. 80.

1378 Hoffmann-Heyden 1954, p. 178; Trenkle 1986, p. 60.

tional ones. This changed with the emergence of a new field of military research, Operational Research.[1379] This scientific discipline remained entirely unknown in Germany until the 1950s.

According to the Gabler Wirtschaftslexikon, Operations Research (OR), also known as Operational Research, is defined as:

> „Die Zielsetzung des OR ist die Entwicklung und der Einsatz von mathematischen Verfahren zur Unterstützung von Entscheidungsprozessen. Im Rahmen eines Entscheidungsprozesses kann auf die Verfahren des OR zur Entscheidungsvorbereitung, Entscheidungsfindung sowie Entscheidungsdurchführung und -kontrolle zurückgegriffen werden."[1380]

Operational research has been defined in many ways. One definition, provided by the UK Operational Research Society, was published on the first page of its in-house journal, Operational Research Quarterly, in 1967, as follows:

> "Operational Research is the application of the methods of science to complex problems arising in the direction and management of large systems of men, machines, materials, and money in industry, business and defence. The distinctive approach is to develop a scientific model of the system, incorporating measurements of factors such as chance and risk, with which to predict and compare the outcome of alternative decisions, strategies or controls. The purpose is to help management determine its policy and actions scientifically."[1381]

The central idea behind this definition is that operational research is based on scientific methods. At a very early stage in its deliberations, the CSSAD recognised the need to research the operational aspects of radar, as radar had rendered existing fighter interception techniques based on continuous patrols obsolete. However, it was essential to demonstrate that their own fighter aircraft could

1379 Crowther and Whiddington 1948, pp. 91–120.

1380 Lübbecke. "The objective of Operations Research (OR) is the development and application of mathematical methods to support decision-making processes. Within a decision-making process, the methods of OR can be used for decision preparation, decision-making, as well as decision implementation and control." (translated by the author).

1381 Kirby 2003, p. 3; Raju 2020, p. 4.

effectively counter incoming attackers, provided that appropriate conditions of altitude, position, and readiness were met.[1382]

The term Operational Research originally referred to studies conducted by the British Air Ministry that involved the statistical analysis of RAF test flights. The aim was to draw conclusions about the number and type of aircraft based on radar data and thereby support the operational use of radar. Historical accounts trace the origins of the term to Bawdsey in the summer of 1938, when Rowe arrived and tasked E. C. Williams and G. A. Roberts with conducting operational research.[1383] Williams, a university scientist, took responsibility for operations in the control room, while telecommunications engineer Roberts focused on the communications system. Rowe not only played a central role in the development of operational research but also gave it its name.[1384] Operational research gained considerable significance during both the Battle of Britain and the Battle of the Atlantic.[1385]

Patrick Blackett[1386], often referred to as the Father of Operational Research, played a pivotal role in the establishment and development of Operational Research.[1387] He was invited to join the Aeronautical Research Committee (ARC) as early as 1935 by its chairman, Henry Tizard, and was also a member of the CSSAD from 1935.[1388] Tizard had been appointed Chairman of the ARC in 1933.[1389]

Having been embedded in two key committees working on the development of an air defence radar system, Blackett was well-

1382 Kirby 2003, p. 70; Kirby and Capey 1997, pp. 560–562.

1383 Kirby 2003, p. 71; McCloskey 1987b, p. 146; Air Ministry 1963, pp. 6–7.

1384 Cunningham et al. 1984, p. 958.

1385 McCloskey 1987a, pp. 453–470; McCloskey 1987c, p. 910–925; Rau 2005, pp. 156–161.

1386 Nye 2011; Lovell 1988; for an extensive biography of P. M. S. Blackett, including his work in physics, his involvement in wartime research, and his political activities after World War II, see Nye 2004.

1387 Ormerod 2003, pp. 187–200; for the American development of OR and its pioneer Philip Morse, see Morse and Kimball 2003; Shrader 2006, p. 47; Thomas 2007, pp. 251–274.

1388 Zimmerman 2003, pp. 110–125.

1389 Collar 1967, pp. 529–538.

versed in the technical challenges from an early stage. At the outset of World War II, he spent some time at the RAE in Farnborough, where he made significant contributions to the development of bombsights. In August 1940, during the Battle of Britain, Blackett became scientific adviser to General Frederick Pile[1390], who commanded Anti-Aircraft Command.[1391] Blackett organised a group of scientists to study the operational use of radar sets, guns, and mechanical calculators, resulting in improved accuracy of anti-aircraft fire during the Battle of Britain.[1392]

In early 1941, Albert Rowe approached the Deputy Chief of the Air Staff, Air Marshal Wilfrid Freeman, and urged that Patrick Blackett be asked to establish a group for operational research at RAF Coastal Command.[1393] Freeman agreed, and the arrangement was made within a few days.

Rowe summarised:

> "I have said that it was the war at sea which was the greatest source of anxiety to me, not so much in my capacity as Head of the Establishment but as a citizen who was interested in being on the winning side. I trust that my many friends in the R.A.F. have forgiven me for pestering them to the subject. One of our troubles at Swanage was that we did not know what was happening in the battles against the submarines, largely because there was nothing at Coastal Command corresponding to the Operational Research Section at Fighter Command. We wanted to know the hours flown per submarine sighting by day and by night, the average ranges of radar locations and, less obviously our business, the results of attacks on submarines. I therefore went to Air Marshal Sir Wilfred Freeman, Vice Chief of the Air Staff, who was ever ready to listen, and urged that P.M.S. Blackett should be asked to form an Operational Research Section at Coastal Command. This was arranged within a few days."[1394]

In March 1941, Blackett joined RAF Coastal Command and established a scientific group focused on anti-submarine warfare. The

1390 Macksey2011.
1391 Air Ministry 1963, p. 40.
1392 Nye 2011, p. 3.; Nye 2004, p. 75.
1393 Waddington 1973, p. 13.
1394 Rowe 2015, p. 105; McCloskey 1987b, p. 148.

group, soon informally known as Blackett's Circus[1395], brought together a multidisciplinary team of scientists, including physicists, mathematicians, statisticians, engineers, as well as psychologists, chemists, and biologists.[1396] Their early work contributed to improvements in the use of airborne radar to detect German submarines threatening Allied shipping. From January 1942 to the summer of 1945, Blackett headed the Operational Research Section (ORS)[1397] at the Admiralty, where he and many of his former Circus members continued to refine radar techniques and other methods that significantly improved anti-submarine warfare effectiveness.

Conrad Waddington, who was in the operational research department of RAF Coastal Command during the war, reported on the OR group with its interdisciplinary composition:

> "A considerable fraction of the Staff of an O.R.S. should be of the very highest standing in science, and many of them should be drawn from those who have had experience at the Service Technical Establishment. Others should be chosen for analytic ability, e.g. gifted mathematicians, lawyers, chess players..."[1398]

Cecil Gordon, a geneticist who developed planned flying and planned maintenance[1399] for RAF Coastal Command, wrote:

1395 Crowther and Whiddington 1948, pp. 96–97; Nye 2011, pp. 3–4; McCloskey 1987b, p. 148; Haider 2021, pp. 145–152.

1396 Air Ministry 1963, p. xix.

1397 The work of the ORS included analyses of submarine attacks and losses, Atlantic convoy systems, artillery deployment, logistics and supply, as well as, where relevant, the impact of air raids and aircraft losses on naval operations, for example in relation to air support for convoys or the use of radar in aircraft detection at sea. While the ORS was primarily focused on maritime warfare, these related aspects of aerial operations were occasionally considered within that context.

1398 Waddington 1973, pp. 8–9.

1399 Planned flying referred to the deliberate and data-driven scheduling of patrol flights to optimise coverage and detection probability, rather than relying on ad hoc decisions. Planned maintenance involved proactive servicing of aircraft at regular intervals based on operational data. This approach reduced breakdowns and increased readiness, and it was a key innovation compared to the traditional method of repairing aircraft only after failures occurred.

> "The complete disregard for frontiers between the different subjects, and the readiness to accept any problems as within their terms of reference, has been a refreshing contrast to the rigid specialization that has developed in all other branches of science. The Operational Research Sections have recaptured the atmosphere of the period of the foundation of the Royal Society."[1400]

Like Gordon, many of the British and Commonwealth scientists who worked in the field of OR were leading experts in their own disciplines yet found themselves applying their skills far outside their original areas of specialisation.

In contrast to Lindemann, who was always in favour of area bombing, Blackett, like Tizard, believed that resources would be better allocated to more effective missions, as his studies had demonstrated the ineffectiveness of area bombing. Blackett gave high priority to fighting German submarines, which were severely hampering the war effort by sinking Allied merchant ships in the Atlantic, a stance that led to conflicts with the military command.

Watson-Watt later explained in his book "Three Steps to Victory":

> "If I was not *the* first and true inventor of Operational Research, which was no less important to victory than was Huff-Duff, and was not incomparably less important than radar, I was at the very least *a* first and true inventor with one or two members of my early staff at Bawdsey. Perhaps I can, as usefully as may be, tell part of the story of O.R. by reproducing something I have written about it in the Proceedings of the Institute of Radio Engineers."[1401]

Although Watson-Watt is primarily known for his developments in radar, his scientific approach to solving military problems indirectly contributed to the emergence and application of Operational Research during World War II.

In Germany, the history of OR did not begin until the mid-1950s.[1402] OR methods, which had already formed the basis for political and military decisions in countries like Great Britain

1400 Shrader 2006, p. 11.
1401 Watson-Watt 1957, p. 200.
1402 Zimmermann 2008, p. 10.

and the USA, were largely unknown in Germany. Additionally, the use of natural and technical sciences for military purposes was largely ignored, which prevented the German leadership from critically examining the limitations of the country's armaments industry.[1403]

9.5 Interaction

The question of whether radar, radio reconnaissance, radio direction finding technology, or operational research contributed most to military success remains open. Radar and radio direction finding technologies were directly associated with the effectiveness of weapons, while the deciphering of German military radio transmissions, for example, was more closely tied to the operational planning and decision-making of the Allied forces.

There is broad consensus that ULTRA significantly shortened the war by providing the Allies with a substantial information advantage. In the first half of the war, it helped compensate for various weaknesses, while in the second half, it enabled more efficient use of resources.[1404] One indication of this impact is a letter of thanks from General Dwight D. Eisenhower to the head of the British Secret Service, General Stewart Menzies, in July 1945. Eisenhower expressed his gratitude for the success in obtaining enemy intelligence through Bletchley Park, which, he stated, had saved thousands of British and American lives and contributed to the quicker defeat and surrender of Germany.[1405] Although there has been some debate over how decisive ULTRA truly was in winning the war, the dominant historical view recognises it as a major strategic advantage. Earlier breaches of ENIGMA security, such as the leak to French intelligence in the early 1930s, laid the groundwork for later Allied codebreaking efforts, but it was the wartime success of Bletchley Park that proved critical.

1403 Ludwig 1974, p. 230.

1404 Pröse 2004, p. 1; Kahn 1979, pp. 40–47.

1405 Winterbotham 1999, p. 2; Winterbotham 1976, p. 14; Rohwer 1969a, p. 368; Ambrose 1981, pp. 153–166.

David Kahn was one of the first to ask why the Allied cipher services were significantly more successful than their German counterparts during World War II. He identifies four technical and five general reasons for this superiority. Among the technical factors, he emphasises the Allies' early understanding of the structure and functioning of the ENIGMA machine, primarily thanks to the work of Polish cryptanalysts before the war. Although ENIGMA was not the only encryption device used by the Germans, with more complex machines such as the Lorenz cipher used for high-level communications, it was by far the most widely employed. In addition, the German keying procedures were flawed, relying on predictable message structures and repetitive phrasing, which made codebreaking easier for the Allies. Among the five general reasons, Kahn primarily cites the fragmentation of the German cipher services and, as a consequence, the lack of concentration of the experts on important tasks. He points out that various military and intelligence agencies operated independently and without centralised coordination. In contrast, the Allies established centralised institutions such as Bletchley Park, where expertise was brought together effectively. Another major factor was the difference in strategic orientation. Britain and France pursued a defensive posture and therefore relied heavily on intelligence and reconnaissance. Nazi Germany, with its offensively driven war strategy, treated intelligence as a secondary concern. The Allies, by contrast, made early and systematic investments in recruiting and training highly qualified personnel, often from universities. Germany also suffered from a rigid bureaucracy in which admitting mistakes was discouraged due to fear of political repercussions under the Nazi regime. Another structural disadvantage was the German reluctance to invest in new and costly technologies. In Britain, the use of vacuum tube technology and the development of electronic decryption machines such as Colossus marked a major breakthrough.

No equivalent efforts were made in Germany. Kahn also highlights the loss of scientific expertise caused by the expulsion of Jewish academics, particularly in the fields of mathematics and physics. Many of these scientists later contributed significantly to Allied research and cryptographic projects. Finally, the Allies benefited from the presence of exceptional individuals such as Alan Turing. The close collaboration between mathematicians, engineers, and linguists enabled the Allies, beginning in 1943, to decipher large volumes of German radio traffic. A similarly interdisciplinary approach was never implemented in Nazi Germany.[1406]

According to Jürgen Rohwer, detailed investigations of the major convoy battles surprisingly revealed that, on the British side, the HF/DF short-wave direction finder had a much greater influence on the majority of operations than centimetre-wave radar, which was regarded as a fundamental element in turning the tide of submarine warfare.[1407]

9.6 Further Aspects

9.6.1 Scientific Intelligence

In addition to the technological measures described above for reconnaissance of enemy operations, a number of other intelligence activities played a significant role. One example is the British physicist Reginald Jones[1408], who was successful both as a scientist and as an employee of the secret service in a wide variety of fields.[1409] After completing his doctorate in 1934 under Professor Lindemann at the Clarendon Laboratory in Oxford, he worked on air defence issues at the research centre of the British Air Ministry from 1936.

On 1 September 1939, he became an employee of the UK's Secret Intelligence Service (SIS).[1410]

The focus of his work was analysing the state of German military research, particularly in the fields of radar, rockets, and infrared technologies, and designing appropriate countermeasures to

1406 Kahn 1979, pp. 42–46.

1407 Bekker 1980, Afterword, J. Rohwer, p. 278.

1408 Martin 2011; Cook 1999; Goodchild 2013; Jones 2017; Jones 2009; Jones 1990.

1409 Jones 1993, pp. 667–684.

1410 Martin 2011, p. 1; Rennert 2021, pp. 27–29; the Secret Intelligence Service (SIS) was officially established in 1909 as the United Kingdom's foreign intelligence agency. The designation MI6 (Military Intelligence, Section 6) was used informally within the British government, particularly during and after World War II, but never served as the agency's official name. Today, MI6 is widely used in public discourse, although the organisation continues to operate under its formal title, SIS (Jeffery 2011).

mitigate the impact of these innovations on the British war effort. His role involved not only gathering intelligence on German scientific developments but also devising strategies to neutralise these advances, often through the use of deception, disruption of enemy technologies, and the development of new defensive systems for the British military.[1411] Additionally, he worked on countering the German Knickebein guidance system by analysing deciphered radio messages from the German Air Force at Bletchley Park. Knickebein was a radio system used for target detection and bombing assistance with great precision.[1412]

Jones emphasised that defence measures against the Knickebein threat were as important as radar development:

> "This had been our only defence; night fighters, until airborne radar was good enough, were powerless, anti-aircraft guns were inaccurate, therefore we could not stop the bombers. Our only hope was to throw them off. Well, we didn't always succeed but we did certainly on a fair number of occasions, with the result that a good many people were alive at the end of the war who otherwise wouldn't have been and a good deal less vital damage was done… If one thinks of what could have happened if with Knickebein all the German Air Force could have bombed accurately. Well, we saw what happened when things went wrong as at Coventry… That could have happened every night."[1413]

Jones had the idea of jamming radar with strips of aluminium foil as early as the late 1930s[1414], a concept that was later further developed and successfully implemented by Joan Curran[1415] and her team in 1942. This technique, known as Window or Chaff, became a crucial part of Britain's wartime strategy, as it allowed the RAF to deceive German radar systems and significantly reduce the effectiveness of enemy bombing raids. Jones was also actively involved in analysing the threat posed by the V-2 rockets developed by Germany during

1411 Jones 2017.

1412 Jones 2017, pp. 92–105; Judkins 2007, p. 287, p. 333; Johnson 2004, pp. 11-61; Johnson 1983, pp. 11–62; Trenkle 1987, pp. 128–132.

1413 Johnson 2004, p. 61; Johnson 1983, p. 62.

1414 Jones 2017, pp. 39–40.

1415 Fletcher 2004; Latham and Stobbs 1999, pp. 196–197; Jones 2017, pp. 291–292; Brown 1999, pp. 295–296.

the war. As an expert in air warfare and new technologies, Jones was a member of the British Cabinet Defence Committee. In this role, he advised on strategies to defend against the V-2 rockets and other advanced German technologies that threatened the British war effort.

9.6.2 Human Intelligence

Another field for gaining knowledge about the enemy was the evaluation of information from non-technical sources, which could be obtained with the help of various groups of people. These included prisoners of war, agents, refugees, and also contacts in neutral countries. A comprehensive overview of British enemy reconnaissance during World War II, with a particular focus on the role of Human Intelligence in British warfare from 1939 to 1945, including case studies, can be found in Falko Bell's dissertation.[1416] The case study on the Battle of Britain in 1940/41, in particular, is an important contribution by Bell to understanding the situation and the resulting consequences for both personnel and material.[1417] Another example of intelligence gathering with far-reaching consequences was the evaluation of the Oslo Report by Jones.

9.6.2.1 Oslo Report

At the beginning of November 1939, the British legation in Oslo received a seven-page anonymous document detailing the latest developments in military research and technology in Nazi Germany. This document, later known as the Oslo Report, contained references to key military research projects, information about new types of aircraft, and details about the previously unknown Air Force test centre at Rechlin. In total, the Oslo Report covered eleven topics, including the Ju 88 program, a large carrier aircraft designed to deploy smaller, remote-controlled gliders, an autopilot system for aircraft, remotely guided rocket-powered weapons, Air Force laboratories and development centres at Rechlin, new methods for

1416 Bell 2016, p. 12; Bell 2015, pp. 556–578.
1417 Bell 2016, pp. 119–194.

attacking bunkers, air raid warning systems and countermeasures, range-finding devices for tracking enemy aircraft, and new types of torpedoes for the German Navy. Finally, the anonymous author also discussed electric fuses for aerial bombs and artillery shells.[1418]

As proof of the credibility of his report, the anonymous author enclosed a glimmer lamp, a component used in the electric detonators being developed by the Germans. The British initially questioned the credibility of the Oslo Report, suspecting it was deliberately misleading. They found the level of detail and the technical range too vast for a single person to have written it, which raised suspicions.[1419]

The document was evaluated by Jones, an intelligence officer who had been with British intelligence since 1 September 1939, and he was quickly convinced of the authenticity of the information.

Jones later reported:

> "...Fred Winterbotham came into my room and dumped a small parcel on my desk and said, 'Here's a present for you!' I asked him what its background was and he said that it had come from our Naval Attaché in Oslo. This was after the Attaché had received a letter, dropped privately through his letterbox, saying that if the British would like to know about various German scientific and technical developments, would we alter the preamble to our news broadcasts in German so as to say, 'Hullo, hier ist London...' instead of the normal preamble, and then our would-be helper would know that it was worth while to give us the information. The change had been duly made, and a package had then been put through the letterbox."[1420]

Despite some errors in the description of general armaments projects and the complete absence of lists of planned military operations by the Germans, which led some British intelligence officers to dismiss the report as largely useless, Jones concluded that the author must be a proven expert, especially in the field of electronics. The author of the Oslo Report never contacted the British authorities or representatives of other Allied countries again.

1418 Jones 2017, pp. 67–71; Jones 1990, pp. 265–275, pp. 333–337; Rennert 2021, pp. 195–201; Johnson 2004, pp. 9–10; Johnson 1983, pp. 9–10; Hagenauer und Pabst 2014, pp. 96–101.

1419 Jones 2017, pp. 69–71.

1420 Jones 1978, p. 68; Johnson 2008, pp. 107–108.

It wasn't until 1954 that Jones managed to find out the identity of the author, thanks to a chain of coincidences. He then met the author a year later at a scientific congress in Munich.[1421] He and Jones agreed that neither the public nor his family should find out who was behind the report. This decision was made to protect the author and his family from potential hostility, which was still quite conceivable in the 1950s.[1422] Jones did not reveal the author's name until 1989 in his work "Reflections on Intelligence".[1423]

So, who was the author of the Oslo Report? His name was Hans Ferdinand Mayer[1424], born on 23 October 1895 in Pforzheim. After studying mathematics, physics, and astronomy from 1916 to 1920, first at the Technical University in Karlsruhe and then at the University of Heidelberg, he obtained his doctorate in 1920 under Professor Philipp Lenard at the Institute of Physics and Radiology.[1425] Following his time as an assistant to Professor Lenard until 1922, Mayer joined Siemens-Werke in Berlin. There, he became a prominent figure in the company's central laboratory, eventually taking over its management in 1936. Mayer played a pivotal role in advancing research in electronics and technology, which laid the foundation for his later contributions to the development of military and aerospace technologies, reflected in the Oslo Report.

On 15 June 1943, Mayer was arrested by the Gestapo for political reasons. He was accused of listening to enemy broadcasts, spreading enemy propaganda, making anti-state statements, insulting the Führer, actively supporting Jews, sabotaging work by misusing the manpower of the central laboratory, and being suspected of contact with enemy powers through secret transmitters. He was also accused of betraying military secrets and was interrogated by the Gestapo until 25 August 1943.[1426] However, at no time was the Oslo Report uncovered, nor was Mayer identified as its author. On 25 August 1943, Mayer was transferred to the Sachsenhausen

1421 Jones 1990, p. 315; Rennert 2021, pp. 179–183.

1422 Rennert 2021; pp. 180–181.

1423 Jones 1990, pp. 320–327.

1424 Jäger und Heilbronner 2010, p. 280; Jones 1990, pp. 320–332; Schoen 1990, pp. 539–540; Hagenauer und Pabst 2014, pp. 14–15; Michel 1994, pp. 85–90.

1425 Jäger und Heilbronner 2010, p. 263.

1426 Hagenauer und Pabst 2014, p. 24.

concentration camp, and on 5 September of the same year, he was moved to the Dachau concentration camp.[1427]

In August 1943, a high-frequency research unit was established at the Dachau concentration camp. At the suggestion of Siemens & Halske, telecommunications and high-frequency specialists from various camps were gathered. Mayer assumed a de facto scientific leadership role and worked with prisoners from twelve nations. As previously mentioned, Himmler, as Reichsführer-SS, had a strong interest in advanced communication technologies and received direct reports from Plendl in October 1943. At that time, the regime had begun to recognise its growing lag in high-frequency research. In July 1943, the Reichsstelle für Hochfrequenzforschung (RHF) was established to centrally coordinate such efforts. In the summer of 1943, around 1,500 engineers and technicians were reassigned to this field as part of the recall campaign.[1428]

Due to the increasing Allied air raids on the Munich area, the high-frequency research unit at Dachau was relocated to the Groß-Rosen concentration camp in Lower Silesia in June 1944, following a decision made in autumn 1943.[1429] Owing to the strategic importance of this wartime operation, the prisoners involved were granted relatively improved conditions within the camp.[1430]

In June 1944, the BHF commando consisted of 30 prisoners with various professional qualifications, under the leadership of Mayer.[1431] In February 1945, as the Soviet army advanced, the prisoners were transferred to the Mauthausen concentration camp in Austria under inhumane conditions, which resulted in numerous deaths among the inmates.[1432] That same month, Mayer managed to secure the further transfer of his commando to the Sachsenhausen concentration camp, where the technical equipment had

---

1427 Ibid., p. 26.

1428 The recall campaign was a Nazi initiative to reallocate skilled engineers and technicians, many of them withdrawn from military service, into high-frequency and radar research in order to accelerate technological development in response to Allied advances; see also Chapter 8.1.

1429 Hagenauer und Pabst 2014, p. 28.

1430 Ibid., p. 29.

1431 Ibid., pp. 27–28.

1432 Ibid., pp. 29–30.

already been sent by mistake. There, the unit was reorganised under the new name Zentralinstitut für Hochfrequenz- und Fernmeldetechnik, Reichsführer SS.[1433] During another forced transfer on 21 April 1945 from Sachsenhausen to Bergen-Belsen, again under horrific conditions, Mayer and 250 other prisoners were released on 27 April. He went into hiding on a farm until 5 May, and after another stopover reached Berlin on 12 May 1945, where he reported back to his former employer, Siemens & Halske.[1434]

From 1946 to 1950, Mayer initially worked as a scientist for the Air Force in Dayton, Ohio, and served as a full professor of communications engineering at Cornell University in Ithaca, New York. After his return to Germany in 1950, he took over the management of research, communications engineering laboratories, and design at Siemens & Halske from 1951 to 1962. His outstanding technical knowledge and associated pioneering work are documented in numerous publications and patents. He received numerous professional honours, including the title of Dr.-Ing. E.H. from the TH Stuttgart in 1956.[1435] Mayer died in Munich on 16 October 1980.

Several attempts by Jones to determine the identity of the author of the Oslo Report after the end of the war were unsuccessful. As previously described, Jones came across the name Mayer by chance in 1954, and he and Mayer first met at a scientific congress in 1955. From that meeting onward, Jones was certain that Mayer was the author of the Oslo Report. Mayer always signed his correspondence with Jones as "The Oslo Person".[1436] It was only two years before his death that Mayer revealed to his wife and son that he was the author of the Oslo Report. Jones also respected the wishes of Mayer's wife, Betti, who requested that the author's identity not be disclosed until after her death. Therefore, the identity of the author of the Oslo Report was not revealed by Jones until after her death in March 1989.[1437]

1433 Ibid., p. 30.

1434 Ibid., p. 30.

1435 Michel 1994, p. 88; Rennert 2021, p. 183.

1436 Jones 1990, p. 323.

1437 Karl-Heinz Ludwig already discussed Mayer and his time in the concentration camp in his 1974 book "Technik und Ingenieure im Dritten Reich", although the Oslo Report had not yet been publicly attributed to Mayer at that time (Ludwig 1974, p. 490).

What was the motivation for writing the Oslo Report? Mayer demonstrated his opposition to the Nazi regime through various personal acts of resistance. He reportedly posted notes with the slogan "Nieder mit Hitler"[1438] on shop windows in Berlin, and in 1942, together with his brother, put up signs reading Adolf Hitler ist unser Untergang[1439] on park benches in Baden-Baden.[1440] At Siemens, he consistently advocated for Jewish and so-calles half-Jewish employees and increasingly expressed his disapproval of the regime openly. This ultimately led to his arrest by the Gestapo on 15 June 1943. His friendship with his British business colleague Henry Cobden Turner was certainly the decisive factor in his decision to write the Oslo Report, alongside his experiences abroad as a German in the aftermath of Kristallnacht.

Mayer had a close relationship with Turner that went beyond their professional life. At a specialist conference in London on 11 December 1938, Turner heard from Mayer about the fate of an 11-year-old girl named Martyl Karweik, who had become a victim of the anti-Jewish legislation.[1441] Her father, an anti-Semitic National Socialist, had separated from his Jewish wife, the girl's mother. The mother was expelled from Germany and emigrated to the USA but was unable to take her daughter with her. With the help of a passport officer from the British embassy, who was also an MI6 secret service agent, Turner finally succeeded in obtaining a visa for Martyl to enter England. She was allowed to travel and join the Turner family in Manchester on 30 January 1939.[1442]

Martyl lived with the Turner family for seven years before joining her mother in New York in 1946, where she later married. Her mother had travelled via France, Spain, and Portugal to New York in 1941. It was through Turner's support and actions during this period that Martyl's life was ultimately saved.[1443]

1438 Down with Hitler.
1439 Adolf Hitler is our downfall.
1440 Hagenauer und Pabst 2014, p. 23.
1441 Rennert 2021, pp. 136–140; Hagenauer und Pabst 2014, p. 22.
1442 Hagenauer und Pabst 2014, p. 22; Rennert 2021, pp. 136–137.
1443 Rennert 2021, pp. 138–139.

10 Innovation Systems in Germany and Great Britain

10.1 Framework

An examination and evaluation of German and British radar development, along with additional technological influencing factors in the context of political events, is always associated with a certain lack of clarity due to its complexity and increasing level of detail and requires a thematic limitation.

Below is an analysis of the respective innovation systems in Germany and Great Britain, focusing on the actions of selected interest groups and actors who were involved either individually or through collaborations in the emergence, dissemination, and application of scientific or technological knowledge.[1444] The promoter model provides an approach for identifying key individuals within an innovation process and for explaining barriers to innovation. However, the model has its limitations in the context of historical analysis, as retrospective empirical verification is challenging.

According to Hauschildt et al.[1445], an innovation system encompasses all persons who actively contribute to the realisation of innovations as participants, but also those who are passively affected by the innovation and contribute to its success through their reaction. Since the relationship between participants and those affected often occurs spontaneously and without regulation, the innovation system must be an open system that allows for the emergence and termination of these relationships.

The innovation system must therefore provide opportunities to initiate, try out, and ultimately institutionalise spontaneous interactions.

1444 Blättel-Mink und Ebner 2020, pp. 1–2.

1445 Hauschildt et al. 2016, pp. 90–93; for further information on modern innovation management, see Tidd and Bessant 2018; Vahs und Brem 2013; Hölzle und Gemünden 2011, pp. 495–512; Mirow et al. 2007, pp. 101–134.

As a consequence, an effective innovation system provides resources for self-organisation; prescriptions or prohibitions are disruptive factors in the structuring of innovation processes. This implies that thinking in terms of connections and relationships is more important than rules or even regulation.[1446]

10.2 Procedure

The attitude of politics, ministries, and the military towards cooperative collaboration with science and industry had far-reaching consequences for the innovation strategy of the participants in Germany and Great Britain, with strongly differing approaches being evident. A key strength of Great Britain lay in the open communication between all parties involved, as well as in the involvement of opponents in technological developments and planned deployment scenarios. At the same time, a small circle of politically responsible individuals made groundbreaking decisions. The approach to innovation barriers and their overcoming differed greatly in both countries, which had profound consequences for technological progress and the superiority of applications during the war years. This is especially evident when comparing the development of radar devices in the centimetre wavelength range, where Great Britain, through a consistent innovation policy, gained an almost uncatchable lead during the course of the war.

A closer look at the elements of a successful innovation culture, using the example of Great Britain, shows that trustful collaboration at all levels, combined with the acceptance of the diversity of those involved and a strong identification with the respective work, formed essential cornerstones for progress. Responsible for this was the continuous creation of an innovation-friendly environment by the respective responsible individuals, who, at a time when only vague ideas about future developments and applications existed, made financial and personnel resources available in an unbureaucratic manner.

Open communication was also reflected in the willingness to engage in trustful collaborations with other countries, even in the

1446 Hauschildt et al. 2016, p. 91.

most secretive areas of technological developments and inventions, in order to accelerate and expand development and production potential. An example of this is the Tizard Mission to the USA in September 1940, which provided a tremendous boost to the further development and production of radar devices in the centimetre-wave range through close collaboration, particularly between British and American scientists.[1447]

In Germany, for the first time, the establishment of the Arbeitsgemeinschaft Rotterdam (AGR) in February 1943, under industrial leadership, with specific development goals and an interdisciplinary team consisting of representatives from industry, research institutions, the Wehrmacht, and other government bodies, as well as an unobstructed flow of information, enabled Germany to largely close the technological gap with the Allies by the autumn of 1944. However, by that time, this had little strategic significance.[1448] Although any historical analysis can only cover specific areas, significant political and military decisions in Germany and Great Britain can still be identified, which had a major impact on the course of World War II. Taking into account the numerous influencing factors in both countries, there is, however, a need for a detailed investigation in many areas, as seemingly unremarkable discussions and decisions had a significant impact on the course of events.

The differing interests and goals of external and internal interest groups played a key role in the respective innovation process, set against the backdrop of innovation culture and strategy. Interest groups are considered to be institutions, communities, and individuals who either had an impact on decisions or were affected by them. Especially in Germany, the involved circles had a variety of diverse self-interests and acted accordingly, whereas in Great Britain, measures were largely unified and focused, both on the political and military as well as the scientific level. In Great Britain, it was consistently possible to resolve emerging conflicts through the intervention of responsible bodies at all levels.

1447 Phelps 2010.

1448 Handel 2002, pp. 250–270.

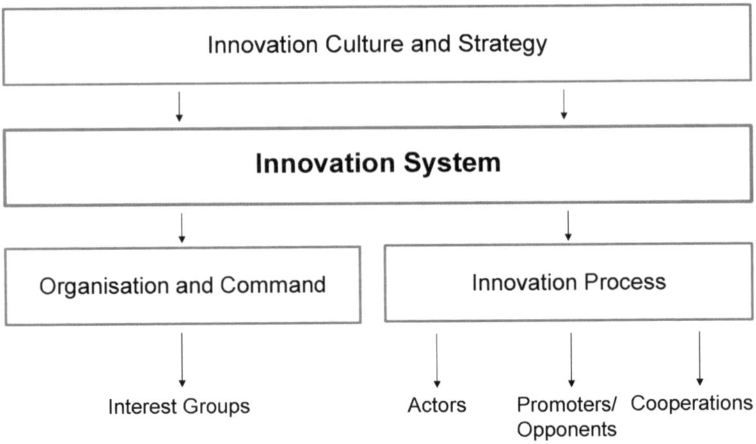

Fig. 35: Elements of the Innovation System

10.3 Elements of the Innovation System

The innovation systems in Germany and Great Britain analysed in the following are primarily described here in accordance with Hauschildt et al.:

- Innovation Culture and Strategy
- Innovation Process
- Organisation and Command
- Interest Groups and Actors
- Cooperations.

Fig. 35 shows the elements of the innovation system considered in this work, in contrast to today's complex systems, which feature expanded communication and interaction structures within the framework of modern innovation management.

While the impact of technical factors on achieving military objectives can be described in relatively simple terms, analysing the institutional and political influences on government actions is considerably more complex, and their effects tend to be indirect. A comparison between Germany and Great Britain clearly shows that the respective political frameworks had a strong influence, either promoting or inhibiting innovation, on technological and operational measures, and were ultimately responsible for the lead or lag in the course of the war. In contrast to Great Britain, where a consistent defence policy was pursued overall despite various political and scientific disputes, Germany, due to the polycratic structure of the Nazi regime, often lacked a structured and coordinated approach. This further complicated the identification and characterisation of key promoters and opponents.

A closer examination of key influencing factors requires further differentiation.[1449] This section does not revisit the fundamental discussion on the Nazi regime's hostility toward science and technology, or the self-mobilisation of science[1450], as these topics have already been discussed in detail in Chapter 8.1. It would also go beyond the scope of this work to attempt a fundamental assessment of the Nazi regime's scientific policy concepts. It has also been shown that the absence of collaboration did not necessarily hinder progress in research and development. In the field of high-frequency technology, R&D activities were carried out not only by numerous fragmented institutions, but also by companies such as GEMA and Telefunken.[1451]

Other large German companies, such as Siemens, AEG, Lorenz, Blaupunkt, and Pintsch, also developed and manufactured their own devices or supported the production activities of the companies mentioned above through their manufacturing capacities.[1452]

When analysing the British innovation system, in addition to the consistent cooperation between politics, ministries, the military, science, and industry, the ongoing dispute between Tizard and Lindemann plays a significant role, particularly in the context of Lindemann's close friendship with Churchill. In this regard, German-speaking literature largely relies on the accounts of C. P. Snow, who portrayed Lindemann and his role in a highly critical manner.[1453] Recent investigations into Lindemann's role present

1449 Maier 2002.

1450 Dinçkal et al. 2010.

1451 In his reports of 16 December 1943 and 7 January 1944, Plendl states that, at the time of his appointment as BHF on 14 November 1942, the German research capacity was divided into approximately 100 different, mostly small, and very small laboratories (BArch, NS 19 / 2057).

1452 Godt 2003, p. 221.

1453 Weintraub 2011; Snow 2013; Snow 1961b.

a more nuanced picture, although the well-documented disputes between Tizard and Lindemann within the CSSAD ultimately led to Lindemann's departure from the committee. However, from today's perspective, his role in the development of radar technology for defence against air attacks requires a more detailed and neutral reassessment.

The following analysis focuses on the factors directly relevant to the innovation system in relation to radar technology and its application. Broader technological and operational measures are also considered, insofar as they were significant to the course of the war in connection with radar development in both countries. Even with the inclusion of these additional framework conditions, this analysis of the respective innovation systems in Germany and Great Britain can only be regarded as another contribution to the ongoing process of examining the innovation steps and barriers under the political conditions in both countries.

On the British side in particular, there are very different interpretations of the actions of individuals who played a key role in the development of radar technology and defence against air attacks. The often-cited statement that radar technology was primarily responsible for the Allied victory must be subjected to critical scrutiny.[1454] This is particularly relevant in light of the additional technological and operational measures, as discussed in Chapter 9, which, in conjunction with the use of radar technology, contributed to a comprehensive defence system against German air attacks.

10.3.1 Innovation Process

While technological developments in the field of radar technology occurred almost simultaneously in both countries in the early stages, the innovation processes differed significantly from the outset. Although the interest groups were similar in structure, their functions and actions diverged considerably. In Great Britain, a highly coordinated approach to technological development was established early on, with radar technology promptly integrated into

1454 Kern 1984, pp. 10–11.

a comprehensive air defence system, supplemented by radio intelligence at Bletchley Park, radio direction finding, and operational research. As early as 1940, there was also a pioneering collaboration with the USA as part of the Tizard Mission.

To gain a deeper understanding of these different innovation processes, the following analysis applies the well-known promoters' model from innovation research. This model allows for the systematic classification of the specific roles and contributions of different promoters in the innovation process.[1455] The primary task of the promoters is to actively drive innovation and transformation processes by reducing or overcoming barriers related to willingness or ability.[1456] This model can be used to analyse the interactions between government, the military, science, and industry before and during the war, involving all parties in the conflict. It was the collaboration of promoters at different levels, each with specific tasks, that had a significant impact on the success of implementing the innovation strategy and overcoming innovation barriers.

10.3.2 Innovation Culture and Strategy

A variety of publications on the development of radar technology before and during World War II primarily focus on the competition among scientists, engineers, and technicians in the context of advancements in device development and their potential applications. The radar principle was simultaneously recognised in various countries, leading to practically identical solutions for its implementation, although military secrecy meant that cooperation occurred only occasionally.

However, it became clear that effective innovation management, under favourable political conditions, could significantly influence the respective lead in research and development. The way in which the innovation culture was shaped by the quality of cooperation among all those involved must not be overlooked. In terms of

1455 Hauschildt et al. 2016, pp. 183–209; Hölzle und Gemünden 2011, pp. 495–512; Gemünden et al. 2007, p. 412; Folkerts 2001, pp. 27–43.

1456 Hauschildt et al. 2016, pp. 188–195; Folkerts 2001, pp. 27–43.

technology and equipment development, Great Britain was initially unable to compete with Germany at the outset of the war.

However, a major advantage of British air defence was the early integration and evaluation of information obtained from a network of radar stations within an air defence system that had already been designed before the start of the war.

The goal was not to develop highly advanced and perfected radar technology, but rather its rapid operational use. A crucial factor was how new insights and inventions were quickly put into practice.[1457] With its defence system, Great Britain took a leading position globally. Additionally, the mental resilience of the population, especially during the Battle of Great Britain, played a decisive role in their endurance.[1458]

Watson-Watt expressed this in retrospect in 1945:

> "It cannot be too often or too firmly stated that the victory over Germany was essentially a victory of the spirit. No technical devices can turn the scale of war save as tools of the spirit. Nor is there, indeed, any real gap between spirit and material, though the very mechanical perfection of our new technical devices may hide the vital fact that they are not only the tools but also the products of the spirit. Radiolocation is perhaps the best of all examples of the interaction of spirit and technique in the forging of a decisive weapon of war. But its origins in the spirit of science may be temporarily lost to view behind its successes. The curiosity-value of its devices, the wealth of its applications to all forms of warfare, and its spectacular victories may dazzle the casual observer."[1459]

The innovation culture and strategy thus formed a connecting element of the innovation system, enhancing innovative strength through positive incentives in both organisation and management.

1457 A key success factor for progress in both the technological and operational areas in Great Britain was the cooperative collaboration between military and civilian scientists, regardless of military rank or civilian status.

1458 In both Germany and Great Britain, the determination of the population to hold out was completely underestimated by the respective opposing side.

1459 Watson-Watt 1945, p. 321.

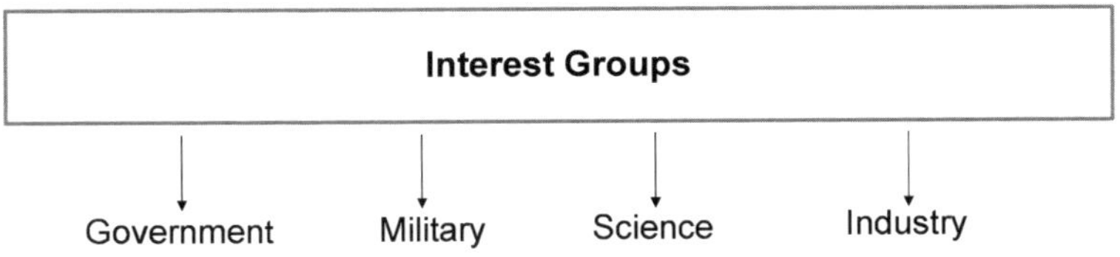

Fig. 36: Interest Groups in Germany and Great Britain

10.3.3 Organisation and Command

A comprehensive analysis of the power structures and interests of the various influencing groups is not possible due to the almost unmanageable variety of individual interests and the changing constellations of cooperation and rivalry. It would also be necessary to analyse smoothly functioning collaborations as counterexamples and to provide a conclusive assessment that takes into account the persistent shortage of skilled workers, which was a particularly pressing issue in Germany. However, this is likewise not feasible.

Therefore, the key interest groups involved in the decision-making processes for significant research projects, as well as their direction and development in the field of radar technology in Germany and Great Britain from the beginning to the end of the war, are presented below based on Fig. 36.[1460]

10.3.4 Promoter Model

10.3.4.1 Approach

Donald A. Schon, in his 1963 publication "Champions for Radical New Inventions", described the driving force behind the realisation

1460 The term Government serves as an umbrella term, encompassing both political institutions and ministries, which together form essential components of a state's governing structure. In addition to these governmental bodies and ministries, numerous committees played a crucial role in coordinating efforts across various sectors. These committees facilitated collaboration between the military services, scientific institutions, and industries involved in the development of radar technology. By bridging the different spheres of governance and expertise, they ensured that the necessary resources, knowledge, and strategies were aligned in support of the wartime technological advancements.

of innovation projects as a person who invests their heart and soul into a new cause.[1461] Schon referred to this person as a Champion. The champion combines several roles, and according to the Great Man Theory, the success of an innovation project is often attributed to this outstanding individual. Schon also derived the concept of the Product Champion as a person who has enough power and influence within the organisation, as well as the necessary expertise, to develop and drive ideas forward.[1462]

While in American research, the champion is primarily seen as an individual, promoter research in the German-speaking world refers to various types of promoters with their specific sources of power and roles.[1463]

This is primarily about the division of labour and collaboration among various types of promoters, each contributing specific types of expertise and having the ability to overcome specific barriers.[1464]

In 1973, Eberhard Witte[1465] developed a model of task division within the innovation process, distinguishing between two types of promoters: the Expert Promoter, who possesses specialised knowledge, and the Power Promoter, who holds influence and authority.[1466] Through this promoter model, Witte saw an opportunity to overcome innovation barriers. He was the first to use the term promoter for individuals who intensively support an innovation process.[1467] Witte's model was later expanded by Hauschildt/Chakrabarti to include the Process Promoter.[1468] Subsequently, Gemünden/Walter further expanded the model to include the Relationship Promoter.[1469]

1461 Schon 1963, p. 84; Gemünden 2010, pp. 86–107.

1462 Schon 1963, p. 85; Gemünden 2010, p. 87; Hölzle und Gemünden 2011, p. 498.

1463 Hauschildt et al. 2016, pp. 186–188, pp. 204–209.

1464 Gemünden 2010, p. 88.

1465 Dowling und Kretschmer 2018.

1466 Witte 1973b; Witte 1973a, p. 17–24.

1467 Hauschildt et al. 2016, p. 189; Hölzle und Gemünden 2011, p. 501.

1468 Hauschildt und Chakrabarti 1988, pp. 378–388; Gemünden 2010, p. 89.

1469 Hauschildt et al. 2016, p. 187; Gemünden later extended the promoter model by adding the roles of the Technological Gatekeeper and the formally institutionalised Project Manager (Gemünden 2010, pp. 90–91).

Especially when considering critical innovation factors within an innovation system, promoters play a central role. They are primarily responsible for shaping and driving innovations. Another success factor is the constructive engagement with opponents, who should be involved in the innovation process to successfully implement new technological developments and operational measures. According to Hauschildt et al., it is not only a matter of identifying individuals or positions, but also of clearly defining and describing functions and roles. In innovation management, there are various role designations that have been assigned to key actors in the innovation process by different authors[1470], starting with Joseph Schumpeter.[1471] For the following analyses, four specific promoter roles are selected from this diversity, and their contributions to the innovation process will be analysed. These are:

- Expert Promoters
- Power Promoters
- Process Promoters
- Relationship Promoters.

1470 Hauschildt et al. 2016, pp. 186–188.

1471 Austrian economist Joseph Schumpeter is regarded as one of the founding figures of innovation theory. In his 1911 book "Theorie der wirtschaftlichen Entwicklung", he described economic development as primarily driven by the implementation of new combinations of production factors by dynamic entrepreneurs. These entrepreneurs, acting as pioneers of economic progress, earn entrepreneurial profits as a result of their early and successful implementation of innovations, thereby triggering economic upswings. Schumpeter understood these innovations mainly as the introduction of new products, new methods of production, the opening of new markets and supply sources, and new forms of organisation. While Schumpeter focused on the entrepreneur as the central figure of innovation, later innovation research increasingly shifted attention to the organisational context and the roles of various actors within it. This development culminated in the promoter model, which emphasises the division of tasks and cooperation among different types of promoters, each contributing to the innovation process in specific ways.

10.3.4.2 Roles of Promoters

10.3.4.2.1 Expert Promoter

The Expert Promoter influences the innovation process through their specific expertise and methodological knowledge, with their hierarchical position being of secondary importance. They are both a learner and a teacher: they expand their knowledge through professional development and share it with others who support the innovation process. At the same time, they use their professional authority to convince opponents and win them over to the innovation. The Expert Promoter does not necessarily have to be a member of the organisation; they can also be embedded informally as an external consultant for the duration of the innovation process.[1472] They recognise opportunities, realise concepts, understand critical details, develop alternatives, and evaluate external solution proposals. With their expert knowledge, they help overcome the barrier of not knowing[1473] and contribute to the successful implementation of innovations. In addition to their technical expertise, the Expert Promoter plays a key role in facilitating knowledge sharing within the team, bridging gaps between influence groups, and aligning them with the innovation's goals. Their contribution is essential not only in solving technical problems, but also in navigating organisational challenges, fostering collaboration, and ensuring that the innovation aligns with the broader strategic objectives of the organisation.

10.3.4.2.2 Power Promoter

The Power Promoter actively uses their high hierarchical position to drive the innovation process forward by providing necessary human and financial resources. Their formal authority enables them to set strategic goals, define priorities, and make key personnel decisions that align with the innovation agenda. They have the capacity to reward support for the innovation and to sanction resistance, thereby creating clear incentives and removing obstacles. A critical function of the Power Promoter is to protect the innova-

1472 Hauschildt et al. 2016, p. 204.
1473 Hölzle und Gemünden 2011, p. 502; Vahs und Brem 2013, pp. 184–185.

tion team from opposition, especially from individuals with significant influence or subject-matter authority. By standing behind the innovation and its actors, the Power Promoter fosters a safe space for experimentation and risk-taking. Through their position of power, they overcome the barrier of unwillingness, which often hinders the adoption of change. Their role is particularly essential when resistance arises at upper management levels or when interdepartmental coordination is required. Ideally, Power and Expert Promoters form a synergistic Promoter Team, combining strategic influence with technical and methodological competence to ensure the success of the innovation across all levels of the organisation. The Power Promoter cannot separate themselves from the content of the innovation; they are responsible for ensuring that its goals and direction are aligned with the broader strategic objectives of the top management.[1474]

10.3.4.2.3 Process Promoter

Hauschildt/Chakrabarti expanded the promoter model by introducing the Process Promoter.[1475] This expansion positions the Process Promoter as a mediator between experts and decision-makers, connecting Power Promoters and Expert Promoters across hierarchical boundaries within the organisation. In particular, the Process Promoter helps overcome organisational and administrative barriers.[1476] They are responsible for managing the entire innovation process, ensuring effective coordination and collaboration between relevant groups within the organisation. Additionally, the Process Promoter ensures that innovative decisions remain aligned with the organisation's strategic objectives and regularly coordinates with the Power Promoter to ensure this alignment.[1477] Typically, they belong to middle management and possess extensive knowledge of the organisation. With strong communication and collaboration skills, they foster cooperation between different departments and individuals. While the Process Promoter is not pri-

1474 Hauschildt et al. 2016, pp. 206–207.
1475 Hauschildt und Chakrabarti 1988, pp. 378–388.
1476 Folkerts 2001, pp. 30–32.
1477 Hauschildt et al. 2016, pp. 205–206.

marily the initiator of innovation processes, they play a key role in coordinating and connecting relevant influence groups within the organisation. In doing so, they help remove the barrier of not being allowed to, particularly in the context of organisational and administrative constraints.[1478] Additionally, the Process Promoter is responsible for planning and managing innovation projects, granting them a high degree of control and influence over the innovation process. Acting as a mediator between Expert Promoters and Power Promoters, they ensure that these two perspectives are connected across hierarchical boundaries within the organisation.

10.3.4.2.4 Relationship Promoter

Gemünden / Walter expanded the Hauschildt / Chakrabarti promoter model to include the Relationship Promoter, who overcomes the barrier of not knowing each other by establishing connections to networks outside the company and to potential cooperation partners such as companies, universities, and research institutes.[1479] In doing so, they initiate new business relationships and foster collaboration beyond the organisation's boundaries. The Relationship Promoter plays a crucial role in facilitating the exchange of knowledge by establishing and maintaining strong relationships with external partners. With their strong negotiation and social skills, they are responsible for identifying potential partners who are critical to the innovation process. This role is essential for supporting the work of Expert Promoters, Power Promoters, and Process Promoters outside the organisation, while the Process Promoter typically operates within the organisation. Often, the Relationship Promoter possesses enough expertise to recognise the technical relevance of ideas from external partners and assess their potential for integration into the organisation's innovation efforts.[1480] The Relationship Promoter holds a unique position in the innovation process because, unlike the other promoters, they

---

1478 Gemünden 2010, pp. 89–90; Folkerts 2001, pp. 30–32; Vahs und Brem 2013, pp. 185–186.

1479 Hölzle und Gemünden 2011, p. 505; Gemünden und Walter 1995, pp. 971–986.

1480 Hauschildt et al. 2016, pp. 204–205.

are not only responsible for specific innovation projects but also for continuously networking organisations and building long-term strategic relationships. Their work is essential for expanding the organisation's external network, thus creating new opportunities for collaboration and knowledge exchange. Although the roles of the Relationship Promoter and Process Promoter can overlap in certain tasks, the distinction lies primarily in the focus of their work: while the Process Promoter operates within the organisation, the Relationship Promoter is focused on external relationships and networks. In the innovation process, the Relationship Promoter helps bridge interpersonal distances, strengthen social ties, and connect the organisation with potential partners, making them an integral part of the innovation ecosystem.[1481]

10.3.4.3 Role Perception and Interaction of Promoters

The roles of individual promoters cannot be sharply delineated in terms of a defined division of labour; instead, success lies in the coordinated interaction of all those involved, with a clear understanding of each role's perception.[1482] Effective coordination ensures that the efforts of the different promoters are not only aligned but also actively complement each other, creating synergy and optimising the innovation process.[1483] A single person may take on different promoter roles at various stages of the process, leading to the creation of different promoter structures. Additionally, a single role can be performed by multiple individuals, depending on the needs of the process at a given time.

Liesa Folkerts differentiates between personal and project-related levels. While the personal level refers to the roles of concrete individuals, the project-related level considers the entirety of the relationships between all persons and their respective roles.[1484] The interaction between promoters on both levels ensures smooth collaboration and facilitates the exchange of expertise and resources.

1481 Gemünden und Walter 1995, pp. 971–986; Hölzle und Gemünden 2011, p. 505.

1482 Hauschildt et al. 2016, pp. 204–207.

1483 Hauschildt et al. 2016, pp. 204–207; Schültz 2016, pp. 23–26.

1484 Folkerts 2001, pp. 40–41.

The flexibility of the promoter roles allows them to adapt to the changing needs of the innovation process, ensuring that the right expertise and leadership are applied at the right time. The importance of each respective promoter role can vary depending on the phase of the process. For example, at the initiation phase, the Power Promoter may play a key role in successfully kick-starting the innovation, while the significance of the Process Promoter or Relationship Promoter steadily increases as the process progresses. At the same time, the importance of the Expert Promoter remains constant throughout the entire process. However, all roles should be filled in every phase to ensure a balanced and effective progression.

As the process evolves, the relative importance of each role changes, necessitating regular reassessments and realignment to maintain momentum and ensure that each phase progresses effectively. This iterative development, where the roles adjust in response to the needs of innovation, highlights the dynamic nature of the process and the key role of the promoters in adapting to these changes. In this context, feedback loops between phases are critical, ensuring that insights gained from earlier stages are incorporated into later stages for continuous improvement.

By maintaining a clear understanding of the roles and coordinating their actions throughout the entire process, promoters ensure that innovation remains flexible and aligned with strategic objectives, thereby contributing to successful outcomes.

10.3.4.4 Barriers in the Innovation Process

When examining barriers to innovation, it is important to take a closer look at the role of opponents. Unlike promoters, opponents are often less visible and harder to identify in their specific roles. While promoters typically act openly and proactively, opponents may exert their influence covertly or indirectly. Nevertheless, opponents can play an important role in innovation processes, especially when their criticism is constructive and helps uncover weaknesses. Just as expert and power promoters often work together as a team, a counterpart can emerge in the form of expert and power opponents. In the best-case scenario, such a team identifies deficits and contradictions in innovation planning and implementation at an early stage, helping to refine and improve the process.[1485] In this sense, opposition does not always signal resistance to change but can be a catalyst for thoughtful reflection and improvement.

Barriers to innovation can arise at both the organisational and process levels. One useful tool to support progress is the definition of clear checkpoints throughout the innovation process.[1486] These milestones enable comparisons between actual progress and predefined goals, providing a degree of planning security and facilitating efficient allocation of resources. However, excessive regulation, inflexible procedures and prolonged decision-making processes can delay or even block innovation. A balance must therefore be found between control and agility.

Barriers may have obvious causes or stem from deeper, less visible factors. Surface-level obstacles are usually easy to identify and resolve, such as lack of funding or outdated infrastructure. However, deeper barriers such as internal resistance, rigid structures or ingrained mental models often remain hidden and only become apparent over time. These can manifest as active or passive resistance from individuals, taking various forms: overt or covert, constructive or destructive, direct or indirect, loyal or opportunistic. It is especially important to distinguish between constructive and destructive opposition. While the former aims to improve or modify the innovation, the latter typically seeks to block or undermine it. Addressing resistance early and identifying its nature can help prevent long-term friction or failure.

In addition to individual resistance, barriers can arise from within the organisation itself. These include bureaucratic hurdles, unclear responsibilities, lack of cross-functional cooperation or silo thinking.[1487] Equally significant are cultural barriers such as risk aversion, lack of psychological safety or an organisational culture that punishes failure instead of treating it as a learning opportu-

1485 Mirow et al. 2007, pp. 104–105.

1486 The progress of modern innovation projects is often assessed using stage-gate models. In this approach, an interdisciplinary team makes structured decisions at pre-defined stages based on a standardised framework. At each stage, the team evaluates whether the project should continue, be modified, or be terminated.

1487 Mirow et al. 2007, pp. 107–108.

nity. Psychological barriers like fear of losing status, uncertainty about new responsibilities or reluctance to leave one's comfort zone can also significantly slow down innovation. From a structural perspective, innovation barriers can be classified as internal or external. External barriers may include market volatility, technological uncertainty, political or legal constraints or customer scepticism. Internal barriers often arise from the organisation's own processes and rules or from insufficient support structures. In contemporary innovation research, a lack of support from top management is considered one of the main causes of innovation failure. If management fails to back an innovation, whether through lack of commitment, resources, or communication, it can demotivate all involved parties.[1488]

This is particularly paradoxical, as it is ultimately top management that decides which innovations are pursued.

Historical examples such as the contrasting wartime leadership of Winston Churchill and Adolf Hitler illustrate how leadership style can significantly influence innovation outcomes, either enabling or suppressing innovation, with wide-ranging consequences.

10.3.4.5 Limitations of the Promoter Model

The promoter model reaches its limits when applied to the assessment of radar development in Germany and the UK, as it can only partially capture the complexity of the processes involved and the nature of the available data. While the promoter model is a valuable framework for understanding modern innovation processes and plays a key role in this work, its application to the historical assessment of radar development in Germany and the UK reveals certain limitations. The complexity of the processes and the nature of the available data extend beyond the scope of the model. Nevertheless, in the context of this work, the promoter model provides valuable insights into the role of key actors in the innovation process, while acknowledging the need for additional frameworks to fully capture the nuances of historical developments.

1488 Ibid., p. 110.

There are significant discrepancies between what can be reconstructed from a historical perspective and what was actually considered relevant at the time. Moreover, there is a risk of idealising certain individuals by portraying them as key drivers of innovation, which may obscure the collective and context-dependent nature of historical innovation processes. The promoter model, by focusing on individuals, may downplay the broader socio-political and technological systems in which innovations occur. This can lead to an oversimplified narrative that fails to capture the full complexity of historical innovation processes. The interplay between key actors often remains ambiguous and continues to be a subject of historical inquiry. Innovation often emerges from a continuous dialogue and negotiation between diverse actors, rather than from the efforts of individual promoters alone. This interaction is often not adequately captured by the promoter model, which tends to emphasise individual action over collaborative processes.

Since the promoter model is primarily used today as a management tool focused on short- to medium-term innovation processes and based on empirical research, it becomes speculative when applied to historical developments. While the model has been adapted for use in diverse fields, including management and technological development, its limitations in historical contexts become more evident when applied to non-linear and multi-faceted innovation processes. Nonetheless, the model allows for the identification of certain characteristics of innovation by highlighting the role of individuals as either drivers or blockers of change. Equally important to understanding innovation are the roles of those who resist or block innovation. The promoter model does not fully account for the significant impact of opposition, which can shape the trajectory of technological and social change.

In conclusion, the promoter model should be regarded as a heuristic framework that can support the analysis of actors, but it does not provide a comprehensive explanation of historically complex innovation processes. With this understanding of its strengths and limitations, the model is now applied to the analysis of key political, military, scientific, and industrial actors involved in the development and implementation of radar technology between 1933 and 1945. Although many of these actors operated in multiple capacities, the focus here is on their respective decisive roles within the innovation process. As shown in Chapter 10.3.4, the promoter model provides a useful lens for identifying different functional roles assumed by individuals over time. However, clear distinctions between promoters and opponents are often blurred and subject to interpretive judgment.

Despite these complexities, it is possible to trace the attitudes and actions of individuals in light of major wartime developments. From an overall perspective, it becomes evident that clear opponents are difficult to identify in terms of a specific role. More commonly, indifference or scepticism toward the new developments in radar technology can be observed.

10.4 Innovation Systems: A Comparative Analysis

The subsequent analysis of the innovation systems in Germany and Great Britain focuses on structures of organisation and command, as well as on the innovation process itself, considered in light of each country's specific innovation culture and strategic priorities. Following an overview of the relevant interest groups, the key actors and their respective functions within the innovation process are examined, with particular attention to their roles as promoters or opponents. The analysis then turns to collaborative efforts, which also had a decisive influence on technological and operational developments.

10.4.1 Germany

10.4.1.1 Organisation and Command

10.4.1.1.1 Interest Groups

As explained in detail in Chapter 8.1, targeted funding for research and development in radar technology appears to have played a rather subordinate role in Germany, in clear contrast to the situation in Great Britain.

This is evident in the Entwicklungsstopp-Befehl dated 3 February 1940 and the Schlüsselkraft-Befehl dated 19 February 1942. Both measures focused on personnel needs for war production in order to ensure a rapid and decisive end to the war. As a result, forward-looking support for science and technology was deemed unnecessary. The prevailing assumption was that the military conflict would be resolved by conventional means. This mindset was shared not only by Hitler and Göring but also by many within the military command structure. Even at the outset of the war, there was a significant shortage of technical and scientific personnel due to their conscription into military service or reassignment to production roles.[1489] Further inhibiting factors included the fragmentation of research and development capacities, resulting from a lack or complete absence of coordination. The strategic importance of radar technology was only gradually recognised over the course of the war, and appropriate measures were eventually taken to improve the situation.

10.4.1.1.1.1 Ministries and Institutions

10.4.1.1.1.1.1 Beauftragter für den Vierjahresplan

At the Nuremberg Rally on 9 September 1936, the Zweiter Vierjahresplan[1490] was announced, and Göring was put in charge of its implementation. The increasing importance of radio communication for operational warfare was gradually being recognised. In 1938, General Erich Fellgiebel was appointed Generalbevollmächtigter für technische Nachrichtenmittel (GBN).[1491] His area of responsibility included the implementation of the Vierjahresplan in the field of radio communications, including securing the necessary raw material quotas. Wilhelm Ohnesorge[1492], Reichspostminister from 1937, also played a key role in this. The Reichspost established its own Reichspostforschungsanstalt[1493], and following the reorganisation of the Reichsforschungsrat, Göring appointed Ohnesorge as a member of its Presidium on 24 July 1942.[1494]

1489 Ludwig 1974, pp. 271–300.
1490 Second Four-Year Plan.
1491 Reuter 1971, p. 49.
1492 Lotz 1999, pp. 494–495; Klee 2011, pp. 443–444; Dittmann 2021, p. 178.
1493 Faensen 2001, pp. 81–83.
1494 Ibid., p. 83.

10.4.1.1.1.1.2 Reichsforschungsrat

The tense military situation of Germany during World War II led to a reorganisation of the Reichsforschungsrat (RFR) in 1942, with further adjustments into 1943. On 9 June 1942, Hermann Göring was tasked by Hitler with transforming the RFR into a legally independent institution.[1495] As a result of this restructuring, the RFR was no longer under the control of the Reichsministerium für Wissenschaft, Erziehung und Volksbildung (REM), but rather under the Reichsministerium für Bewaffnung und Munition (RMfBuM), headed by Albert Speer. This restructuring reflected the prioritisation of military research and development efforts and aimed to ensure that innovative technologies and weapons could be made available to the armed forces quickly and efficiently. The RFR's new status allowed for greater flexibility and independence in research, which was crucial given the ongoing war effort. Göring himself became president of the RFR on 9 June 1942 and appointed Government Commissioners for critical areas of war-related research. These Government Commissioners were tasked with intensifying specialised research efforts in their respective fields and commissioning relevant institutes and industrial companies. The goal was to avoid duplication of efforts and ensure that research was directly aligned with the needs of the war.

On 14 November 1942, Hans Plendl was appointed Bevollmächtigter für die Hochfrequenzforschung (BHF), with direct responsibility to Göring in this position. On 16 July 1943, the Reichsstelle für Hochfrequenzforschung e. V. (RHF) was established as an administrative authority to carry out the tasks of the BHF. The task of the RHF was to manage research contracts and ensure that they aligned with the war effort and state-controlled economic priorities. The allocation of funds to the RHF was carried out by the Reichsforschungsrat under Göring's presidency, which gave him significant influence over high-frequency research. After only about a year, Plendl was ultimately replaced by the president of the PTR, Abraham Esau.

1495 Flachowsky 2005, p. 208.

10.4.1.1.1.1.3 Reichsministerium für Bewaffnung und Munition

From 1942, the Kommission für Nachrichtenmittel was established under the Reichsleiter für Rüstung und Kriegsproduktion.[1496] Karl Küpfmüller was appointed head of the development group within the commission, responsible for the development and standardisation of military communication technology, particularly in the areas of high-frequency and radar technology. This placement reflects Speer's party function as Reichsleiter, distinct from his state office as Minister für Bewaffnung und Munition (RMfBuM), which he had assumed earlier in February 1942.

After a meeting between Göring, Dönitz, and Speer in early January 1944, responsibility for consolidating all high-frequency technologies was transferred to Speer's Reichsministerium für Rüstung und Kriegsproduktion (RMfRuK). Göring attempted to transfer control of the entire radio measurement industry, including developments for the navy, to the Air Force. However, Dönitz and Speer vehemently opposed this plan. Finally, Göring agreed to consolidate and manage the development and production of radar under the RMfRuK.[1497]

Friedrich Lüschen[1498] of Siemens & Halske, who served as chairman of the Hauptkommission Elektrotechnik[1499] within Speer's ministry, was appointed Beauftragter für Elektrotechnik and thereby exerted considerable influence on research activities. In May 1944, the Entwicklungsgruppe Funkmeß was dissolved and replaced by the Sonderkommission für Funkmeßtechnik (SKFM), a subcommittee of the Hauptkommission Elektrotechnik, led by Karl Rottgardt of Telefunken.[1500] The Arbeitsgemeinschaft Rotterdam (AGR) was also involved in this work, and Leo Brandt was tasked with continuing its responsibilities.[1501] At an AGR meeting on 25 February 1944, Brandt had already announced Speer's intention to

---

1496 From September 1943 Reichsministerium für Rüstung und Kriegsproduktion (RMfRuK).

1497 Reuter 1971, p. 195.

1498 Blumtritt 1987, pp. 475–476; Klee 2011, p. 383.

1499 Reuter 1971, p. 195.

1500 BArch-MA, RL 3 / 42.

1501 See also Chapter 10.5.1.1.

establish the SKFM.[1502] On 5 April 1944, AGR was again tasked with continuing the previous work.[1503] In addition to the SKFM, further special commissions were established by Speer's decree of 12 May 1944.[1504] Members of the Special Commission included Dr. Heyne as a representative of the Air Force and Küpfmüller as a representative of the Navy.[1505]

The representation of both military services was intended to ensure that planned developments were coordinated early between the Air Force and the Navy. The goal was to transfer full responsibility for the development and production of radar technology to industry, as Speer believed that only industry could identify urgent needs and manage them efficiently. In doing so, Speer aimed to respond to the increasingly critical war situation by fostering collaboration between industry and the military.[1506] During meetings with Hitler from 1 to 4 January 1944, Lüschen and Küpfmüller assured him through a presentation and technical briefings that it was still possible for German research to catch up with the British.[1507]

10.4.1.1.1.2 Air Force

On 11 March 1935, the existence of the German Air Force was officially made public, signaling a clear breach of the Treaty of Versailles. On 1 May 1935, Hermann Göring was appointed Commander-in-Chief of the Air Force. In addition to his military role, he also assumed authority over civilian aviation, thereby consolidating control over all aspects of German air policy.[1508]

10.4.1.1.1.2.1 Reichsluftfahrtministerium

On 30 January 1933, Hermann Göring was appointed Reichsminister ohne Geschäftsbereich in Hitler's cabinet and, effective the same

1502 AGR-Protokoll vom 25. Februar 1944, p. 2.
1503 AGR-Protokoll vom 5. April 1944, p. 1; Reuter 1971, p. 195.
1504 BArch, R 26 III / 192; Albert Speer sought to streamline and centralise armaments efforts by establishing various commissions.
1505 Reuter 1971, pp. 195–196; Hagenauer und Pabst 2014, p. 18; Maier 2007a, p. 1036.
1506 Reuter 1971, pp. 195–196; Weiss 2002, pp. 137–138.
1507 BArch, R 3 / 1509.
1508 See also Chapter 2.5.2.1; Benz et al. 2001, p. 682.

day, Reichskommissar für den Luftverkehr. On 2 February he oversaw the establishment of the Reichskommissariat für die Luftfahrt which was reorganised into the Reichsluftfahrtministerium (RLM) on 5 May 1933, with Göring as its head.[1509] The RLM was now responsible for organising and overseeing the creation of the Air Force. Due to the provisions of the Treaty of Versailles, particularly Article 198, which prohibited the German Reich from establishing and operating air forces, the creation of the Air Force initially proceeded covertly.[1510]

The Technisches Amt within the RLM was responsible for awarding development and procurement contracts to industry, including radio and navigation equipment. The Technisches Amt was established in 1933 within the RLM structure, initially under the direction of Colonel Wimmer, and from 1936 under Colonel Ernst Udet.[1511] The Technisches Amt reported to the Staatssekretär der Luftfahrt, Erhard Milch[1512], who was also Generalinspekteur der Luftwaffe from 1939.

On 1 February 1939, Udet[1513] was appointed Generalluftzeugmeister (GL) and, in this new role, continued to serve as the head of the office in the RLM. His duties and responsibilities were clearly defined in the service instructions for the GL of 6 February 1939:

„1. Der Generalluftzeugmeister ist dem Staatssekretär der Luftfahrt und Generalinspekteur der Luftwaffe unterstellt; er hat die Dienststellung eines Kommandierenden Generals der Luftwaffe.

Er hat das Recht des unmittelbaren Vortrags beim R.d.L. u. Ob.d.L. u. hat den Staatssekretär der Luftfahrt und Generalinspekteur der Air Force über den Inhalt dieser Vorträge zu unterrichten.

2. Er ist im Krieg und Frieden verantwortlich für die Forschung, Entwicklung, Erprobung und Beschaffung des gesamten Luftwaffengerä-

1509 Völker 1968, pp. 131–133; Rieger 2018a.

1510 Völker 1967, pp. 61–70; a comprehensive overview of the development of military aviation in Germany from 1920–1933 can be found in: Völker 1962, pp. 121–230.

1511 Potempa 2016, pp. 514–515; Hümmelchen 2011b, pp. 258–264; Baeumker 1971, p. 28.

1512 Hümmelchen 2011a, pp. 171–177.

1513 from 1 November 1938 Lieutenant General.

tes, sowie für die Geräteausstattung und Bevorratung aller Dienststellen der Luftwaffe.

Er hat Inspektionsbefugnis in sämtlichen Nachschubeinrichtungen der Luftwaffe, sowie in den Waffenmeister- und Feuerwerkerschulen der Luftwaffe.

Er überwacht alle für die Luftwaffe im Verantwortungsbereich des Heeres und der Marine laufenden Aufgaben auf dem Gebiet der Technik und hat das Recht, sich durch direkte Vorträge der zuständigen Dienststellen unterrichten zu lassen.

3. Er leitet ferner verantwortlich den Aufbau und Ausbau der Industrie, sowie ihre Finanzierung und wirtschaftliche Betreuung; er vertritt in diesen Fragen die Belange der Luftwaffe gegenüber den anderen Wehrmachtdienststellen und Reichsbehörden.

4. Der Bedarf an Werk- und Rohstoffen, sowie an Industrie-Personal wird durch ihn im Benehmen mit dem O.K.W. sichergestellt, wobei er die Forderungen der gesamten Luftwaffe vertritt.

5. Nach den Weisungen des R.d.L. u. Ob.d.L. überwacht und steuert er die Ausfuhr von Luftwaffengerät.

6. Zur Durchführung seiner Aufgaben unterstehen ihm der Chef des Stabes und der Leitende Chef-Ing. Ihre Aufgaben sind durch bes. Dienstanweisungen festgelegt.

7. Ferner sind ihm unmittelbar unterstellt:

a) das Technische Amt

b) das Nachschubamt

c) die Amtsgruppe „Technische Wirtschaft und Haushalt".

8. Der Generalluftzeugmeister ist der oberste Fachvorgesetzte der Ing.-Offiziere, der Angehörigen des Ing.-Korps und der Technischen Angestellten und hat maßgebenden Einfluß auf deren Verwendung im Benehmen mit dem Personalamt.

9. Er ist befugt, der Truppe unmittelbar technische Weisungen auf dem Truppendienstwege zu erteilen. Soweit diese die Einsatzbereitschaft der Luftwaffe berühren, können solche nur im Einvernehmen mit dem Generalsstab gegeben werden."[1514]

1514 Völker 1968, p. 156; Hentschel 1989, pp. 11–12. The Generalluftzeugmeister is subordinated to the State Secretary for Aviation and the Inspector General of the Luftwaffe; he holds the position of a Commanding General of the Luftwaffe. He has the right to present matters directly to the R.d.L. (Reich Minister for Aviation) and Ob.d.L. (Commander-in-Chief of the Luftwaffe)

From 1933 to 1941, however, aviation research became increasingly fragmented within the overall structure of the RLM, making it impossible to provide unified and targeted leadership to achieve the relevant goals.[1515] As a result, incorrect decisions were made in the context of the extensive and complex tasks, progressively weakening the operational readiness of the Air Force as the war went on. Among the series of misappointments to technical leadership positions, the appointment of Udet as head of the Technisches Amt, probably the most important office in the RLM after the General Staff, stands out.[1516] After Udet's suicide on 17 November 1941, Milch resumed the duties of the GL once again.[1517]

---

and must inform the State Secretary for Aviation and the Inspector General of the Luftwaffe about the contents of these presentations. In both peace and war, he is responsible for the research, development, testing, and procurement of all Luftwaffe equipment, as well as for the equipment and stockpiling of all Luftwaffe units. He has inspection authority over all Luftwaffe supply facilities, as well as the Luftwaffe's weapons and pyrotechnic schools. He supervises all technical tasks related to the Luftwaffe within the Army and Navy's responsibility and has the right to receive direct reports from the relevant offices. He is also responsible for directing the establishment and expansion of industry, as well as its financing and economic management; he represents the Luftwaffe's interests in these matters to other branches of the Wehrmacht and Reich authorities. The demand for materials and raw materials, as well as for industrial personnel, is ensured by him in coordination with the O.K.W. (Oberkommando der Wehrmacht), representing the needs of the entire Luftwaffe. Following the instructions of the R.d.L. and Ob.d.L., he monitors and controls the export of Luftwaffe equipment. To carry out his duties, he is supported by the Chief of Staff and the Chief Engineer. Their tasks are specified in special service instructions. He is directly subordinated to: a) The Technisches Amt b) The Supply Office c) The Office Group 'Technical Economy and Budget.' The Generalluftzeugmeister is the highest-ranking superior of the engineering officers, members of the engineering corps, and technical staff, and has a decisive influence on their deployment in coordination with the Personnel Office. He is authorised to issue technical instructions directly to the troops through the troop service channel. As far as these instructions affect the operational readiness of the Luftwaffe, they can only be given in agreement with the General Staff (translated by the author).

1515 Baeumker 1971, pp. 28–30.

1516 Boog 1982, pp. 36–76.

1517 For selected minutes of GL meetings specifically concerning radar technology in relation to operations against the Royal Air Force between April 1942 and June 1944, see Hentschel 1989, pp. 120–137.

10.4.1.1.1.2.2 Forschungsführung des R.d.L. and Ob.d.L.

In February 1942, Erhard Milch, in his roles as Staatssekretär der Luftfahrt of the RLM and Generalinspekteur der Luftwaffe, established the research leadership under the authority of Hermann Göring, who remained Reichsminister der Luftfahrt (R.d.L.) and Oberbefehlshaber der Luftwaffe (Ob.d.L.). Its main responsibilities were to plan and oversee aeronautical research and armament, and to manage the allocation of resources, facilities, and personnel for aviation research.

Additional responsibilities involved coordinating with scientific institutions, industry, and the combat forces at the front.[1518] However, this structure became obsolete with the establishment of the second Reichsforschungsrat on 9 June 1942, headed by Göring. He appointed so-called Government Commissioners to coordinate efforts in key research areas. On 20 June 1944, responsibility for air armament was transferred to the RMfRuK under the leadership of Albert Speer, and Milch's positions as Secretary of State for Aviation and Generalluftzeugmeister were terminated.[1519]

10.4.1.1.1.2.3 Oberkommando der Luftwaffe; Chef des NVW

From the end of 1933, Wolfgang Martini initiated the build-up of the Nachrichtenverbindungswesen (NVW) and the Luftnachrichtentruppe within the RLM, a position he held until the end of the war. His responsibilities gradually expanded, and in 1938 he was appointed head of the Air Force's communications system.

In September 1941, he was promoted to General of the Luftnachrichtentruppe, and in November 1941, he was appointed Sonderbeauftragter für Funkmeßgeräte, overseeing the development of radar, radio, and electronic systems across the German armed forces. Drawing on his expertise and authority, he coordinated efforts between the Air Force and the Navy, bringing together military representatives and industry specialists for joint technical planning. This centralised management significantly enhanced the efficiency and effectiveness of these crucial systems, while his posi-

1518 Reuter 1971, p. 191.
1519 Hümmelchen 2011a, p. 175.

tion ensured the necessary resources and support were secured for their development.[1520]

On Martini's initiative, the Entwicklungsgruppe Funkmeß, along with other development groups for various areas of aviation technology, was established in late 1941. It was placed under the leadership of Leo Brandt, head of equipment development at Telefunken.[1521]

From 1944 onward, Martini held the position of the Generalnachrichtenführer, managing all military communication systems, including signals, radio, and radar operations.[1522] He consistently sought to foster cooperation between the various branches of the armed forces in radar development, particularly between the Navy and the Air Force. In February 1943, he was responsible for establishing the Arbeitsgemeinschaft Rotterdam (AGR). The Military Communication Network and the Air Communications Corps of the German Air Force were crucial for communications and surveillance technology during World War II. The NVW primarily organised and set up telecommunications links between various air force units, headquarters, and other military branches. The Luftnachrichtentruppe was responsible for the technical implementation and operation of these connections and played a key role in air surveillance and defence within the German Reich and occupied territories. Within the Air Force, the Luftnachrichtentruppe handled flight control, radio navigation, and radio direction finding, all essential for the orientation and protection of friendly aircraft. The development and deployment of radar technologies were particularly important, supporting both air defence and fighter control coordination. The Luftnachrichtentruppe also conducted jamming and deception measures to disrupt enemy communications, as well as radio reconnaissance to intercept and analyse enemy signals. With its wide range of technical and operational tasks, the Luftnachrichtentruppe was one of the four major services within the Air Force and played a crucial role in the Air Force Command's defence strategy. Martini recognised the significance of radar tech-

1520 Reuter 1971, p. 48–49, p. 194; Flachowsky 2005, p. 211.

1521 Reuter 1971, p. 194; Flachowsky 2005, p. 211.

1522 Hoffmann 1990, Rieger 2018b.

nology early on and was one of the driving forces behind the development and production of radar systems by GEMA.

10.4.1.1.1.3 Navy

The most important naval organisations involved in radar development were:

10.4.1.1.1.3.1 Nachrichtenmittel-Versuchskommando

The Nachrichtenmittel-Versuchskommando (NVK) and the Amtsgruppe Technisches Nachrichtenwesen were the principal Navy bodies responsible for radar development. From December 1939, for example, the NVK was engaged in developing a radar system intended for use on fighter aircraft. According to von Willisen of GEMA, the NVK had established an excellent reputation in both scientific and naval circles through its contributions to target location, radio direction finding, underwater acoustic devices, and self-guiding torpedoes.[1523]

10.4.1.1.1.3.2 Amtsgruppe Technisches Nachrichtenwesen und Marinenachrichtendienst

At the Oberkommando der Marine (OKM), the Amtsgruppe Technisches Nachrichtenwesen was responsible for communications and reconnaissance technologies, including radar development. However, due to initial personnel shortages, the knowledge and practical experience gained could often only be passed on to the tactical command posts inadequately and with delays. In June 1941, the Amtsgruppe Technisches Nachrichtenwesen was renamed Marinenachrichtendienst (MND) to emphasise the growing importance of its role in naval warfare. This renaming reflected the expansion of its responsibilities. The MND was tasked with maintaining and improving radio and radar systems, developing direction-finding techniques, and jamming enemy communications.[1524] It was only with the formal establishment of the MND under the command of Admiral Erhard Maertens that clear reporting lines

1523 Willisen ca. 1984, p. 25.

1524 Reuter 1971, p. 76.

were introduced for the first time, regulating the flow of information and communication with the OKM. In this capacity, Maertens was responsible for the military-tactical deployment and coordination of naval detection and reconnaissance systems.[1525] During the Nazi era, the MND operated as the intelligence service of the German Navy and collaborated with other intelligence organisations, including the Abwehr, the German military intelligence service, and the German Air Force's signals intelligence units. Maertens was promoted to Vice Admiral in September 1942 and served as the Chief of the MND until May 1943. Among the initial problems faced were, among other things, secrecy regulations that hindered the training of personnel, as well as delays in the training of soldiers to install and operate highly sensitive equipment. It was not until 1943 that clear and comprehensive instructional materials became available for the further training of soldiers.[1526] In May 1943, following the submarine crisis, Dönitz grew increasingly concerned about the growing threat posed by the enemy's advantages in detection technology and lost confidence in Maertens, appointing Rear Admiral Ludwig Stummel as his successor. Stummel was tasked with reorganising the navy's entire technical intelligence system and was instructed to improve both personnel and material resources to better address the future challenges of submarine warfare, particularly in the field of high-frequency research. As part of this effort, the Organisation der deutschen Schwingungsforschung für den U-Boot-Krieg was established, bringing together naval, industrial, and independent research facilities into a total of nine scientific working groups. Its goal was to develop new methods of camouflage, deception, jamming, and warning, particularly in the fields of radio, light, and sound detection.[1527]

10.4.1.1.1.3.3 Wissenschaftlicher Führungsstab beim Oberkommando der Kriegsmarine (WFM)

The WFM was established in December 1943, and Karl Küpfmüller was personally appointed by Dönitz as its head. From January 1944

1525 Ibid., p. 159.
1526 Ibid., p. 190.
1527 Ibid., p. 159, p. 191.

to May 1945, Küpfmüller led this important office.[1528] The WFM's role was to consolidate the Navy's research capabilities in high-frequency technology and to allocate corresponding contracts to institutions and companies. Close cooperation with the BHF ensured coordination between the Air Force and the Navy in the field of high-frequency research.[1529]

10.4.1.1.1.4 Science

In the field of high frequency technology, a wide range of research tasks was carried out by both academic institutions and industrial companies in Germany.[1530]

In addition to major universities and large firms, a number of specialised research institutes also played an important role. Institutions involved in various aspects of high frequency research included, for example:

- Deutsche Versuchsanstalt für Luftfahrt (DVL)
- Flugfunk-Forschungsinstitut Oberpfaffenhofen (FFO)
- Ernst-Lecher-Institut
- Ernst-Orlich-Institut
- Ferdinand-Braun-Institut
- Technische Hochschulen in Berlin, Braunschweig, Breslau, Danzig, Darmstadt, Dresden, Hannover, München
- Universität Jena.

As already explained in Chapter 8.1, Plendl noted in his accountability reports dated 16 December 1943 and 7 January 1944, which were written in connection with his dismissal as head of the BHF, that at the time of his appointment on 14 November 1942, Ger-

---

1528 Hagenauer und Pabst 2014, p. 18; Reuter 1971, p. 159, p. 191; BArch, R 26 III / 91.

1529 Reuter 1971, p. 194.

1530 A good overview of the research commissioned by the BHF in radar technology for the year 1944 can be found in file BArch, R 26 III / 6. Additionally, file BArch-MA, RM 24 / M / 697 [E] provides an overview of individual research contracts awarded by the Navy between 1943 and 1944. BArch, R 26 III / 91 also contains research tasks related to the Navy's work in communications and detection.

many's research capacity in this field was significantly inferior to that of the Allies:

> „Die deutsche Forschungskapazität auf diesem Gebiet war der feindlichen etwa im Verhältnis von 1:10 unterlegen und ausserdem noch auf rd. 100 verschiedene, meist kleine und kleinste Laboratorien aufgeteilt…"[1531]

In addition to the well-known industrial companies and universities, various independent research institutes were also active in the field of radar technology. Notably, the Drahtlostelegraphische und Luftelektrische Versuchsstation Gräfelfing (DVG), founded in 1908, and its subsidiary, the Flugfunk-Forschungsinstitut Oberpfaffenhofen (FFO), founded in 1937, made significant progress in transmitter technology for radiolocation and navigation.[1532]

Tests were carried out in the wavelength range from 5 to 10 centimetres. For example, in August 1937, reflection experiments with a steamboat on Lake Ammersee at a wavelength of 6 centimetres and a distance of 1 kilometre were successfully conducted. The velocity of the steamboat was determined by analysing the Doppler effect.[1533] Moreover, employees of the FFO conducted backscatter measurements on aircraft models to determine the frequency dependence of the radar backscatter cross-section, providing important groundwork for radar signature analysis.[1534] A key figure in this field was physicist Max Dieckmann.[1535] After studying mathematics and physics at the universities of Göttingen, Leipzig, Strasbourg, and the Technical University of Munich, he earned his doctorate in Strasbourg in 1907 and completed his habilitation in 1912 with a thesis on experimental studies at the intersection of wireless telegraphy and atmospheric electricity. With private funding,

1531 BArch, NS 19 / 2057; Reuter 1971, pp. 192–193. "The German research capacity in this area was inferior to that of the enemy at a ratio of about 1:10 and, what is more, it was still distributed across about 100 different, mostly small and very small laboratories…" (translated by the author).

1532 Dieckmann 1953, pp. 49–55.

1533 Keydel und Röde 2004, p. 209; Röde 1997, pp. 10–11.

1534 Keydel und Röde 2004, pp. 210–212; Röde 1997, p. 11; Klausing und Holpp 2000, pp. 14–15.

1535 Jäger und Heilbronner 2010, pp. 101–102; Zetzmann 1983, pp. 46–48.

Dieckmann established the DVG in Gräfelfing near Munich, where he conducted pioneering experiments with students. His work in atmospheric electricity laid the foundation for the application of wireless telegraphy in airship communication.

During World War I, Dieckmann focused on the wireless transmission of simple images from aircraft to the ground. His research also extended to flight navigation, antenna technology, and early radar-like systems. In 1936, he was appointed as an extraordinary professor at the Technical University of Munich. A year later, the FFO was founded to intensify research on radio-based systems for aviation, and in 1944, the DVG was formally integrated into this new institute. The FFO became a key centre for the development of location, communication, and navigation technologies using electromagnetic waves.

Following World War II, the FFO was incorporated into the Institut für Flugfunk und Mikrowellen within the Deutsche Versuchsanstalt für Luftfahrt (DVL) in Oberpfaffenhofen. In 1969, the DVL was reorganised and became the Deutsche Forschungs- und Versuchsanstalt für Luft- und Raumfahrt (DFVLR), which later became the Deutsches Zentrum für Luft- und Raumfahrt (DLR).[1536]

Dieckmann emigrated briefly to the United States in 1947 but returned to Germany the following year for health reasons. He remained active in the scientific community until his death in 1960.

10.4.1.1.1.5 Industry

Among the industrial companies engaged primarily in radar research, development, and production were:

- AEG
  Allgemeine Elektricitäts-Gesellschaft (AEG) was one of the leading manufacturers of electronics and electrical equipment. In radar technology, AEG developed high-frequency components, transmitters, receivers, and system elements used in military radar installations, and was deeply involved in armament projects.

1536 DLR n.d..

- Siemens & Halske
  Siemens & Halske, one of Germany's largest technology companies, worked on high-frequency technology, antenna development, and the production of electronic components for various military radar systems. The company was broadly engaged in communication technology, antenna systems, and radar components.
- GEMA
  Gesellschaft für elektroakustische und mechanische Apparate mbH (GEMA) was a key provider of radar systems, especially for the navy. GEMA developed the first German radar system for naval applications (Seetakt) and specialised in receiver technology, signal processing, and radar antennas, and was central to naval radar technology.
- Telefunken
  Telefunken was a leader in high-frequency technology and radar systems. The company manufactured transmitters and receivers and pioneered innovations in signal processing, and was known for the Würzburg radar family. Telefunken's innovations in high-frequency technology and radar systems played a central role in advancing German radar capabilities during World War II.
- Rohde & Schwarz
  Rohde & Schwarz specialised in measurement technology and communication electronics. In radar, the company contributed to the development of testing equipment and high-frequency components, and specialised in measurement technology for high-frequency systems. The company's expertise supported the development and testing of advanced radar systems.
- J. Pintsch AG
  Pintsch was primarily active in power supply and signalling technology but expanded its portfolio during the war to include radio and radar development, particularly for maritime applications. Although a smaller company, it was involved in signalling technology for both naval and aviation applications.
- Hagenuk/Blaupunkt-Werke/Zeiss Jena
  Hagenuk, Blaupunkt-Werke, and Zeiss Jena had diverse technical focuses, ranging from radio technology and early elec-

tronic products for private use, up to optical devices. Notably, Blaupunkt was already active in the 1930s as a pioneer in manufacturing in-car radio receivers, marking one of the earliest developments in automotive audio equipment. During the war, these companies shifted their production capacities towards radar and navigation systems, with Zeiss Jena standing out for its expertise in precision optics. These companies had diverse technical focuses, ranging from consumer electronics and radio technology to optical devices. Notably, Blaupunkt was a pioneer in car radios from the 1930s onward. Zeiss Jena was indispensable in optics, especially for telescopic sights, periscopes, and radar beam guidance.

10.4.1.1.1.6 Forced Labour in High-Frequency Research

In August 1943, a high-frequency research unit was reportedly established at the Dachau concentration camp, initiated by Siemens & Halske in cooperation with the Reichsstelle für Hochfrequenzforschung.[1537] For this purpose, telecommunications and radio engineers who had been imprisoned in various concentration camps were brought together.[1538] Technical equipment was supplied by the BHF.

On 5 September 1943, Hans Ferdinand Mayer[1539], a former director at Siemens and head of its central laboratory since 1936, was transferred to Dachau and is believed to have supervised the technical work of a multinational group of highly qualified prisoners from twelve countries.[1540] The unit was considered significant for the war effort and aligned with BHF priorities.[1541] Himmler, who showed personal interest in advanced communication technolo-

---

1537 See also Chapter 9.6.2.1.

1538 Hagenauer und Pabst, p. 26; Ludwig 1974 p. 490; a high-frequency research outpost was established at Sudelfeld above Bayrischzell in the Bavarian Alps and remained in operation until 1945.

1539 The author of the Oslo report.

1540 Hagenauer und Pabst 2014, p. 26; before his transfer to Dachau, Mayer was sent to Sachsenhausen concentration camp on 25 August after three months in pre-trial detention.

1541 Hagenauer und Pabst 2014, pp. 27.

gies, reportedly received a direct report on the project from Plendl in October 1943.[1542]

10.4.1.2 Innovation Process

10.4.1.2.1 Historical Context

In the context of ongoing debates about armaments research during the Nazi era, particularly concerning rearmament and autarky, the interplay between government, military, science, and industry played a crucial role. The actions of key actors and the allocation of vital resources were central to this dynamic. Nevertheless, this interaction did little to overcome significant barriers to innovation, shaped by fragmented power structures and intense competition among government bodies and branches of the armed forces. These factors profoundly impacted the course of World War II, especially in the field of radar technology.

Hitler exhibited limited understanding of the natural sciences but had a more nuanced perception of technology, viewing inventions as the achievements of individuals rather than collective scientific efforts. He regarded technology not as an application of scientific knowledge but as an imitation of nature. As a result, there was no coherent or integrated science and technology policy.[1543] Those in power did not pursue a comprehensive innovation strategy aligned with political and military objectives. Ignorance of technological innovation, even after the early successes of the Blitzkrieg, meant Germany lost its initial advantage in research and development as the war progressed. This was exacerbated by a shortage of skilled labour and the persecution of scientists for racial and political reasons. Physicists at universities and research institutions were particularly affected, as physics often conflicted with Nazi ideology.[1544] In addition to well-known figures in historical accounts, physicists also played an important role as technical experts in companies, particularly in high-frequency research.[1545]

1542 Ludwig 1974, p. 490.
1543 Kern 1984, pp. 135–136.
1544 Beyerchen 1980.
1545 Kern 1984, p. 137.

The shortage of skilled labour was a recurring issue in radar development. It worsened with the decline of scientifically trained engineers at technical universities.[1546] Recall actions in 1943 aimed at addressing this shortage had limited success. The ban on amateur radio operators and the confiscation of their equipment further reduced the pool of technically skilled individuals in high frequency technology, while the Allies fully exploited this resource.

Strict secrecy regulations within the Wehrmacht created additional barriers to innovation. An analysis of key figures shows there were no outright opponents of radar development. Rather, a general lack of interest and limited understanding among political and military leaders, shaped partly by experiences in World War I, influenced key decisions on radar technology. The military events of 1941 and 1942, marked by large campaigns, triggered a shift in scientific policy. Scientists and engineers began to push back against state control, taking research promotion into their own hands. Examples include Plendl's appointment as director of the BHF in November 1942 and the founding of the AGR in February 1943.

In addition to structural barriers, technical misjudgements further hindered progress. Decisions by experts and military leaders delayed or even prevented important advances, sometimes with fatal consequences. Notable examples include the halt in centimetre-wave development, which caused Germany to lose its technological lead, and the failure to introduce alternative frequencies for anti-aircraft radars, which enabled effective Allied jamming during the July 1943 Hamburg attack. Strategically, Germany also lacked a long-term vision for the use of electromagnetic waves to detect ships and aircraft, unlike the UK, which had secured funding for radiolocation research well before the war. Although the AGR achieved a high level of research in centimetre waves by the war's end, these results came too late to affect the outcome.

This lack of foresight was compounded by institutional obstacles. The political and military leadership did not fully recognise radar's operational potential and failed to establish a coordinated technological strategy. Conflicts of interest and the passive stance of sceptics within the Wehrmacht went largely unchallenged, further hindering innovation.

1546 Ludwig 1974, p. 281, p. 290.

At the highest level, Hitler and Göring did not act as decisive promoters of innovation. At the beginning of the war, attention remained focused on improving proven radar technologies rather than developing a holistic technical and operational defence strategy. Their limited technical understanding meant that support for science and research was often delayed or absent, despite occasional demonstrations of radar progress. Measures were usually isolated and tied to the Vierjahresplan, Hitler's 1936 four-year economic and armaments programme led largely by Göring. While it aimed to prepare the economy for war and supported rearmament and autarky, it lacked a coherent scientific research strategy. Technological innovation was subordinated to short term production goals, and support for basic research remained fragmented and reactive.

10.4.1.2.2 Key Actors and Their Roles

For the analysis of the development of radar technology and air defence systems in Germany, selected key actors are considered, focusing on individuals directly involved in the technical and operational implementation of these systems. Contributions from other politicians or government members are included only when relevant, but the primary emphasis remains on the core actors.

Many of these actors already possessed professional expertise, influential positions, or networks prior to their involvement. This analysis, however, focuses on the roles they actively assumed within the framework of the Promoter Model, rather than on their general expertise, position, or connections. The model identifies five key roles: a Power Promoter leverages authority to support initiatives; an Expert Promoter applies technical knowledge to advance projects; a Process Promoter organises and drives workflows; a Relationship Promoter builds and mobilises networks to influence outcomes; and an Opponent resists or favours alternative approaches. Emphasising active influence and actions helps ensure clarity.

Each actor's contribution is analysed in terms of the type of influence they exerted, and the results are presented in a table, which focuses on actors whose roles can be clearly categorised, including both Promoters and Opponents. The table lists the actors alphabetically, regardless of their relative importance or influence.

Describing the roles of Hitler and Göring within the Promoter Model is inherently difficult, as neither assumed a fully supportive role as Power Promoters nor can they be convincingly considered visible Opponents. Their influence was largely indirect, exerted through delegation of operational responsibilities and the setting of political and strategic priorities. Hitler occasionally displayed aspects of a situational Opponent when initiatives conflicted with his strategic or ideological preferences, while Göring exercised partial influence as a Power Promoter by managing resources, production, and coordination across military and industrial actors. Due to this indirect and complex influence, they are discussed in the text but are not included in the table.

The chronology of the most important innovations and obstacles in radar technology highlights the wide-ranging impacts of these decisions on technological development and operational measures, showing that even small changes could have far-reaching consequences. The table instead focuses on key actors, including scientists, engineers, and technical experts, whose contributions can be directly attributed and categorised within the Promoter Model, reflecting their active role in advancing projects, managing processes, and coordinating resources.

Werner Beckmann, a department head in the Technisches Amt (LC 4) of the RLM, played an important yet often overlooked role in advancing Germany's radar capabilities.

In October 1939, he proposed awarding a contract for a short-range aircraft detection device, even though the device did not yet exist, and simultaneously put forward the prospect of a large production order. This development contract later led to the Würzburg radar family at Telefunken.

Beckmann's primary roles were as an Expert Promoter, recognising the technical potential of the system, and as a Process Promoter, initiating and coordinating the development process. His involvement as a Power or Relationship Promoter was limited, which is why these roles are not reflected in the table.

His initiative enabled significant progress in design, production, and testing, ultimately laying the groundwork for the Würzburg radar to become the standard anti-aircraft system. Beckmann's example demonstrates how individual expertise and initiative can drive technological advancement even within hierarchical and bureaucratic constraints.

Leo Brandt played a central role in the development of the Würzburg radar family, which became standard equipment for German anti-aircraft artillery. While Brandt had technical knowledge of the system, he acted as an Expert Promoter only in a secondary capacity, with his primary role being a Process Promoter, overseeing project organisation, coordinating workflows, and ensuring that development milestones were met efficiently.

Through his leadership of the Arbeitsgemeinschaft Rotterdam, a pioneering collaboration between the Wehrmacht, industry, and research institutions, Brandt also fulfilled a Relationship Promoter function. In this capacity, he linked military, industrial, and research stakeholders, reconciling differing interests, fostering trust, and accelerating radar development. While the Relationship Promoter role was important, his primary influence remained in the technical and process domains, which ensured that the Würzburg radar moved from prototype to operational deployment.

Brandt's combination of technical expertise, organisational skill, and ability to connect key actors significantly advanced the operational deployment and strategic integration of radar into German air defence, exemplifying how coordinated promotion across multiple roles can drive technological innovation in a complex military-industrial environment.

Max Dieckmann was a pioneering figure in German high-frequency and radar research. He founded the Drahtlostelegraphische und Luftelektrische Versuchsstation Gräfelfing (DVG), conducting groundbreaking work in wireless communication and aviation radio systems. Later, together with Claudius Dornier, he was involved in the establishment and early development of the the Flugfunk-Forschungsinstitut Oberpfaffenhofen (FFO) at the Dornier Works' airfield, advancing radio navigation and radar technologies. Throughout his career, Dieckmann actively collaborated with both military and industrial partners, helping to translate theoretical research into practical applications. He also mentored a generation of engineers and scientists, ensuring the continuity of expertise within German high-frequency and radar programs.

Within the promoter framework, Dieckmann acted as an Expert Promoter, applying his scientific knowledge to radio and radar systems; as a Process Promoter, organising research programmes and coordinating technical development; as a Relationship Promoter, linking military, industry, and academic partners through the FFO; and as a Power Promoter, ensuring resources and strategic direction for both the DVG and FFO.

Dieckmann's work laid crucial foundations for Germany's wartime and post-war developments in radar, navigation, and high-frequency communication. He retired in 1958 but continued to influence the research community until his death on 28 July 1960.

Karl Dönitz, commander of the German submarine forces, played a key role in shaping technological priorities within the U-boat fleet. By 1939, his stance towards radar was cautious. He did not actively oppose it, but operational concerns and a strong preference for sonar slowed its adoption. Early radar systems were technically complex and required highly skilled operators, whose training only began after the devices were introduced, further delaying effective use. These limitations partly explain why Dönitz initially prioritised sonar over radar for operational deployment.

Nevertheless, Dönitz occasionally supported experimentation with radar and encouraged evaluation of new detection technologies, recognising their potential to improve submarine tactics. From 1943, he acted as a limited Power Promoter, facilitating selected innovations and allocating resources, although radar deployment remained constrained by personnel and technical limitations. Prior to the war, some inter-service cooperation occurred, notably through figures such as Wolfgang Martini, though conflicting priorities within the Technisches Amt limited meaningful coordination.

Dönitz's cautious yet pragmatic approach illustrates how senior military leaders could simultaneously promote certain technologies while constraining others, shaping innovation through strategic choices, operational experience, and institutional culture. His decisions had a lasting impact on the pace and direction of submarine radar development during the war. To further support radar development, Dönitz occasionally intervened to prioritise experimental projects and ensure resources were available, demonstrating his targeted but significant influence on innovation.

Paul-Günther Erbslöh and *Hans-Karl Freiherr von Willisen* were technically skilled entrepreneurs who co-founded GEMA in January 1934, following initial work in Tonographie. They were introduced to Rudolf Kühnhold at the NVA in 1932 to 1933, developing high-powered acoustic impulse generators for acoustic detection, which led to the Navy's interest in active radio detection. Encouraged by this success, they established GEMA to focus on early radar prototypes, including what later became the Seetakt system, laying the groundwork for German naval and aerial radar systems such as Freya, Mammut, and Wassermann.

From the founding of GEMA in 1934 until the company's dissolution in 1945, Erbslöh and von Willisen acted primarily as Expert Promoters, applying their technical skills to advance radar development. They also served as Process Promoters, organising experiments, coordinating teams, and structuring project workflows. Through their interactions with the Navy, NVA, research institutions, and industry partners, they functioned as Relationship Promoters, building and maintaining crucial networks. As owners and managers of GEMA, they exercised Power Promoter functions, controlling resources, setting strategic priorities, and steering the company's role in military radar innovation throughout the war.

Their combination of technical expertise, organisational skill, networking, and authority ensured that GEMA became a central player in German radar development from its inception to the end of World War II.

Abraham Esau was a prominent German physicist and a pioneer in ultrashort wave technology. In the early 1930s, he conducted research on centimetre and millimetre waves, laying the scientific foundation for radar and microwave technologies. In 1939, he became president of the Physikalisch-Technische Reichsanstalt (PTR), coordinating high-frequency research across academic, industrial, and military institutions. In December 1943, following Johannes Plendl's removal, he was appointed Bevollmächtigter für die Hochfrequenzforschung (BHF), overseeing military-relevant projects and steering radar research priorities. Despite wartime disruptions, including Allied air raids and laboratory relocations, Esau ensured continuity of critical research. His strategic guidance and coordination of personnel and projects ensured that high-frequency and radar research remained aligned with military priorities, facilitating the development of operational systems despite wartime constraints. Esau's efforts also strengthened collaboration between academic, industrial, and military stakeholders, ensuring that research outcomes could be effectively translated into practical radar systems.

Within the promoter framework, Esau acted as an Expert Promoter, applying his technical knowledge to radar and high-frequency innovations; as a Process Promoter, structuring research programmes and managing laboratories; as a Relationship Promoter, linking universities, military agencies, and industry; and as a Power Promoter, overseeing appointments and research priorities within the PTR and BHF.

Hans E. Hollmann was one of the pioneers of high-frequency and ultra-short wave research in Germany. He developed the first centimetre- and decimetre-wave transmitter–receiver systems at the Technical University of Darmstadt and earned his doctorate with a thesis on generating very high-frequency oscillations. At the Heinrich-Hertz-Institut in Berlin, he achieved fundamental advances in the generation and detection of decimetre waves, providing the scientific foundation for later radar applications. Hollmann also contributed as a scientific consultant to GEMA, supporting the development of early radar prototypes and providing guidance on technical approaches.

Within the promoter framework, Hollmann acted as an Expert Promoter, applying his scientific expertise to advance radar technology; as a Process Promoter, linking academic research with industrial development; and as a Relationship Promoter, connecting institutions, companies, and military stakeholders. His influence was primarily in the scientific and advisory domains rather than in direct operational implementation. After a period of active involvement, he distanced himself from radar work due to increasing military control and censorship, focusing instead on medical applications and later post-war high-frequency research. Hollmann's work laid essential theoretical and technical foundations that enabled subsequent practical radar developments in Germany.

Rudolf Kühnhold was a pioneering figure in the early development of radar and underwater acoustics in Germany. After joining the Torpedo-Versuchsanstalt (TVA) of the Reichsmarine in 1929, he became scientific director of the Nachrichtenmittel-Versuchsanstalt (NVA) in early 1933. In the autumn of that year, he began experimenting with the reflection of electromagnetic waves to detect ships, laying the groundwork for German radar research.

Though not a commanding figure, Kühnhold facilitated technical cooperation between the Navy and industry. At the end of 1933, he met with Wilhelm Runge of Telefunken to explore support for further radar development, but the company showed little interest at the time. Kühnhold then continued the project with other partners and initiated the establishment of GEMA in January 1934, which became a key centre for radar and acoustic detection research, although he was not a co-founder.

Kühnhold guided early German radar experiments, including work on centimetre-wave detection and radio navigation systems such as Knickebein. From the late 1930s until 1943, he conducted microwave research at the NVK, the successor to the NVA. In the final war years, he returned to underwater acoustics and collaborated closely with ELAC in Kiel. After the war, he resumed research in commercial radar at ELAC, leading to several patents, including one registered in the United States in 1954.

Within the promoter framework, Kühnhold acted primarily as an Expert Promoter, advancing the scientific foundation and technical feasibility of radar systems. He also held Process Promoter functions, overseeing research teams and experimental programmes, and contributed in a minor Relationship Promoter capacity, facilitating collaboration with industrial and military partners.

Despite technical and bureaucratic challenges, Kühnhold's expertise and strategic foresight laid groundwork for Germany's operational radar systems and post-war radar development.

Karl Küpfmüller was a leading German electrical engineer whose work had lasting impact on telecommunications and radar technology. From 1921, he worked at Siemens & Halske and later became a professor at the Technical Universities of Danzig and Berlin. From 1937, he led Siemens' telecommunications development, focusing on high-frequency and radar systems crucial for the Navy. In December 1943, Dönitz appointed him head of the Wissenschaftlicher Führungsstab, coordinating naval research and liaising closely with the BHF. In this role, Küpfmüller oversaw research contracts, the integration of developments between the Navy and Air Force, and the advancement of radar and communications technologies, including systems for U-boat operations. His expertise also extended into system theory, where his work on signal transmission and control engineering influenced both contemporary and later generations of electrical engineers.

Within the promoter framework, Küpfmüller acted primarily as an Expert Promoter, applying his technical knowledge to drive innovation. He also served as a Process Promoter, organising research programmes, structuring projects, and coordinating laboratories. In addition, he functioned as a Relationship Promoter, fostering cooperation between Navy, industry, and research institutions. His Power Promoter role was limited and mainly consisted of coordinating research priorities within the strategic framework set by senior naval command.

Küpfmüller's leadership illustrates how technical expertise, process management, and strategic coordination shaped German radar development during World War II.

Wolfgang Martini, head of the Nachrichtenverbindungswesen, played a central role in the development of radar and military radio detection. Within the armed forces' innovation network, he acted as a Power, Process, and Relationship Promoter, coordinating technical development, building networks, and advocating for radar's strategic importance while mediating interservice rivalries between the Air Forces and the Navy. As a Process Promoter, he encouraged standardisation and improved communication between military branches and industry. In his role as a Relationship Promoter, he forged links across branches, companies, and scientific institutions, fostering collaboration in a hierarchical and fragmented environment. Despite resistance from other branches, particularly the Navy, which pursued independent projects, Martini's leadership advanced the acceptance and operational use of radar within the Air Forces. He also played a crucial role in advising operational units on the tactical application of the technology, ensuring that innovations were effectively translated into practical military benefits. His work demonstrates how strategic networking, technical insight, and mediation between competing interests were essential for successful innovation, and his ability to navigate bureaucratic hurdles and persuade sceptical commanders further ensured that radar became an integral part of Air Forces operations.

Werner Osenberg occupied a unique position in German radar development by focusing on key technical and scientific personnel. As head of the Planungsamt within the Reichsforschungsrat from 1939 to 1945, he advocated recalling important researchers and specialists from frontline service to contribute to research and development in high-frequency technology, communications, and armaments. He organised a central personnel database of around fifteen thousand staff essential for innovation projects. Osenberg acted primarily as an Expert Promoter, using his expertise in assessing personnel needs rather than direct technological research. He also functioned as a Relationship Promoter, mediating between scientific institutions, the military, the ministry, and research policy. To a lesser extent, he served as a Process Promoter, structuring personnel management and improving project workflows. By ensuring that qualified staff were available for essential projects, Osenberg indirectly influenced the pace and success of radar and high-frequency research, making his contributions indispensable despite limited formal authority over technological development.

Hans Plendl, appointed by Hermann Göring as Bevollmächtigter für Hochfrequenzforschung (BHF) in November 1942, was already a recognised expert in high-frequency and radar technology. In his role as BHF, he played a central part in consolidating and coordinating high-frequency research in Nazi Germany, functioning primarily as an Expert Promoter, applying his technical expertise to evaluate and advance complex projects. He also acted as a Process Promoter, organising and structuring research programmes. A distinct Relationship Promoter role is not assigned to Plendl, as networking activities played only a secondary role compared with his technical and organisational responsibilities. To a limited extent, he exercised influence over research priorities and project approvals. Plendl oversaw experimental trials, facilitated communication between laboratories and operational units, and ensured that research outcomes could be translated into deployable radar systems.

Although his tenure was brief, ending in December 1943 when Abraham Esau succeeded him, his work laid important organisational foundations for the centralisation and systematic management of radar research, enabling subsequent technological progress under more coordinated leadership. His meticulous approach to project evaluation and documentation helped create a professional framework for later radar research, while his guidance ensured that personnel and technical resources were used efficiently across multiple sites. Plendl's contributions, though short in duration, shaped a more structured and effective research environment for high-frequency and radar developments in Germany.

Erich Raeder, Commander-in-Chief of the Navy from 1935 to January 1943, played a decisive role in shaping technological priorities before and during the war. In September 1935, he attended a successful demonstration of GEMA radar, which convinced the Navy's technical leadership and secured initial financial support. Nevertheless, Raeder prioritised sonar as the Navy's most urgent requirement, leaving radar underfunded and initially sidelined. Early research into centimetre-wave began in January 1936 but was later discontinued. Formal requirements for radar installation on ships and submarines followed later in 1936, yet the fleet command remained cautious. As the war progressed, Raeder increasingly recognised radar's value and provided stronger backing, commissioning experiments and allocating resources where feasible. Although his support was selective and limited, these actions helped maintain continuity in radar research and laid the groundwork for later operational deployment. Within the promoter model, Raeder acted primarily as a Power Promoter, providing strategic direction and resources. Raeder did not act as an Opponent; his involvement was limited, and he showed no active opposition to radar development. He had limited influence on coordination between research teams and naval leadership and did not directly manage project workflows; he is therefore not listed as a Process Promoter in the table. His decisions illustrate how senior military leaders could simultaneously delay and promote innovation, shaping the pace of technological adoption in a hierarchical and resource-constrained environment.

Wilhelm Runge of Telefunken played a pivotal role in German radar development from the mid-1930s. In 1935 he conducted measurement tests with a Ju 52, and in 1936 he became head of the high-frequency laboratories, a position he held until 1942. As an Expert Promoter from 1935 to 1942, Runge advanced key radar systems such as Würzburg and Lichtenstein and contributed to high-frequency transmitter technologies, enhancing performance and reliability. From 1936 to 1942, he also acted as a Process Promoter, coordinating projects within Telefunken and liaising with military agencies. While he maintained some contact with military authorities, this function was secondary and did not constitute a primary promoter role. He also exercised limited project-level influence from 1936 to 1942, steering project priorities and ensuring progress without formal authority over broader military decisions.

After 1942, Runge continued to work at Telefunken, focusing on optimisation and implementation of radar systems. His cautious stance on centimetre-wave research delayed certain important innovations, illustrating the balance between technical ambition and practical constraints. Nevertheless, his foundational work significantly advanced German radar systems and left a lasting impact.

Theodor Schultes was a leading German engineer in high-frequency and radar technology. He studied electrical engineering at the Technical University of Darmstadt and later conducted research at the Heinrich-Hertz-Institut in Berlin, completing his doctorate while focusing on room acoustics and high-frequency systems.

Schultes joined GEMA as head of the research laboratory, collaborating closely with Hans E. Hollmann. He played a central role in early radar development, combining theoretical innovations with engineering expertise, and was instrumental in creating high-performance transmitter tubes and experimental radar devices. His work contributed to systems that evolved into Seetakt, Freya, Wassermann, and Jagdschloß, shaping both naval and air-defence radar capabilities.

Within the promoter framework, Schultes acted as an Expert Promoter, advancing technical knowledge and practical radar applications; as a Process Promoter, coordinating and structuring development projects at GEMA and later at Siemens & Halske; and as a Relationship Promoter, linking industrial, military, and research stakeholders to facilitate collaboration. After 1945, when Allied restrictions limited radar research in Germany, Schultes worked from 1948 to 1952 at Albiswerk Zürich, a Siemens & Halske subsidiary, and from 1952 to 1964 at Siemens & Halske in Munich, contributing to the radio and high-frequency sector and leaving a lasting legacy in high-frequency and radar engineering.

Albert Speer was a central figure in Nazi Germany, initially as Hitler's preferred architect and from 1942 as Minister for Armaments and War Production. Following Fritz Todt's death, Speer assumed broad authority over Germany's war industry and the Todt Organisation. He reorganised armaments production, implemented rationalised workflows, increased efficiency, coordinated resources, and prioritised critical technologies, thereby exerting substantial influence on Germany's war effort. His policies led to significant increases in tank, aircraft, and munitions output despite Allied bombing campaigns. Within the promoter framework, Speer acted primarily as a Power Promoter, controlling resources, setting strategic priorities, and directing the war economy. He also functioned as a Process Promoter, implementing organisational structures, standardisation, and efficient production workflows. As a Relationship Promoter, he mediated between Hitler, the military, the Rüstungsministerium, and industry to ensure execution of production plans. His role as an Expert Promoter was limited to strategic and organisational knowledge rather than technical expertise in weapons or radar development.

Ernst Udet, appointed Generalluftzeugmeister of the Air Force on 1 February 1939 and a celebrated World War I fighter ace, was highly respected among pilots for his skill and operational experience. In the summer of 1938, he was shown Telefunken's radar prototype in Darmstadt, a precursor to the Würzburg system. Initially sceptical, Udet viewed external radar guidance as limiting pilot autonomy, reflecting broader cultural resistance within the Air Force. Although Udet held the powerful position of Generalluftzeugmeister, he did not exercise formal authority over radar projects. Unlike institutional promoters such as Wolfgang Martini, who exercised direct authority over radar development and prioritisation, Udet's influence was primarily symbolic and advisory: through personal experience and credibility among pilots, he acted as a Relationship

Promoter, easing resistance and mediating between pilots and technical developers. Over time, he recognised radar's tactical value, lending credibility among pilots and easing resistance to its use. His influence on broader policy was limited, as key decisions remained with the RLM's Technisches Amt. Udet's death by suicide in November 1941 ended his personal involvement, but his evolving attitude provides an example of how operational experience influenced pilot acceptance of radar technology.

10.4.1.2.2.1 Summary of Key Roles

Assessing the actions and decisions of the top political and military leadership using the Promoter Model is challenging, particularly when analysing the roles of Power Promoters and their opponents. With few exceptions, those in charge fulfilled their roles as promoters of innovation processes inadequately, and no one can convincingly be classified as a clear destructive opponent. While this lack of support did not entirely prevent innovations, it significantly delayed them, often with serious consequences for the development of radar technology. A major reason for this undifferentiated behaviour was the experience of World War I, when conflicts were expected to be resolved through conventional means, resulting in a limited understanding of emerging radar technology.

Numerous actors were involved in the development of radar technology, often assuming multiple roles within the innovation network. For the present analysis, only the roles primarily evident in the context of radar are considered, in order to highlight the main driving forces and obstacles in technological development and operational deployment. Since responsibilities and influences frequently overlapped and were not clearly documented, the classification of roles is inherently imprecise. This applies not only to political and military actors but also to research institutions and industrial participants.

The contributions of these actors are presented in Table 7 and Table 8 according to the Promoter Model. Table 7 focuses on political, military, and ministerial actors, while Table 8 covers research, industry, and institutional actors from 1933 to 1945. To account for multiple roles, only the most significant role of each actor is considered, so some ambiguity in the assignments is unavoidable.

Power Promoter											Dönitz[*1]	Dönitz[*1]	Dönitz[*1]
		Martini	Martini	Martini	Martini	Martini	Martini	Martini	Martini	Martini	Martini	Martini	
			Raeder[*2]	Raeder[*2]	Raeder[*2]	Raeder[*2]	Raeder[*2]	Raeder[*2]	Raeder[*2]	Raeder[*2]			
										Speer	Speer	Speer	Speer
Expert Promoter							Beckmann						
							Osenberg[*3]	Osenberg[*3]	Osenberg[*3]	Osenberg[*3]	Osenberg[*3]	Osenberg[*3]	Osenberg[*3]
										Plendl	Plendl		
Process Promoter							Beckmann						
		Martini	Martini	Martini	Martini	Martini	Martini	Martini	Martini	Martini	Martini		
							Osenberg[*4]	Osenberg[*4]	Osenberg[*4]	Osenberg[*4]	Osenberg[*4]	Osenberg[*4]	Osenberg[*4]
										Plendl	Plendl		
										Speer	Speer	Speer	Speer
Relationship Promoter							Esau	Esau	Esau	Esau	Esau	Esau	Esau
		Martini	Martini	Martini	Martini	Martini	Martini	Martini	Martini	Martini	Martini	Martini	Martini
							Osenberg[*4]	Osenberg[*4]	Osenberg[*4]	Osenberg[*4]	Osenberg[*4]	Osenberg[*4]	Osenberg[*4]
										Speer	Speer	Speer	Speer
						Udet[*5]	Udet[*5]	Udet[*5]	Udet[*5]				
Year	1933	1934	1935	1936	1937	1938	1939	1940	1941	1942	1943	1944	1945

Opponent									Göring*6				
							Dönitz*7						
			Raeder*8										
						Udet*9	Udet*9						
Year	1933	1934	1935	1936	1937	1938	1939	1940	1941	1942	1943	1944	1945

*1 Dönitz: limited Power Promoter; radar not a priority, influence selective.
*2 Raeder: limited Power Promoter; primarily focused on sonar, radar development delayed.
*3 Osenberg: Mainly handled human capital; no direct technological expertise.
*4 Osenberg: Coordinated project workflows and personnel; indirect influence on radar-related innovation.
*5 Udet: Symbolic Relationship Promoter; legitimised radar among pilots; limited direct influence.
*6 Göring: indirect opposition; tolerated delays in adoption of new technologies.
*7 Dönitz: radar not a priority; no active opposition.
*8 Raeder: limited involvement; no active opposition.
*9 Udet: Radar was considered a low priority; he did not actively oppose its development.

Table 7 D: Roles of actors in political and military leadership and ministries

Power Promoter	Dieckmann	Dieckmann	Dieckmann	Dieckmann	Dieckmann	Dieckmann	Dieckmann	Dieckmann	Dieckmann	Dieckmann	Dieckmann	Dieckmann	Dieckmann
		Erbslöh	Erbslöh	Erbslöh	Erbslöh	Erbslöh	Erbslöh	Erbslöh	Erbslöh	Erbslöh	Erbslöh	Erbslöh	Erbslöh
											Esau	Esau	Esau
											Küpfmüller*[1]	Küpfmüller*[1]	Küpfmüller*[1]
		Willisen	Willisen	Willisen	Willisen	Willisen	Willisen	Willisen	Willisen	Willisen	Willisen	Willisen	Willisen
Expert Promoter	Brandt	Brandt	Brandt	Brandt	Brandt	Brandt	Brandt	Brandt	Brandt	Brandt	Brandt	Brandt	Brandt
	Dieckmann	Dieckmann	Dieckmann	Dieckmann	Dieckmann	Dieckmann	Dieckmann	Dieckmann	Dieckmann	Dieckmann	Dieckmann	Dieckmann	Dieckmann
		Erbslöh	Erbslöh	Erbslöh	Erbslöh	Erbslöh	Erbslöh	Erbslöh	Erbslöh	Erbslöh	Erbslöh	Erbslöh	Erbslöh
					Esau	Esau	Esau	Esau	Esau	Esau	Esau	Esau	Esau
	Hollmann	Hollmann	Hollmann	Hollmann									
	Kühnhold	Kühnhold	Kühnhold	Kühnhold	Kühnhold	Kühnhold	Kühnhold	Kühnhold	Kühnhold	Kühnhold	Kühnhold	Kühnhold	
					Küpfmüller	Küpfmüller	Küpfmüller	Küpfmüller	Küpfmüller	Küpfmüller	Küpfmüller	Küpfmüller	Küpfmüller
			Runge	Runge	Runge	Runge	Runge	Runge	Runge	Runge			
		Schultes	Schultes	Schultes	Schultes	Schultes	Schultes	Schultes	Schultes	Schultes	Schultes	Schultes	Schultes
		Willisen	Willisen	Willisen	Willisen	Willisen	Willisen	Willisen	Willisen	Willisen	Willisen	Willisen	Willisen
Year	**1933**	**1934**	**1935**	**1936**	**1937**	**1938**	**1939**	**1940**	**1941**	**1942**	**1943**	**1944**	**1945**

*[1] Power Promoter role limited; acted within senior command framework.

Process Promoter							Brandt	Brandt	Brandt	Brandt	Brandt	Brandt	Brandt
	Dieckmann	Dieckmann	Dieckmann	Dieckmann	Dieckmann	Dieckmann	Dieckmann	Dieckmann	Dieckmann	Dieckmann	Dieckmann	Dieckmann	Dieckmann
		Erbslöh	Erbslöh	Erbslöh	Erbslöh	Erbslöh	Erbslöh	Erbslöh	Erbslöh	Erbslöh	Erbslöh	Erbslöh	Erbslöh
							Esau	Esau	Esau	Esau	Esau	Esau	Esau
		Hollmann	Hollmann	Hollmann									
	Kühnhold	Kühnhold											
					Küpfmüller	Küpfmüller	Küpfmüller	Küpfmüller	Küpfmüller	Küpfmüller	Küpfmüller	Küpfmüller	Küpfmüller
				Runge	Runge	Runge	Runge	Runge	Runge	Runge			
		Schultes	Schultes	Schultes	Schultes	Schultes	Schultes	Schultes	Schultes	Schultes			
		Willisen	Willisen	Willisen	Willisen	Willisen	Willisen	Willisen	Willisen	Willisen	Willisen	Willisen	Willisen
Relationship Promoter							Brandt	Brandt	Brandt	Brandt	Brandt	Brandt	Brandt
					Dieckmann	Dieckmann	Dieckmann	Dieckmann	Dieckmann	Dieckmann	Dieckmann	Dieckmann	Dieckmann
		Erbslöh	Erbslöh	Erbslöh	Erbslöh	Erbslöh	Erbslöh	Erbslöh	Erbslöh	Erbslöh	Erbslöh	Erbslöh	Erbslöh
							Esau	Esau	Esau	Esau	Esau	Esau	Esau
		Hollmann	Hollmann	Hollmann									
					Küpfmüller	Küpfmüller	Küpfmüller	Küpfmüller	Küpfmüller	Küpfmüller	Küpfmüller	Küpfmüller	Küpfmüller
		Schultes	Schultes	Schultes	Schultes	Schultes	Schultes	Schultes	Schultes	Schultes			
		Willisen	Willisen	Willisen	Willisen	Willisen	Willisen	Willisen	Willisen	Willisen	Willisen	Willisen	Willisen
Opponent										Runge*2			
Year	**1933**	**1934**	**1935**	**1936**	**1937**	**1938**	**1939**	**1940**	**1941**	**1942**	**1943**	**1944**	**1945**

*2 Runge opposed the continuation of centimetric-wave radar development, not out of active resistance, but due to a lack of vision.

Table 8 D: Roles of actors in research, industry, and institutions

10.4.1.2.3 Innovation Steps and Barriers

From 1933 onwards, a chronological overview is provided of the most important events that either promoted or hindered innovation in radar technology until the end of the war. In addition to the clearly identified barriers to innovation, it is evident that deep-rooted mistrust of new developments in many circles led to radar technology being abandoned or put on hold in favour of conventional methods as soon as problems arose during the early test phases. This unfortunate tendency cast a long shadow over the years leading up to and during the war.

Table 9 presents the key innovation steps and barriers relating to government, ministries, and the military, while Table 10 focuses on science, research, and industry. It should be noted that the boundaries between the two tables are not always sharply defined.

Year	**Innovation Steps**	**Barriers to Innovation**
1933–1945		
1933–1945		Government and Wehrmacht leaders did not promote a coherent science and technology policy.
1933–1945		Lack of understanding of radar technology by Hitler, Göring and Wehrmacht authorities.
1933–1945		Competition between different authorities to gain Hitler's attention, hindering promotion of technology and research.
1933–1945		Rivalries within the Wehrmacht and secrecy regulations limiting collaboration.
1936–1940		Sharp decline in electrical engineering student numbers, reducing availability of skilled personnel.

1937		
16 March	Foundation of the Reichsforschungsrat RFR.	Power struggles within government and ministries influenced the promotion and direction of research under the RFR.
1939		
1939		Recruitment of technical and scientific personnel for military service, limiting available talent for radar projects.
1939		Amateur radio banned and equipment confiscated, restricting experimentation and technical development.
1940		
3 February		Entwicklungsstopp-Befehl; resource allocation and strategic priorities limited radar development.
1941		
End of the year		Only 400 engineers, technicians and mechanics still working in the German radar industry.
1942		
Beginning	Recall of experts by General Martini.	Request initially met with little success.
19 February		Schlüsselkraft-Befehl; coordination challenges among research units, limiting effectiveness of retaining key specialists.
9 June	Founding of the second RFR as a reorganisation to combine all existing forces in the interests of the state and concentrate them on the goals to be achieved.	Fragmented research capacity; Germany's radar research still behind Allied development.

14 November	Appointment of Hans Plendl as Bevollmächtigter für Hochfrequenzforschung (BHF) with the aim of coordinating all research work in the field of high frequency technology.	Research capacity heavily inferior to Allies; divided among numerous small and micro laboratories
1943		
22 February	Foundation of the Rotterdam Working Group (AGR). Close and trusting cooperation between industry, research institutions and the military for the first time.	
May	Recall actions for high-frequency specialists.	
Summer	1500 high-frequency specialists recalled from the front.	
2 December		Replacement of Plendl as BHF due to lack of rapid successes.
End of the year	Secret decree for the repatriation of 5000 scientists from the Wehrmacht due to the efforts of Werner Osenberg.	Only 2,000 scientists repatriated by July 1944; a total of 3,430 by November 1944.

Table 9 D: Steps and barriers in government, ministries, and the military

Year	**Innovation Steps**	**Barriers to Innovation**
1933		
Autumn	Application of the acoustic detection principle to electromagnetic waves, marking the first use of radio waves instead of sound waves for location purposes in the NVA.	
Autumn		Unsuccessful reflection (backscatter) tests in the centimetre-wave range due to inadequate performance of existing tubes. The development of high-power tubes was estimated to take approximately ten years.

Autumn	Six magnetrons with a continuous wave output of 40 watts at a wavelength of around 50 centimetres were purchased from Philips in Eindhoven, and the first test transmitters were built at Tonographie in Berlin in December 1933.	
End of the year		Unsuccessful negotiations between the NVA and Telefunken regarding cooperation. Decades of Telefunken's R&D expertise in radio technology remained unutilised.
1934		
January	After unsuccessful negotiations with the Telefunken company, GEMA was founded on 16 January 1934 at Kühnhold's instigation and subsequently became a major developer and manufacturer of radar equipment. The co-founders and, from the beginning of 1935, sole shareholders of GEMA, Erbslöh and von Willisen, were highly motivated, and there was a trusting and unbureaucratic cooperation between the NVA and GEMA.	
1934		No further talks took place between the NVA and Telefunken. In the following period, there was no support or guidance from political or military authorities, unlike in Great Britain, to consolidate the expertise and decades of experience of major companies such as Telefunken, AEG, Siemens, and Lorenz.
	Two proven high-frequency technology experts, Hollmann and Schultes, were recruited by GEMA and contributed significantly to its success.	

Autumn	Further successful trials against ships and aeroplanes with improved transmission and reception technology at GEMA. TVA provides funding for further trials.	However, strict secrecy was initially imposed, even within the Navy itself, because at that time it was not yet possible to measure range and azimuth, both essential components for effective radar operation.
End of the year	Development from pure reflection (backscatter) measurement to radar measurement was successfully realised at GEMA.	
1935		
Early	Kühnhold was convinced of the effectiveness of centimetric waves for achieving greater accuracy in locating and imaging targets. Initial work in the 4 to 10 cm wavelength range had already begun, despite the lack of results in generating the necessary power with Pintsch's tube development. Practical trials were scheduled for January 1936.	Insufficient power generation in the centimetre-wave range led to discontinuation of further development at this wavelength.
Early	For the first time, C. Lorenz AG undertook development work in the field of radar technology without being commissioned by official authorities. The advanced state of tube technology proved advantageous for the further development of higher-performance radar sets, and the pulse method was also investigated.	
Summer	Successful tests at wavelengths of approximately 2 metres and 52 centimetres led to the adoption of the 2-metre range for subsequent developments at GEMA, forming the basis for the Freya radar set.	No further focus was placed on shorter wavelength ranges, which would prove to be a serious mistake in later years. Furthermore, there was no development of frequency-agile antennas.

Summer	First attempts by Runge at Telefunken to detect aeroplanes. The company's own Ju 52 was successfully located at an altitude of 5,000 metres.	The hoped-for enthusiasm from Telefunken's management failed to materialise. Runge continued work on location using radio waves without the knowledge of Telefunken's board. The work was also kept secret from Telefunken's parent companies, AEG and Siemens, for fear that this field of research would be taken away from Telefunken.
September	After successful demonstrations of devices by the NVA and GEMA to the Navy, the military command recognised the importance of radar technology and promised funding for further developments.	Despite these impressive results, however, Admiral Raeder saw the detection of underwater targets with sound waves as the most urgent military task and thus also the work within the NVA, as this was where equipment for the front could be expected in the foreseeable future.
1936		
April	The NVA was ordered to press ahead with the urgently needed development of a smaller radar system to equip destroyers, torpedo boats, and submarines. Notably, the technical staff recognised the importance of radar technology for the Navy at an early stage.	The German Navy did not pay much attention to the development of radar equipment, but the Torpedo- und Artillerie-Waffenamt Offices now intervened regarding their requirements for increasing the accuracy of firing values for torpedo and, above all, anti-aircraft artillery. This meant that the ordnance offices initially had a greater influence on the development of radar technology than the naval authorities responsible for tactics and operational deployment.
Autumn		GEMA was only allowed to provide information about ongoing trials to the Navy. Although the Air Force showed interest in the work, contact could not be established.

End of the year	The Freya device was presented to Wolfgang Martini, head of the Nachrichtenverbindungswesen in the RLM, at Pelzerhaken. Martini then ordered the delivery of 12 devices.	
1937		
Summer		Implementation of an NVA test programme to investigate pulse technology at short wavelengths in the 10 to 14 cm range. Research and further development at these short wavelengths in the centimetre wave range were eventually discontinued again. As a result, it was not until 1943 that tests and developments in the field of centimetre waves were carried out again, after a radar device with an operating wavelength of approx. 9 cm was found in an English bomber shot down near Rotterdam. Until then, it had been assumed that no developments would be made in this wavelength range in Great Britain either. A mistake with fatal consequences.
1938		
Summer		The Heereswaffenamt ordered 10 prototypes of Telefunken's Darmstadt device, which already had altitude measurement, for anti-aircraft artillery trials. Here too, however, there was no visible pressure for rapid delivery. In the summer of 1938, the Darmstadt device was presented to Generalluftzeugmeister Erich Udet. His reaction was characterised by a deep-seated rejection of this pioneering development.

July		In July 1938, GEMA's Freya device was demonstrated to Hitler and high-ranking officers. Göring was annoyed that the head of the Air Force's NVW, General Martini, had not been invited. The background to this was that the Navy did not want to have any competition from the Air Force as a customer for radar equipment.
July		Göring demanded the production of 200 Freya devices per month, but nothing happened at first. Even Martini's order for 12 Freya devices, which had been in place for over a year, was temporarily cancelled, until Martini, now a general, managed to enforce it.
November		At a meeting between the Navy and the Technisches Amt at the RLM, the Navy was casually informed that the companies Lorenz and Telefunken had been commissioned to develop radar equipment in the 50 and 60 cm wavelength range.
1939		
August	Shortly before the outbreak of the war, the Technisches Amt of the RLM gave a presentation to Hitler, Göring, Milch, Udet and other leading figures on the state of German air armament. During the same occasion, Telefunken demonstrated its Darmstadt device.	The Technisches Amt focused on planning and development before the war and did not implement operational measures until after its outbreak.

October	Leo Brandt, head of device development at Telefunken, was commissioned to develop and manufacture the Nahflugmeldegerät, capable of measuring distance, lateral position and altitude.	
	The Technisches Amt showed great foresight in the development work at Telefunken, marking the birth of the Würzburg radar devices. However, it should be noted that this commissioning of a device, which essentially did not yet exist, was largely due to the initiative of a single individual within the RLM and was not part of an overarching strategic plan of the military command. An order was placed for 4,000 devices, which enabled the successful trialling of new processes in development, design, and production, with these stages occurring almost simultaneously. At the same time, the head of the NVW ordered the delivery of 200 Freya devices for the air warning service.	There was no overall strategic concept on the part of the military command.
1942		
27 November		Development of centimetre waves at Telefunken was halted at Runge's suggestion.
1943		
22 February	From 22 February 1943, the AGR, under Leo Brandt, coordinated the development and production of radar devices, beginning with the analysis of the captured British H_2S radar.	Despite significant technical advances, the AGR was unable to close the technological gap with the Allies before the end of the war.

Table 10 D: Steps and barriers in science, research, and industry

The development of German radar technology during World War II was significantly hindered by governmental, military, industrial, and technical obstacles.

Key factors included political restrictions, lack of authorisation, and incorrect technical decisions, all of which contributed to delays in technological progress. Top leaders, particularly Hitler and Göring, failed to actively support the developments, leading to further setbacks. The failure to recognise new technologies such as radar, coupled with insufficient permissions and restrictive secrecy rules, exacerbated blockages in the research process.

One of the most decisive mistakes was the decision to discontinue research into centimetre waves in 1942. This became particularly evident when the Germans discovered a British radar with a wavelength of about 9 centimetres in February 1943, underscoring Germany's technological lag in radar technology.

Another critical mistake was the 1936 decision to restrict research to fixed frequencies and abandon further developments in frequency agility. This had dramatic consequences: during the 1943 air raid on Hamburg, German air defences were unable to counter the British Window jamming technique because their radar systems could not switch frequencies, rendering them ineffective.

Additionally, the lack of skilled personnel due to military service and the prohibition on amateur radio operators further hindered progress. The hostile attitude toward physicists and the scarcity of students also made it difficult to recruit qualified personnel for radar development.

Overall, these political and technical errors led to a significant lag in German radar technology compared to the Allies. The lack of innovative momentum and organisational stagnation not only delayed technological progress but also hindered Germany's ability to develop effective countermeasures, significantly weakening its war effort.

10.4.2 Great Britain

10.4.2.1 Organisation and Command

10.4.2.1.1 Interest Groups

In contrast to Germany, political decision-makers played a decisive role in the development of radar-based air defence systems in Great Britain. The political landscape was characterised by close and confidential relationships between key politicians, ministries, the military, scientific institutions, and industry, and was shaped by various committees.[1547] However, as in Germany, the complexity of the overall situation allows only limited insight into the precise roles and actions of the different actors. Judkins[1548], Zimmerman[1549], and Burns[1550] offer deeper perspectives on the positions of key figures, highlighting their specific interests and personal conflicts within the broader context of radar-based air defence development, supported by extensive source material.

10.4.2.1.1.1 Government

The Government played a key role in the development of a comprehensive air defence system in Great Britain. Political leaders such as Churchill were pivotal in driving technological innovation and securing personnel and material resources. Based on Fig. 17 in Chapter 7.1.2.2, the following chapters analyse the key political governing bodies and committees that were particularly important for the development of radar technology and the establishment of an integrated defence system against aerial attacks.[1551] Among several leading politicians, there was a remarkable awareness that information technologies such as radar and radio direction finding could only be effective as part of a broader complementary system. Only such an integrated approach would result in an effective concept for national air defence. However, before any broad agreement on

1547 Klausing 2022.
1548 Judkins 2007.
1549 Zimmerman 2013.
1550 Burns 2013.
1551 A comprehensive overview of the almost unmanageable number of committees and their members can be found in Rogers 2014.

a radar-based system was reached, there were intense controversies within politics, the ministries, and the scientific community. Churchill and his scientific adviser Lindemann played a decisive role in shaping this discourse. Although Churchill held no ministerial office between 1929 and 1939, his outspoken criticism of appeasement and his warnings about the Nazi regime gained increasing public attention. Lindemann passionately supported this stance. From late 1934 onwards, he worked actively to promote his ideas for defending against German air raids.

He believed that future attacks would resemble those of World War I, but with far greater destructive power, especially from nighttime bombing, which would lead to significant damage to British cities.[1552] An article by Lindemann expressing these concerns appeared in The Times on 8 August 1934:

> "SCIENCE AND AIR BOMBING
>
> TO THE EDITOR OF THE TIMES
>
> Sir,
>
> In the debate in the House of Commons on Monday on the proposed expansion of our Air Force, it seemed to be taken for granted on all sides that there is and can be no defence against bombing aeroplanes and that we must rely entirely upon counter-attack and reprisals. That there is at present no means of preventing hostile bombers from depositing their load of explosives, incendiary materials, gases, or bacteria upon their objectives I believe to be true; that no method can be devised to safeguard great centres of population from such a fate appears to me profoundly improbable.
>
> If no protective contrivance can be found and we are reduced to a policy of reprisals, the temptation to be "quickest on the draw" will be tremendous. It seems not too much to say that bombing aeroplanes in the hands of gangster Governments might jeopardize the whole future of our Western civilization.
>
> To adopt a defeatist attitude in the face of such a threat is inexcusable until it has definitely been shown that all the resources of science and invention have been exhausted. The problem is far too important and too urgent to be left to the casual endeavors of individuals or departments. The whole weight and influence of the Government should

1552 Zimmerman 2013, pp. 97–98.

be thrown into the scale to endeavour to find a solution. All decent men and all honourable Governments are equally concerned to obtain security against attacks from the air, and to achieve it no effort and no sacrifice is too great.

I am, Sir, yours faithfully.

F. A. LINDEMANN.

Christ Church, Oxford, Aug. 2."[1553]

Further correspondence and articles on the air defence debate between 1934 and 1939, with contributions from Churchill, Lindemann, Wimperis, and others, can be found in The Washington Quarterly.[1554] This was a critical situation for the Secretary of State for Air, Lord Londonderry[1555], as Churchill and Lindemann signalled their willingness to attack government policy through the press and radio. Londonderry attempted to keep air defence research and development secret by informing Lindemann on 20 December 1934 about his decision to establish the CSSAD, advising him to contact Tizard directly. However, this only increased Lindemann's mistrust of the Air Ministry. Lindemann believed that the Tizard Committee could never be effective and that he would have to appeal over Londonderry's head. Londonderry's aim was to ensure that there would be no parliamentary debate on secret air defence research projects. On 7 January 1935, Lindemann composed a letter signed by Churchill and Austen Chamberlain[1556] to Prime Minister MacDonald[1557], requesting a review of the correspondence between Londonderry and Lindemann regarding the demand to establish a C.I.D. subcommittee. MacDonald succumbed to the pressure and agreed to the establishment of a C.I.D. subcommittee. It is likely that MacDonald was initially unaware of the

1553 Lindemann 1934; Burns 2013, p. 96; Fisher 1989, pp. 81–82; Birkenhead 1961, p. 175, The Air Defense Debate 1985, pp. 200–201.

1554 The Air Defense Debate 1985, pp. 199–216.

1555 Jackson 2008.

1556 Not to be confused with Neville Chamberlain, Austen Chamberlain and Churchill were outspoken critics of the British government's foreign policy in Parliament in response to the growing threat of the Nazi regime. They called for rearmament and a stronger stance against German aggression (Dutton 2013).

1557 Marquand 2015.

latest initiatives of the Air Ministry, such as the existence of the CSSAD, which was not mentioned in the letter from Churchill and Chamberlain.[1558] MacDonald received the letter on 9 January and replied to Chamberlain on 10 January, expressing his support for a committee to be overseen by the C.I.D.:

> "...I quite agree with you that it should not be a departmental Committee, and have already made up my mind that it is something that the C.I.D. should take in hand..."[1559]

After MacDonald was informed about the existence of the CSSAD, he attempted to revise his hasty decision with a letter dated 15 January, asking A. Chamberlain to disregard the previous correspondence. However, Chamberlain had already been briefed on the CSSAD by Lindemann.[1560] MacDonald informed A. Chamberlain that the CSSAD had already been staffed with Tizard as chairman, along with Wimperis, Blackett, and Hill, and that Lindemann had been informed about this by Londonderry.[1561] This led to strong criticism from both Churchill and Lindemann towards MacDonald, with Lindemann also directing frequent criticism at the CSSAD. On 28 January, A. Chamberlain wrote to MacDonald, requesting a meeting with him and Churchill to discuss air defence research. On 30 January, MacDonald responded with an attempt to co-opt the critics. He offered Lindemann full membership in the CSSAD on behalf of the Air Ministry. Lindemann initially withheld his response because A. Chamberlain had warned him that he would be muzzled. However, unwilling to be silenced, he replied to the offer as follows:

> "I thank you for your letter of the 30th January. I understand the whole question is soon to be debated in Parliament, and I think you will agree that in the circumstances it is best for me to defer giving a definite answer for the time being."[1562]

1558 Zimmerman 2013, pp. 98–99.
1559 Birkenhead 1961, p. 179.
1560 Zimmerman 2013, p. 99.
1561 Birkenhead 1961, pp. 179–180; Zimmerman 2013, p. 99.
1562 Zimmerman 2013, p. 100.

On 15 February, a personal meeting took place between place between MacDonald, and reportedly also attended by Churchill, Lindemann, and A. Chamberlain, during which MacDonald explained that he was convinced of the necessity of establishing a subcommittee of the C.I.D.. In return, MacDonald and Londonderry hoped for political support for the rearmament policy and a reduction in public criticism. However, their hopes were soon dashed. Just three days later, on 18 February 1935, Lindemann delivered a public speech highlighting the threat from the air and revealed the existence of the CSSAD. He justified his actions by citing a lack of trust in the Air Ministry. Instead of cooperating with the CSSAD, he called for the creation of a powerful body that would be accountable to Parliament. The government, however, did not initially respond by creating a new institution as Churchill and Lindemann had hoped. Instead, it officially announced the existence of the CSSAD, including a list of its members. Once again, the competence of the committee was emphasised, and it was stated that Lindemann had already been invited to join the body some time earlier.[1563] The press took a keen interest in the public announcement of the CSSAD. In contrast to the controversial political debates surrounding rearmament, there was broad consensus on the need for an effective air defence system. One example is an article published in The Manchester Guardian Weekly.

Although the paper had initially opposed increased military spending, it now expressed full support for the government's efforts:

> "One cannot envy the committee of distinguished scientists who have been appointed by the government 'as a matter of urgency' to investigate new possibilities of countering air defence. People often say that an effective 'reply' to the bomber is bound to be discovered ... It may be so but there are no grounds for any immediate optimism ... We wish the eminent scientists well, but there is no denying they have taken on a full size job."[1564]

In the period that followed, several parliamentary debates took place concerning the progress of the CSSAD's work and the possible

1563 Ibid., p. 101.

1564 The Manchester Guardian Weekly 1935, p. 183; Zimmerman 2013, p. 102.

appointment of an additional committee under the C.I.D. These discussions were set against the backdrop of growing public and political pressure on the government to take concrete measures for air defence. In response, MacDonald announced the formation of a new subcommittee of the C.I.D. tasked specifically with defence against air attacks. At the same time, the continued existence of the CSSAD was confirmed, as the committee had already begun developing concrete solutions and had initiated the first experimental work. The new committee was named the Air Defence Research Sub-Committee (ADRC) of the C.I.D.[1565]

The ADRC met for the first time on 10 April 1935 under the chairmanship of Philip Cunliffe-Lister, Secretary of State for the Colonies.[1566] Tizard was a member of this committee from the very beginning. Dowding was also a member, along with other prominent figures from politics, ministries, and the military; however, Lindemann was not included. At this time, Tizard was already a long-standing member of the Aeronautical Research Committee (ARC) and had been its chairman since 1933.[1567] As there was a risk of overlapping responsibilities between the ADRC and the CSSAD, a pragmatic solution was found. While the CSSAD focused on direct research tasks in the field of air defence, the ADRC concentrated on the political and military decisions required to implement the results of these research and development programmes into practical air defence measures.[1568] Churchill subsequently moderated his criticism of the government. The establishment of the ADRC was well received by the press. For example, The Manchester

1565 Zimmerman 2013, p. 103.

1566 Ibid., p. 105.

1567 Collar 1967; Rogers 2014, p. 245; as already explained, close cooperation between government, science, and industry was established through a large number of bodies, research institutions, and committees. Two important organisations were the Aeronautical Research Committee (ARC) and the Radio Research Board (RRB). Close contact between the armed forces and scientists was facilitated through the ARC and the RRB. Endowed with far-reaching powers, the ARC was able to advise both the government and industry on aviation matters. Its aim was to support and promote aviation research within the Air Ministry and other ministries, as well as to foster aviation training (Zimmerman 2013, pp. 56–57).

1568 Farren 1961, p. 332; Judkins 2007, pp. 108–109; Zimmerman 2013, p. 103.

Guardian Weekly highlighted the importance of this new body, especially in light of Stanley Baldwin's speech that the bombers would always get through. The article expressed hope that the new committee would help to effectively counter the threat of hostile air raids.[1569] However, the work of the ADRC was viewed differently by some. From a positive perspective, the ADRC was seen as a bridge between politics, the military, and science. On the other hand, critics argued that, in the absence of additional resources for air defence, the committee served more as a shell for an illusion of progress.[1570]

In the period leading up to the outbreak of World War II on 3 September 1939, Churchill repeatedly warned against Hitler's aggression and was a staunch opponent of the policy of appeasement towards Germany. Following the annexation of Austria to the German Reich and the Sudetenkrise in 1938, as well as the breach of the Münchener Abkommen, the appeasement policy was widely seen as a failure, and Churchill's stance gained increasing public recognition.[1571] After the invasion of Poland on 1 September 1939 and the subsequent declaration of war on Germany on 3 September, Churchill was appointed to the cabinet by Chamberlain and, on the same day, was made First Lord of the Admiralty.[1572]

On 10 May 1940, following Neville Chamberlain's resignation, Churchill became Prime Minister in the newly formed War Cabinet, a position he held until 23 May 1945.[1573] He also took on the role of Minister of Defence. Churchill's appointment significantly strengthened Lindemann's position, while Tizard's influence began to wane. As a result of Lindemann's growing influencc, Tizard resigned from all his duties at the Air Ministry in June 1940, except for his role at the ARC.

On Churchill's suggestion, a mission was organised in September 1940 to the USA and Canada, under Tizard's leadership, to advance joint research and development in radar technology and explore manufacturing capacities. With the cavity magnetron in his

1569 Zimmerman 2013, p. 103.
1570 The Air Defense Debate 1985, p. 208.
1571 See also Chapter 2.
1572 Roberts 2019, pp. 462–493; Addison 2018, pp. 31–32.
1573 Roberts 2019, pp. 494–514; Addison 2018, pp. 32–33.

possession, Tizard and the six scientists accompanying him succeeded in convincing the USA of the high standard of British research, leading to extensive collaboration with American scientists and industry. This secret mission, known as the Tizard Mission, was a major success for Tizard.[1574] After returning in the autumn of 1940, Tizard became a freelance scientific adviser to the Ministry of Aircraft Production and was appointed to the Air Council by the Royal Air Force.[1575]

In 1920, Tizard became Deputy Secretary at the Department of Scientific and Industrial Research (DSIR), where he organised the coordination of scientific work between defence and civilian departments.[1576] His administrative abilities quickly led him to rise through the ranks, and in 1927, he took on the position of Secretary. He had a major influence on the department's policy and founded the Chemical Research Laboratory in Teddington, which later became known as the National Chemical Laboratory. Tizard left the Department of Scientific and Industrial Research in 1929 and became the Rector of the Imperial College of Science and Technology in London. He held this post until 1942.[1577] Afterward, he served as President of Magdalene College, Oxford, from 1942 to 1946, and was briefly chairman of a policy committee that dealt with the growth in the use of radar and radio jamming techniques. However, internal battles between Tizard and Lindemann, now Lord Cherwell, led Tizard to threaten resignation. Churchill persuaded him to become a member of the new Radio Board.[1578]

In 1942, a renewed disagreement arose between Tizard and Lord Cherwell, also known as Frederick Lindemann, who served as Winston Churchill's close scientific adviser. Although Cherwell was not a formal member of the War Cabinet, he held the position of Paymaster-General and had significant influence on strategic decisions due to his close personal relationship with Churchill. In

1574 For the Tizard Mission see also Chapter 10.5.2.1.
1575 Snow 2013, p. 40; Snow 1961b, p. 43.
1576 Piller 2008, p. 3.
1577 Ibid., pp. 3–4.
1578 Piller 2008, p. 7; Clark 1965, pp. 317–318.

a now well-known memorandum called the Dehousing Paper[1579], Cherwell argued that the morale of the German population could be undermined through systematic area bombing of working-class residential neighbourhoods. He presented highly optimistic estimates regarding the level of destruction such a campaign could cause. Tizard and other critics questioned the validity of these assumptions, and later assessments confirmed that Cherwell's estimates were significantly overstated and that the actual effects of the bombing campaign on German morale and industrial capacity were far more limited than anticipated.[1580]

10.4.2.1.1.2 Ministries and Institutions

10.4.2.1.1.2.1 Air Ministry

The Air Ministry played a key role in the development of British radar technology through the efforts of Albert Rowe, who conducted an analysis of existing detection systems for air defence in 1934, and Harry Wimperis, the Director of Scientific Research at the Air Ministry. Rowe was Wimperis' personal assistant. Based on Rowe's findings, Wimperis proposed the establishment of a research committee for air defence at the Air Ministry.[1581] Henry Tizard was appointed chairman of this CSSAD.[1582] In the following

---

1579 The Dehousing strategy was proposed in a memorandum dated 30 March 1942 by Frederick Lindemann, who was the British Prime Minister's chief scientific adviser. This document, often referred to as the Dehousing Paper or Dehousing Memorandum, outlined a plan focused on systematically bombing working-class residential areas in German cities. Lindemann believed this strategy would allow the Allies to avoid a full-scale invasion of Europe. The strategy was based on the effects of the Area Bombing Directive issued in February 1942 and was endorsed by the War Cabinet. The Dehousing Paper became the foundational document for this approach. Maurice Kirby, a respected historian and expert on operations research during World War II, notes that while the Dehousing Paper was presented early in 1942, its strategic implications and detailed operational assessments by British Operations Research teams became more prominent towards the end of 1942. Later analysis showed that the estimates of destruction and impact on German morale were overly optimistic (Kirby 2003, p. 139).

1580 Snow 2013, pp. 43–44; Snow 1961b, p. 47.

1581 Judkins 2007, p. 8.

1582 See also Chapter 7.1.2.2.

years, Churchill and Lindemann launched severe criticism against the Air Ministry, accusing it of inactivity and delays in developing an effective air defence research program. Churchill and Lindemann wanted to set the agenda for the research programme. However, the research programme had the continued support of the Baldwin and Chamberlain governments, as well as that of Air Marshal Hugh Dowding.

10.4.2.1.1.2.2 Committee of Imperial Defence (C.I.D.)

The Committee of Imperial Defence (C.I.D.) was composed of leading figures from the political and military spheres and functioned as the principal advisory body for the strategic defence of the British Empire.

Established in the early twentieth century, the C.I.D. developed into a key institution for coordinating imperial defence policy. In the interwar years, its responsibilities encompassed a broad spectrum of issues, including political, diplomatic and military planning. Following periods of inactivity, the committee reconvened in 1922 in response to emerging security challenges, particularly those concerning defence policy.

From 1935 onwards, the C.I.D. increasingly concentrated on the rapid rearmament of Britain, prompted by the escalating threat of German militarisation. In the years leading up to World War II, its core functions included the integrated planning of national defence and the logistical coordination of military supply across the three armed services.

In order to relieve the Prime Minister of certain responsibilities, the office of Minister for Co-ordination of Defence was created in 1936 under Prime Minister Stanley Baldwin. This ministerial post, which did not entail the leadership of a government department, entailed the role of deputy chairman of the C.I.D. The first to hold the office was Thomas Inskip.[1583]

When Winston Churchill was appointed Prime Minister in 1940 and simultaneously assumed the newly consolidated position of Minister of Defence, the role of the Co-ordination Minister became redundant. During the war, the War Cabinet assumed

1583 Robbins 2006, p. 2.

the core strategic functions previously carried out by the C.I.D.[1584] The committee was not reconstituted following the end of World War II.[1585]

A detailed analysis reveals that the various committees of the Air Ministry and the Committee of Imperial Defence (C.I.D.) were intricately interconnected through shared subcommittees, with key individuals such as Tizard actively participating across multiple bodies. The Royal Air Force, notably represented by Air Marshal Hugh Dowding, played a pivotal role by providing substantive support for the development of the air defence system, including experimental work conducted at Biggin Hill. It is significant that, despite recurrent setbacks during development and testing phases, confidence in the system's eventual success remained unwavering, and institutional support persisted unabated even under critical circumstances, reflecting a firm conviction that an effective air defence apparatus would secure national defence capability.

A fundamental strength of the British approach was the concurrent development and evaluation of the Chain Home radar network alongside the integrated air defence system. Had Tizard's efforts been confined solely to the technological advancement of radar equipment without parallel strategic planning for air defence, it is probable that Britain would have been less adequately prepared to counter the German aerial threat in 1940.

While scholarship on British radar development has predominantly concentrated on the structure and operations of the CSSAD, there remains an important lacuna regarding the interrelation and coordination among the multiple committees from the perspective of the actors involved.

10.4.2.1.1.2.3 Air Defence Research Sub-Committee (ADRC) of the C.I.D.

The Air Defence Research Committee (ADRC) was established on 10 April 1935 and subsequently incorporated as a subcommittee within the C.I.D. Political responsibility and oversight of the ADRC were vested in Cunliffe-Lister, who assumed the office of Secretary

1584 Probert 1993, pp. 790–791.
1585 McKercher 2008, pp. 391–442.

of State for Air in June 1935.[1586] Scientific leadership, however, remained under the direction of Henry Tizard.

Following Cunliffe-Lister's elevation to the peerage as Viscount Swinton in November 1935, the ADRC became commonly known as the Swinton Committee.[1587] The primary objective of the committee was to expedite technological advancements in air defence and to secure the necessary financial appropriations to support such developments. It remains uncertain to what extent Tizard was apprised of the government's underlying motives for establishing the ADRC, particularly in light of prior lobbying efforts spearheaded by Frederick Lindemann. Nevertheless, Tizard was aware of Lindemann's intent to displace the Committee for the Scientific Survey of Air Defence (CSSAD).[1588]

Tizard's response to the formation of the ADRC was characterised by both disappointment and frustration, as he perceived a considerable duplication of responsibilities between the ADRC and the CSSAD. Moreover, the modalities of political and operational cooperation between the two bodies were unclear. In correspondence dated 21 March 1935, Air Marshal Hugh Dowding assured the continued independence of the CSSAD's work. By early April, Tizard was offered membership of the ADRC and was once again assured that neither the Prime Minister nor government departments would intervene in the CSSAD's activities.

Dowding, a staunch advocate of the CSSAD, also joined the ADRC. The committee's membership included senior civil servants and military officials at ministerial rank, as requested by Lindemann, albeit excluding Lindemann's closest supporters. Coordination between the ADRC and the CSSAD was further institutionalised through the appointment of Rowe as joint secretary to both committees.[1589]

In the period that followed, both the slow functioning of the ADRC and its erroneous assessment of German aircraft production which was considered underestimated were subject to criticism.

1586 Cunliffe-Lister was Secretary of State for Air between 1935 and 1938 and Minister for Civil Aviation from 1944 to 1945.

1587 Robbins 2008, pp. 4–5.

1588 Zimmerman 2013, p. 105.

1589 Ibid., p. 105.

The ADRC operated in a more formal manner than the CSSAD and primarily relied on the scientific and technical investigations carried out by the CSSAD. The fact that the ADRC convened only twice a year led Churchill to insist that air defence research be included in government debates.

During the debate on 6 June 1935, Churchill reiterated the considerable efforts he, Lindemann, and Austen Chamberlain had made to persuade the government to address the threat of aerial bombing. He further accused the government of acting with undue slowness and argued that had a truly effective scientific committee been established earlier, significant progress could have been achieved. Prime Minister MacDonald, however, firmly rejected Churchill's accusations, maintaining that Churchill had largely misunderstood the ADRC's role. He emphasised that the ADRC's task was to coordinate research activities and that it worked closely with the CSSAD, which comprised distinguished scientists. Based on the initial results of the CSSAD, MacDonald expressed optimism regarding advances in air defence. A Chamberlain shared this view and advised Lindemann to seek membership in the CSSAD.

On 7 June 1935, MacDonald resigned as Prime Minister. Stanley Baldwin succeeded him on 7 June and formed a new cabinet. Cunliffe-Lister replaced Lord Londonderry as Secretary of State for Air. Baldwin's primary aim was to strengthen political oversight of air defence. This period also saw intense debate over the financial resources allocated by the government to establish an effective air defence system. Against the backdrop of Germany's increasing rearmament, discussions regarding the expansion of squadron numbers were contentious and resulted in compromises on various points.

Due to long standing concerns about the disclosure of sensitive information on technological developments and operational concepts in air defence and in light of significant press interest the government sought to coopt its most vocal critics Churchill and Lindemann. Churchill insisted that his scientific adviser Lindemann be included in the CSSAD.[1590]

On 11 June 1935, Lindemann finally agreed to join the CSSAD but continued to pursue his fundamental goal of influencing and

1590 Snow 2013, p. 27; Snow 1961b, pp. 30–31.

controlling the agenda of the air defence research programme. Even before being informed of the secret research, he decided to set out his views on air defence in an eight-page memo to Tizard. In this way, he wanted to ensure that he could disclose his positions before becoming subject to confidentiality requirements.[1591] He vehemently advocated the use of aerial mines and parachute bombs or bombs attached to balloons. He also proposed the use of heavy cables attached to a parachute at one end and to a weight at the other. Another suggestion was the use of acoustic proximity fuses, designed to detonate bombs upon detecting the sounds of approaching enemy aircraft. The use of infrared detectors to locate aircraft was also considered a possible defence measure. Since he was not familiar with Watson Watt's work, he additionally suggested using reflected electromagnetic waves. However, Lindemann regarded the development and deployment of aerial minefields as his top priority.[1592]

Lindemann's proposals were evaluated by Rowe and, with one exception, were found to be unrealisable or lacking in impact.[1593] In a meeting with Tizard, Lindemann indicated that he would continue to discuss air defence research with outsiders and once again expressed deep distrust of the employees at the Air Ministry. Tizard then issued an ultimatum: he reminded Lindemann of the authority of state secrecy and demanded that Lindemann cooperate openly and without hesitation or otherwise stay away from the CSSAD.[1594]

Lindemann eventually gave in and attended a CSSAD meeting for the first time on 25 July. Earlier in July, Cunliffe-Lister had suggested appointing Churchill to the ADRC, primarily to calm him down regarding his criticism of air defence.

On 6 July, Churchill accepted the offer on the condition that he "must be free to debate all of the general issues of strength, air policy and air programmes."[1595]

1591 Zimmerman 2013, p. 109.

1592 Ibid., p. 110.

1593 However, it should be noted at this point that the Daventry Experiment conducted by Watson Watt and Wilkins in February 1935 was only four months old and radar development was still in its infancy.

1594 Zimmerman 2013, p. 110

1595 Ibid., p. 111.

Churchill quickly tried to put his stamp on the ADRC with Lindemann's support. Together, they prepared a paper on air tactics and air defence for the ADRC's first meeting.

Churchill expressed doubt about the effectiveness of early warning systems for air attacks. Instead, he proposed the use of very fast, radio-equipped aircraft to meet attacking bomber formations over the sea and report their strength and direction by radio. At the ADRC meeting on 25 July 1935, it was decided to examine this proposal further, while also drawing attention to the CSSAD's work on defence against enemy aircraft.[1596]

As Churchill and Lindemann remained firmly convinced that, given the increasing German threat, insufficient efforts were being made for air defence, Churchill emotionally proposed establishing a further organisation to report to the ADRC and ensure the implementation of its concepts. On 18 August, Cunliffe Lister replied to Churchill, informing him that the Air Ministry already had a system for following up useful concepts and updating him on major advances in detection and location technologies. This reassured Churchill and Lindemann for the time being. Meanwhile, the radar group's work at Orfordness had been in full swing since May 1935, and the many political and committee disputes had no negative impact on the progress of the air defence research programme.[1597]

10.4.2.1.1.2.4 Committee for the Scientific Survey of Air Defence (CSSAD)

The CSSAD, also known as the Tizard Committee, is the best-known committee and was described in detail in chapter 7.1.2.2. Much has been reported about the conflict between Tizard and Lindemann. C. P. Snow interpreted the conflict against the backdrop that Lindemann wanted nothing to do with radar and believed it should be placed low on the list of urgent priorities, while his own ideas should take precedence.[1598] Watson-Watt contradicted Snow's statements, reporting that Lindemann was indeed in favour of developing radar, although he was fully aware of the possibility

1596 Ibid., p. 111.
1597 Ibid., p. 112.
1598 Snow 2013, p. 29; Snow 1961b, p. 33.

of enemy countermeasures. The actual basis of the conflict, however, was the status of the CSSAD, which Lindemann considered to have insufficient executive authority. He personally saw no way to carry out the necessary lobbying in the public and political spheres through his contacts. Lindemann always endeavoured to influence the committee's research agenda. The members of the CSSAD also found it unacceptable that Lindemann reported to Churchill in the ADRC. After Blackett and Hill resigned from the CSSAD due to the impossibility of working with Lindemann, the committee was quickly re-staffed without him, with Appleton joining as a new member. In a letter dated 2 November 1936, Swinton informed Lindemann that he had re-staffed the CSSAD under Tizard's chairmanship and expressed regret that no constructive collaboration had taken place, especially since he had been the one to invite Lindemann to join the CSSAD in the first place.[1599] After his dismissal from the CSSAD, Lindemann apparently was not given the opportunity to work with the ministries and the Admiralty on his research ideas, receiving only brief and sometimes dismissive responses.[1600] Lindemann was pleasantly surprised when he was invited to join the ADRC by the new Secretary of State for Air, Kingsley Wood[1601], with Churchill and Tizard also being members. This news came as a real shock to Tizard, who had hoped for a fruitful collaboration with scientists with whom he could work well. As a consequence, he offered his resignation from the ADRC, seeing a continuation of the disputes between himself and Lindemann within the committee. Ultimately, he was persuaded to continue working in the ADRC, though not without first expressing his concerns once again.

10.4.2.1.1.3 Royal Air Force (RAF)

Unlike in Germany, military leadership in Great Britain, particularly within the Royal Air Force (RAF), was involved from the out-

1599 Birkenhead 1961, pp. 207–208.
1600 Ibid., p. 208.
1601 As successor to Swinton.

set and played a key role in advancing both radar development and the establishment of an integrated air defence system.[1602]

This was especially true of Air Marshal Hugh Dowding, who as early as February 1935 called for a practical demonstration of using radio waves for defence purposes. This marked an important development step in the British air defence system, following the successful Daventry Experiment by Watson Watt and Wilkins.

Further important steps were the experiments initiated by Tizard and Dowding to develop a system that would analyse all available information in order to ensure the optimal deployment of forces. During this process, Dowding assumed increasing responsibility within the RAF. It was thanks to the foresight of both men that, despite resistance from senior officials in the Air Ministry and officers in the RAF, the experiments moved forward. Both used their influence to bring scientists, engineers, and operational users together in the practical development of the system.

By 1939, an air defence concept was already in place, later known as the Dowding System, which required significant specialist personnel and financial resources. Equally important was the fact that RAF leadership did not allow setbacks in technical or operational areas to halt progress but continued to support the project consistently. Dowding can be regarded as the architect of victory in the Battle of Britain. General Frederick Pile said of him:

> "A difficult man, a self-opinionated man, a most determined man, and a man who knew more than anybody about all aspects of aerial warfare."[1603]

10.4.2.2 Innovation Process

10.4.2.2.1 Historical Context

A closer examination of key actors reveals, unlike in Germany, a clear interconnection and mutual dependence between various institutions and individuals across politics, government ministries, the military, science, and industry, spanning multiple levels of hier-

---

1602 For an overview of the command of the Royal Air Force during World War II, see Probert 1993, pp. 786–807.

1603 RAF Museum, London.

archy. From the outset, the tasks necessary for the development of an integrated air defence system were consolidated in such a way that parallel work was enabled and lengthy decision-making processes or jurisdictional disputes could be resolved relatively quickly by the responsible authorities. This integration resulted in a significant boost to innovation. A more detailed analysis of decision-making processes, particularly at the political and military levels, also reveals substantial disagreements regarding the assessment of the German threat. Among the most frequently cited disruptions to otherwise efficient cooperation were the influence of Churchill's close friendship with Lindemann and the ongoing conflicts between Tizard and Lindemann over the direction of radar development.

On closer inspection, the strength of the British approach becomes apparent, particularly in the cooperation among all parties involved, which led to solutions achieved through courageous decisions by a few key individuals. In contrast to Germany, the roles and responsibilities of the parties involved were often less rigidly defined. Also noteworthy is the deep trust placed by the government, especially the Air Ministry, and the Royal Air Force in the work of the research group at Bawdsey. High ranking officials from the Air Ministry and Royal Air Force officers went to great lengths to promote radar technology as a key component of an air defence system.

A clear counterexample to Germany's approach, exemplified by figures such as Goering, can be found in Air Marshal Dowding, who worked closely and trustfully with scientists at all levels. Parallel to the further development of radar at Bawdsey, the effectiveness of radar was tested through two major air defence programs with a strong operational focus. These included experiments at Biggin Hill aimed at developing effective interception tactics by the Royal Air Force, and the integration of all available airspace surveillance information at Bentley Priory to coordinate interceptor operations effectively.

Starting in 1938, under the designation Operational Research, staff at Bawdsey conducted statistical analyses of Royal Air Force aircraft test flights, drawing conclusions from radar data about the number and types of aircraft, thus gaining insights for the operational use of radar. Watson Watt later described Operational Research, along with radar and radio direction finding, as one of the three critical steps leading to the Allies' victory.[1604]

Alongside this highly successful collaboration among participants from diverse fields, one must also consider disruptive interruptions in radar development caused by the failed relocations of radar teams, first to Dundee and then to other, often unsuitable sites. These disruptions particularly affected the development of effective airborne radars for night operations. The comprehensive radar system, integrating ground and airborne radar to guide interceptors toward attacking aircraft accurately at night and in poor weather, was developed late and contributed to significant losses early in the Blitzkrieg in 1940, a notable innovation barrier.

Besides radar-based air defence, the Committee also explored other methods of combating attacking bombers. Although radar was seen as a key element of future air defence, practical and quickly implementable solutions were needed given radar's early developmental stage.[1605] Lindemann, as Churchill's scientific adviser, strongly advocated for such alternatives. He considered aerial mines, and parachute bombs the most effective means against enemy aircraft.[1606] The Committee also considered these measures feasible and conducted thorough studies assessing their likelihood of success. Although such methods had been proposed during World War I, the low probability of direct hits rendered them largely impractical. Another idea was acoustic proximity fuses, designed to detonate bombs upon detecting the sound of approaching enemy aircraft.[1607] Lindemann also viewed infrared technology as an effective radar alternative, though he did not reject radar development outright.

It is worth noting that many reports on the Committee's work tended to downplay the importance of other technologies, as Lindemann's push to postpone radar development often took precedence over other items on the agenda of urgent priorities.[1608]

1604 Gießler 1971, p. 89.

1605 Zimmerman 2013, p. 141.

1606 Snow 2013, p. 29; Snow 1961b, p. 33.

1607 Zimmerman 2013, p. 139.

1608 Snow 2013, p. 29; Snow 1961b, p. 33.

10.4.2.2.2 Key Actors and Their Roles

Key actors who contributed to the successful development of both radar technology and the air defence system are selected. With the exception of Churchill, the focus is on individuals directly involved in the technical and operational implementation of the air defence system. Contributions from other politicians and Cabinet members who were influential in their respective areas of responsibility are not considered here. Some senior politicians influenced air defence indirectly; their roles are analysed when relevant. A further distinction is also necessary between support for the technical and operational aspects of air defence and broader political debates, such as the discussion around appeasement.

Many key actors were already experts in their respective fields, held influential positions, or possessed strong networks prior to their involvement in radar and air defence projects. In this analysis, however, the emphasis lies on the roles they actively assumed within the Promoter Model.

This definition applies to all types of promoters and opponents: a Power Promoter must actively deploy authority to support initiatives; an Expert Promoter must actively apply knowledge to advance projects; a Process Promoter must manage and drive processes to enable progress; a Relationship Promoter must deliberately use networks to influence outcomes; and an Opponent is defined by actively resisting or preferring alternative approaches. Focusing on active influence and actions rather than inherent status and skills ensures clarity and avoids misunderstandings.

Each key actor's contribution is analysed in terms of the type of influence they exerted, whether as a Power Promoter, Expert Promoter, Process Promoter, Relationship Promoter, or, in some cases, as an Opponent. The results are then presented in a table, which lists the promoters alphabetically, regardless of their relative importance or influence, in order to ensure clarity and comparability.

Winston Churchill's role in the development of an integrated radar-based air defence system must be analysed within a broader historical and political context. Churchill assumed different roles at different times, acting both as a promoter of technological developments and, occasionally, as an opponent. His position was not one-dimensional but multifaceted and context-dependent.

In the 1930s, prior to 1939, Churchill often acted as a situational opponent showing scepticism or caution regarding early radar and other military innovation projects. He prioritised conventional defence measures or questioned their strategic value, without fundamentally opposing technological development.

After becoming Prime Minister in 1940, Churchill assumed the role of a visionary Power Promoter. He recognised the strategic importance of radar and actively supported and prioritised its development, most notably through his support of the Chain Home radar network, which proved decisive during the Battle of Britain in 1940. At the same time, he acted as a Relationship Promoter, leveraging his close connection with his scientific adviser Frederick Lindemann to critically evaluate and foster technological innovations.

Thus, Churchill's role evolved over time: from a situational Opponent in the 1930s to a flexible Power Promoter and Relationship Promoter from 1940 onward, with occasional phases of opposition depending on resource constraints and strategic priorities.

Edward Appleton, a distinguished British physicist, was noted for his pioneering work on the ionosphere. In 1924, he confirmed the existence of the Kennelly-Heaviside layer, showing that radio waves could be reflected by the upper atmosphere, a discovery fundamental for long-distance radio communication and later radar development. Between 1924 and 1929, he developed methods to measure critical radio frequencies, laying the groundwork for practical radar applications.

From 1936 to 1940, Appleton served on the CSSAD, acting as an Expert Promoter by advising on radar development, and as a Relationship Promoter by facilitating coordination between scientists, military planners, and government officials.

From 1939 to 1949, as Secretary of the Department of Scientific and Industrial Research, he oversaw projects crucial to the war effort, including radar and the Tube Alloys nuclear programme, further strengthening links between scientists and policymakers.

Awarded the Nobel Prize in Physics in 1947, Appleton later returned to academia as Principal of the University of Edinburgh, promoting international scientific collaboration until his death in

1965. His work bridged theoretical research and practical military applications, securing radar as a key component of Britain's defence.

Patrick Blackett and *Archibald Hill* were members of the CSSAD from its foundation in 1934, contributing significantly through their scientific and military expertise.

Blackett, already serving as an Expert Promoter in the CSSAD, contributed significantly to radar and air defence research during its early years, providing scientific guidance and helping to coordinate efforts among scientists and military planners. He later played a central role in developing operational research, particularly for military applications during World War II. As an Expert Promoter, he advanced operational research through his scientific knowledge, and as a Process Promoter and Relationship Promoter, he led the Operational Research Section at the Admiralty, guiding the multidisciplinary team known as Blackett's Circus. Through his leadership, he applied rigorous scientific methods to improve military efficiency and operational effectiveness.

Hill, Nobel Prize winner in Physiology or Medicine in 1922, acted as an Expert Promoter by providing analytical and scientific foundations for operational research through physiology and mathematics. He also served as a Relationship Promoter, using his reputation to secure legitimacy for operational research and facilitate its acceptance in both academic and military circles.

Together, Blackett and Hill bridged the gap between theory and practice, promoting the implementation of research findings into operational decision-making. Their combined roles as Experts and Promoters of processes and relationships were crucial for the successful development and practical application of operational research during the war. Their work set a lasting standard for the integration of scientific analysis into military planning.

Alan Blumlein was one of Britain's most versatile electronics engineers of the twentieth century, with pioneering work in telecommunications, sound recording, television, and radar technology. During World War II, he played a crucial role in advancing radar systems, applying his technical expertise to develop innovative solutions essential for British air defence.

Within the promoter model, Blumlein acted as an Expert Promoter, driving key technological innovations and solving complex engineering challenges. He also served as a Relationship Promoter, collaborating closely with leading research institutions, including the Telecommunications Research Establishment (TRE), and coordinating with other prominent scientists and engineers to integrate diverse technologies into operational systems. His work ensured that radar and related equipment could be effectively deployed for both air and naval operations.

Tragically, Blumlein's career was cut short in 1942 when he died testing the H_2S radar system. Nevertheless, his visionary approach, technical skill, and ability to connect multiple disciplines left a lasting impact on wartime technology and postwar developments in communication and recording systems. His contributions illustrate the vital role of Expert and Relationship Promoters in translating scientific knowledge into practical, operationally effective innovations.

Edward Bowen played a central role in the development, testing, and operational deployment of airborne radar during World War II. He was instrumental in adapting ground-based radar for aircraft, greatly improving the detection and tracking of enemy submarines and aircraft. His work on Airborne Interception (AI) radar was critical for RAF night fighter operations, while his development of Air to Surface Vessel (ASV) radar enhanced anti-submarine warfare.

Within the promoter model, Bowen acted as an Expert Promoter, bringing exceptional knowledge in high-frequency engineering and radar electronics. He understood both the physics and the operational demands of airborne systems and contributed decisively to the miniaturisation and stabilisation of radar equipment. As a Process Promoter, he translated laboratory innovations into operational procedures, organising testing programmes and coordinating scientists, engineers, and service units to convert prototypes into effective combat systems.

Bowen also served as a Relationship Promoter, bridging scientists, engineers, pilots, and military planners to integrate radar into active service. This was particularly evident during the Tizard Mission to the United States in 1940, where he shared technical knowledge and fostered Anglo-American cooperation. While he never held formal office, Bowen exercised some influence akin to a Power Promoter through his scientific authority and collaboration with decision-makers such as Watson-Watt and Tizard.

His technical brilliance, organisational skill, and ability to mediate between diverse groups made him pivotal in transforming radar from an experimental concept into an indispensable tool for Allied air and naval operations.

Joan Curran played an important role in British electronic warfare during World War II. From 1941 onwards at the TRE, she developed and refined the radar countermeasure Window, also known as chaff, standardising its design and operational deployment to conceal Allied bomber formations during major air raids. Her meticulous work ensured that the technique could be applied reliably on a large scale, significantly reducing aircraft losses.

Within the promoter model, Curran acted as an Expert Promoter by applying her deep knowledge of physics to innovate and optimise radar countermeasures. She also functioned as a Process Promoter, systematising deployment procedures and integrating Window into coordinated operational plans. In addition, she was a Relationship Promoter, collaborating closely with engineers, military planners, and operational personnel to ensure its successful application in the field. Her expertise and leadership were also evident in her contributions to the development of proximity fuses and her later work with the British delegation to the Manhattan Project, where she applied her skills to the electromagnetic separation of uranium isotopes.

Curran's ability to translate scientific knowledge into practical and strategic outcomes exemplifies the multifaceted impact of her work and highlights her role in advancing both technology and the position of women in science.

Hugh Dowding, promoted to Air Marshal in 1933, immediately used his senior position to support emerging scientific work on air defence. By backing radar research at an early stage and securing institutional recognition, he already acted as a Power Promoter, ensuring that the scientists involved received resources and official attention.

His role expanded significantly in July 1936, when he became the first Air Officer Commanding-in-Chief of the newly established RAF Fighter Command. In this position, Dowding was able to combine his influence with operational responsibility. From 1937, he focused on his operational responsibilities as Process Promoter and Relationship Promoter. As a Process Promoter, he oversaw the practical introduction of radar into Britain's defence system, translating laboratory progress into operational capability. This role primarily involved strategic oversight and coordination of radar integration across the RAF, rather than direct operational process management. At the same time, he acted as a Relationship Promoter, building trust and cooperation between the RAF, government, and scientific community.

Central to his achievement was the Dowding System, an integrated network of radar detection, ground control, and rapid communication that enabled fighters to be directed with unprecedented efficiency. Despite resistance from conservative voices, he persistently advocated the centrality of radar and air defence. His strategic vision and steadfast support of science provided the foundation for Britain's success in the Battle of Britain, and his influence continued to shape the RAF's defensive thinking long after his departure.

Robert Hanbury Brown advanced the development and operational deployment of radar technology in Britain during World War II. From 1936, when he joined Robert Watson-Watt's team at the Radio Research Station in Slough and later at Bawdsey Manor, Hanbury Brown acted as an Expert Promoter, applying his knowledge of electronics and radar systems to practical development and testing. He contributed to the installation of the first Chain Home coastal defence radar systems and early airborne radar prototypes, directly influencing Britain's air defence capabilities.

At the same time, he served as a Relationship Promoter, fostering collaboration between civilian scientists, engineers, and military personnel. This was particularly evident in 1940 during the establishment of fighter interception units at Northolt and Tangmere, where he coordinated radar teams and operational pilots. Later, his work with Edward Bowen on airborne radar systems and with J.W.S. Pringle's group on the Rebecca-Eureka transponder system further integrated scientific innovation with operational needs.

Although his flying career ended in 1941 due to an accident, Hanbury Brown continued to promote radar development internationally, including work in the USA from 1942 onwards. His combined contributions as Expert and Relationship Promoter transformed radar from an experimental concept into a fully operational system, significantly strengthening Britain's wartime defence.

Maurice Hankey's role in British strategic and military decision-making was often less recognised. Serving as Secretary of the Committee of Imperial Defence from 1912 onwards, he advised the Prime Minister and Cabinet for over two decades, influencing strategic decisions during World War I and the interwar period. His exceptional organisational skills and deep understanding of governmental and military structures made him a key figure in ensuring coherent policy development.

Within the promoter model, Hankey was primarily a Relationship Promoter, using his ability to manage communication between ministers, military leaders, and government officials, facilitating collaboration at the highest levels. He also acted as a Process Promoter, organising and structuring the work of the committee to ensure complex strategic issues were addressed efficiently and decisions coordinated across departments. His careful management of information and meetings enabled the smooth integration of multiple perspectives into strategic planning.

Although his influence effectively ended with his retirement from the C.I.D. in 1938, Hankey's earlier work laid essential groundwork for the coordination and integration of military and governmental efforts, contributing to the successful development and implementation of British defence policy during World War II. He is often remembered as a key figure working quietly behind the scenes, whose meticulous approach ensured that high-level policies were effectively translated into action.

Reginald Jones was a British physicist and a key figure in scientific intelligence during World War II. After studying at Wadham College, Oxford, he joined the Royal Aircraft Establishment in 1936, focusing on air defence technologies. From September 1939, Jones served as Assistant Director of Intelligence Science at the Air Ministry, the first post dedicated to evaluating enemy scientific and technical developments. He analysed German radar, radio navigation, and rocket technologies, developing effective countermeasures such as the disruption of the Knickebein navigation system and advocating the use of Window to deceive enemy radar. He also assessed critical intelligence such as the Oslo Report, recognising its strategic significance for Britain's defensive planning.

Within the promoter model, Jones acted as an Expert Promoter, applying his scientific knowledge to practical intelligence problems and devising innovative technical solutions. He was also a Process Promoter, coordinating the collection, analysis, and application of intelligence to ensure that countermeasures were deployed efficiently and effectively. Additionally, he functioned as a Relationship Promoter, fostering collaboration between scientists, military units, and intelligence services to align expertise with operational needs. Through this combination of technical skill, organisational coordination, and networking, Jones ensured that technological threats were understood and countered, playing a crucial role in Britain's air defence during World War II.

Frederick Lindemann's role in the development of radar technology and the air defence system was highly multifaceted, combining scientific expertise with strategic influence. As Churchill's close scientific adviser, he acted as a central Promoter of radar technology form 1938 to 1945, combining Expert Promoter from 1938 to 1945 and Relationship Promoter from 1938 to 1945 roles, while also advocating alternative approaches such as aerial mines, infrared devices, and acoustic fuzes, and occasionally displaying situational opposition.

Prior to 1938, Lindemann was already a leading physicist with significant influence in scientific circles, but his direct involvement in radar development was limited. From 1938 onwards, he increasingly assumed the role of Expert Promoter, applying his knowledge to assess and guide technological projects, particularly radar, and advising Churchill on their strategic potential. Simultaneously, he acted as a Relationship Promoter, leveraging his personal rapport with Churchill to advance research priorities and facilitate communication between scientists, military planners, and government officials.

While Lindemann occasionally preferred alternative approaches or questioned the feasibility of certain projects, he did not fundamentally oppose radar development. His influence was decisive in shaping the government's research agenda and ensuring that innovative projects received political support.

Thus, Lindemann's role evolved over time from a respected scientific expert to a central Promoter of radar technology, combining technical expertise with strategic guidance and personal influence to support Britain's air defence efforts. His guidance and advocacy were instrumental in ensuring that radar and related innovations received the attention and resources necessary to become decisive elements of Britain's defensive strategy.

Bernard Lovell played an important role in the development and operational implementation of radar technology during World War II. Initially recruited by Patrick Blackett in 1939, he shifted from cosmic ray research to support military innovation, training operators on the Chain Home radar system and quickly gaining practical expertise. His early work laid the foundation for the effective use of radar in detecting and tracking enemy aircraft, contributing directly to Britain's air defence.

Within the promoter model, Lovell acted as an Expert Promoter, applying his technical knowledge to the H_2S airborne radar system. He addressed critical engineering challenges, including miniaturising equipment and improving antennas to enhance radar performance. He also served as a Process Promoter, coordinating development and testing, training operators, and ensuring that innovations were translated into operational procedures. As a Relationship Promoter, he worked closely with scientists, engineers, and military personnel to guarantee the successful deployment of radar systems.

After the war, Lovell continued to promote scientific and technological advancement as the founding director of the Jodrell Bank Observatory, overseeing the construction of the large radio telescope and fostering collaboration between academic researchers, engineers, and international partners. His career combined technical skill, organisational capability, and collaborative leadership, making him a key figure in both the wartime and postwar development of radar and radio astronomy.

John Randall and *Henry Boot* developed the cavity magnetron in 1940 at the University of Birmingham, representing a major breakthrough in radar technology. The magnetron enabled unprecedented power at centimetre wavelengths, forming the basis for high-resolution airborne radar systems. These systems allowed precise targeting of enemy positions and more effective operations against German submarines, significantly influencing the course of World War II. The success of the magnetron also accelerated the development of centimetre-wave radar in both Britain and the United States, highlighting its strategic importance.

Within the promoter model, Randall and Boot were primarely Expert Promoters, using their technical knowledge and problem-solving skills to invent, refine, and stabilise the magnetron for practical use in aircraft radar systems. Their achievements, supported by Professor Mark Oliphant at the University of Birmingham, were central to the Allied success in radar technology and established Randall and Boot as leading figures in wartime electronics research.

Albert Rowe's role in the development of radar technology and the air defence system was multifaceted, combining scientific expertise with operational implementation. From June 1934, when he began investigating existing detection systems for air defence, Rowe played a pioneering role in shaping Britain's radar capabilities. At the inaugural meeting of the CSSAD in January 1935, he was appointed secretary, marking the first significant step towards the systematic development of an air defence system in which radar would play a key role.

Rowe can be seen as an Expert Promoter, applying his knowledge to develop and refine detection systems, as well as a Process Promoter, managing and coordinating early research efforts. He also acted as a Relationship Promoter, facilitating collaboration between scientists, engineers, and government officials to ensure that innovative ideas were effectively communicated and implemented.

Throughout the war, Rowe continued his critical work, later contributing as a leading scientist at the TRE, where he oversaw further development and practical implementation of radar and communication systems. His sustained influence across expertise, process management, and network facilitation made him a central figure in the operationalisation of Britain's radar-based air defence.

His expertise and leadership were instrumental in ensuring that Britain's radar systems became fully operational and effective, laying the groundwork for the country's successful air defence during World War II.

Henry Tizard was a central figure in British air defence and science policy before and during World War II. In 1934, he was appointed chairman of the newly established CSSAD. As a scientist, government official, and World War I veteran, he possessed broad expertise and valuable contacts, making him ideally suited to lead the committee. In this role, Tizard primarily acted as a coordinator, facilitating collaboration between scientists, the military, and policymakers. Within the promoter model, Tizard served as a Process Promoter, managing and coordinating CSSAD activities to ensure that research, development, and operational needs were effectively integrated. He was also a Relationship Promoter, leveraging his network to mobilise resources, foster cooperation among scientific, military, and governmental stakeholders, and advance key technological innovations such as radar.

During his tenure as chairman from 1934 until 1940, Tizard occasionally disagreed with Frederick Lindemann, Churchill's scientific adviser. While Tizard strongly advocated radar development and the integration of scientific expertise into defence planning, Lindemann sometimes preferred alternative approaches. Despite these challenges, Tizard's steady leadership maintained progress in radar research and strengthened the committee's effectiveness.

Tizard was also instrumental in organising the Tizard Mission in 1940, which shared British innovations including radar, jet engines, and atomic research with the United States, accelerating the integration of new technologies into Britain's air defence and demonstrating the strategic value of scientific collaboration.

Alan Turing's contribution to Britain's war effort and the development of cryptographic technology was pivotal during World War II. Beginning on 4 September 1939, he worked full-time at Bletchley Park for the Government Code and Cypher School, joining Hut 8, responsible for deciphering German Naval ENIGMA messages. His mathematical expertise and innovative thinking enabled him to design the British Bombe, an electromechanical device that significantly accelerated ENIGMA decryption, providing the Allies with vital intelligence.

Within the promoter model, Turing was primarily an Expert Promoter from 1939 to 1945, applying his technical knowledge to solve complex cryptographic problems. He also acted as a Process Promoter, designing and coordinating the workflow of decryption activities, optimising interactions between cryptanalysts and machines, and ensuring the efficient operation of Hut 8. Additionally, he served as a Relationship Promoter, fostering collaboration among mathematicians, linguists, engineers, and military personnel to align diverse teams toward timely intelligence production.

Turing's work extended to breaking the Lorenz cipher and improving other cryptographic systems, demonstrating his versatility and operational impact. By the end of World War II, his innovative designs, leadership in process optimisation, and coordination of interdisciplinary teams had firmly established him as a central figure in wartime technological innovation. His efforts not only ensured crucial intelligence for the Allies but also laid the conceptual groundwork for modern computing, artificial intelligence, and automated problem-solving.

Robert Watson-Watt, a central figure in British air defence and science policy before and during World War II, presented proposals for an effective air defence system in February 1935 in his memorandum "Detection and Location of Aircraft by Radio Methods". This laid the foundation for radar development in Britain as aerial threats in Europe increased.

Watson-Watt had already established himself in the 1920s and early 1930s as a pioneer in radio wave propagation and lightning detection. His expertise and innovative thinking enabled him to recognise radar as a practical solution for aircraft detection. In 1935, he led the team that successfully demonstrated aircraft detection using radio waves in the Daventry Experiment, confirming radar's feasibility for early warning and surveillance.

In the late 1930s and during World War II, Watson-Watt coordinated the rapid expansion of Britain's radar network, known as Chain Home. Within the promoter model, he acted as an Expert Promoter, providing technical guidance and innovative solutions; a Process Promoter, overseeing the development and implementation of radar systems; and a Relationship Promoter, fostering collaboration between scientific institutions, the Air Ministry, and military units.

Despite challenges such as extending detection range, improving signal processing, and integrating radar data into operational command, Watson-Watt and his teams continuously refined the technology. The Chain Home system became a cornerstone of British air defence and was instrumental in the victory during the Battle of Britain.

Arnold Wilkins played a key role as a radar pioneer alongside Watson-Watt from the very beginning. In February 1935, they conducted the first field experiments on the reflection of electromagnetic waves from aircraft in Daventry. Wilkins was crucial not only in practical testing but also in establishing the theoretical basis of radar, demonstrating that radio waves could be reflected by aircraft and used for detection. By presenting detailed reports and convincing government and military officials of radar's strategic importance, he acted as an Expert Promoter, helping secure funding and institutional support.

He also contributed to technical innovations, improving signal processing and detection range, and worked closely with the Royal Air Force to ensure radar met operational needs. In this capacity, he functioned as a Relationship Promoter, fostering collaboration between engineers, scientists, and military personnel. Furthermore, Wilkins played a central role in developing the Identification Friend or Foe system in 1938, reinforcing his position as an Expert Promoter.

Wilkins was instrumental in translating radar research into practical military applications by developing operational procedures and training programmes for radar operators. His work bridged the gap between laboratory experiments and real-world use, making radar a reliable component of Britain's air defence. His analytical skills and ability to communicate complex concepts helped align stakeholders around the common goal of effective air defence.

Harry Wimperis was instrumental in the development of Britain's air defence system, combining strategic leadership with organisational and scientific expertise. As Director of Scientific Research at the Air Ministry from 1932 to 1937, he laid the institutional and scientific groundwork for air defence innovation, recognising the need for new technological approaches to aircraft detection and reinforcing research capacity.

In November 1934, Wimperis proposed to the Minister of Aviation and senior officials the establishment of a defence research committee within the Air Ministry, leading to the creation of the CSSAD. Through this initiative, he consolidated scientific efforts and created a framework that facilitated interdisciplinary collaboration.

Within the promoter model, Wimperis can be seen as a Power Promoter, mobilising institutional support and influencing research agendas; as a Relationship Promoter, leveraging networks across government, military, and science to secure resources; and as a Process Promoter, establishing structures that enabled effective collaboration between stakeholders.

From the outset, Wimperis engaged with Robert Watson-Watt and other researchers to advance studies on electromagnetic detection of aircraft. His early involvement and strategic vision ensured that innovative technologies, including radar, were integrated effectively into Britain's air defence. Through his leadership, Wimperis bridged the gap between scientific research and operational application, creating the foundation for rapid development and deployment of radar technology.

10.4.2.2.2.1 Summary of Key Roles

In Germany, the roles of actors can be relatively well divided into Political and Military Leadership and Ministries on the one hand, and Roles of Actors in Research, Industry and Institutions on the other. In Britain, however, such a distinction is difficult. Although the roles of individual promoters can be considered separately from their areas of responsibility, unambiguous assignment to politics, government, the military, or research and industry is also difficult, as their activities were closely intertwined.

Power Promoter								Churchill	Churchill	Churchill	Churchill	Churchill	Churchill
	Dowding	Dowding	Dowding	Dowding									
		Wimperis	Wimperis	Wimperis	Wimperis								
Expert Promoter				Appleton	Appleton	Appleton	Appleton	Appleton					
		Blackett	Blackett	Blackett	Blackett	Blackett	Blackett	Blackett	Blackett	Blackett	Blackett	Blackett	Blackett
							Blumlein	Blumlein	Blumlein	Blumlein			
			Bowen	Bowen	Bowen	Bowen	Bowen	Bowen	Bowen	Bowen	Bowen		
									Curran	Curran	Curran	Curran	Curran
				Hanbury B.	Hanbury B.	Hanbury B.	Hanbury B.	Hanbury B.	Hanbury B.	Hanbury B.			
		Hill	Hill	Hill	Hill	Hill	Hill	Hill	Hill	Hill	Hill	Hill	Hill
				Jones	Jones	Jones	Jones	Jones	Jones	Jones	Jones	Jones	Jones
						Lindemann	Lindemann	Lindemann	Lindemann	Lindemann	Lindemann	Lindemann	Lindemann
							Lovell	Lovell	Lovell	Lovell	Lovell	Lovell	Lovell
								Randall/ Boot	Randall/ Boot	Randall/ Boot	Randall/ Boot	Randall/ Boot	Randall/ Boot
		Rowe	Rowe	Rowe	Rowe	Rowe	Rowe	Rowe	Rowe	Rowe	Rowe	Rowe	Rowe
							Turing	Turing	Turing	Turing	Turing	Turing	Turing
			Watson-Watt	Watson-Watt	Watson-Watt	Watson-Watt	Watson-Watt	Watson-Watt	Watson-Watt	Watson-Watt	Watson-Watt	Watson-Watt	Watson-Watt
			Wilkins	Wilkins	Wilkins	Wilkins	Wilkins	Wilkins	Wilkins	Wilkins	Wilkins	Wilkins	Wilkins
Process Promoter		Blackett	Blackett	Blackett	Blackett	Blackett	Blackett	Blackett	Blackett	Blackett	Blackett	Blackett	Blackett
			Bowen	Bowen	Bowen	Bowen	Bowen	Bowen	Bowen	Bowen	Bowen		
									Curran	Curran	Curran	Curran	Curran
Year	**1933**	**1934**	**1935**	**1936**	**1937**	**1938**	**1939**	**1940**	**1941**	**1942**	**1943**	**1944**	**1945**

					Dowding	Dowding	Dowding	Dowding					
	Hankey	Hankey	Hankey	Hankey	Hankey	Hankey							
							Jones	Jones	Jones	Jones	Jones	Jones	Jones
							Lovell	Lovell	Lovell	Lovell	Lovell	Lovell	Lovell
			Rowe	Rowe	Rowe	Rowe	Rowe	Rowe	Rowe	Rowe	Rowe	Rowe	Rowe
		Tizard	Tizard	Tizard	Tizard	Tizard	Tizard	Tizard					
							Turing	Turing	Turing	Turing	Turing	Turing	Turing
			Watson-Watt	Watson-Watt	Watson-Watt	Watson-Watt	Watson-Watt	Watson-Watt	Watson-Watt	Watson-Watt	Watson-Watt	Watson-Watt	Watson-Watt
		Wimperis	Wimperis	Wimperis	Wimperis								
Relationship Promoter				Appleton	Appleton	Appleton	Appleton	Appleton					
		Blackett	Blackett	Blackett	Blackett	Blackett	Blackett	Blackett	Blackett	Blackett	Blackett	Blackett	Blackett
							Blumlein	Blumlein	Blumlein	Blumlein			
			Bowen	Bowen	Bowen	Bowen	Bowen	Bowen	Bowen	Bowen	Bowen		
								Churchill	Churchill	Churchill	Churchill	Churchill	Churchill
									Curran	Curran	Curran	Curran	Curran
					Dowding	Dowding	Dowding	Dowding					
				Hanbury B.	Hanbury B.	Hanbury B.	Hanbury B.	Hanbury B.	Hanbury B.	Hanbury B.			
	Hankey	Hankey	Hankey	Hankey	Hankey	Hankey							
		Hill	Hill	Hill	Hill	Hill	Hill	Hill	Hill	Hill	Hill	Hill	Hill
							Jones	Jones	Jones	Jones	Jones	Jones	Jones
						Lindemann	Lindemann	Lindemann	Lindemann	Lindemann	Lindemann	Lindemann	Lindemann
							Lovell	Lovell	Lovell	Lovell	Lovell	Lovell	Lovell
			Rowe	Rowe	Rowe	Rowe	Rowe	Rowe	Rowe	Rowe	Rowe	Rowe	Rowe
		Tizard	Tizard	Tizard	Tizard	Tizard	Tizard	Tizard					
Year	**1933**	**1934**	**1935**	**1936**	**1937**	**1938**	**1939**	**1940**	**1941**	**1942**	**1943**	**1944**	**1945**

							Turing	Turing	Turing	Turing	Turing	Turing	Turing
			Watson-Watt	Watson-Watt	Watson-Watt	Watson-Watt	Watson-Watt	Watson-Watt	Watson-Watt	Watson-Watt	Watson-Watt	Watson-Watt	Watson-Watt
			Wilkins	Wilkins	Wilkins	Wilkins	Wilkins	Wilkins	Wilkins	Wilkins	Wilkins	Wilkins	Wilkins
					Wimperis	Wimperis	Wimperis	Wimperis					
Opponent*													
Year	**1933**	**1934**	**1935**	**1936**	**1937**	**1938**	**1939**	**1940**	**1941**	**1942**	**1943**	**1944**	**1945**

*Although Churchill and Lindemann occasionally acted as opponents, particularly on strategic decisions, this ultimately had no significant adverse effect on the overall outcome.

Table 11 GB: Roles of actors and their interconnected responsibilities

10.4.2.2.3 Innovation Steps and Barriers

Table 12 presents the key innovation steps and barriers. In contrast to Germany, in Great Britain all relevant actors are closely interlinked, meaning that there can be no sharp demarcation between politics and the other actors involved.

Year	Innovation Steps	Barriers to Innovation
1933–1945		
	Strong sensitisation of the political and military leadership regarding the threat to the country from German air attacks.	
	Clear vision by the political leadership; forward-looking, comprehensive co-operation across ministries, the military, science and industry, with clear targets.	
	Bundling of all necessary work to develop an effective air defence system from the outset.	
		Through their friendship, Lindemann influenced Churchill with his ideas on air defence; in return, Churchill stayed up to date on technical developments for his committee work.
1934		
June	Unofficial analysis of detection systems for air defence by Albert Rowe, found conventional methods like acoustic devices insufficient.	

November	Official formation of the Committee for the Scientific Survey of Air Defence (CSSAD) in November 1934, commonly known as the Tizard Committee, to investigate how scientific and technological advances could strengthen existing air defence methods.	
1935		
Beginning	Watson-Watt shared Kühnhold's 1933 view that conventional acoustic detection methods were insufficient for effective early warning of air attacks; Arnold Wilkins mathematically disproved Tesla's Death Rays theory.	
28 January	First meeting of the CSSAD to examine suitable defence methods against air attacks.	
26 February	The Daventry Experiment on 26 February 1935 was the first successful demonstration of aircraft detection by radio waves and showed for the first time the feasibility of radar for air defence. These first elementary results were sufficient to secure immediate funding for further investigations.	

27 February	Second Watson-Watt memorandum on air defence using radiolocation, which outlined in twenty points how effective air defence could be achieved through radio detection and became a key foundation for Britain's radar system.	
13 May	Move to Orfordness and start of development of new tracking methods for air defence under the leadership of Watson-Watt, developments that later led to the Chain Home system.	
11 June		Lindemann becomes a member of the CSSAD at Churchill's instigation. In the period that followed, there were fierce controversies between Lindemann and Tizard and the other members of the CSSAD.
24 July	First reception of echo signals from a formation of aircraft, with determination of the number of aircraft.	
September	Further successful measurements with increasing range at different wavelengths.	

16 September	At the ADRC meeting, progress on the coastal radar system was reported to far exceed expectations, achieving in five months what was originally estimated to take five years. The committee approved construction of a chain of radar stations and recommended additional ones to ensure reliable detection, despite no fully operational prototype yet existing.	
24 October	First official conference under Air Marshal Hugh Dowding addressing the challenges of military operations involving radar and discussing the integration of technical and operational requirements.	
1936		
May	Completion of move to Bawdsey Manor (Bawdsey Research Station).	
September	Bowen and a small team began investigating the feasibility of installing radar equipment in aircraft for air combat.	
15 July		Fierce controversy between Lindemann and members of the CSSAD over priorities in air defence technologies. Lindemann's attacks on Tizard were so severe that members Blackett and Hill resigned.

	CSSAD is quickly reorganised. Tizard remains chairman, Hill and Blackett rejoin as members, and Appleton becomes a new member. Lindemann is no longer a member.	
4 August	Start of experiments at Biggin Hill to test radar detection of aircraft under operational conditions, providing crucial data for air defence development.	
1938		
September	Five Chain Home stations start 24-hour operation, marking the beginning of continuous early warning coverage along the British coastline and significantly enhancing air defence readiness.	
1939		
August	From August 1939, development of the first radar for installation in aircraft; first fitted ASV Mk. I in Lockheed Hudson and Short Sunderland in January 1940 for maritime surveillance. Improved successor system, the ASV Mk. II, entered service in late 1940.	
End of August	Rowe's group at Bawdsey consisted of about 250 personnel, including roughly 100 scientists.	

2 September	Relocation from Bawdsey Manor to Dundee.	Radar development delayed due to poor laboratory facilities and inadequate infrastructure.
3 September	18 Chain Home radar stations were operational and connected to a central evaluation unit known as the Filter Room.	
September	Due to lack of space at the University of Dundee, Bowen's AI and ASV radar teams were relocated to RAF Scone near Perth.	Radar development delayed due to poor laboratory facilities, inadequate infrastructure, and suboptimal working conditions.
September	Bletchley Park's concentrated decoding of German radio traffic, combined with radio direction finding, complemented early aircraft radar, providing crucial data for maritime surveillance and air warning.	
Winter 1939/40		Due to unsuitable conditions at Scone, Bowen and his team were relocated to St Athan, where work on airborne radar was carried out in winter 1939/40 under difficult conditions. Maintaining contact with headquarters in Dundee proved challenging due to long communication and postal routes.

1940		
21 February	Randall and Boot finalised the basic design of the cavity magnetron, producing 400 watts at 9.8 cm wavelength, a breakthrough for centimetric radar.	
Spring	Relocation of Bowen's radar team from St Athan to Christchurch near Worth Matravers; team tasked with installing AI radars on aircraft, training personnel, and exploring new applications; Radar Test and Research Station of the TRE.	Confusion between Coastal Command and the Navy over prioritising AI and ASV radar installations increased the team's workload, causing delays in fighter radar fitting, setbacks in ground control procedures, and limited progress on tactical AI radar applications.
May	Relocation of remaining radar group from Dundee to Worth Matravers, Dorset; around 200 personnel moved to a site with good vantage over the English Channel; Radar Test and Research Station of the TRE.	Isolated location with limited infrastructure and occasional adverse weather conditions.
May	Bowen's team from Christchurch merged with the remaining radar group, which was based at Worth Matravers, at Malvern, consolidating radar development and forming a central hub for advanced radar research and testing.	Resistance from the local residents; difficulties with accommodation and catering.

May – June	Air Defence System, later Dowding System, fully operational; integrated Chain Home radar, radio, Filter and Operations Rooms, and RAF air defence control, forming the world's first radar-based air interception network.	
August / September	The Tizard Mission was a British delegation led by Henry Tizard that travelled to the United States in August 1940 to share critical technologies, including the cavity magnetron. Bowen played a key role in facilitating radar collaboration with MIT, laying the foundation for joint US and UK radar development and accelerating the deployment of centimetric radar systems.	
September	Magnetrons manufactured at GEC with a peak output of 100 kW, operating at a wavelength of approximately 10 cm.	
1942		
Early	Window was developed at the TRE as a radar countermeasure using strips of aluminium foil.	
March	AI Mk. VII, was the first centimetric radar designed for low-flying aircraft interception.	

December	AI Mk. VIII consolidated the advances in centimetric radar, providing RAF night fighters with unprecedented precision.	
1943		
March	ASV Mk. III Centimetric radar with extended detection range for maritime patrols.	

Table 12 GB: Steps and barriers across sectors

10.5 Innovation Cooperations

In Germany, the polycratic structure of the Nazi regime often led to uncoordinated cooperation between research and development, especially in radar technology. In contrast, Great Britain adopted a more structured approach from the start, both in strategic objectives and in research, development, and resource planning.

Moreover, Germany faced major obstacles such as the Entwicklungsstop-Befehl of 3 February 1940 and the Schlüsselkraft-Befehl of 19 February 1942. Subsequent recall campaigns failed to maintain Germany's lead over the UK due to delays in returning specialists from the front.

The efforts of Hans Plendl as BHF to coordinate research in this area and the measures taken by Werner Osenberg to provide qualified personnel ultimately did not enable Germany to catch up with Great Britain's technological lead in the field of radar technology. Added to this were the wrong decisions made when research on centimetre waves was halted in Germany. After parts of a centimetre-wave radar were found in the wreck of a British bomber shot down near Rotterdam on 3 February 1943, Germany realised the backwardness of its own research and development in this area. As a consequence, the Arbeitsgemeinschaft Rotterdam (AGR) was hastily founded on 22 February 1943 at the suggestion of General Martini.[1609]

With the founding of the AGR, Germany achieved significant progress in research and development under immense time pressure. While collaboration in basic and applied research had existed despite disputes between competing interest groups, the AGR was the first institution to adopt a holistic approach to a common development goal.

The type and form of cooperation seen in the AGR had already existed in Great Britain. The next major step in gaining an advantage over Germany was the so-called Tizard Mission to the United States and Canada in September 1940, which, against the backdrop of the Battle of Great Britain, was primarily aimed at intensifying and accelerating cooperation between the countries in the field of war-related research. A key factor in the success of the mission was the full disclosure of war-related developments by Great Britain, which led to close and trusting cooperation between the countries, despite occasional tensions between the parties involved.

10.5.1 Germany

10.5.1.1 Arbeitsgemeinschaft Rotterdam

Just two weeks after General Martini's request to resume work in the centimetre wave range was rejected again by the Technisches Amt of the RLM on 15 January 1943 due to low chances of success[1610], the British used the H_2S search and navigation radar with a wavelength of about 9 centimetres and a pulse power of 10 kW for the first time on board an aircraft during an attack on Hamburg in the night from 30 to 31 January 1943.[1611] After parts of the H_2S device were found in the wreckage of a British Stirling bomber shot down on 3 February, the German side was initially unclear

1609 Handel 2002, pp. 257–261; Handel 1999, pp. 59–60; Brandt 1952, Foreword.

1610 Since 8 August 1942, General Martini had submitted a request to resume work on centimetre waves.

1611 Reuter 1971, pp. 113–114; Lovell 1991, p. xvii.

about its purpose.[1612] It was named the Rotterdam device after the place where it was downed.[1613] This device was examined at Telefunken in Berlin-Zehlendorf, and it was quickly discovered that it operated under the designation Experimental 6, with an acceptance stamp dated 24 December 1942[1614], at a wavelength of 9.15 centimetres.[1615] Afterwards, the German side quickly recognised the connection with the increasing British successes in fighting German U-boats.[1616]

The aim of the Arbeitsgemeinschaft Rotterdam (AGR) was to consolidate all the experience gained in industry and research institutions in the field of centimetre waves as a countermeasure to the H_2S device. Leo Brandt, a senior engineer at Telefunken, took over the chair. Already in mid-1942, Fellgiebel assigned Brandt responsibility for the Entwicklungsgruppe Funkmeß, in addition to his work at Telefunken.[1617] Through his involvement in these various tasks, Brandt gained valuable experience in working with government institutions, the military, industry, and research organisations.[1618]

1612 The H_2S device was originally designed to search for submarines, until it was realised that with a surprisingly long range of around 70 km, the rotating antenna could be used to develop a panoramic sighting device that could be used over land and depict the surfaces of coastlines, lakes and rivers as well as the outlines of cities similar to a map. The British quickly recognised the enormous importance of this technology for the bombing war against Germany (Rowe 2015, pp. 116–120; Brandt 1962c, p. 70; Brandt 1962a, pp. 85–86; Reuter 1971, pp. 113–114).

1613 Reuter 1971, p. 114.

1614 After the great importance of this radar device for bombing raids was recognised in Britain, however, there were voices that warned against using it too early for reasons of secrecy over land, but these were ignored (Brandt 1962c, p. 70).

1615 Brandt 1962a, p. 86; AGR Protokoll, 17 March 1943, p. 2.

1616 Brandt 1962a p. 88.

1617 The Entwicklungsgruppe Funkmeß under the leadership of Leo Brandt was dissolved in May 1944. Instead, in the Reichsministerium für Rüstung und Kriegsproduktion (RMfRuK) the Sonderkommission für Funkmeßtechnik (SKFM), headed by Director Karl Rottgardt of Telefunken, was established under the Hauptkommission für Elektrotechnik. The AGR was assigned to it, and Leo Brandt remained its head (Reuter 1971, p. 195).

1618 In 1952 and 1953, a summary of the minutes of the AGR meetings was published with Leo Brandt as editor (Brandt 1952; Brandt 1953a).

The first meeting of the AGR took place at Telefunken on 23 February 1943.[1619] Representatives from the military, industry, and research institutions, such as the Physikalisch-Technische Reichsanstalt with Esau and Plendl in his capacity as BHF, were present.[1620]

The following resolutions, among others, were passed and set out in the minutes of the first meeting of the AGR on 23 February 1943:

Item 1: Goals:

„Der Arbeitsgemeinschaft wurde die Aufgabe gestellt, durch Zusammenfassung aller Erfahrungen von Forschung und Industrie auf dem Zentimeterwellengebiet schnellstens die notwendigen Gegenmassnahmen gegen das Verfahren „Rotterdam“ zu schaffen.“[1621]

Item 2: „Rotterdam“ device:

„Es wurde beschlossen in achttägigen Zusammenkünften jeweils einen Bericht über den Stand der Ermittlungen bzgl. Technik und Verfahren der Anlage „Rotterdam“ entgegenzunehmen. Der Nachbau von 6 Anlagen wird einstimmig, als schnellstmögliche Massnahme zur Erarbeitung der Verfahren und ihrer Anwendungsmöglichkeiten angesehen. Telefunken hat die Erstellung übernommen.“[1622]

Item 3: “Naxos I” device:

„Nach Beratung der Möglichkeit zur schnellsten Erstellung von Warnempfängern gegen Anlage „Rotterdam“ wurde der Bau von „Naxos I“-Empfängern als einzige Lösung ermittelt. Um die besten Empfangs-

1619 All minutes of the AGR from 1943 to 1944 are available for free download on the website CDVandT.org (CDV&T) operated by Arthur Bauer (URL: https://www.cdvandt.org/agr_protocols.htm).

1620 AGR-Protokoll, 23 February 1943, p. 1; Reuter 1971, p. 114.

1621 AGR-Protokoll, 23 February 1943, p. 1. “The working group was given the task of creating the necessary countermeasures against the „Rotterdam“ method as quickly as possible by combining all the experience of research and industry in the centimetre wave field.” (translated by the author).

1622 AGR-Protokoll, 23 February 1943, p. 1; Reuter 1971, p. 114. “It was decided to hold eight-day meetings to receive a report on the status of investigations into the technology and processes of the „Rotterdam“ plant. The reproduction of 6 systems is unanimously considered to be the fastest possible measure to develop the processes and their possible applications. Telefunken has taken over the construction.’ (translated by the author).

> Detektoren vorzusehen wurde beschlossen, dass die PTR einige Detektoren von Lorenz, Prof. Dieckmann und eigener Herstellung, vergleicht. Der für diesen Zweck optimale Detektor wird von der Fa. Lorenz in den benötigten Stückzahlen (einige 100 Stück) schnellstens hergestellt.“[1623]

In addition to the repair of the captured device, the reproduction of six units of the Rotterdam device[1624] and the rapid development of warning receivers were the main priorities. The warning receiver for the Rotterdam device, Naxos, was primarily intended to protect submarines from fighter aircraft, and later also to be used aboard night fighters for protection against Allied bombers. These projects were given high priority.[1625] Additionally, there were topics such as overlay receivers and HF cables.[1626] The chairmanship of the AGR by Leo Brandt was appreciated by many engineers working in industry, as concrete research and development results had to be achieved here, along with their implementation into production. In contrast, BHF Plendl was seen more as a representative of basic research.[1627] Ultimately, it was decided that representatives from Siemens & Halske and GEMA should also attend the second AGR meeting.[1628]

In the second meeting on 17 March 1943, the wavelength of the Rotterdam device was determined to be approximately 9.15 centimetres, with the magnetron power ranging from 20 to 30 kW. Furthermore, it was concluded that the radar was a navigation device, and that so-called Pathfinder aircraft were equipped with it

1623 AGR-Protokoll, 23 February 1943, p. 2. "After discussing the possibility of quickly developing warning receivers to counter the „Rotterdam“ device, the construction of "Naxos I" receivers was determined to be the only solution. To ensure the best receiving detectors, it was decided that the PTR should compare several detectors from Lorenz, Prof. Dieckmann and from its own production. The optimal detector for this purpose will be quickly produced by the company Lorenz in the required quantities (several 100 pieces)." (translated by the author).

1624 Later expanded to 20 units.

1625 The development and testing of the Naxos warning receiver was always a key topic in the subsequent meetings of the AGR.

1626 AGR-Protokoll, 23 February 1943, pp. 2–3.

1627 Handel 1999, p. 59.

1628 AGR-Protokoll, 23 February 1943, p. 3.

to drop marker bombs for subsequent bombers.[1629] A key topic was the fundamental improvement of tube technology and the development of more efficient detectors for the Naxos warning receivers.[1630] In a development meeting chaired by Martini, topics such as the reproduction of the Rotterdam device, jamming devices, and tube developments were addressed.[1631]

After the third AGR meeting on 8 April, which focused on further findings regarding the Rotterdam device and the Naxos technology, fully functional replicas of the Rotterdam device were already demonstrated in one of the subsequent meetings on 19 May 1943.[1632]

At the meeting on 1 June 1943, a newly captured device[1633] and the first prototype of a series model in a reconstruction, equipped exclusively with German components, were presented. After that, work began on installing the repaired original device in an aircraft.[1634]

In the meeting on 22 June 1943, a report was presented on the commissioning of the repaired Rotterdam device aboard an aircraft, with only those parts from Telefunken being reproduced that could not be restored.[1635]

In the report from the AG Rotterdam on 23 July 1943, regarding the first test flights with the repaired Rotterdam device, the following conclusion was stated:

> „Man erkennt, dass das Gerät „Rotterdam“ ein wertvolles Gerät zur Blindnavigation darstellt. Man erhält das Geländebild auf dem Braunschen Rohr allerdings sehr unvollkommen, erkennt aber verschiedene Navigationshilfspunkte insbesondere grössere Wasserflächen, so dass das Blindanfliegen einzelner Ziele unter Zuhilfenahme

1629 AGR-Protokoll, 17 March 1943, pp. 2–3.

1630 AGR-Protokoll, 17 March 1943, pp. 3–5.

1631 BArch-MA, RL 2 V / 75.

1632 AGR-Protokoll, 19 May 1943, p. 2.

1633 It was probably a Rotterdam device which was on board a Halifax bomber that was shot down during an attack on Düsseldorf on the night of 26 May 1943 (Reuter 1971, p. 114).

1634 AGR-Protokoll, 1 June 1943, p. 2.

1635 AGR-Protokoll, 22 June 1943, p. 2.

einer Karte bis auf eine Genauigkeit von etwa 1–2 km ohne weiteres möglich ist."[1636]

The first replicas were delivered to various locations at the beginning of September, such as the BHF, the Navy and the Air Force. It quickly became clear that disruption would be very difficult. It was recognised that the lakes in and around Berlin were easily detectable with the radar, which aided navigation for British aircraft. Therefore, experiments were carried out with mirrors of various sizes, which were mounted on rafts or placed on land, to investigate whether it would be possible to camouflage entire lake complexes and thereby prevent their detection in the aircraft for navigation purposes.[1637] These camouflage and decoy attempts did not achieve the desired success. For example, the Havel River, camouflaged with mirrors, could still be clearly traced along its course, and the camouflaged Müggelsee remained easily recognisable.[1638] In his accountability reports of 16 December 1943 and 7 January 1944, Plendl refers in Appendix 3 to the possibility of approaching the Reich capital with the help of the Rotterdam device and to the possibilities of defence.[1639] Another requirement for AGR was the development of warning receivers and jammers. The provision of powerful detectors was one of the main requirements for the Naxos warning receiver. At the AGR meetings, regular reports were given on the progress of research in this area.[1640] Despite the insights gained from the reconstruction of the Rotterdam device, the development of suitable countermeasures progressed only slowly. The

1636 AGR-Protokoll, 23 July 1943, p. 2. "It is evident that the „Rotterdam" device is a valuable tool for blind navigation. Although the terrain image from the Braun'sche Röhre is quite incomplete, various navigation reference points, particularly larger bodies of water, can be identified, making it possible to approach individual targets blindly using a map with an accuracy of about 1 to 2 kilometres without difficulty." (translated by the author)

1637 AGR-Protokoll, 2 September 1943, pp. 2–3.

1638 AGR-Protokoll, 3 March 1944, pp. 3–4; Handel 1999, p. 62.

1639 BArch, NS 19/2057, Annex 3.

1640 For the history of semiconductor research and development in Germany, see the dissertation by Handel 1999. For the development of detectors and the companies and institutions embedded in the AG Rotterdam, see Handel 2002, pp. 260–270.

Task of the passive Naxos device[1641], which was developed and built by the company Telefunken in various versions, was to receive the transmitted signals from the H_2S device in the enemy aircraft and then determine the direction.

The Korfu receiver was developed by the company Blaupunkt as a land-based receiver. In the AGR meeting on June 22, 1943, reports were presented on the first trials of the Korfu receiver, during which electromagnetic signals were already detected at distances of up to 200 kilometres.[1642]

In autumn 1943, the first Naxos Z device had been installed in a Ju 88 and tested. 15 ground stations[1643] had already been experimentally equipped with the Korfu receiver.[1644] During attacks on Berlin, the Korfu receiver was able to detect around 50 British airborne radars operating at wavelengths between 8.8 centimetres and 9.2 centimetres.[1645]

The AGR repeatedly addressed the development of a jammer in its meetings, but with limited success.[1646] As already stated, it was decided in the first AGR meeting to include the best reception detectors. The PTR was tasked with comparing the detectors from Lorenz, Telefunken, and Prof. Dieckmann from FFO[1647], and the optimal detector should be quickly produced in the required quantities by Lorenz.[1648] In addition to representatives from Lorenz and Telefunken, who had already attended the first meeting, it was decided to invite representatives from Siemens & Halske and GEMA to the upcoming meetings.

Subsequently, Telefunken and other institutions carried out intensive research and development of suitable detectors, since

1641 Bauer 2004c.

1642 AGR-Protokoll, 22 June 1943, p. 5.

1643 Stations were set up on the Brocken, the Zugspitze, and the Eiffel Tower (Reuter 1971, p. 134).

1644 The devices of the ground warning systems for the air reporting service were given the code names Naxos W and Korfu W, while they were called Naxos U for use on submarines. Naxos Z and Korfu Z were target acquisition devices for fighter aircraft, with which Rotterdam devices could be targeted and then attacked (Reuter 1971, p. 115; Brandt 1953a, p. 8).

1645 AGR-Protokoll, 2 September 1943, p. 3.

1646 Reuter 1971, p. 115; Brandt 1962a, pp. 92–94; Brandt 1953a, p. 7.

1647 FFO: Flugfunk-Forschungsinstitut Oberpfaffenhofen.

1648 AGR-Protokoll, 23 February 1943, p. 2; Handel 2002, pp. 260–261.

these were needed in large quantities.[1649] Highly qualified experts such as Herbert Mataré[1650] and Karl Seiler[1651] from the Telefunken laboratory and Heinrich Welker from the Physical-Chemical Institute of the University of Munich contributed to the success in the field of developing high-performance detectors.[1652]

Starting in 1943, the first prototype of a replica of the British H_2S device, with the code name Berlin A, developed by Telefunken, was available as an airborne radar system. It operated at a wavelength of 9 centimetres with a peak pulse power of 20 kW.[1653] One of the major challenges was adapting it to the limited space compared to the British heavy bombers, particularly in the smaller German aircraft types, such as the Me 110 and Ju 88.[1654]

While the developments of 1943 were primarily focused on offensive warfare, by the end of the year, the advantages of centimetre-wave radar for anti-aircraft operations and night fighting had been recognised. Until then, it had been assumed that the existing defence measures were sufficient.[1655]

During the Christmas period of 1943, Telefunken planned to develop new radar devices with wavelengths of 9 centimetres, 3 centimetres and 1 centimetre.

In his report on 8 February 1944, Brandt presented the planning to several high-ranking officials. These included the Chief of the Nachrichtenverbindungswesens (NVW) of the Air Force, the Chief of Marinenachrichtendienstes (MND) in the RLM, Generals of the Air Force, Admiral Stummel, and his colleagues from the Navy, as well as representatives from industry. As a result of this presentation, the transition to centimetre waves had practically been achieved with the Berlin device.

1649 Handel 1999, p. 63.
1650 Handel 1999, pp. 64–65; Handel 2002, pp. 262–266.
1651 Handel 2002, pp. 263–268.
1652 A detailed description of the development of high-performance detectors can be found in Handel 1999, pp. 64–93.
1653 AGR-Protokoll, 25 February 1944, pp. 2–4; Trenkle 1979, p. 116.
1654 Brandt 1962a, p. 97.
1655 Reuter 1971, p. 117.

This laid the foundation for the subsequent Berlin series for the Air Force and Navy.[1656]

The original assumption was that centimetric waves were well suited for mapping the Earth's surface due to their short wavelength, but not for air surveillance, night fighter control, or anti-aircraft artillery. For these applications, the well-known radar sets such as Freya, Würzburg, Würzburg-Riese, and Lichtenstein SN2 continued to be used.[1657]

On 22 December 1943, high-frequency parts of a captured Rotterdam set were installed in a Würzburg set as part of a general test. The achievable range was estimated at 8 kilometres, but it turned out that a range of up to 30 kilometres could be achieved. This realisation led to a drastic rethink in the assessment of the suitability of centimetre waves.[1658]

During the AGR meeting on 5 April 1944, reports were made on the first test flights with the modified Rotterdam device Wiesbaden. Compared to the replica radar device Rotterdam, the Wiesbaden device had a higher sensitivity, which allowed test flights to be conducted at higher altitudes. The long range was particularly noticeable.[1659]

Fig. 37 shows a photograph taken with the Wiesbaden device from Lake Müritz. The image was captured from an altitude of 4,000 metres. At the edge of the image, Stralsund, 85 kilometres away, appears as an arc. Additionally, Wittstock can be seen at a distance of 48 kilometres, and both Lake Müritz and Lake Plauer are depicted.[1660]

In the meantime, American and British bombers were already in use with radar equipment with a wavelength of 3 centimetres. Probably at the end of 1943, an American captured device called Meddo[1661], which was later called Rotterdam in German, fell into

---

1656 AGR-Protokoll, 31 May 1944, pp. 8–9; Brandt 1962a, pp. 99–100; Reuter 1971, p. 118, pp. 239–243.

1657 Brandt 1962a, pp. 97–98; Reuter 1971, p. 117.

1658 Reuter 1971, p. 117–118.

1659 AGR-Protokoll, 5 April 1944; pp. 17–18; Ergänzungsprotokoll zu den AGR-Sitzungen, 1 September 1944, pp. 47–56.

1660 Ergänzungsprotokoll zu den AGR-Sitzungen, 1 September 1944, pp. 50–51.

1661 Named after the location where it was found, near the Dutch village of Meddo, close to the Dutch-German border.

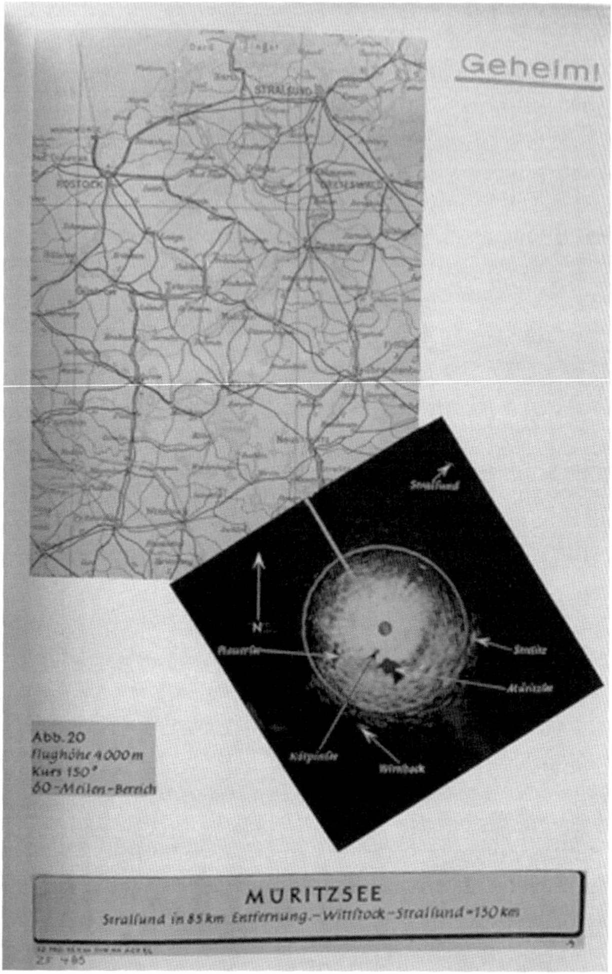

Fig. 37: Photograph taken with the Wiesbaden device[1662]

German hands. However, only parts of the Meddo device were usable. At the AGR meeting on 25 February, the company Telefunken was tasked with reproducing 10 units.[1663]

1662 Ergänzungsprotokoll zu den AGR-Sitzungen, 1 September 1944, p. 50, Fig. 37; Image credit: Helmut Klausing.

1663 AGR-Protokoll, 25 February 1944, p. 7; Bauer 2009.

Right at the start of the 25 February meeting, Brandt reported that Speer had established the Sonderkommission für Funkmeßtechnik, under the leadership of Karl Rottgardt from Telefunken.[1664]

At the AGR meeting on 13 March 1944, further progress in the reproduction of the Meddo device was reported.[1665] Initially, only parts of a Meddo device had been secured, but following the AGR meeting on 5 April 1944, another device was recovered. In early April, a nearly undamaged Meddo device was retrieved from a US B24 Liberator that had made an emergency landing near Calais. The radar system operated at a wavelength of 3.2 centimetres and was functionally equivalent to the Rotterdam device.[1666]

During the AGR meeting on 31 May 1944, a comprehensive report was presented on the current state of development in radio measurement technology.[1667] Additionally, ground observation images from the Berlin device, the original Rotterdam device, the Rotterdam replicas, and a Meddo device were analysed.[1668]

At the AGR meeting on 26 July 1944, technological issues were discussed in detail, including the topic of Düppel, known in English as Window.[1669]

The impressive test results of 22 December 1943, when a captured Rotterdam device was integrated into a Würzburg set and achieved an extraordinary range, had already prompted intensified work on centimetre-wave technology. This area was regarded as extremely urgent by the highest decision-makers. Around the turn of 1943/44 and in early January 1944, several meetings took place at the highest political and military levels with all leading representatives. As part of the consolidation of research and development, the ongoing rivalry for leadership in high-frequency and radio measurement technology between the Navy and the Air Force was also addressed.

1664 AGR-Protokoll, 25 February 1944, p. 2.

1665 AGR-Protokoll, 13 March 1944, p. 13.

1666 AGR-Protokoll, 5 April 1944; pp. 18–24; AGR-Protokoll, 26 April 1944, pp. 6–11.

1667 AGR-Protokoll, 31 May 1944; pp. 3–41; Reuter 1971, pp. 239–243.

1668 AGR-Protokoll, 31 May 1944; pp. 8–17.

1669 AGR-Protokoll, 26 July 1944; pp. 49–75.

On 3 January 1944, a discussion between Dönitz, Göring, and Speer took place regarding the consolidation of the radio measurement industry under Speer's ministry. Göring emphasised that the communications chiefs of the military branches shared responsibility for the backlog in the development of centimetre-wave technology.[1670]

10.5.1.1.1 Summary

With the founding of the AGR, developments were carried out for the first time under industrial leadership and with a specific goal. Until the beginning of 1943, political and military decision-makers had paid little attention to high-frequency research, especially in the field of centimetre waves. Until then, the existing and technically optimised radar equipment was considered sufficient and was continuously improved. The limited resources were divided between the ongoing development areas, and at the end of 1942, research in the field of centimetre waves was completely discontinued.

With the discovery of a radar device aboard a British bomber shot down near Rotterdam on 3 February 1943, operating in the centimetre wave range, the gap in German research and development in this frequency range became dramatically clear.

This was particularly significant in light of the increasing successful bombing of German cities and industrial facilities by Allied bombers, as well as the fight against German submarines. Alongside science and research, the available capacity of development and manufacturing facilities, as well as their personnel, increasingly came into focus against the backdrop of the growing bombings of German cities and industrial sites. With the AGR, all available resources were strategically concentrated on a common goal, and as a result, the technological backlog was largely caught up in a relatively short time.

Collaboration and an unhindered flow of information also took place within different working groups, crossing the boundaries of the different services of the armed forces, especially be-

1670 Reuter 1971, p. 200; BArch, R 3 / 1739.

tween the Navy and the Air Force.[1671] Alongside the briefly outlined different research and development tasks of the AGR, there were additional efforts in industry and institutions. With the appointment of Plendl as BHF on 14 November 1943, research in radar technology for air and coastal defence was assigned high priority. In addition, Plendl should work together with the communications chiefs of the military branches and incorporate the high-frequency research in industry and institutions into the resulting research tasks.

The aim was to create a central body for coordinating research in the field of radar technology for the first time.[1672] Between April 1943 and February 1945, the BHF awarded nearly 540 research projects to various research institutes, universities, military branches, and industrial entities.[1673]

10.5.2 Great Britain

10.5.2.1 Tizard Mission

Another example of effective innovation cooperation was the secret trip by Henry Tizard, known as the Tizard Mission[1674], who travelled to the United States in August and September 1940, accompanied by six scientists.

Among the members of this mission were John Cockcroft and Edward Bowen, both technical experts in radar.[1675] Tizard travelled to Canada by plane as early as August to discuss possible Canadian contributions with representatives of the Canadian government. The other members arrived in Halifax, Canada, by ship on 6 September and travelled on to Washington, D.C., a few days later.[1676]

In America, Tizard first held talks with Vannevar Bush, the senior U.S. scientific administrator, who, in many ways, was Ti-

1671 Handel 2002, pp. 257–270.
1672 Flachowsky 2005, pp. 212–213.
1673 Flachowsky 2005, p. 218; CIOS 1946, pp. 4–80.
1674 Zimmerman 1996; Phelps 2010; Bowen 1988b, pp. 296–307; Blackett 1960, pp. 647–653.
1675 Redhead 2001, p. 326.
1676 Boot and Randall 1976, p. 726.

zard's counterpart in America at the time.[1677] He was one of the most important figures in U.S. war effort during World War II. After the Battle of Great Britain had begun in the summer of 1940, and with the attacks on London starting on 7 September, Great Britain appeared to be without allies, and Tizard realised that the country's capacity in research, development, and production in the field of military armaments was insufficient to withstand Germany.[1678] Therefore, Great Britain sought cooperation with the United States.[1679] Prime Minister Winston Churchill agreed with U.S. President Franklin D. Roosevelt that British scientists should bring their secret research and development findings to the U.S. in order to make use of American capacities in research, development, and manufacturing.

They carried samples, blueprints, and reports on almost all new British war-related developments, including radar, fire control systems, sonar detection devices, and other military innovations, as well as a sample of perhaps the most spectacular invention, the cavity magnetron, which became the key component and centrepiece of all future radar devices.[1680]

James Phinney Baxter 3rd[1681], Official Historian of the Office of Scientific Research and Development (OSRD) said about the magnetron:

> "When the members of the Tizard Mission brought one to America in 1940, they carried the most valuable cargo ever brought to our shores. It sparked the whole development of microwave radar and constituted the most important item in reverse Lease-Lend."[1682]

After arriving in Washington D.C., the Tizard Committee sought contacts with American scientists. However, it was not the universi-

---

1677 Phelps 2010, p. 67.

1678 The United States entered World War II on 8 December 1941, following the Japanese attack on Pearl Harbor. Germany and Italy declared war on the USA on 11 December 1941.

1679 Eldridge 2000, pp. 1–20.

1680 Blackett 1960, p. 649.

1681 Phelps 2010, p. 289.

1682 Baxter 3rd 1968, p. 142; Watson-Watt 1957, p. 297; Buderi 1997, p. 27; Snow 2013, p. 39; Snow 1961b, p. 42.

ties, industry, or military that showed interest in the British guests, but rather a wealthy private individual named Alfred Loomis.[1683] Loomis had set up a private laboratory in Tuxedo Park, about 35 miles northwest of New York City, where he, along with other scientists such as Karl Compton, conducted research in the field of microwave technology.[1684] When he heard about the British delegation, he invited them to his laboratory and realised that British scientists were far ahead of their American colleagues in the field of radar. This was particularly true for radar devices to be installed in aircraft, but also for ground-based and ship-based systems.[1685] After the presentation of the magnetron, he was willing to participate in the development of a radar system in the microwave range.

In view of the threat posed by German submarines and attacks on supply ships en route to Great Britain, there was enormous pressure to accelerate the development of a microwave radar system. In October 1940, the MIT Radiation Laboratory (RadLab) was founded with Loomis' support and placed under the control of the National Defence Research Committee (NDRC).[1686]

The NDRC was established after a conversation between Vannevar Bush and President Roosevelt on 27 June 1940. Roosevelt immediately recognised the value of such an organisation when he signed the decree approving the NDRC. Bush became chairman of the NDRC. The NDRC's mission was to organise, coordinate and develop, in strict secrecy, technologies relevant to warfare, including radar. On 28 June 1941, the NDRC was incorporated into the OSRD, with Vannevar Bush as chairman.[1687] The task of the OSRD was to coordinate research for military purposes during World War II. The RadLab, which focused on microwave and radar research, expanded quickly and had around 30 to 40 employees within the NDRC organisation as of November 1940 and almost 3,900 employees at the OSRD as of summer 1945. Alfred Loomis performed important functions at both the NDRC and the

1683 Alvarez 1980, pp. 329–330.

1684 Phelps 2010, pp. 153–159.

1685 Bowen 1988b, p. 297.

1686 Stewart 1948, Chapter 2, pp. 7–32.

1687 Stewart 1948, Chapter 3, pp. 35–51; the NDRC existed until 1947 and still had an advisory function.

OSRD.[1688] At the end of 1945, the RadLab was formally closed, and the research work was continued by MIT's Research Laboratory of Electronics (RLE).[1689] At a meeting of the British team with Loomis on 19 September 1940, who was at that time chairman of the Microwave Committee[1690] within the NDRC, the British presented technical details of their Chain Home radar system, which was very similar to a long-wave radar of the US Navy. After that, the results of the research by Compton and Loomis in the field of microwaves were presented. The Americans did have an experimental device operating at a wavelength of 10 centimetres, but with insufficient transmission power. After the British had presented their cavity magnetron, the collaboration went smoothly. Everyone had recognised the value of the magnetron in the development and production of radar equipment, for example for installation in night fighters to locate surfaced submarines or as a valuable aid in bomber navigation. RadLab was not alone in its research efforts. Other centres, such as Harvard and Columbia Universities, also conducted work on microwave and radar technology with NDRC funding.

The RadLab also signed contracts with seventy industrial companies as part of a collaboration on the mass production of radar equipment, such as General Electric, Raytheon, RCA, Westinghouse, Philco, Sperry, and Western Electric. In October 1941, the RadLab established its own company, the Research Construction Company (RCC), to manufacture limited quantities of microwave radar systems and components that were not immediately available from industry.[1691] As a result, the quantity of research, development and production quickly exceeded years of work in Great Britain. From 1943, the RadLab provided more and more improved devices that the industry could produce in large quantities. Another important milestone was the development of LORAN, a long-range radio navigation system for aviation and shipping. The forerunners of this navigation system were Loomis' 1940 works at RadLab.

1688 Buderi 1997, p. 50.
1689 *RLE currents* 1991, p. 7.
1690 Buderi 1997, p. 50.
1691 *RLE currents* 1991, p. 6.

When Tizard returned to England, he was not given any further central role in radar development.[1692]

10.5.2.1.1 Summary

The close and trusting cooperation between British and American experts in the Tizard Mission ultimately led to the mass production of new radar equipment, primarily for installation in aircraft. This helped the Allies in their fight in Europe and the Pacific to a considerable extent and contributed significantly to their military superiority in the second half of the war.

Just one year before the USA entered the war, Blackett said:

> "This imaginative act of trust, which Tizard and A. V. Hill first envisaged and finally forced through Whitehall, had immensely beneficial effects on the scientific aspects of the Allied war effort. Cockcroft reminds us that the Mission was magnificently organized by Tizard and that he had the inspiration to bring a mixed team of serving officers and scientists. For the first time our American friends heard civilian scientists discussing authoritatively the instruments of war and then heard the Service people following on with practical experience."[1693]

10.5.2.1.2 Remark

In the September 2022 issue of The Journal of Defence Electronics History Society (DEHS), Phil Judkins, a leading British researcher on the history of radar, drew on original sources to examine the collaboration between government research, industrial production and military users in creating effective capabilities, complementing the already familiar background of the Tizard Mission (Judkins 2022, p. 1 and pp. 3-28). His analysis highlights the tensions and conflicts between individuals and organisations over scientific, industrial and military priorities in Great Britain, the United States and Canada during this intense phase of radar development. A closer look at the interaction between political leaders, the armed forces, scientists and industry reveals how differing priorities, par-

1692 Cockcroft became head of the British mission on 5 October after Tizard returned to England (Baxter 3rd 1968, p. 121).

1693 Blackett 1960, p. 649; Snow 2013, p. 39.

ticularly in the United Kingdom and the United States, frequently led to disputes and delays. However, unlike in Germany, these conflicts did not result in fundamental misjudgements that undermined technological progress or jeopardised Britain's defensive capabilities.

10.5.2.2 Sunday Soviets

The so-called Sunday Soviets[1694] were a distinctive feature of the innovation process at the TRE. These weekly discussion sessions brought together government officials, scientists, leaders, and RAF pilots in an informal setting. Employees of all ranks were encouraged to contribute their ideas freely, as the meetings were deliberately held without formal ranks or titles ("no ranks, no titles"). When a good idea was proposed, it could be implemented immediately, since all relevant experts and decision-makers were present. These meetings played a key role in securing acceptance for radar technology.

The key figure and founder of these meetings was Albert Rowe, who had already conducted an analysis of existing location systems for air defence purposes at the British Air Ministry in 1934. He became Secretary of the CSSAD in 1935 and Chief Superintendent of the TRE in 1940, giving him oversight of all research projects. Rowe emphasised that the meetings provided a unique opportunity for users of radar equipment, such as RAF personnel and other military staff, to openly discuss practical problems they encountered with the engineers and designers responsible for developing and fabricating the technology. This direct dialogue helped rapidly identify issues and implement improvements, fostering close collaboration between users and creators.

While at Worth Matravers near Swanage in 1940, Rowe conceived the idea of inviting senior military officers to visit the TRE on Sundays to meet informally with research engineers and scientists. These meetings, held in a completely informal and trusting atmosphere, were unique. A strong sense of connection developed between experts and high-ranking military officers. Initially, the venue was the Grosvenor Hotel in Swanage, but when the TRE

1694 Rowe 2015, pp. 83–86; Air Ministry 1963, p. 39; Burns 2013, p. 365.

relocated to Malvern, the meetings were held on-site at Malvern College and continued there until the end of the war.[1695]

By the end of the war, Rowe reported that these meetings were attended by Cabinet Ministers, Henry Tizard, Frederick Lindemann, Commanders, Admirals, Air Marshals, university professors, service personnel returning directly from the front, decision-makers from headquarters, and notably, scientists working in or around the laboratories.[1696] Air Chief-Marshal Joubert de la Ferté's remark, "You can say anything in Rowe's office on a Sunday", epitomised the frank exchange of views on the progress of the war and discussions on how radar could best support the fighters.[1697]

The concept of the Sunday Soviets gradually evolved alongside the development of operational research. For Rowe, the connection between scientists and users was of paramount importance. The term Sunday Soviets originated from a comment made by a senior government visitor shortly before the war. Attending one of the informal Sunday meetings at Bawdsey, the visitor was surprised by the open and free exchange of ideas, without formal ranks or restrictions, and remarked that it was "just like a Soviet".[1698] The nickname stuck and came to symbolise the unique atmosphere of cooperation and trust between scientists and military leaders at these sessions.

Originally, Sunday had been intended as a day of rest. Although work continued seven days a week in Swanage after the German attack on France in May 1940, it soon became clear that this was not effective. After staff were given the option to take one day off per week, which proved insufficient, Sunday was chosen as the most suitable day.

While Swanage's location offered some recreational relief, many participants still found the meetings exhausting. On the other

1695 Hutchinson 2022, p. 132.

1696 Rowe 2015, p. 83; Burns 2013, p. 365.

1697 Rowe 2015, p. 84.

1698 Ibid., p. 86; Judkins explains, referring to a famous photograph of a meeting of the Sunday Soviets with Lord Cherwell, RAF officers, and scientists from the TRE, that this image had a positive effect on public opinion. However, it does not show any manufacturers, even though it cannot be claimed that British industry lacked personnel who were on par with the scientists at the TRE (Judkins 2022, p. 23).

hand, friendships were formed among participants from a wide variety of circles. Rowe recalled that the visitor's comparison with the Soviets aptly highlighted the stark contrast between Great Britain and Germany in terms of cooperation and trust.

11 Appendix

11.1 Key Individuals

11.1.1 Germany

11.1.1.1 Politics and Military

Adolf Hitler

Adolf Hitler (1889–1945)[1699] was appointed Reichskanzler of Germany on 30 January 1933. Following the death of Reichspräsident Hindenburg on 2 August 1934, he assumed the roles of Reichskanzler, Reichspräsident, and Oberbefehlshaber der Wehrmacht, consolidating all power in a single individual. On 26 February 1935, the establishment of the Air Force as the third branch of the Wehrmacht was announced, and on 16 March compulsory military service was introduced alongside the abolition of the military restrictions imposed by the Treaty of Versailles.

Hitler's understanding of technology remained largely shaped by experiences in World War I. Plans for war preparations in the summer of 1939 assumed a conflict conducted along traditional lines, similar to World War I, with the only adjustment being the increased personnel demands for the armaments industry. He exhibited limited understanding of the natural sciences, and his approach to technology was often influenced more by ideology than by scientific knowledge.

Hitler's approach to technological advancement was heavily influenced by his ideological beliefs, often at odds with modern scientific developments. He prioritised traditional methods of warfare and paid little attention to the innovations such as radar. Although the Air Force and some generals recognised the importance of radar and other modern technologies, Hitler remained largely

1699 As with Churchill, Hitler has been the subject of extensive biographical literature.; e.g., Kershaw 2009; Kershaw 1998; Kershaw 2000; Pyta 2024; Wirsching 2016, pp. 387–418.

indifferent to their development. This contributed to a fragmented and undercoordinated approach to technological development, as the Nazi regime focused primarily on military strength rather than technological superiority.

Despite holding near-unlimited power as dictator, radar development remained a secondary concern in Hitler's eyes, as reflected in his minimal engagement with or approval of this modern technology.

Adolf Hitler died by suicide on 30 April 1945.

Hermann Göring (1893–1946)[1700] was a successful fighter pilot during World War I and held numerous prominent positions after the Nazi Party came to power. On 30 August 1932, he was elected President of the Reichstag. Following Hitler's appointment as Chancellor on 30 January 1933, Göring became Reich Minister without Portfolio, Reich Commissioner for Aviation, and provisional Prussian Minister of the Interior. On 11 April 1933, he was appointed Prime Minister of Prussia, and on 5 May 1933, Minister of Aviation.

On 26 February 1935, he assumed command of the newly formed Air Force. On 18 October 1936, he was appointed Beauftragter für den Vierjahresplan, responsible for preparing the German economy for war. Although Göring often presented himself as a supporter of modern technologies, his technical knowledge, particularly of the then novel radar, was limited. During a 1941 demonstration of radar for night fighters, he dismissively stated: "Ich will kein Kintopp (Kino) im Flugzeug".[1701] His interest in technical developments was largely politically motivated and superficial, unlike figures such as Albert Speer, who, as Armaments Minister, played a key role in organising industrial war production and fostering technological innovation.

After the defeat in the Battle of Britain in 1940, Göring's reputation steadily declined. The Allied bombing of German cities and the failure of the airlift to supply the encircled 6th Army at Stalingrad in the winter of 1942/43 further undermined his prestige. From that point on, he no longer played a significant role in the leadership

1700 Stoltenberg 1964, pp. 525–527; Overy 2012.

1701 "I don't want a Kintopp (cinema) in the aircraft!" (translated by the author).

of the Nazi regime. Despite holding numerous offices and formal power, Göring remained primarily a politician and military leader who largely avoided involvement with technical matters such as radar.

In April 1945, Hitler dismissed Göring by telegram, considering him incapable of acting. In his political testament[1702], later that month, Hitler appointed Karl Dönitz as his successor as President of the Reich and Commander-in-Chief of the German Armed Forces. Göring was captured by the Allies shortly thereafter. At the Nuremberg Trials, he was sentenced to death but committed suicide by ingesting cyanide on 15 October 1946.

Karl Dönitz

Karl Dönitz (1891–1980)[1703] was a German naval officer who played a key role in building and strategically developing the German U-boat force. During World War I, he gained his first experience as a U-boat commander and remained active in the navy throughout the interwar period. In the 1930s, Admiral Erich Raeder tasked him with building a modern U-boat fleet. From 1935, Dönitz served as Führer der Unterseeboote (FdU), directing the development and operations of Germany's submarine force. In 1939, he was officially appointed Befehlshaber der U-Boote (BdU), the formal title within the German Navy.

Dönitz is credited with developing and implementing the Wolfsrudeltaktik, a strategy in which groups of submarines coordinated their attacks on Allied convoys, significantly increasing the effectiveness of U-boat operations during the early years of the Battle of the Atlantic. Initially, he showed little interest in radar technology, focusing instead on acoustic systems such as hydrophones and sonar. Only after heavy losses in 1943 did he recognise the strategic importance of electronic detection and countermeasures.

On 11 December 1943, the Wissenschaftlicher Führungsstab beim Oberkommando der Kriegsmarine (WFM) in Berlin was established to coordinate technological development and countermeasures for the U-boat fleet. Despite these efforts, the situation re-

1702 Bradsher 2015; pp. 16–25; Ryan 2015.

1703 Hillmann 2022; Kraus 2011, pp. 316–325; Dönitz 2018.

mained dire: of roughly 41,000 German submariners, only about 15,000 survived the war.

On 30 January 1943, Dönitz was promoted to Grand Admiral and Commander-in-Chief of the Navy. In Hitler's political testament of April 1945, he was named Reich President and Supreme Commander of the Armed Forces. He assumed leadership of the so called Flensburg Government until his arrest by British forces on 23 May 1945.

At the main Nuremberg trial, he was sentenced on 1 October 1946 to ten years in prison. After his release in 1956, he lived a quiet life near Hamburg, remaining a controversial symbolic figure in naval circles. He consistently portrayed the Kriegsmarine as politically neutral in his memoirs.

Karl Dönitz died on 24 December 1980.

Erich Raeder

Erich Raeder (1876–1960)[1704] played a key role in rebuilding and leading the German Navy between the world wars and during the early years of World War II. He joined the Imperial Navy in 1894, remained in service after 1918, and in 1928 became Chief of the Naval Command. Shaped by the traditions of the pre-1918 fleet, Raeder aimed to rebuild a disciplined navy loyal to the state, regardless of its political system. He sought to reconstruct the German Navy within the constraints of the Treaty of Versailles and placed great emphasis on officer training and naval discipline. Despite his experience and careful planning, he was slow to push technological modernisation, particularly in expanding the U-boat fleet and integrating new systems such as radar.

From 1935 onwards he oversaw the development of the Kriegsmarine, pursuing a balanced surface fleet of battleships, cruisers and smaller vessels. His commitment to surface warfare brought him into conflict with Karl Dönitz, who prioritised submarine strategy, and his assumption that war with Britain would not come before 1943 quickly proved wrong. Raeder was cautious in adopting new technologies such as radar, recognising its potential but failing to prioritise its development, which left the navy at a disadvantage compared with the Allies.

1704 Raeder 1957; Salewski 2003, pp. 104–106; Fischer 2011, pp. 185–194.

Raeder remained in command until early 1943. After the failed surface engagement in the Barents Sea and Hitler's decision to withdraw the capital ships, he resigned and was succeeded by Dönitz, who focused fully on the U-boat war. Raeder was sentenced to life imprisonment at the Nuremberg Trials on 1 October 1946, but released on 27 September 1955 due to illness. He died on 6 November 1960. His legacy is that of a capable naval administrator whose traditional outlook limited the effectiveness of Germany's naval strategy in World War II.

Ernst Udet

Ernst Udet (1896–1941)[1705] was, like Hermann Göring, one of the most successful German fighter pilots of World War I with 62 aerial victories. After the war he gained international fame as a stunt pilot, aircraft manufacturer, and adventurer, performing spectacular flights and expeditions that enhanced his public reputation. In 1933 he joined the Nazi Party and became Vice Commodore of the German Air Sports Association, with the regime exploiting his popularity and wartime record to support the establishment of the Air Force.

From 1936 Udet headed the Technisches Amt in the RLM, responsible for technical development and procurement of aircraft and equipment. In February 1939 he was promoted to Generalluftzeugmeister, giving him supreme responsibility for aircraft development, production, and armament within the Air Force. Despite his rapid promotions to Colonel General, he lacked administrative experience and technical expertise, which frequently led to conflicts with Erhard Milch. Udet was initially sceptical of emerging technologies such as radar, famously dismissing its value for flying, but he later recognised the Würzburg-A system's importance for night fighting after 1940.

Although his position gave him considerable influence, Udet's leadership weaknesses and inability to manage complex armaments projects undermined the Air Force's organisational efficiency. His public profile and status as a war hero often outweighed his administrative abilities, making him more of a symbolic figure than a fully effective manager.

1705 Potempa 2016, pp. 514–515; Hümmelchen 2011b, pp. 258–264.

After a final serious conflict with Göring, Udet committed suicide on 17 November 1941. Officially declared an accident, his death was followed by a state funeral. He remained Generalluftzeugmeister until his death, after which his responsibilities were redistributed, mainly under Erhard Milch. Udet's career illustrates the Air Force's reliance on charismatic World War I pilots whose prestige sometimes exceeded their practical capabilities.

Power Without Progress:
Neither Hitler, Göring, Dönitz, Raeder, nor Udet fully recognised the strategic potential of radar, or they did so only belatedly. None of them developed a coherent vision for its systematic development or deployment. As the highest leaders of the Nazi regime, they could have significantly advanced radar technology, yet their World War I experiences, and conventional thinking led to underestimation and delayed engagement. While they were not outright opponents, their interest was sporadic and largely symbolic, resulting in minimal impact on the innovation process. It was only later that technically skilled and strategically minded figures such as Wolfgang Martini actively promoted radar, recognised its full potential, and implemented it deliberately in military research and development.

Wolfgang Martini
Wolfgang Martini (1891–1963)[1706] was born on 20 September 1891 in Lissa, Province of Posen. He joined the German Army in 1910 and served in various radio communication roles during World War I. After the war, he continued in the signal corps and from 1928 worked as a technical officer in the Reich Ministry of Defence.

In 1933, Martini joined RLM, playing a leading role in developing the communication and reporting structures that would later form the Luftwaffe's signal corps. From 1934 to 1939 he headed the Abteilung Nachrichtenverbindungswesen in the RLM, which from 1937 to 1939 was attached to the General Staff. From 1939 to 1944 he served as Chef des Nachrichtenverbindungswesens and Inspekteur

1706 Hoffmann 1990, pp. 301–302; Rieger 2018b; Watson 2009, pp. 253–254; Traces of War n.d.

der Luftnachrichtentruppe, receiving promotion to General in 1941 and appointment as Sonderbeauftragter für Funkmeßgeräte.

Martini was an early advocate of radar in Germany. After witnessing the Seetakt naval radar in 1938, he initiated the development of the Freya system for the Air Force and promoted cooperation between military branches and academic institutions. He recognised the potential of centimetre-wave radar following the recovery of a British H_2S device in 1943, leading to the development of systems such as the Berlin radar. Martini played a leading role in founding research groups, including the Arbeitsgemeinschaft Rotterdam, aimed at consolidating radar development and improving production efficiency.

He was instrumental in collecting and preserving wartime knowledge in radar and navigation, helping to facilitate Germany's postwar reintegration into international developments in both civilian and military communication and navigation technology.

Wolfgang Martini died on 6 January 1963.

Werner Osenberg

Werner Osenberg (1900–1974)[1707] was a German materials scientist and a central organiser of armaments research during World War II. After studying engineering and materials science, he joined research institutions in Germany and became known for his organisational skills and technical expertise. In June 1943, Göring issued a decree establishing the Planungsamt within the RFR, appointing Osenberg as its head (with responsibility for identifying and allocating specialist personnel to critical wartime research projects.

Between June 1943 and 1944, Osenberg built up the Planungsamt, created a central card index of approximately 15,000 scientists and engineers, which became the basis for a large-scale recall operation in late 1944. Around 5,000 specialists were withdrawn from military service, while a further around 10,000 were declared indispensable to prevent conscription. This effort ensured that Germany retained vital technical expertise for armaments and research projects despite manpower shortages. Following heavy bombing of Hannover in 1943, he relocated the Planungsamt to Lindau in the Harz re-

1707 Federspiel 2002, pp. 72–105.

gion, where the office expanded to nearly 300 staff. In addition to personnel management, Osenberg's office also coordinated armaments development projects, including experimental rocket warheads and specialised industrial research.

In 1944, Osenberg further organised armaments research within the RFR as head of the Planungsamt, demonstrating his ability to mobilise scientific expertise efficiently. His work had a lasting impact on German wartime innovation and the efficient deployment of scientific expertise under extreme conditions.

After the war, Osenberg was interned by the Americans until 1947. His card index reportedly informed the recruitment of German scientists for programmes such as Operation Paperclip. He returned to Germany in 1948, resumed his academic career at the Technical University of Hannover, and retired in 1970.

Werner Osenberg died on 14 December 1974.

Johannes (Hans) Plendl

Hans Plendl (1900–1991)[1708] was a German physicist and a leading developer of military radio navigation systems during the Third Reich. After earning his doctorate under Jonathan Zenneck, he initially worked at Telefunken and from 1927 at the Deutsche Versuchsanstalt für Luftfahrt (DVL). From 1934, he led the Funkforschungsreferat in Rechlin, where he developed the so-called X-method, a radio beam guidance system that enabled the Air Force to conduct precise bombing attacks. The X-method was employed, for example, during the air raids on Coventry during the night of 14–15 November 1940, when pathfinder aircraft used it to mark targets.

In the early phase of the war, Plendl and his deputy, Walter Dieminger, initiated a system to predict shortwave radio connections for the German Air Force based on ionospheric observations.

In November 1942, Hermann Göring appointed Plendl as Bevollmächtigter für Hochfrequenzforschung (BHF), a newly created position intended to coordinate the previously uncoordinated research in high-frequency and radar technology. Plendl systematically documented ongoing projects, established new working

1708 Seiler 2007, pp. 54–55, pp. 114–117, pp. 133–136; Handel 1999, pp. 57–58; Reuter 1971, pp. 192–193; Watson 2009, pp. 254–255.

groups, and structured the field with a long-term perspective. His methodical approach emphasised quality and interdepartmental cooperation, although it did not deliver immediate military results.

Following setbacks suffered by the Air Force in the summer of 1943, including the Allied attacks on Hamburg, Plendl lost Göring's support and was dismissed on 2 December 1943. He was succeeded by Abraham Esau, President of the Physikalisch-Technische Reichsanstalt (PTR). Despite his dismissal, Plendl's initiatives had already begun to bear fruit: between April 1943 and February 1945, the BHF awarded around 540 research contracts to state institutions and industrial laboratories.

After the war, Plendl moved to the United States, where he worked for the Air Force. He later turned his focus to solid-state physics rather than continuing with radio navigation research.

Hans Plendl died on 10 May 1991.

Albert Speer

Albert Speer (1905–1981)[1709] was a central figure in Nazi Germany, initially as an architect and later as Minister for Armaments and War Production. He studied under Heinrich Tessenow in Berlin and quickly became Hitler's preferred architect, designing monumental buildings, stages for mass events such as the Nuremberg Party Rallies, and the New Reich Chancellery in Berlin. After the death of Fritz Todt in 1942, Hitler appointed Speer as Minister for Armaments and War Production, giving him broad authority over Germany's war industry and the Todt Organisation. Speer reorganised armaments production and increased efficiency despite intensifying Allied bombing. Contemporary research critically examines Speer's role, often described as the Speer Myth, emphasising the contrast between the purported Rüstungswunder after his appointment and the largely peacetime economy that persisted from 1939 to 1941. Allied air raids increased year by year, doubling the tonnage of bombs dropped on Germany, but until 1943 many industrial sites and transport routes were only sporadically hit and quickly repaired. Towards the end of the war, Speer attempted to prevent the implementation of the Nero Decrees, which aimed to

1709 Speer 2013; Schroeter 2019; Herbst 2010, pp. 644–646.

destroy German infrastructure. Following Hitler's death, he served briefly in the Dönitz administration, was arrested, and sentenced to 20 years' imprisonment at the Nuremberg Trials. Released in 1966, he lived near Heidelberg and wrote memoirs that remain important sources on the inner workings of the Nazi regime.

Speer was one of the most influential figures in the Nazi regime, as he not only shaped its architecture but also significantly increased armaments production, thereby prolonging the war effort. His memoirs, despite his self-portrayal as an innocent technocrat, continue to provide valuable insights into the structures and functioning of the Nazi state.

Albert Speer died on 1 September 1981.

Werner Beckmann (?)
In October 1939, Telefunken was awarded a contract to develop a short-range aircraft detection device, marking an early stage in the Würzburg radar system family. Senior Staff Engineer Werner Beckmann, head of a department at the RLM, initiated the development of radar prototypes at a time when the RLM had no centralised strategy for radar research.

Beckmann's department handled the production and testing of several radar systems, including short-range devices for anti-aircraft artillery and air surveillance, as well as the long-range Freya radar. These developments took place with limited involvement from Hermann Göring, reflecting a degree of decentralisation in German radar research during the early war years.

The work of Beckmann's department contributed to the operational deployment of radar systems in Germany, providing a technical foundation for later improvements and broader use in air defence during World War II.

11.1.1.2 Science and Industry

Leo Brandt
Leo Brandt (1908–1971)[1710] was a prominent German radio and radar technology expert, known for his significant contributions

1710 Mittermaier und Rusinek 2009; Mauel 1983a, pp. 25–27; Weinreich 2001d, pp. 109–110.

during World War II. He began his career at Telefunken in 1933 and quickly rose through the ranks. In 1935, Brandt became head of the Empfängerlabor for radio equipment at Telefunken. By 1937, he had taken on the management of Geräteentwicklung in the areas of radio receivers, direction-finding systems, and targeting devices. In 1939, he was given responsibility for the radio network and radio measurement technology.

Brandt played a crucial role in the development of the Würzburg radar family, starting in 1939. The Würzburg radar became standard equipment for the German anti-aircraft artillery in 1940 and played a critical role in German air defence. This development marked a milestone in German radar and radio technology, and Brandt's leadership was key to its success. By 1941, he was appointed head of the Entwicklungsgruppe Funkmeß, reporting to the Generalbevollmächtigter für technische Nachrichtenmittel (GBN), General Erich Fellgiebel.

In this role, Brandt highlighted the critical shortage of skilled workers in the radar industry, a concern that led to significant organisational and personnel improvements. His ability to respond swiftly and effectively to the pressing needs of the radar development sector cemented his reputation as one of the leading figures in the field of radio and radar technology at the time.

In February 1943, Brandt became chairman of the Arbeitsgemeinschaft Rotterdam (AGR), a groundbreaking collaboration between the Wehrmacht, industry, and research institutions. This cooperation enabled open and trusting dialogue, which significantly accelerated the development of radar technologies. As chairman, Brandt's leadership skills were crucial in reconciling the different interests of military and civilian sectors, ensuring that radar technology development could meet the strategic needs of both.

During the war, Brandt's contributions were not limited to the Würzburg radar systems. He also played a pivotal role in the development of radar and radio technology more broadly, including innovations that were key for German military efforts. His organisational and leadership skills, particularly in the development of radar systems, were vital to the success of the projects in which he was involved.

After the war, Brandt continued to contribute to radar and radio technologies, holding various responsibilities that helped shape the post-war technological landscape. He was appointed Staatssekretär im Ministerium für Wirtschaft und Verkehr des Landes Nordrhein-Westfalen and contributed to reconstruction projects, including railway electrification and the establishment of research institutions.

Brandt can ultimately be seen as one of the key figures in radar development, both for his direct involvement with radar technologies and his leadership in strategic collaboration. His ability to guide complex projects, such as the Würzburg radar system, and his efforts to build bridges between military and civilian research had lasting effects on radar and radio technology during and after the war.

Leo Brandt died on 26 April 1971.

Max Dieckmann

Max Dieckmann (1882–1960)[1711] played a central role in advancing radio measurement technology through his work at the Drahtlostelegraphische und Luftelektrische Versuchsstation Gräfelfing (DVG) and the Flugfunk-Forschungsinstitut Oberpfaffenhofen (FFO). Founded in 1908 and 1937 respectively, these specialised research institutes made significant contributions to the development of transmitter technologies for radio location and navigation, effectively bridging the gap between academic theory and practical application. Dieckmann was instrumental in both institutions, contributing significantly to advances in wireless communication and atmospheric electricity.

Born in Hermannsacker, Germany, in 1882, Dieckmann studied mathematics, chemistry, experimental physics, and electrical engineering at the universities of Leipzig, Göttingen, and Strasbourg. After completing his doctorate in 1907 under Ferdinand Braun in Strasbourg, he focused on the intersection of wireless telegraphy and atmospheric electricity. In 1908, he established the DVG near Munich, which became a hub for research on high-frequency radio waves, atmospheric electricity, and the practical application of wireless communication for aviation, particularly for airships.

During World War I, Dieckmann worked on radio communication and navigation projects for military purposes, including the

1711 Jäger und Heilbronner 2010, pp. 101–102; Zetzmann 1983, pp. 46–48.

wireless transmission of drawings from aircraft to the ground. This research laid the foundation for future advancements in aviation communication.

In 1927, he joined the Technical Advisory Board of the Deutsche Versuchsanstalt für Luftfahrt (DVL) in Berlin-Adlershof. Although the DVL had attempted to recruit him earlier for a high-frequency research division in 1926, the Technical University of Munich, where he was based, did not release him. He remained at the university and, in 1936, was appointed extraordinary professor for radio technology. That same year, the Institute for Radiotechnics and Aeronautical Radio Communication was established at the TU Munich to support his work.

In 1937, together with Claudius Dornier, Dieckmann was involved in the establishment of the Flugfunk-Forschungsinstitut Oberpfaffenhofen (FFO) at the Dornier Works' airfield. The institute focused on aviation radio communication, navigation systems, and radar technology. While Dieckmann's leadership was crucial, Dornier's support was also key to the institute's successful establishment.

In 1942, the DVG was fully integrated into the FFO, which continued to advance radio communication, radar, and navigation technologies, particularly for aviation. In 1956, the FFO was integrated into the Deutsche Versuchsanstalt für Luftfahrt (DVL) as an institute in Oberpfaffenhofen. Following a reorganisation in 1969, the DVL was incorporated into the Deutsche Forschungs- und Versuchsanstalt für Luft- und Raumfahrt (DFVLR), which later became the German Aerospace Center (DLR).

Dieckmann's research and leadership left a lasting impact on radio communication and radar technology. In 1947, he spent a year at Wright-Patterson Airfield in Ohio, USA, before returning to Germany for health reasons. He became emeritus in 1958 but continued to influence the research community.

Max Dieckmann died on 28 July 1960.

Paul-Günther Erbslöh (1905–2002) and Hans-Karl Freiherr von Willisen (1906–1966) were technically skilled and entrepreneurially minded pioneers of early German radar development.[1712] In the

1712 Kroge 2000; Kroge 1998.

early 1930s, they worked in a small workshop in Berlin-Charlottenburg, developing an electric cutter for wax discs used in record production. Their company, Tonographie, founded in 1931, financed itself through commercial recording services for record labels, political organisations, and advertising clients.

A pivotal moment came in late 1932 when a mutual acquaintance introduced them to the Nachrichtenmittel-Versuchsanstalt (NVA) in Kiel. The newly appointed scientific director, Rudolf Kühnhold, was seeking high-powered acoustic impulse generators for submarine detection using sonar. Erbslöh and von Willisen successfully developed a prototype, which was demonstrated in mid-1933. Encouraged by this success, they founded the Gesellschaft für elektroakustische und mechanische Apparate m.b.H. (GEMA) on 16 January 1934 in Berlin, establishing it separately from Tonographie to limit financial risk.

Starting with only two employees and support from freelance collaborators such as Hans E. Hollmann and Theodor Schultes, Erbslöh and von Willisen began developing early radar prototypes, including the system that would later become Seetakt for naval use. GEMA quickly became a key player in the field of radar technology, presenting its first working prototypes in 1935 and becoming an official military contractor in 1937. The company moved to larger facilities in Berlin-Köpenick and significantly expanded its workforce. By 1945, GEMA employed several thousand people.

Erbslöh and von Willisen promoted close collaboration between scientific research, industry, and the military, accelerating technological development. Their work laid the foundation for numerous innovations that continued to influence radar technology after the war. Following World War II, GEMA was dissolved, its facilities either dismantled, destroyed, or placed under Soviet administration. The remaining site in Berlin later became part of the VEB Funkwerk Köpenick, one of East Germany's leading communications technology companies.

The legacy of Paul-Günther Erbslöh and Hans-Karl Freiherr von Willisen lies in their technical ingenuity, entrepreneurial vision, and their ability to translate scientific expertise into practical radar systems. They are recognised as key figures in the early development of radar in Germany.

Hans-Karl Freiherr von Willisen died on 26 January 1966, and Paul-Günther Erbslöh on 18 February 2002.

Abraham Esau

Abraham Esau (1884–1955)[1713] was a prominent German physicist and pioneer in ultrashort wave technology. After completing his studies in physics at the Technical University (TH) of Berlin and the TH of Danzig, where he earned his doctorate in 1908, he began his early career at Telefunken. In 1913, on behalf of Telefunken, he established the Kamina radio station in Togo as part of building a wireless communication network between the German colonies and Germany. He was caught there by the outbreak of World War I in 1914 and remained in French captivity until 1919.

In 1925, Esau joined the University of Jena, where he advanced research on ultrashort waves, a key development that later contributed to the creation of VHF technology. While he also explored medical applications of electromagnetic waves, such as shortwave therapy, these played a minor role in his scientific and political significance.

Esau's major pioneering contributions were in centimetre and millimetre wave technology. He investigated the propagation, generation, and application of wavelengths in the centimetre range and developed powerful magnetrons capable of producing wavelengths below two millimetres. These breakthroughs laid the foundation for modern radar and microwave technologies, which became crucial during World War II and in the postwar period.

During the Nazi era, Esau increasingly assumed political roles. He served twice as rector of the University of Jena, joined the Nazi Party in 1933, and became a member of the Reich Cultural Senate in 1935. In 1939, he was appointed president of the Physikalisch-Technische Reichsanstalt (PTR) in Berlin and professor at the Technical University of Berlin. In 1942, he was named Bevollmächtigter des Reichsmarschalls für Kernphysik but handed over this position to Walther Gerlach at the end of 1943. Despite relinquishing this role, Esau remained president of the PTR until the end of the war in 1945.

1713 Jäger und Heilbronner 2010, p. 120; Schröter 1959, pp. 640–641; Hoffmann und Stutz 2003, pp. 136–179; Mauel 1983b, pp. 52–53; Seiler 2007, pp. 136–139.

In December 1943, following the removal of Plendl, Esau was appointed Bevollmächtigter für die Hochfrequenzforschung (BHF), overseeing military-relevant research in high-frequency technologies. His focus shifted increasingly to radar and microwave technologies, particularly in the centimetre and millimetre wave ranges, which were of high military significance during World War II.

Due to Allied air raids, the PTR was relocated to multiple sites across Germany, being effectively dismantled by 1945.

After the war, Esau was imprisoned but later denazified. He resumed academic work as an honorary professor at Aachen University in 1949. From 1953 until his death in 1955, he headed the Institute for High-Frequency Technology at the Deutschen Versuchsanstalt für Luftfahrt (DVL) in Mülheim an der Ruhr.

Esau's career exemplifies the close connection between scientific innovation and National Socialist politics, particularly in the fields of high-frequency technology and nuclear research.

Abraham Esau died on 12 May 1955.

Hans Erich Hollmann

Hans E. Hollmann (1899–1960)[1714] was a pioneering German scientist in high-frequency and radio measurement technology. In 1927, at the Physics Institute of the Technische Hochschule Darmstadt, he developed the first very high-frequency transmitter and receiver for centimetre and decimetre waves. He received his doctorate in 1928 with a thesis on the generation of very high-frequency electrical oscillations.

Hollmann continued his research at the Deutsche Forschungsgemeinschaft (DFG) and later at the Heinrich-Hertz-Institut in Berlin. There, together with Gustav Leithäuser, he succeeded in generating and receiving decimetre waves, a crucial milestone for later radar development. In 1933, he completed his Habilitation at the Technische Hochschule Berlin and began working at Telefunken, while also consulting for Berlin Tonographie GmbH, contributing to early high-frequency devices and concepts for electromagnetic echo location.

1714 Frühauf 1959, pp. 141–143; Küpfmüller 1972, p. 543; Watson 2009, pp. 252–253.

These efforts contributed to the founding of the Gesellschaft für elektroakustische und mechanische Apparate m.b.H. (GEMA) in 1934. Hollmann, together with Hans-Karl Freiherr von Willisen, Paul-Günther Erbslöh, and Theodor Schultes, was among the first scientific staff, laying the foundations for early German radar devices, including the Seetakt naval radar and the Freya radar.

In 1936, he published "Die ultrakurzen Wellen in der Technik", exploring the concept of "Sehen mit elektrischen Wellen" and demonstrating the potential of electromagnetic detection. Due to military secrecy regulations, much of his work was censored, prompting Hollmann to end his cooperation with GEMA. He then established the Laboratorium für Hochfrequenztechnik und Elektromedizin in Berlin-Lichterfelde, collaborating with his brother on technical and medical applications of high-frequency research.

In 1942, Hollmann was appointed Scientific Director of the Forschungsgesellschaft für Funk- und Tonfilmtechnik e. V. in Berlin, but he was soon prohibited from working on short-wavelength research due to state restrictions, reflecting the strict military control over radar-related work.

After the war, he was appointed Associate Professor of specialised Hochfrequenztechnik at the Universität Jena. Shortly thereafter, he emigrated to the United States, where he worked for the US Navy and the company Hydro-Aire in California, continuing his pioneering research and gaining international recognition.

Hans Erich Hollmann died on 19 November 1960.

Rudolf Kühnhold

Rudolf Kühnhold (1903–1992)[1715] was a pioneering figure in the early development of radar and sonar technology in Germany. After joining the Torpedo-Versuchsanstalt (TVA) of the Reichsmarine in 1929, he rose to become scientific director of the Nachrichtenmittel-Versuchsanstalt (NVA) of the Reichsmarine in early 1933. In the autumn of that year, he began experimenting with the reflection of electromagnetic waves to detect ships, laying the groundwork for German radar research.

Though not a commanding figure, Kühnhold was instrumental in facilitating cooperation between the Navy and industry. At the end

---

1715 Weinreich 2001g, pp. 108–109; Surhone 2010, pp. 1–2.

of 1933, he held a meeting with Wilhelm Runge of Telefunken to explore support for further radar development. However, the company showed little interest at the time, which prompted Kühnhold to continue the project with other partners. His close collaboration with Paul-Günther Erbslöh and Hans-Karl Freiherr von Willisen, both key figures in the emerging electroacoustic industry, facilitated the establishment of the Gesellschaft für elektroakustische und mechanische Apparate (GEMA) in January 1934. This company played a pivotal role in the following period as a dedicated research and development centre for radar and acoustic detection technologies, becoming instrumental in advancing Germany's early radar capabilities.

By 1935, Kühnhold was already advocating the use of centimetre waves for higher accuracy in target detection, and early experimental work began at Pintsch, though technical limitations delayed further progress. Between 1936 and 1937, he contributed to radio navigation systems, including work on the Knickebein beam guidance system. From just before the war until 1943, Kühnhold conducted microwave research at the NVK, the successor to the NVA. Despite his expertise in magnetron technology, he, like Hans Erich Hollmann, was not involved with the Arbeitsgemeinschaft Rotterdam group.

In the final war years, Kühnhold returned to underwater acoustics and worked closely with ELAC in Kiel. After the war, he resumed research in commercial radar at ELAC, leading to patents including one registered in the United States in 1954. His career there ended in the 1960s following ELAC's financial decline and the sale of its nautical division.

Rudolf Kühnhold died in 1992.

Karl Küpfmüller

Karl Küpfmüller (1897–1977)[1716] was a prominent German electrical engineer whose work had a lasting impact on telecommunications, radar, and system theory, particularly during and after World War II. Born in Nuremberg, he studied electrical engineering from 1913 to 1919 before joining the Telegraphen-Versuchsamt in Berlin. In 1921, he moved to Siemens & Halske AG, contributing to signif-

1716 Jäger und Heilbronner 2010, pp. 250–251; Schoen 1994, pp. 96–102; Mielert 1982; Mauel 1983c, pp. 101–102; Hagenauer und Pabst 2014.

icant advancements in telecommunications. By 1928, Küpfmüller was appointed professor at the Technical University of Danzig and later at the Technical University of Berlin.

In 1937, he returned to Siemens as head of the telecommunications development department. That same year, he joined the NSDAP, and in 1942, he became head of the Kommission für Nachrichtenmittel at the office of the Reichsleiter für Rüstung und Kriegsproduktion, overseeing the development and standardisation of military communication technology, with a focus on high-frequency and radar systems critical to the war effort. In January 1944, Küpfmüller was appointed head of the newly established Wissenschaftlicher Führungsstab beim Oberkommando der Kriegsmarine (WFM) by Dönitz, where he coordinated the Navy's research capabilities, particularly in high-frequency technology, allocating research contracts to institutions and companies.

Küpfmüller played a central role in coordinating radar and communication technology efforts between the Air Force and the Kriegsmarine. His close cooperation with the BHF ensured that technological advances, especially in radar and radio equipment, were shared and optimised across both branches. This coordination was crucial for improving U-boat detection and targeting systems, enhancing Germany's operational effectiveness.

From May 1944, he also became a member of the Hauptkommission Elektrotechnik within the RMfRuK, overseeing electrical engineering technologies with military applications. Later that year, he headed the Technical Advisory Board to the Chief of Naval Armaments and joined the Wissenschaftlicher Führungsstab of the Wehrforschungs-Gemeinschaft.

After the war, Küpfmüller continued to influence electrical engineering in Germany, although public recognition declined due to his wartime associations with the Nazi regime. Nevertheless, his technical contributions left a lasting impact on the development of German electrical engineering in the early Federal Republic.

Karl Küpfmüller died on 26 December 1977.

Wilhelm Tolmé Runge

Wilhelm Runge (1895–1987)[1717] was a German physicist and electrical engineer who played a key role in the development of radar technology in Germany. After studying mathematics, physics, and chemistry in Göttingen, followed by electrical engineering at the Technical University of Darmstadt, he began working in 1923 at the receiver laboratory of the Telefunken Gesellschaft für drahtlose Telegraphie m.b.H. In the autumn of 1924, he became head of the laboratory. He is considered a pioneer in receiver technology, radar, and radio relay systems.

In the summer of 1935, Runge successfully carried out his own aircraft detection experiments, identifying a Ju 52 at an altitude of 5,000 metres. From 1936 to 1942, he served as chief engineer, heading Telefunken's high-frequency laboratories, where he was responsible for nearly all technical device development, with the exception of radio receivers.. He made significant contributions to the development of the Würzburg and Lichtenstein radar systems.

During the war, Runge not only played a leading role in radar development but also contributed to the advancement of essential components, which significantly enhanced radar performance. These innovations were critical to improving the accuracy and efficiency of German radar systems in military applications. Following a company restructuring, Runge was relieved of his active duties at Telefunken and, on 1 November 1944, assumed leadership of the Institut für Elektrophysik der Deutschen Versuchsanstalt für Luftfahrt (DVL) in Berlin-Adlershof. In 1943, he took part in four meetings, and in 1944 in two meetings, of the Arbeitsgemeinschaft Rotterdam as Telefunken's technical representative.

After the war, Runge played a leading role in rebuilding Telefunken. As head of the company's Technical Central Department, he reorganised development efforts and, in 1955, established a new Telefunken research institute in Ulm focused on modern communication and radar technologies. Over the course of his career, Runge held more than one hundred patents in the field of electrical engineering.

Wilhelm Runge died on 9 June 1987.

1717 Jäger und Heilbronner 2010, pp. 371–372; Rangnick 2005, pp. 261–262; Weinreich 2001e, pp. 105–106; Nestel 1983, pp. 151–153.

Theodor Jakob Joseph Schultes

Theodor Schultes (1901–1981)[1718] was a leading German engineer in high-frequency and radar technology. He studied electrical engineering at the Technical University of Darmstadt from 1921 to 1926 and completed his doctorate in 1934 while working at the Heinrich-Hertz-Institut for Oscillation Research in Berlin, specialising in room acoustics and electroacoustic applications. In the early 1930s, he also acted as a consultant for Tonographie GmbH, a company developing electric cutters for records, where he first collaborated with Hans E. Hollmann.

In June 1934, Schultes joined the newly founded Gesellschaft für elektroakustische und mechanische Apparate (GEMA) as head of the research laboratory. Together with Hollmann, he played a crucial role in the early development of radar-like technologies, combining theoretical innovation with practical engineering. Their work laid the foundation for the Seetakt naval radar and the Freya mobile aircraft detection system, which significantly advanced German radar capabilities prior to World War II.

By 1937, Schultes had been a key contributor to the development of the Freya system, capable of detecting aircraft at ranges of up to 100 kilometres. During the war, he continued to drive radar development at Siemens & Halske in Berlin, leading the creation of the Wassermann long-range radar in 1940, which extended detection ranges to 250 kilometres, and the Jagdschloß panoramic radar system in 1943, providing comprehensive air warning.

Following World War II, Allied restrictions initially prohibited radar research in Germany, and Schultes focused on electromedicine and electroacoustic research. From 1948, he gradually returned to radar-related projects, first at Siemens' subsidiary Albiswerk in Zurich, and from 1952 at Siemens in Munich, eventually serving as plant manager and technical director at the Industrieanlagen-Betriebsgesellschaft mbH (IABG) until his retirement in 1971.

Throughout his career, Schultes' technical expertise, collaboration with Hollmann, and close cooperation with the founders of GEMA were central to advancing German radar technology before and during the war, laying important foundations for post-war de-

1718 Jäger und Heilbronner 2010, pp. 391–392; Weiher 1983b, pp. 171–173; Weinreich 2001a, pp. 106–108.

velopments. His work continued to influence radar and high-frequency engineering long after the war, bridging the gap between wartime innovation and peacetime technological progress.

Theodor Schultes died on 16 March 1981.

11.1.1.3 Actors and Perspetives

Key figures from politics, the military, government ministries, industry, and scientific research drove the development of Germany's rearmament and military technologies in the 1930s. To this day, historians debate the influence and motivations of these actors, reflecting how different perspectives and historical contexts shape their portrayal. Some highlight Adolf Hitler and leading Nazi officials as the main drivers of policy and strategic decisions, including Hermann Göring, who oversaw the Air Force and later the Vierjahresplan, and Albert Speer, who directed armaments production as Reichsminister für Bewaffnung und Munition. Others focus on the military, engineers, and technical experts who implemented rearmament plans, designed weapons systems, and organised logistics. Interpretations of these roles often depend on the sources consulted, ranging from official government and military documents to personal memoirs and post-war analyses. These discussions raise broader questions of responsibility, initiative, and the balance between personal ambition and institutional constraints. The range of perspectives illustrates the complex interplay of political, ideological, technological, and administrative forces that shaped Germany's rapid militarisation in the years leading up to World War II.

11.1.2 Great Britain

11.1.2.1 Politics, Military, Science, and Industry

Winston Leonard Spencer Churchill
Winston Churchill (1874–1965)[1719] was the most prominent British politician between the two World Wars and during World War II. He played a central role in developing countermeasures against German air raids, both politically and technically. Churchill worked closely with his scientific adviser Frederick Lindemann to influence the government's research agenda in air defence. Initially sceptical of radar and other technologies, he quickly recognised their potential, particularly through the early results of the Chain Home radar stations, which enabled timely detection of enemy air raids.

Churchill's support was crucial for the further development of Britain's air defence system. He warned early about the unpredictability of the Nazi regime and opposed the government's appeasement policy, emphasising the need for a strong defence state. Between 1930 and 1939, Churchill held no ministerial office but remained politically active as a Member of Parliament. As a member of the Air Defence Research Committee (ADRC) from 1935 to 1938, he was able to influence defence policy and promote cooperation between the military, political authorities, and scientists. He supported technological innovations such as air mines, parachute bombs, and infrared detectors and later became a strong advocate for radar, which underpinned the implementation of the Chain Home network.

Together with other key figures such as Harry Wimperis and Hugh Dowding, Churchill laid the foundation for an effective air defence system. Under his leadership, Britain achieved a decisive technological advantage, which proved crucial during the Battle of Britain in 1940. By combining political foresight with support for new technologies, Churchill not only strengthened Britain's air defence but also set standards for modern warfare that provided strategic advantages for the Allies.

Winston Churchill died on 24 January 1965.

1719 Addison 2018; Roberts 2019.

Edward Victor Appleton

Edward Appleton (1892–1965)[1720] was a distinguished British physicist, widely recognised for his pioneering work on the ionosphere and its role in radio-wave propagation. Born in Bradford, West Yorkshire, he demonstrated an early talent for science and mathematics at Hanson Grammar School. Awarded a scholarship to St John's College, Cambridge, he excelled in the Natural Sciences Tripos, earning first-class honours, and began research under W. L. Bragg at the Cavendish Laboratory, one of the leading centres of modern physics.

During World War I, Appleton served with the Royal Engineers as a signals officer. This experience introduced him to thermionic valves and the challenges of radio-wave propagation, knowledge that would prove vital in his later work on long-distance communication and radar technology. In 1924, he experimentally confirmed the existence of the Kennelly-Heaviside layer, an ionised layer in the upper atmosphere that reflects radio waves. This discovery was fundamental to understanding long-distance radio transmission and laid the foundation for the practical development of radar. Between 1924 and 1929, he developed systematic methods to measure critical radio frequencies and investigate the electrical structure of the ionosphere, directly influencing early radar signal analysis.

In 1936, Appleton joined the CSSAD, replacing Frederick Lindemann, where he contributed to radar research and helped coordinate efforts between scientists, military planners, and government officials until the committee's dissolution in 1940. From 1939 to 1949, he served as Secretary of the DSIR, overseeing wartime projects, including radar development and the Tube Alloys nuclear programme, acting as a crucial link between research teams and government decision-makers. He also encouraged interdisciplinary collaboration, recognising the importance of combining physics, engineering, and operational planning for national defence, particularly in the optimisation of radar networks.

Appleton's research not only advanced the scientific understanding of the ionosphere but also had lasting practical applications in radio communication, radar detection, navigation, and defence

1720 Ratcliffe and Hutchins 2012; Clark 1971.

systems. His mentorship of younger scientists ensured the continued development of atmospheric physics in the UK for decades, and his insights influenced both civilian and military technologies.

In recognition of his groundbreaking contributions to physics and radar-related research, Appleton was awarded the Nobel Prize in Physics in 1947. After the war, he returned to academia as Principal and Vice-Chancellor of the University of Edinburgh, where he remained an inspiring leader, fostering international scientific collaboration and continuing his lifelong study of the ionosphere and radar propagation.

Edward Appleton died on 21 April 1965.

Patrick Maynard Stuart Blackett, Baron Blackett

Patrick Blackett (1897–1974)[1721] studied physics at the University of Cambridge and served as an officer in the Royal Navy during World War I. After the war, he began research at the Cavendish Laboratory under Ernest Rutherford in 1920, specialising in cosmic radiation and nuclear physics. He completed his doctorate in 1924 and later held professorships at Birkbeck College from 1933 and at the University of Manchester from 1937, where he became known for his innovative approach combining theory and experiment.

Already in the 1930s, Blackett combined scientific work with military practice. In 1935, he became a member of the CSSAD, emphasising his role as a scientist with military experience and a practical understanding of warfare, as well as a visionary approach to applying science to defence challenges.

In August 1940, Blackett became scientific adviser to the Commander-in-Chief of the Anti-Aircraft Command. He contributed significantly to strategic planning and the enhancement of British air defence, particularly through innovative use of radar and the application of operational research techniques that integrated statistical methods into military decision-making.

In March 1941, Blackett joined Coastal Command, leading a team that recalculated depth settings for anti-submarine explosives. Using mathematical methods, including Poisson's distribution, he resolved tactical and strategic disputes. His scientific ap-

1721 Nye 2004; Kirby 2003.

proach greatly improved anti-submarine warfare and cemented operational research as an essential military tool.

Simultaneously, Blackett led an interdisciplinary working group at RAF Coastal Command, later known as Blackett's Circus, which optimised military tactics using statistical analysis and improved radar systems. From 1942 to 1945, he headed the Operational Research Section at the Admiralty, where his analyses shaped wartime and post-war military research.

During the war, Blackett criticised the strategic bombing of German cities, advocated targeted strategies, and influenced government military thinking with his rigorous scientific arguments and data-driven assessments.

In 1948, he was awarded the Nobel Prize in Physics for his work on the Wilson cloud chamber, enabling major progress in nuclear physics and cosmic radiation studies. After the war, Blackett promoted peaceful atomic energy use and warned against nuclear arms escalation.

From 1965 to 1970, as President of the Royal Society, he emphasised the link between science, politics, and social responsibility. His legacy endures, including at the Blackett Laboratory at Imperial College London, demonstrating that scientists should engage both in research and societal responsibility, influencing generations of physicists and policymakers alike.

Patrick Blackett died on 13 July 1974.

Alan Dower Blumlein

Alan Blumlein (1903–1942)[1722] was one of Britain's most brilliant and versatile electronics engineers of the 20th century. Educated at Highgate School and Imperial College London, he graduated in 1923 with first-class honours in electrical engineering. Over just 18 years of professional life, he was awarded 128 patents across diverse fields, including telecommunications, sound recording, television, and especially radar technology.

Blumlein began his career at International Western Electric, where he developed key innovations in telephony. Later, at Columbia Graphophone, he created a superior monaural recording

1722 Burns 2014; Jäger und Heilbronner 2010, p. 54.

system and pioneered stereophonic sound with a landmark 1931 patent, decades ahead of its time. After the formation of EMI, he contributed to the world's first high-definition television service, inventing components ranging from wide-band amplifiers to camera tubes.

His most critical work came during World War II. In 1939, he and colleague E. L. C. White privately developed a 60 MHz radar system, anticipating wartime needs. Blumlein played a key role in the advancement of Air Interception (AI) radar, solving the short-range detection problem and helping engineer the AI Mark IV system, which was used from 1941 to counter German night bombers. He also co-developed the AI Mark VI, adapted for single-seat fighter aircraft.

Blumlein was central to the development of H_2S radar, the first airborne ground-scanning radar, used for navigation and bombing accuracy. Working with Bernard Lovell at the TRE, he helped adapt H_2S into the Air to Surface Vessel (ASV) radar, which proved vital in detecting German U-boats during the Battle of the Atlantic. His delay-line circuits, automatic strobe-following systems, and other innovations also enhanced systems like the GL Mark III gun-laying radar, used in British air defence.

On 7 June 1942, Blumlein was killed in a plane crash over the Wye Valley while testing an H_2S radar system, at the age of just 38. Despite his early death, his legacy endures. As EMI's research director Isaac Shoenberg said, "There was not a single subject to which he turned his mind that he did not enrich extensively."

Alan Blumlein remains a central figure in the history of electronics and radar development. His technical innovations were crucial to Britain's air defence, submarine detection, and strategic bombing during World War II. Beyond their wartime significance, his ideas laid the foundation for modern communication and sound recording technologies.

Edward George Bowen

Edward Bowen (1911–1991)[1723] was one of the most important pioneers of radar technology in the 20th century and played a key role

1723 Minnett 2015; Bowen 1987; Watson 2009, p. 47.

in the development of airborne radar. After completing his physics studies in 1930, he initially conducted research on X-rays and the structural analysis of metals. In 1931, he earned his Master of Science (MSc) and later completed his PhD under Professor Edward Appleton at King's College London.

In 1933 and 1934, Bowen worked at the Radio Research Station in Slough, developing a cathode ray direction-finding device. This work brought him to the attention of Robert Watson Watt, who recruited him into his team in 1935. That year, Bowen joined Watson Watt's team at Orfordness, working on the technical implementation of early radar systems for detecting aircraft. In June 1935, the team achieved the first successful radar detection of an aircraft under realistic conditions. These experiments led to the development of the Chain Home radar network starting in 1936, which by 1939 had become fully operational as part of Britain's air defence.

After relocating to Bawdsey Manor in 1936, Bowen focused on developing airborne radar, aiming to miniaturise the large and heavy equipment for fighter aircraft. By autumn 1937, a demonstration proved that airborne radar could effectively detect enemy aircraft and ships, laying the foundation for its operational use in World War II. Bowen's work enabled night-fighter and coastal patrol aircraft to detect enemy planes and vessels at a distance, giving the Royal Air Force a vital technological advantage.

With the outbreak of World War II, Bowen's unit moved to St Athan in South Wales, where he continued developing radar for ship and submarine detection. A breakthrough came in February 1940 with the cavity magnetron, developed by John Randall and Henry Boot, which allowed the creation of compact and powerful airborne radar systems. Bowen was also part of the British Tizard Mission to North America in 1940, sharing advanced radar technologies with the United States and Canada.

From 1941, Allied aircraft increasingly used airborne radar, giving them a decisive advantage in the Battle of the Atlantic. After the war, Bowen worked in Australia for CSIR, heading the Division of Radiophysics from 1946. He advanced civilian applications of radar, conducted particle physics research, and promoted scientific education.

Edward Bowen was not only an outstanding scientist but also a bridge-builder between research, the military, and international collaboration. His work on airborne radar, in particular, proved decisive for Britain's wartime air defence.

Edward Bowen died on 12 August 1991.

Hugh Caswall Tremenheere Dowding, 1st Baron Dowding

Hugh Dowding (1882–1970)[1724] was an officer in the Royal Air Force (RAF) who played a pivotal role in Britain's air defence during World War II. In 1933, he was promoted to Air Marshal, and in July 1936, he became the commander of the newly created Fighter Command. Dowding's leadership was marked by his trust in scientists and technological innovation, particularly in radar, which he saw as essential to modern air defence.

He actively supported early experiments led by Henry Tizard in 1935–1937, known as the Biggin Hill Experiments, aimed at integrating all available information for optimal use of defensive forces. Dowding's collaboration with Tizard and other scientists led to the creation of the Dowding System, the world's first integrated air interception network. This system connected coastal radar stations with a central command, enabling near-continuous monitoring of Britain's airspace and precise coordination of fighter interceptors. Its effectiveness was fully demonstrated during the Battle of Britain, where early radar warnings allowed the RAF to successfully repel German air attacks.

Despite his critical role in Britain's defence, Dowding was cautious about committing full RAF support to France during the German invasion in 1940, believing that resources needed to be concentrated for the forthcoming Battle of Britain. This decision caused disappointment among the French and remains a controversial aspect of his career.

Following Britain's successful defence, Dowding relinquished command on 24 November 1940. Winston Churchill then assigned him to lead a British mission to the United States focused on procuring new aircraft. He subsequently took on administrative roles within the RAF until his retirement in 1942.

Dowding is widely regarded as the architect of the victory in the Battle of Britain. His foresight in integrating technology, particu-

1724 Orange 2008.

larly radar, into military strategy, and his emphasis on collaboration between scientists, the military, and political leadership, were decisive in Britain's success. His career exemplifies visionary leadership and the effective application of scientific innovation in warfare. He was also admired for his calm demeanour under pressure, which helped maintain morale among pilots and staff during Britain's darkest hours. Dowding's emphasis on rigorous training and disciplined coordination of Fighter Command ensured that pilots could respond swiftly and effectively to enemy raids. His support for innovation encouraged the adoption of new aircraft, tactics, and technologies, which collectively strengthened Britain's air defence.

His career left a lasting legacy as a strategist who combined operational acumen with trust in technological progress.

Hugh Dowding died on 15 February 1970.

Robert Hanbury Brown

Robert Hanbury Brown (1916–2002)[1725], known to his friends as Hanbury, showed an early interest in science and electrical engineering. He was born on 31 August 1916 in Aruvankadu, India, where his father served in the British Indian Army. After returning to England, he studied at Tonbridge School and then Brighton Technical College, earning a BSc with first-class honours in 1935. He began postgraduate research at Imperial College London, focusing on oscillators, before being recruited in 1936 by Henry Tizard to join Robert Watson-Watt's radar project at the Radio Research Station in Slough.

Hanbury Brown was involved in installing the first Chain Home coastal radar at Dunkirk, Kent, which played a crucial role during the Battle of Britain. He later worked with Edward Bowen on airborne radar, demonstrating the first operational system capable of detecting ships and aircraft in poor visibility in 1937. His expertise in flight testing and radar operational techniques helped train pilots and improve interception procedures. He was widely respected by both scientists and military personnel for his practical ingenuity and meticulous attention to detail.

A 1941 flying accident ended his flying career, after which he joined J.W.S. Pringle's group to develop the Rebecca-Eureka

1725 Graham-Smith 2015; Davis and Lovell 2003, pp. 83–106.

transponder system, later vital for guiding D-Day airborne operations. He also contributed to extending the system for all Allied forces in North America.

After the war, Hanbury Brown returned to the TRE, applying navigation aids to civil aviation. In the late 1940s, he shifted to radio astronomy, collaborating with Bernard Lovell at Jodrell Bank to develop radio interferometry, a technique for measuring the angular size of radio sources. From 1964 to 1981, he was Professor of Physics and Astronomy at the University of Sydney, making significant contributions to intensity interferometry. He also served as President of the International Astronomical Union from 1982 to 1985.

Hanbury Brown's work transformed radar from experimental technology to an operational system, greatly contributing to Britain's air defence and the advancement of both military and civil radar techniques. His pioneering efforts in radio astronomy laid the foundation for many modern observational methods. His combination of theoretical insight, practical engineering, and dedication ensured that his contributions had lasting impact across both military and scientific communities. He was admired not only for his technical skill but also for his collaborative spirit.

Robert Hanbury Brown died on 16 January 2002.

Maurice Pascal Alers Hankey, 1st Baron Hankey

Maurice Hankey (1877–1963)[1726] was a British civil servant who became one of the most influential figures in government during both World Wars. Born in Biarritz, France, as the third son of R.A. Hankey, he was educated at Rugby School before joining the Royal Marine Artillery. In 1908, he was appointed Naval Assistant Secretary to the Committee of Imperial Defence (C.I.D), becoming Secretary of the C.I.D in 1912, a post he held until 1938. This role placed him at the centre of Britain's defence planning and gave him vital experience during World War I.

During the World War I, Hankey coordinated military efforts and liaised between political and military leaders, earning the trust of Prime Minister David Lloyd George. He was instrumental in estab-

1726 Naylor 2008.

lishing the War Cabinet and was appointed its Secretary, playing a central role in government decision-making. His efficiency and organisational skill earned him a reputation as one of Britain's most capable civil servants.

In the interwar years, Hankey continued to influence defence policy, advising successive Prime Ministers and coordinating strategy with the British dominions and colonies. He maintained close contact with scientific advisers such as Frederick Lindemann, recognising the importance of modernising Britain's air defence.

Hankey retired in 1938 but returned in 1939 when Neville Chamberlain appointed him Minister without Portfolio and member of the War Cabinet. He oversaw intelligence activities and ensured that scientific and technological developments supported the war effort. Despite increasing tensions with Winston Churchill, Hankey remained a respected adviser. He was appointed Paymaster General in 1941 but was removed from the Cabinet in 1942 due to dissenting views.

After the war, Hankey became a critic of the Allied war crimes trials, arguing that the Allies had no right to convict German and Japanese leaders. His extensive diaries, documenting his work throughout his career, are now held at the Churchill Archives Centre in Cambridge and are a valuable resource for historians.

Hankey's career was marked by his ability to navigate complex political landscapes, his administrative talent, and his influence on Britain's defence policy. He was raised to the peerage in 1939 and became a member of the Privy Council. His work ensured that Britain maintained coherent strategy and scientific foresight in both world wars, leaving a lasting legacy in 20th-century British governance.

Maurice Hankey died on 26 January 1963.

Archibald Vivian Hill

Archibald Hill (1886–1977)[1727] was a British physiologist, mathematician, and pioneer of biophysics. He gained international recognition not only for his research on heat production in muscles but also for his contributions to military science and air defence. Educated in mathematics at Trinity College, Cambridge, Hill turned to physiology under the influence of Walter Morley Fletcher and Fred-

1727 Katz 2009.

erick Gowland Hopkins. By applying physico-chemical methods to biological problems, he laid the foundation for modern biophysics.

Even before World War I, Hill had made theoretical contributions to muscle and nerve physiology and developed precise thermal measurement techniques. When the war broke out in 1914, he volunteered for the army and soon led experimental anti-aircraft research under the Ministry of Munitions. There, he headed an interdisciplinary team, applying scientific expertise to improve British air defence.

After the war, Hill returned to academic life. In 1922, he was awarded the Nobel Prize in Physiology or Medicine, shared with Otto Meyerhof, for their discoveries in muscle physiology. Hill remained interested in applying science to public needs. In 1935, he became a member of the CSSAD, chaired by Henry Tizard. The committee evaluated and improved Britain's air defence capabilities. Hill scientifically validated early warning systems and conceptualised air defence as an integrated structure involving radar stations, command centres, interceptor aircraft, and anti-aircraft artillery.

While Robert Watson-Watt led the technical development of radar, Hill contributed through mathematical modelling and empirical analysis, helping to optimise deployment and coordination. His work within the CSSAD was essential in shaping the Dowding System, the world's first fully integrated air defence network.

During World War II, Hill continued to serve in key positions. He was an independent Member of Parliament for Cambridge University, undertook scientific missions to the United States, and advised the Indian government on postwar reconstruction. He also worked to support displaced scholars under the Nazi regime as a founding member of the Academic Assistance Council.

Hill's contributions to air defence and radar science were crucial to Britain's preparedness for aerial warfare. His ability to bridge scientific research with military planning made him a key figure in science-based national security policy. His humanitarian efforts and dedication to the practical application of science exemplified the wider social responsibility of scientists in the twentieth century. Hill's work influenced both science and military strategy for decades.

Archibald Hill died on 3 June 1977.

Reginald Victor Jones

Reginald Jones (1911–1997)[1728] was a British physicist and one of the most important scientific intelligence experts of World War II. He was born in Herne Hill, South London, and studied Natural Sciences at Wadham College, Oxford, graduating with first-class honours in 1932. He completed his doctorate in 1934 at the Clarendon Laboratory under Professor Frederick Lindemann. After a research fellowship in astronomy at Oxford, Jones joined the Royal Aircraft Establishment in Farnborough in 1936, focusing on air defence.

On 1 September 1939, he took up a newly created position at the British Air Ministry as Assistant Director of Intelligence Science, the first role specifically dedicated to systematic scientific and technical intelligence. Throughout the war, Jones analysed German weapons and guidance technologies, particularly radar, radio navigation, and rockets, and developed effective countermeasures. He played a key role in detecting and disrupting the German Knickebein system during the Battle of the Beams and recognised early the potential to deceive enemy radar using metal foil strips known as Window, later called Chaff.

Jones was known for his pragmatic and unconventional approach, often using wit and ingenuity to solve complex problems. He also assessed the so-called Oslo Report, an anonymous warning about German high technology, recognising its strategic significance. His work encompassed defending against V-weapons and orchestrating deliberate misinformation within the Double-Cross system. His brief period at Bletchley Park strengthened the connection between scientific analysis and codebreaking efforts, fostering collaboration essential to Allied success.

After the war, Jones declined a career in intelligence and accepted a chair in Natural Philosophy at the University of Aberdeen in 1946, which he held until 1981. There, he focused on developing highly precise scientific instruments. He was appointed Commander of the Order of the British Empire (CBE) in 1942, Companion of the Order of the Bath (CB) in 1946, and Companion of Honour in 1994. In 1965, he was elected a Fellow of the Royal Society.

Reginald Jones is remembered as a brilliant scientist and pioneer of scientific intelligence. His analytical mind, intuition for techni-

1728 Cook 1999, pp. 239–254; Martin 2011.

cal developments, and ability to translate scientific findings into strategic action made him a key figure in Britain's defence during World War II. His work ensured Britain could respond to Nazi technological threats and preserved operational flexibility. Jones's legacy profoundly shaped modern military intelligence and technology assessment, illustrating how scientific clarity and personal courage can influence history.

Reginald Jones died on 17 December 1997.

Frederick Alexander Lindemann, 1st Viscount Cherwell

Frederick Lindemann (1886–1957)[1729], later Baron (1941) and Viscount (1956) Cherwell , was one of Britain's most influential scientific advisers during World War II and a close confidant of Winston Churchill. He earned his doctorate in physical chemistry in 1910 under Professor Walther Nernst in Berlin. During World War I, Lindemann joined the Royal Flying Corps, gaining recognition for his research on aircraft spin recovery, establishing his reputation in aeronautical engineering.

In 1919, he was appointed to the Chair of Physics at the University of Oxford and became director of the Clarendon Laboratory, marking the start of a distinguished academic career. Though he did not make groundbreaking discoveries, his analytical mind, problem-solving skills, and versatility made him a sought-after scientific adviser.

Lindemann met Winston Churchill in 1921, forming a lifelong friendship. By the 1930s, he had become Churchill's most trusted scientific adviser. When Churchill became Prime Minister in May 1940, Lindemann was appointed his personal scientific adviser, influencing key strategic and technical wartime decisions, including matters of military technology and air defence.

During the war, Lindemann played a complex role in the development of radar. While he recognised its potential, he approached the technology with pragmatic caution, concerned about its technical challenges and resource demands. He supported alternative methods, such as aerial mines and experimental infrared sensors, and often clashed with Henry Tizard within the Committee for the

1729 Blake 2008; Birkenhead 1961; Burns 2013.

Scientific Survey of Air Defence (CSSAD), leading to temporary restructuring of the committee. Despite his scepticism, Lindemann did not oppose radar; he evaluated it realistically and ensured that resources were allocated efficiently within Britain's wartime economy.

Lindemann's influence helped bridge military leadership and scientific research, ensuring that multiple avenues of technological innovation were pursued. His advice was instrumental in securing a strategic advantage for Britain, and British radar ultimately became decisive during the Battle of Britain.

In 1941, he was elevated to the peerage as Baron Cherwell and, in 1956, promoted to Viscount Cherwell. Lindemann remained a trusted adviser to Churchill throughout the war and contributed to the integration of science into British military and political strategy. His ability to understand and communicate complex technical issues made him indispensable, even when his positions were controversial.

His legacy is that of a pragmatic, influential scientific adviser whose guidance helped shape Britain's wartime technological and strategic decisions.

Frederick Lindemann died on 3 July 1957.

Alfred Charles Bernard Lovell

Bernard Lovell (1913–2012)[1730] was an English physicist and radio astronomer, known for his pioneering work in radar during World War II and for founding the Jodrell Bank Observatory. He is professionally known by his second given name, Bernard Lovell.

Lovell was born in Oldland Common, Bristol, in 1913. He studied physics at the University of Bristol, earning a BSc in 1934 and completing a PhD in 1936 on the electrical conductivity of thin metal films. Before World War II, he conducted cosmic ray research at the University of Manchester.

At the outbreak of the war, Patrick Blackett recruited him to contribute to radar development. Lovell trained operators on the Chain Home early warning radar system and quickly became familiar with its operation. On 3 September 1939, he observed unusual echo sig-

1730 Murdin 2016; Davies et al. 2016, pp. 323–344.

nals, initially attributed to the ionosphere, but he suspected they were related to cosmic rays, drawing on his earlier research.

Lovell's wartime work focused on the H_2S airborne ground scanning radar, which enabled British bombers to navigate and locate targets in poor visibility. He helped miniaturise the equipment and improve the parabolic antennas for higher resolution. In June 1942, after a Halifax bomber equipped with H_2S crashed, he assisted in recovering the highly secret cavity magnetron. Despite the tragic loss of all eleven crew members, including key colleagues such as Alan Blumlein, Lovell continued his work, as H_2S was considered vital for the war effort.

After the war, Lovell returned to astronomy and established the Jodrell Bank Observatory in Cheshire, converting a former wartime searchlight station into a centre for radio astronomy. Using ex-military radar equipment, he detected echoes from daytime meteor showers, confirming their solar orbits. With university funding, he built the Lovell Telescope, then the world's largest fully steerable radio telescope, which remains a key instrument in radio astronomy and part of interferometric networks.

Lovell's radar expertise and project leadership advanced radio astronomy significantly. In 1951, he became Professor of Radio Astronomy at the University of Manchester and founded the Nuffield Radio Astronomy Laboratories. He collaborated internationally and advised the British Ministry of Supply and Ministry of Defence on technical and defence matters. Between 1969 and 1971, he served as President of the Royal Astronomical Society, contributing to the planning of major UK astronomical projects.

With his unique ability to transform wartime technologies into scientific tools, Bernard Lovell bridged military innovation and peaceful exploration of the heavens. His contributions laid the foundation for modern radio astronomy and demonstrated the broader value of radar technology.

Bernard Lovell died on 6 August 2012.

John Turton Randall and Henry Albert Howard Boot
John Randall (1905–1984)[1731] and Henry Boot (1917–1983)[1732] are historically recognised as the inventors of the Cavity Magnetron, a groundbreaking device developed in 1940 at the University of Birmingham in the laboratory led by Professor Mark Oliphant. He played a crucial role by supporting and guiding the research team, helping to provide the necessary resources that enabled Randall and Boot to achieve their breakthrough. Their invention revolutionized radar technology by dramatically increasing power output in the centimetre-wave frequency range, which was vital for high-resolution radar systems during World War II. The cavity magnetron overcame previous limitations, contributing significantly to the Allied war effort.

John Randall was born on March 23, 1905, in Newton-le-Willows, Lancashire, England. After attending grammar school in Ashton-in-Makerfield, he studied physics at the University of Manchester, graduating with first-class honours in 1925 and earning his MSc in 1926. He began his career at the General Electric Company in Wembley, focusing on luminescent materials and phosphorescence. By 1937, Randall was recognised as a leading British researcher in luminescence and received a Royal Society Fellowship that brought him to the University of Birmingham. There, he worked on electron trap theory related to phosphorescence. With the outbreak of World War II, Randall shifted to radar research and, together with Boot, invented the cavity magnetron in 1940. For this achievement, he received multiple honours, including election as a Fellow of the Royal Society in 1946, the Hughes Medal, and a knighthood in 1962. After the war, Randall turned to biophysics, leading a multidisciplinary research group at King's College London and contributing to important discoveries about biological structures such as collagen molecules.

John Randall died on 16 June 1984.

Henry Boot was born in 1917. Less is documented about his early life and education compared to Randall. It is known that he worked

1731 Wilkins 2011; Wilkins 1987, pp. 491–535; Watson 2009, p. 143.
1732 Nicholls 2011; Watson 2009, p. 143.

alongside Randall at the University of Birmingham and co-invented the cavity magnetron. Boot was a brilliant physicist whose technical skills and experimental expertise were crucial to the magnetron's successful development. During the war, Boot dedicated himself to advancing radar technology, a critical military innovation. After the war, he continued his scientific career but remained less publicly known than Randall. Boot was recognised with awards such as the Thomas Gray Memorial Prize (1943), the John Price Wetherill Medal (1958), and the John Scott Award (1959) for his contributions.

Henry Boot died on 8 February 1983.

The invention of the cavity magnetron by Randall and Boot proved to be one of the most significant technological breakthroughs of World War II. By enabling high-power, high-frequency radar systems, it greatly improved the detection and tracking of enemy aircraft and submarines, directly influencing the outcome of key battles and ultimately contributing to the Allied victory.

Albert Percival Rowe

Albert Rowe (1898–1976)[1733] was a pivotal figure in the development of British radar technology, which played a decisive role in air defence during World War II. Born on 23 March 1898 in Bush Park, Launceston, Cornwall, Rowe attended the Royal Dockyard School in Portsmouth and later the Royal College of Science in London. In 1922, he joined the scientific staff of the Air Ministry and, in 1925, became assistant to Harry Wimperis, the first director of scientific research. Between 1927 and 1937, he also lectured part-time at Imperial College London while contributing to military research.

In 1934, Rowe began analysing existing air defence detection systems and recognised that conventional technologies like acoustic devices and constant patrols were insufficient. Reviewing all available files, he found a significant gap in practical solutions, motivating the pursuit of novel technologies that ultimately led to radar development. In 1935, following initial discussions about a "Death Ray" with Robert Watson Watt, Rowe helped develop radar systems, including tests using the Daventry transmitter.

1733 Lovell 2004a; Rowe 2015; Putley 2009, pp. 45–48.

In January 1935, Rowe was appointed secretary to the CSSAD, later known as the Tizard Committee, which investigated methods to defend Britain against hostile aircraft. He collaborated closely with Watson Watt and other scientists to develop Britain's first operational radar. By May 1938, Rowe succeeded Watson-Watt as Head of the Bawdsey Research Station, a key facility in radar research. During World War II, the station was relocated multiple times to avoid German attacks, eventually becoming the Telecommunications Research Establishment (TRE) in Malvern, Worcestershire. Rowe oversaw the growing organisation, coordinating scientists, engineers, and military personnel to rapidly deploy radar technology.

Rowe's leadership ensured that scientific advancements were aligned with military needs, contrasting with the German model, where research was often isolated from operational requirements. TRE became a centre of excellence under his guidance, producing groundbreaking devices critical to the war effort. He remained Superintendent until 1941, playing a central role in organisational development before John Cockcroft assumed scientific leadership.

After the war, Rowe served as Deputy Controller of Research and Development at the Admiralty. In 1947, he moved to Australia, becoming Chairman of the Defence Advisory Committee and scientific adviser to the government. He published *One Story of Radar* in 1948 and briefly served as Vice-Chancellor of the University of Adelaide before returning to Britain in 1958.

Rowe received numerous honours for his pioneering work in radar and his ability to integrate scientific research with military strategy.

Albert Rowe died on 25 May 1976.

Henry Thomas Tizard

Henry Tizard (1885–1959)[1734] was a distinguished British chemist and scientist whose leadership played a central role in the development of radar and air defence technologies during World War II. He studied at Magdalen College, Oxford, focusing on mathematics and chemistry, and gained early experience in aeronautical research. During World War I, Tizard served in the Royal Flying Corps as

1734 Piller 2008; Clark 1965; Watson 2009, pp. 42–43.

an Experimental Equipment Officer, acquiring practical experience with aircraft and aviation technology.

In 1933, Tizard became chairman of the Aeronautical Research Committee (ARC), and in 1934 he took the lead of the Committee for the Scientific Survey of Air Defence (CSSAD). Under his guidance, the committee explored new technologies, particularly radar, to strengthen Britain's air defence. He worked closely with Air Marshal Hugh Dowding and scientists such as Robert Watson-Watt to integrate technical innovations into operational defence systems. Notably, he oversaw the Biggin Hill experiments in 1936, which tested radar detection and interception procedures. Follow-up analyses and refinements in 1937 helped integrate radar into the emerging Dowding System, the world's first operational air defence network.

Tizard was a determined advocate for radar, despite initial scepticism from some military officials, including Frederick Lindemann. Thanks to his persistence, radar became a central component of Britain's air defence network by the late 1930s, providing critical early warning capabilities during the Battle of Britain. His work contributed significantly to the development of the Dowding System, linking radar stations with central command for coordinated interception of enemy aircraft. He also played a key role in promoting the sharing of scientific knowledge between allied nations, ensuring that British innovations could be applied effectively in multiple theatres of the war.

In 1940, Tizard led the top-secret Tizard Mission to the United States to present British developments in radar and air defence. After his return, he had no direct responsibility for the implementation of these technologies but continued to serve as an adviser on radar and air defence matters. His work, together with Watson-Watt, was instrumental in establishing radar as a cornerstone of modern air defence and laid the foundation for many post-war technological advances.

Tizard's career exemplifies the successful integration of scientific innovation and military strategy. Under his guidance, Britain achieved a decisive technological advantage crucial in defending against German air attacks. His contributions, alongside those of Watson-Watt and Dowding, provided the basis for modern air defence systems and post-war technological developments.

Henry Tizard died on 9 October 1959.

Robert Alexander Watson-Watt

Robert Watson-Watt (1892–1973)[1735] was born in 1892 in Brechin, Angus. He graduated in engineering from University College Dundee, then part of the University of St Andrews, in 1912. His early interest in radio waves set the stage for his pioneering work in radar technology.

During World War I, Watson-Watt worked as a meteorologist for the Royal Flying Corps and later at the Royal Aircraft Factory in Farnborough, developing methods to detect thunderstorms using radio waves to provide early warnings for airmen. In 1916, he proposed using cathode ray oscilloscopes to display radio signals, though these became available only in 1923.

From the early 1920s, Watson-Watt held a senior position at the Department of Scientific and Industrial Research (DSIR). In 1924, he transferred to the newly founded Radio Research Station (RRS) in Slough, and by 1933 he was head of the new radio department at the National Physical Laboratory in Teddington.

In early 1935, he submitted a memorandum titled "Detection and Location of Aircraft by Radio Methods" to the CSSAD, detailing how radio waves could form the basis of an effective air defence system. The day before, he and his assistant Arnold Wilkins conducted a successful field test near the BBC transmitter in Daventry, demonstrating that an illuminated aircraft reflected enough radio energy to be detected.

In September 1936, Watson-Watt was appointed head of the Bawdsey Research Station (BRS) at Bawdsey Manor near Felixstowe. Work there led to the creation of the radar network along England's east and south coasts, including Chain Home for high-altitude early warning and Chain Home Low for low-flying aircraft. These systems were operational by the outbreak of World War II in 1939 and were decisive in detecting German bomber formations during the Battle of Britain.

1735 Watt 2017; Watson-Watt 1957; Ratcliffe 1975, pp. 548–568; Watson 2009, pp. 39–40; Jäger und Heilbronner 2010, pp. 452–453; Weinreich 2001f, pp. 103–104.

Watson-Watt oversaw both technical development and team co-ordination, resolving early organisational conflicts. When he left Bawdsey in 1938, the station was well-prepared under his successor Albert Rowe. He then moved to the Air Ministry as Director of Communications Development, overseeing radar, radio, and other detection technologies. From 1940, he advised the Ministry of Aircraft Production and, in 1941, the United States on air defence weaknesses exposed by Pearl Harbor. He was knighted in 1942 and received the US Medal of Merit in 1946.

In 1966, he married Dame Katherine Jane Trefusis Forbes, former Director of the Women's Auxiliary Air Force (WAAF).

His work laid the foundations not only for wartime radar defence but also for post-war advances in radio and electronic detection technologies.

Robert Watson-Watt died on 5 December 1973.

Arnold Frederic Wilkins

Arnold Wilkins (1907–1985)[1736] was a British physicist whose theoretical work significantly influenced the development of radar technology. From 1931 he closely collaborated with Robert Watson-Watt at the Radio Research Station in Slough, an important facility of the National Physical Laboratory. While Watson-Watt served as project leader and public figure, Wilkins was responsible for the essential theoretical calculations and experimental approaches.

Although Watson-Watt initially conceived the idea of detecting aircraft by reflected radio waves, it was Wilkins who demonstrated the physical feasibility through detailed analysis and developed practical testing methods. His contributions laid the foundation for Britain's air defence system.

In February 1935 Wilkins conducted the Daventry Experiment, which for the first time proved that radio waves could be reflected by an aircraft up to eight miles away. This experiment marked a critical milestone in radar development. He also debunked technical claims surrounding Nikola Tesla's so-called Death Rays in 1934 with precise calculations, helping to steer British research away from impractical projects.

1736 Latham and Stobbs 2011.

Following the Daventry test Wilkins and his team relocated to Orfordness on the North Sea coast where they laid the groundwork for the Chain Home radar network. This early warning system played a decisive role in defending Britain against German air raids during the Battle of Britain. In 1938 he also contributed to the development of the Identification Friend or Foe system enabling forces to distinguish friendly from enemy aircraft.

Throughout World War II Wilkins remained a key figure in refining radar technology, which gave Britain a crucial technological advantage during the conflict. His work enabled early detection of enemy aircraft and was instrumental in Britain's air defence success. Despite his fundamental contributions, Wilkins received little public recognition during and immediately after the war, often overshadowed by Watson-Watt. Historians later emphasised his crucial behind the scenes role, acknowledging his deep theoretical insight and relentless dedication that were vital to radar's success.

After the war Wilkins focused on expanding radar's civilian applications such as meteorology and civil aviation, demonstrating its broad utility beyond military purposes. He was awarded the Order of the British Empire for his achievements. Today he is remembered as one of the true pioneers of radar, whose theoretical skill and practical experimentation shaped modern technology and wartime defence alike.

Arnold Wilkins died on 5 August 1985.

Harry Egerton Wimperis

Harry Wimperis (1876–1960)[1737] was born in 1876 in London and educated at Cambridge University. Before his work in air defence, he had a distinguished career in science and engineering, contributing to advancements in aeronautics, navigation, and military technology. He served in various capacities in the Royal Air Force (RAF) and the Air Ministry, establishing himself as a key figure in integrating scientific innovation into military strategy.

Wimperis played a critical role in the development of radar and air defence technologies essential to Britain's success in World War II. In the early 1930s, he recognised that existing air de-

1737 Serby 2020.

fence systems were inadequate. In 1934, he proposed the formation of the Committee for the Scientific Survey of Air Defence (CSSAD), established under the leadership of Henry Tizard. The committee aimed to explore new technologies, particularly radar, to strengthen air defence. Wimperis was instrumental in forming the committee and was a staunch advocate for radar as a key component of Britain's defence strategy.

The committee's work initially faced scepticism from military officials. However, thanks to the persistence of Wimperis and leading scientists such as Tizard and Robert Watson-Watt, radar technology became a central part of Britain's air defence network. This collaboration between the Air Ministry, military officials, and scientists was crucial in establishing the radar-based system that played a pivotal role in the Battle of Britain.

Wimperis also faced challenges in his interactions with Frederick Lindemann, later Viscount Cherwell, a scientific adviser to Winston Churchill. Lindemann, known for his pragmatic focus on immediate solutions, was initially sceptical of radar and other long-term technological advancements. Despite these tensions, Wimperis remained steadfast in his belief that radar was essential to Britain's defence. His persistence led to the operational use of radar by the late 1930s, greatly enhancing Britain's ability to detect and intercept enemy aircraft.

In 1937, Wimperis retired as Director of Scientific Research at the Air Ministry but continued to contribute to military research, advising both the British and Australian governments on aeronautical issues. He played a key role in establishing the Aeronautical Research Laboratory in Melbourne, Australia.

Wimperis' legacy is one of scientific foresight and leadership in integrating new technologies into military defence. His advocacy for radar development was a cornerstone of Britain's ability to defend its skies during World War II. His contributions to radar and air defence remain a vital part of Britain's wartime success and helped shape modern air defence strategies.

Harry Wimperis died on 16 July 1960.

The development of radar technology was driven by numerous scientists and engineers. Equally important, however, were advances in computer science and cryptanalysis, which laid essential foundations for modern technologies. Alan Turing is a key figure in this context. Although he was not directly involved in radar technology, his work at Bletchley Park during World War II in codebreaking German encrypted messages was of great significance. This work was an important factor in the success of the Allies. At the same time, Turing established fundamental principles of information processing that are also applied in radar and computer technology.

Parallel to these developments, other scientists made crucial contributions directly to radar countermeasures. Joan Curran, a Welsh physicist, played a pivotal role in enhancing radar technology by inventing the Window technique also known as chaff which involved dispersing metallic strips from aircraft to confuse enemy radar signals. This innovation significantly reduced Allied bomber losses and became a cornerstone of electronic warfare. Curran's work not only demonstrated practical ingenuity but also highlighted the vital role women scientists played in wartime technological advances, complementing efforts in fields like cryptanalysis and computing.

Alan Mathison Turing

Alan Turing (1912–1954)[1738] was a British mathematician, logician and one of the founders of modern computer science. After his studies at Cambridge he became a Fellow of King's College and in 1936, during a research stay at Princeton, published his groundbreaking work on computable numbers. In it he introduced the theoretical Turing machine, a model that laid the foundations of computer science.

At the outbreak of the World War II Turing joined the Government Code and Cypher School at Bletchley Park. There he worked in Hut 8, the section responsible for deciphering ENIGMA messages used by the German Navy. These posed a serious threat to Allied shipping in the Atlantic. To overcome the challenge, Turing designed the concept of the Bombe, an electromechanical device based on earlier Polish work. The first prototype, built by Harold Keen, became decisive, especially after Gordon Welchman's improvement through the diagonal board.

Turing's work made it possible to analyse German signals on a systematic basis. Together with an interdisciplinary team he played

1738 Hodges 2012; Hodges 2017.

a crucial role in breaking the ENIGMA and later contributed to the decryption of the Lorenz cipher used by the German High Command. Many historians regard these achievements as one of the key factors behind the Allied victory, as they shortened the war significantly and saved countless lives.

After the war Turing returned to academic and practical computing. At the National Physical Laboratory he designed the Automatic Computing Engine (ACE), one of the earliest blueprints for a digital computer. Later at Manchester he contributed to the development of early stored programme computers and pursued pioneering ideas in artificial intelligence. In 1950 he proposed the Turing Test, which remains a reference point in debates about machine intelligence.

Despite his extraordinary contributions Turing faced personal hardship. In 1952 he was prosecuted for homosexuality, then a criminal offence in Britain, and forced to undergo hormone treatment that had severe effects on his life. Two years later he died in tragic circumstances, most likely by suicide.

Recognition came only decades later. In 2009 the British government issued an official apology and in 2013 he was posthumously granted a royal pardon. Since 2021 his portrait has appeared on the £50 banknote, symbolising both his scientific legacy and the fight for human dignity.

Alan Turing is remembered as a brilliant thinker, a pioneer of computer science and a central figure in the wartime code breaking effort that shaped the course of history.

Alan Turing died on 7 June 1954.

Joan Elizabeth Curran, Lady Curran

Joan Curran (1916–1999)[1739] was a Welsh physicist whose career was closely tied to key technological developments during World War II. Born in Swansea, she attended Swansea Girls' High School and won a scholarship in 1934 to Newnham College, Cambridge. There, she completed a degree in physics with distinction, although women were not formally awarded degrees at the time. She later joined the Cavendish Laboratory under Philip Dee, specialising in

1739 Jones 2009, pp. 291–292; Fletcher 2004.

experimental physics. Curran was noted for her skill in constructing and operating scientific instruments, a talent crucial in wartime work. She was married to Samuel Curran, also a physicist, yet maintained an independent career, notably through her invention of the Window technique, which had a profound impact on radar operations.

The idea of disrupting radar using aluminium foil strips was strongly promoted by R. V. Jones. Curran developed and refined the concept from 1941 at the Telecommunications Research Establishment, standardising the size and deployment of the strips, which could be dropped from aircraft to create radar echoes that concealed bomber formations. This countermeasure, codenamed Window and also known as chaff, became one of the most effective tools in Allied electronic warfare, influencing air operations throughout the European theatre.

The first large-scale operational use of Window occurred during Operation Gomorrah in July 1943, reducing Allied bomber losses. In Operation Taxable in June 1944, Window simulated a phantom invasion fleet, deceiving German radar and diverting attention from the Normandy landing zone. The Germans had a similar technique, Düppel, but it was rarely deployed or used systematically.

During the war, Curran also contributed to developing the proximity fuse and later joined the British delegation to the Manhattan Project, assisting with uranium enrichment. Following the end of World war II, she co-founded ENABLE Scotland in 1954 to support families of children with disabilities, inspired by her own daughter. In 1987, she received an honorary Doctor of Laws from the University of Strathclyde, recognising both her scientific achievements and her humanitarian efforts.

Joan Curran's legacy combines technical brilliance, practical ingenuity, and social responsibility. Her work on Window advanced electronic warfare, saved countless lives, and highlighted the crucial but often overlooked contributions of women in science. Her pioneering role in radar countermeasures remains a milestone in twentieth century military science, showing how scientific innovation directly influenced the course of the war.

Joan Curran died on 10 February 1999.

11.1.2.2 Actors and Perspectives

The development of radar technology in Great Britain involved key figures from politics, government, scientific research, and the military. To this day, historians continue to debate the roles and motivations of these individuals, showing how their contributions are interpreted differently depending on perspective and context. Some emphasise the collaboration among scientists and engineers, highlighting the teamwork and innovation behind the technology. Others focus on political leaders and military officials who coordinated resources and set strategic priorities. Assessments of these roles rely on a range of sources, from official government documents and technical reports to personal accounts and later historical analyses. These discussions and studies raise questions of recognition, responsibility, and the influence of individual ambition within institutional frameworks. The variety of viewpoints illustrates the complex social, political, and technological forces that shaped Britain's pioneering work in radar during the 1930s and early 1940s, as well as the international political and strategic situation of the time.

11.2 Pioneers of Radio Technology

James Clerk Maxwell

James Clerk Maxwell (1831–1879)[1740] developed his significant differential equations in 1864/65 on the interaction of the electric and magnetic fields and the phenomena caused by them.[1741] Based purely on a mathematical analysis, Maxwell concluded that the electromagnetic field must propagate in waves, like light. The equations named after him, the Maxwell equations, which summarized the results of his work on electromagnetic field theory, were published in 1873 in his two-volume work "A Treatise on Electricity and Magnetism".[1742] The wave theory of light had already been formu-

1740 Harman 2009; Jäger und Heilbronner 2010, pp. 278–280; Wehner 1987.

1741 Forbes and Mahon 2014; Peruzzi 2000; Tricker 1975; Klausing und Holpp 2000, pp. 19–54.

1742 Maxwell 1954a; Maxwell 1954b; Weiss 1986, pp. 71-82.

lated by Augustin Jean Fresnel after conducting extensive experiments starting around 1815. Maxwell proved, purely from theoretical considerations, that electromagnetic waves propagate at the same speed as light, even in a vacuum, and that light occupies only a small portion of the electromagnetic spectrum at very short wavelengths, i.e., very high frequencies.[1743]

1743 The 19th century brought significant scientific discoveries, culminating in Heinrich Hertz's (1857–1894) demonstration of electromagnetic waves and their properties in 1887/88, and the mathematically exact formulation of the electromagnetic field theory by James Clerk Maxwell (1831–1879) in 1864/65. The wave theory of light was already formulated by Augustin Jean Fresnel (1788–1827) after extensive experiments he conducted starting around 1815. In 1820, Hans Christian Ørsted (1777–1851) discovered the phenomenon of electromagnetism by observing that a magnetic needle was deflected near a current-carrying conductor (Jäger und Heilbronner 2010, pp. 315–316). Based on Ørsted's work, André-Marie Ampère (1775–1836) described how two conductors carrying electric current exert forces on each other due to the magnetic field generated around each conductor. This discovery led to the definition of the terms electric current and electric voltage (Jäger und Heilbronner 2010, pp. 20–21). He established the foundation for the theory of electrodynamics with his law of electrodynamics, Ampère's law. It was Michael Faraday (1791–1867) who discovered electromagnetic induction in 1831, showing that a changing magnetic field induces an electric current in a conductor. This was a major milestone in the field of electromagnetism (James 2011; Jäger und Heilbronner 2010, pp. 124–125). Faraday also developed the concept of the electric and magnetic fields, although it was James Clerk Maxwell who later mathematically formulated the theory of electromagnetism in the mid-19th century, unifying and expanding upon the ideas of Faraday and others. Up until this point, the concept of force lines was still influenced by Newton's mechanical laws, with the visual or tangible representation of forces. The few electrical and magnetic phenomena known at the time were understood as mechanical forces and movements; the current flowing through a wire was still conceivable. By analogy to magnetic lines of force, Faraday introduced electric lines of force and field lines. In particular, the conclusion that electromagnetic waves, unlike sound waves, can propagate in a vacuum was difficult for many scientists to understand. Another key step on the path to Heinrich Hertz's discoveries was the work of William Thomson (1824–1907), who later became known as Lord Kelvin (Smith 2011). In 1853, based on his findings in the field of electrical oscillations, he published his formula for calculating the frequency of oscillations resulting from the interaction between inductance and capacitance. The accuracy of this formula was confirmed by the German physicist Wilhelm Feddersen (1832–1918) in 1859 through experiments in which he studied electri-

Karl Ferdinand Braun

Karl Ferdinand Braun (1850–1918)[1744] studied mathematics and physics in Marburg and Berlin. After completing his doctorate at the Friedrich-Wilhelms-Universität in Berlin, he was appointed director of the Institute of Physics and full professor at the University of Strasbourg in 1895, following several stops as a professor in Marburg, Strasbourg, Karlsruhe and Tübingen.

In addition to technological developments in the field of wireless transmission and reception circuits, Braun made further groundbreaking contributions to electronics and radio technology, including the discovery of the rectifying properties of a metal-semiconductor junction in 1874. From this time onwards, he was involved in the investigation of cathode rays and their properties for the graphic visualisation of electrical processes, which led to the invention of the cathode ray tube, later named after him as the Braun'sche Röhre, which was first presented to the public in 1897.[1745] It was of fundamental importance for measurement technology and formed the basis for the cathode ray oscilloscope. The Braun'sche Röhre was also of outstanding importance for later television technology and radar technology for target indication. In 1898, Braun also succeeded in operating a transmitter with a closed-circuit oscillator, thereby enabling the possibility of delivering larger amounts of energy per discharge, which reduced the attenuation of the radiated waves. Braun's further efforts to replace damped oscillations with undamped ones were only successful in 1905 with the development of the extinguishing radio transmitter by Max Wien[1746], which replaced the noisy spark-gap radio transmitter and Braun's other experimental setups for generating undamped oscillations. From 1895, Braun also worked on wireless telegraphy and improved the separation between the oscillating and antenna circuits, which led to the transmitters he developed in 1899/1900 having a significantly greater range than Mar-

---

cal discharges and oscillations. He conducted experiments that investigated the properties of electrical oscillations and experimentally confirmed the theoretical predictions regarding the oscillation frequency (Wißner 1961).

1744 Jäger und Heilbronner 2010, pp. 63–64; Hars 1999; Kurylo 1965; Russer 2009, pp. 547–554, Fuchs 2004, pp. 69–83; Goetzeler 1983a, pp. 28–30.

1745 Hars 1999, pp. 114–116; Jäger und Heilbronner 2010, p. 63.

1746 Jäger und Heilbronner 2010, pp. 463–464; Goetzeler 1983b, pp. 195–196.

coni's.[1747] Braun's progress in the field of transmitter technology was later utilised by Marconi in his transmission attempts across the Atlantic.[1748] On 24 September 1900, Braun was involved in early radio transmission experiments between Cuxhaven and Helgoland over a distance of approximately 65 kilometres, demonstrating the potential of wireless communication for maritime services.[1749] On 16 November 1900, Braun gave a public lecture on wireless telegraphy for the first time, followed by further lectures, scientific publications and newspaper articles.[1750]

On 10 December 1909, Braun received the Nobel Prize for Physics in Stockholm together with Marconi.[1751]

Heinrich Rudolf Hertz

Between 1886 and 1889, Heinrich Rudolf Hertz (1857–1894), who was a professor of physics at the Technical University of Karlsruhe from 1885 to 1889, carried out systematic experiments to prove the existence of electromagnetic waves and investigate their properties. As a result, he was able to experimentally confirm Maxwell's theory, which stated that electromagnetic waves not only exist in the frequency range of visible light but are also present at lower frequencies and have the same properties as light waves.[1752] Hertz also confirmed that electromagnetic waves have the same properties as light waves in terms of other propagation phenomena such as polarisation, reflection and refraction. He reported these findings in December 1888 in his lecture "Über Strahlen elektrischer Kraft"[1753] at the Berlin Academy of Sciences.[1754] The apparatus used by Hertz with a wavelength of 66 centimetres revealed a crucial finding for the development of radar technology, namely that electromagnetic waves are reflected by metallic and dielectric bodies. These findings

1747 Jäger und Heilbronner 2010, p. 64.
1748 Braun 1901b, p. 547.
1749 Kurylo 1965, p. 186; Fuchs 2004, pp. 78–80.
1750 Kurylo 1965, p. 189; Braun 1901a, pp. 258–260; Braun 1901b, pp. 469–470.
1751 Braun 1909.
1752 Heinrich-Hertz-Institut 1988, pp. 41–62; Hertz 2002a, pp. 64–79; Hertz 1923.
1753 On the Rays of Electrical Force (translated by the author).
1754 Hertz 2002b, pp. 80–96.

were the essential basis for the radar technology to come. Further detailed literature on the life and work of Heinrich Hertz can be found in various sources.[1755] Major advances in the development of radar technology were also closely connected with the Russian physicist Alexander Stepanovich Popov, the German physicist Karl Ferdinand Braun and the Italian radio engineer Guglielmo Marconi.

Alexander Stepanovich Popov

Alexander Stepanovich Popov (1859–1905)[1756] became aware of Hertz's experiments in 1894 and repeated them. He presented the results at a meeting of the Russian Academy of Sciences in St. Petersburg in May 1895. Popov constructed a device for receiving electrical discharges in the atmosphere, which were generated by thunderstorms. He used the already known coherer for detection and a dipole as a transmitting device. In March 1896, he succeeded in sending the world's first wireless radio telegram over a distance of about 250 metres. In the spring of 1897, Popov, with the support of the Russian admiralty, equipped the ships of the Baltic Fleet with radio devices that could already bridge distances of about 5 kilometres. After Popov had failed to patent his ideas, Guglielmo Marconi patented an invention in June 1896, which in principle represented Popov's published receiving device and subsequently led to disputes regarding priority rights.

Guglielmo Marconi

In early 1895, Guglielmo Marconi (1874–1937)[1757] began carrying out experiments at his father's estate in Pontecchio, Italy, to transmit Morse code using electromagnetic waves, based on the work of Heinrich Hertz.

He was inspired early on by the idea of setting up a radio network using wireless telegraphy that would enable communication

1755 Heinrich-Hertz-Institut 1988, pp. 63–67; Wiesbeck 1988; Hertz 1957, pp. 21–34; Fölsing 1997; Eckert 2010; Kniestedt 1989, pp. 41–57; Schwab und Fischer 1998; pp. 1312–1318; Darricau and Blanchard 2002, pp. 16–30.

1756 Jäger und Heilbronner 2010, pp. 343–344; Fuchs 2004, pp. 64–69.

1757 Dragoni 2011; Jacot und Collier 1937; Solari 1942; Jäger und Heilbronner 2010, pp. 273–274; Isted 1991a, pp. 45–56; Isted 1991b, pp. 110–122; Schönborn 2003, pp. 13–35; Fuchs 2004, pp. 60–64.

between ships at sea and between ships and land stations. At that time, ships could no longer be reached with messages once they had left the harbour.[1758] He was thus also one of the first to work on the practical application of Heinrich Hertz's findings on the transmission of information by means of electromagnetic waves, i.e. wireless telegraphy.[1759]

He attended the lectures of Professor Augusto Righi[1760] at the University of Bologna and was also able to observe Righi's experiments with Hertzian waves.[1761] Professor Righi carried out numerous experiments to prove Maxwell's theory on the basis of Heinrich Hertz's findings and built effective devices for this purpose, which were then very helpful to his student Marconi in experiments on wireless telegraphy. Marconi then carried out his own experiments in 1895 with a modified setup in which he connected one pole of the transmitter to the ground and the opposite pole to a wire suspended vertically, achieving a range of up to 2.5 kilometres.[1762] He used an experimental set-up designed by Professor Righi and the coherer constructed by Edouard E. Branly[1763] in 1890.[1764] Marconi offered his invention to the Italian postal and telegraph service but received neither interest nor support for his work.[1765] In February 1896, Marconi went to England where, with the help of connections from his English mother Annie, who came from an Irish-Scottish

1758 Fuchs 1998, p. 116; Wessel 2004, p. 178.

1759 At this time, there was already a worldwide network of cable connections for Morse telegraphy, which was dominated by powerful telegraph companies. As early as 1866, the Atlantic was permanently bridged between Europe and North America and the International Telegraph Union was founded in 1865 (Wessel 2008, pp. 74–75; Fuchs 1998, p. 113). The further development of wireless telegraphy, especially starting with Marconi's experiments, was initially only an extension of wired Morse telegraphy, but with far greater flexibility, which ultimately led to strong competition with the major telegraph companies.

1760 Jäger und Heilbronner 2010, pp. 358–359.

1761 Gerwin 1957, pp. 28–32.

1762 Blumtritt 1988, p. 65; Solari 1942, pp. 19–21.

1763 Jäger und Heilbronner 2010, p. 62; independently of Branly, the Englishman David E. Hughes also constructed a coherer in 1879 (Blumtritt 1988, p. 65; Jäger und Heilbronner 2010, pp. 206–207).

1764 Zenneck und Rukop 1925, pp. 329–333.

1765 Solari 1942, p. 23; Gerwin 1957, p. 45.

whisky dynasty, he quickly found support from the chief engineer of the English telegraph authority, William Preece.[1766]

On 2 June 1896, he applied to the London patent office for a patent for wireless telegraphy using electric waves, having withdrawn his first application from March 1896. This first British patent for wireless telegraphy was granted on 2 July 1896 under number 12,039, entitled "Improvements in transmitting electrical impulses and signals and in apparatus thereof"'. On 13 July an American patent was granted for the same specification. These patent applications were for a radio system with an apparatus that included an aerial antenna and an earth connection.[1767] On 27 July 1896, Marconi successfully carried out the first attempt to transmit a telegram over a distance of approximately 1 kilometre from the roof of the telegraph authority to the roof of another government building under the supervision of officials from the telegraph authority. As a result of this success, Marconi was asked to conduct further experiments over longer distances.[1768] In September 1896, additional experiments were carried out with the goal of covering greater distances, under the supervision of officials from the telegraph authority and representatives from the navy and army. During the tests in September 1896, distances of about 1 3/4 miles were achieved. This was followed by a range of about 4 miles in March 1897, and finally, in May of the same year, by about 9 miles, while the Morse code for V was transmitted.[1769] Adolf Slaby was also present as an observer at Marconi's experiments in May[1770]. These successes were publicised by Preece to a wide audience in lectures and Marconi received a great deal of public attention for the first time.[1771] Following the successful tests in 1896 and 1897,

---

1766 Jäger und Heilbronner 2010, pp. 347–348; above all, the letter of recommendation from Alan A. Campbell-Swinton, a well-known electrical engineer and friend of the Marconi family, enabled Marconi to demonstrate his equipment to Preece and other officials of the English telegraph authority (Solari 1942, p. 24; Schönborn 2003, p. 26).

1767 Dragoni 2011, p. 2; Jäger und Heilbronner 2010, p. 274; Solari 1942, pp. 23–24.

1768 Solari 1942, pp. 24–25.

1769 Marconi 1909, p. 198; E.A. 1897, pp. 505–507; Gerwin 1957, pp. 64–65.

1770 Slaby 1908, pp. 158–160; Dragoni 2011, p. 2.

1771 Solari 1942, p. 25; Jacot und Collier 1937, p. 42.

Preece gave a lecture at the Royal Institution, London, on 4 June 1897, where he concluded with the remark:

> „Mag man sich über die Neuheit der Einzelheiten der Erfindung streiten, eines steht jedoch fest, Marconi gab uns ein neues elektrisches Sinnesorgan, empfindlicher als alle bekannten elektrischen Instrumente.“[1772]

These statements by Preece again generated a strong echo in the press and attracted the attention of many countries to Marconi's findings.[1773] In Germany in particular, the importance of these findings for the German navy was recognised early on. Alongside his scientific work, Marconi began to commercialise his results early on by founding his own company, which was registered on 20 July 1897 under the name Wireless Telegraph & Signal Company, Ltd. and renamed in 1900 as Marconi's Wireless Telegraph Company, Ltd.[1774] On 22 November 1899, an American subsidiary, the Marconi Wireless Telegraph Company of America, was founded in New York. Another company was founded to protect the rights in the shipping sector, alongside the establishment of additional subsidiaries in the USA and Canada. After founding the Marconi International Marine Communication Company in 1900, Marconi travelled to Newfoundland in the winter, believing that atmospheric interference would be less at that time of year.[1775] Ship-to-ship and ship-to-shore communication could be operated for a long time without competition.[1776] Marconi thus combined science, inventiveness and entrepreneurship in his person.[1777] On 26 April 1900,

1772 Preece 1897, pp. 431. "One may debate the novelty of the details of the invention, but one thing is certain, Marconi gave us a new electrical sense organ, more sensitive than all previously known electrical instruments." (translated by the author).

1773 Solari 1942, p. 26.

1774 Jäger und Heilbronner 2010, p. 274; Dragoni 2011, p. 3.

1775 Solari 1942, p. 47.

1776 Teuteberg und Neutsch 1998, p. 116.

1777 However, Marconi did not succeed in gaining a worldwide monopoly because, on the one hand, Braun's transmitters were much more powerful and, on the other, the Gesellschaft für drahtlose Telegraphie m.b.H., System Telefunken was founded in Germany in 1903, which emerged from the merger of the two competing groups around A. Slaby and F. Braun (Jäger und Heilbronner 2010, p. 274; Thiele 2003, p. 19). This subsequently led to a considerable competitive situation with Marconi.

Marconi was granted the famous patent no. 7777, which described a device that allowed for better tuning of the transmitting and receiving devices, thus achieving improved selectivity of individual frequencies and a greater range.[1778]

This patent gave Marconi a significant head start but also resulted in numerous and long-standing disputes with other inventors. Braun, for example, had already applied for a patent in Germany two years earlier, which corresponded to Marconi's invention in essential points.[1779] On 27 March 1899, following experiments in Italy and England, Marconi bridged the English Channel for the first time over a distance of 40 miles between Wimereux in France and South Foreland in England.[1780] The construction of the first radio transmitter at Poldhu, Cornwall, took place from October 1900 to January 1901, with the aim of establishing the first transatlantic radio connection.[1781]

In addition to the challenges of transmission and reception technology, it was still completely unclear at this time how the curvature of the earth, the time of year and day and atmospheric discharges could affect radio transmission.[1782] Marconi travelled to Newfoundland in winter because he suspected that there would be less atmospheric interference at that time of year.[1783]

Furthermore, it was still widely believed that electromagnetic waves propagated in the manner of Hertzian waves, and that the curvature of the Earth could thus present an obstacle to their propagation.[1784] Initially, in January 1901, some successful experiments were carried out between The Lizard near Poldhu and St Catherines' Point on the Isle of Wight over a distance of 186 miles.[1785] Valuable knowledge was gained about tuning and improving the transmission and reception technology, but not about the propagation conditions with regard to the curvature of the earth. The

1778 Marconi 1909, pp. 202–205; Bussey 2000, p. 13; Blumtritt 1988, pp. 69–70.

1779 Blumtritt 1988, p. 70.

1780 Gerwin 1957, pp. 81–88; Marconi 1909, p. 200; Hinnen 1987, p. 10.

1781 Bussey 2000, pp. 15–52; Isted 1991b, p. 110.

1782 Marconi 1909, p. 208; Solari 1942, p. 46.

1783 Solari 1942, p. 47.

1784 Marconi 1909, pp. 205–206.

1785 Ibid., p. 207.

transmission attempts were interrupted by several setbacks, which jeopardised the entire undertaking. On 17 September 1901, for example, the antenna system at Poldhu, which had only been erected in August, was almost completely destroyed by a violent storm. Just one week later, the antenna system was rebuilt, but with a different design.[1786] In Newfoundland, balloons and kites were used to replace the masts of the receiving antennas, with the kites proving more effective in strong winds.[1787]

After the start of the experiments at the beginning of December 1901 and many efforts, the first Morse code signal, the simple letter S (...), was received from Poldhu in Cornwall on 12 December 1901 in St. John's, Newfoundland.[1788] By bridging a distance of almost 2,000 miles, Marconi proved that the curvature of the Earth was not an obstacle to radio transmission, but that the time of day did have an influence on reception quality.

On 18 January 1903, the first transatlantic wireless message was sent between the South Wellfleet Station in Cape Cod, Massachusetts, USA, and Poldhu in Cornwall, England. The message was a congratulation from President Theodore Roosevelt to King Edward VII of England:

> "The President of the United States to his Majesty King Edward the Seventh:
>
> In taking advantage of the wonderful triumph of scientific research and ingenuity which has been achieved in perfecting the system of wireless telegraphy, I extend on behalf of the American people my most cordial greetings and good wishes to you and the people of the British Empire.
>
> Theodore Roosevelt"[1789]

The British side, represented by King Edward the Seventh, replied to the greetings on 19 January:

1786 Bussey 2000, pp. 30–37.

1787 Jacot und Collier 1937, pp. 75–80.

1788 Marconi 1922, p. 215; Solari 1942, p. 48; Bussey 2000, pp. 46–52; Dragoni 2011, pp. 3–4; Hinnen 1987, p. 10.

1789 Bussey 2000, p. 83.

> "The King of England to the President of the United States:
>
> I thank you most sincerely for the kind message which I have just received from you through Signor Marconi's transatlantic wireless telegraphy. I sincerely reciprocate, in the name of the people of the British Empire, the cordial greetings and friendly sentiment expressed by you on behalf of the American nation, and I heartily wish you and your country every possible prosperity.
>
> Edward R."[1790]

Fig. 38 shows the commemorative plaque on the east coast at Wellfleet, Cape Cod, Massachusetts, USA.

After his successes, Marconi initially focused on the development and commercialisation of radio connections from ship to ship and from ship to shore. However, one obstacle was that, in 1870, communication services in the United Kingdom had been nationalised, and, for example, Lloyd's, the ship insurer responsible for commercial shipping, had no interest in Marconi's devices, as the equipment could only be operated by trained personnel. The British military first placed an order for the delivery of wireless telegraph equipment for the Boer War in South Africa in 1898. Marconi recognised that marketing individual devices to coastal services and ship insurers could only lead to sustainable economic success if trained operators were also available. As a consequence, the devices were rented out together with trained personnel, e.g., to shipping companies, with the personnel being strictly required to make connections only with Marconi's on-board or shore stations. Only in emergencies was an exception allowed. Marconi was able to maintain his economic supremacy for a long time.[1792]In 1906, the International Radiotelegraph Conference took place in Berlin, with 27 nations participating, and it was the first major step towards a global agreement on the use of radio frequencies and the coordination of international radio communication. This meant that coastal and on-board stations were obliged to pass on radio messages without discrimination, regardless of the respective radio sys-

1790 Ibid., p. 84.

1791 Image credit: Helmut Klausing.

1792 Blumtritt 1988, p. 69; Teuteberg und Neutsch 1998, p. 116.

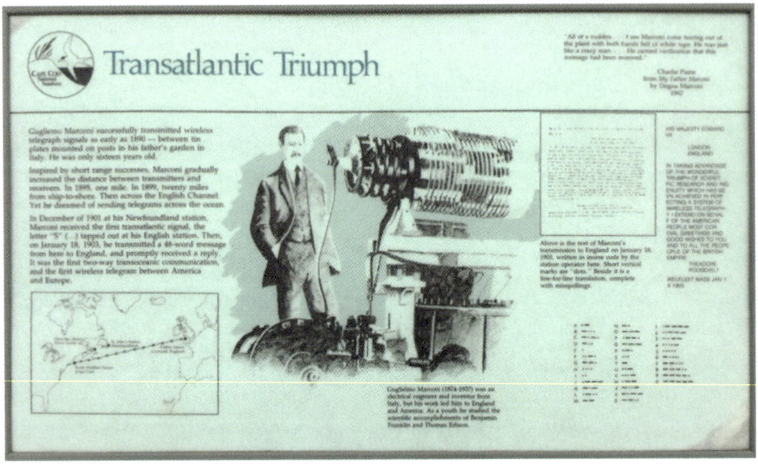

Fig. 38: Commemorative plaque at Wellfleet, Massachusetts, USA[1791]

tem.[1793] The introduction of the SOS distress signal was also an important step for safety at sea. When these obligations came into force in 1908, Marconi's monopoly was finally broken. Until that time, Marconi's system remained technically dominant and was still capable of hindering communication between coast and ship.[1794]

Marconi was seen as a benefactor of humanity after Marconi radio operators transmitted an SOS message that saved the lives of about 700 people on board the sinking SS Titanic.[1795]

Later, in 1922, Marconi recognised the potential of using shortwave electromagnetic radiation to develop a device for measuring the reflection of metallic bodies and using this to locate ships.[1796]

On 10 December 1909, Marconi received the Nobel Prize for Physics in Stockholm together with Braun.[1797]

1793 Wessel 2008, pp. 89–90; Hinnen 1987, p. 11.

1794 Blumtritt 1988, p. 69; Fuchs 1998, p. 116.

1795 Dragoni 2011, p. 4.

1796 Marconi 1922, p. 237.

1797 Marconi 1909.

Adolf Slaby

Marconi's work did not go unnoticed in Germany. In March 1897, Kaiser Wilhelm II learned of Marconi's work on telegraphy without wires from a German newspaper report titled "Das Telegraphieren ohne Drähte".[1798] Wilhelm II saw great potential in the navy, which he particularly regarded as his favourite child.[1799] He instructed Adolf Slaby (1849–1913)[1800], professor of theoretical mechanics and electrical engineering at the Technische Hochschule in Berlin, to make contact with Marconi and to continue his own experiments, since Slaby was already experimenting with Hertzian waves and had increasingly dedicated himself to wireless telegraphy since the mid-1890s.[1801] Slaby was considered an expert in the field of electrical engineering and enjoyed high esteem among companies and the military.[1802]

In January 1893, more than 30 representatives from electrical engineering associations, societies, and companies gathered to establish the Verband Deutscher Elektrotechniker (VDE). At the first meeting, Slaby was elected chairman, and Wilhelm von Siemens became his deputy.[1803] This gave him access to Kaiser Wilhelm II, to whom he had to regularly report on the latest technological developments, leading to a close relationship.

However, Slaby's experiments were only moderately successful. Through the mediation of Gisbert Kapp[1804], Slaby managed to take part in Marconi's experiments in the Bristol Channel in May 1897.[1805] Kapp had worked with Preece in previous years, and so Slaby was able to attend the experiments in great detail, much to Marconi's displeasure, at the invitation of Kapp through Preece.[1806] Back in

1798 Heilbronner n.d., p. 2; Heilbronner was unable to find out the name of the newspaper, but an article by William Preece was published in Germany in July 1897 in the Elektrotechnische Zeitschrift under the title Die Telegraphie ohne Drähte (Preece 1897, pp. 430–431).

1799 Slaby 1908, pp. 189–213.

1800 Jäger und Heilbronner 2010, p. 409; Mauel 1983d, pp. 176–177.

1801 Slaby 1902a, pp. 165–169; Slaby 1902b, pp. 254–258.

1802 Fuchs 1998, p. 117.

1803 Dittmann 2021, p. 21.

1804 Jäger und Heilbronner 2010, p. 220; Dittmann 2021, p. 37.

1805 Heilbronner n.d., p. 3; Slaby 1898, pp. 867–874.

1806 Slaby 1908, pp. 151–159; Heilbronner 2014, p. 3.

Germany, Slaby rebuilt Marconi's experimental apparatus. Slaby undoubtedly gained valuable insights for his own research work and, much later in 1919, Marconi once again made it apparent how damaging he considered Slaby's invitation to be for his own work.[1807]

Slaby and Braun

In Germany at this time, there were essentially two groups that carried out experiments in the field of wireless communication in parallel with Marconi. One was Slaby and his assistant Georg Graf von Arco[1808] at the Technische Hochschule in Berlin and the other was Braun with his employees Jonathan Zenneck[1809] and Mathias Cantor[1810] at the University of Strasbourg in Alsace-Lorraine.[1811]

The AEG company supported the work of Slaby and Graf von Arco, while the Siemens & Halske company supported Braun's work in Strasbourg.[1812] There was intense competition between the two groups, which led to numerous patent disputes.[1813] The systems developed by Braun and Slaby were technically nearly identical, and their performance was roughly equivalent.[1814] Braun, in collaboration with Siemens & Halske, primarily worked for the army, while Slaby, working with AEG, mainly focused on the navy. At the same time, Marconi tried to maintain his monopoly by preventing radio telegrams from being forwarded by Siemens & Halske or AEG from Marconi stations.

Marconi's aim was to maintain his monopoly position in wireless telegraphy, so that his operators did not forward radio telegrams from stations supplied by the AEG/Slaby/Arco-group; even for the German Kaiser no exception was made.[1815] The intervention of Wilhelm II in the business policies of the two electricity companies

1807 Heilbronner n.d., p. 3.

1808 Jäger und Heilbronner 2010, pp. 23–24; Zennek 1953, pp. 337–338; Fuchs 2004.

1809 Jäger und Heilbronner 2010, pp. 473–474; Kind und Mühe 2006a, pp. 193–199; Goetzeler 1983c, pp. 199–201.

1810 Hars 1999, p. 111.

1811 Thiele 2003, p. 18.

1812 Ibid., p. 18.

1813 Hars 1999, pp. 170–177.

1814 Fuchs 2004, p. 83.

1815 Thiele 2003, p. 18.

finally settled the patent disputes and led to an attempt to combine forces by introducing standardised radio systems for the army and navy.[1816]

In the commemorative publication "25 Jahre Telefunken", it is written:

> „Es war erheblich leichter, die beiden Häuser, A.E.G. und Siemens & Halske, in gemeinschaftlicher Arbeit zu vereinigen, als die beteiligten Gelehrten."[1817]

On 27 May 1903, the company Gesellschaft für drahtlose Telegrafie m.b.H, System Telefunken, was founded in Berlin.[1818] The systems of Braun and Slaby were incorporated into this new company, each with a 50 % share.[1819] Graf von Arco became the first technical managing director.[1820]

On 17 April 1923, the company's name was changed to Telefunken Gesellschaft für drahtlose Telegraphie m.b.H. by a shareholders' resolution.[1821]

With the establishment of Telefunken in 1903, the foundation was laid for one of the most significant companies in the fields of radio and radar technology.

Additional Note

A very good overview of the pioneers of radio technology has been provided by Peter Butcher of DEHS.[1822]

1816 Hars 1999, pp. 162–170.

1817 Telefunken 1928, p. 16. "It proved considerably easier to unite the two companies, A.E.G. and Siemens & Halske, in joint work than to bring together the scholars involved." (translated by the author).

1818 Thiele 2003, p. 19; Strunk 1999, p. 32.

1819 Telefunken 1928, p. 17; Fuchs 2004; p. 94.

1820 Thiele 2003, p. 19; Telefunken 1928, pp. 16–18; Fuchs 1998, p. 119; Arco 1909a, pp. 535–540; Arco 1909b, pp. 561–567.

1821 Thiele 2003, p. 379.

1822 Butcher 2020, pp. 1–20.

11.3 Microwave Generation Into A Key Military Technology in World War 2: A Comparative Study

TRANSMISSION LINES

THE JOURNAL OF THE DEFENCE ELECTRONICS HISTORY SOCIETY
www.dehs.org.uk; ISSN 1362-3834

Volume 28 No.4 Editor: Peter Butcher December 2023

Fig. 1. The British prototype E1189 10cm resonant cavity magnetron.

MICROWAVE GENERATION INTO A KEY MILITARY TECHNOLOGY IN WORLD WAR 2: A COMPARATIVE STUDY

Prof Helmut Klausing and Dr Phil Judkins

Fig.2. H2S ground-scanning radar in the crowded cockpit of an RAF Mosquito aircraft.

EDITORIAL

Peter Butcher

INDEX

'Transmission Lines' DEHS Contacts

DEHS Chairman: Dr Phil Judkins, 18, St John's Square, Wakefield, West Yorkshire, WF1 2RA; philjudkins@btinternet.com or 0797 144 9451.

Transmission Lines Editor: Peter Butcher, Meadowcroft House, Churchstanton, Taunton TA3 7RN; petermbutcher@outlook.com

MICROWAVE GENERATION INTO A KEY MILITARY TECHNOLOGY IN WORLD WAR 2: A COMPARATIVE STUDY

Prof Helmut Klausing and Dr Phil Judkins

Abstract

During World War II, Allied radar technique made a technological leap to centimetric wavelengths, based in part on resonant cavity magnetron technology first demonstrated by Randall and Boot in Birmingham, UK, developed further by GEC and shared with US scientists during the 1940 Tizard Mission. German technique made no such leap. This paper examines the comparative developments in both countries to determine possible reasons for these different outcomes and identifies issues in both Britain and Germany which have resonance in R&D and equipment procurement today.

1. Introduction

In war fighting, the development of sensors to detect the approach of an enemy and to guide counterattacks are of obvious importance. During the Second World War, on land, in the air and at sea, much attention was devoted to development of radar technology as a significant new sensor in these fields of conflict. In Britain and, later, the USA, the development of means to generate high power at microwave frequencies (centimetric wavelengths) led to compact radars with smaller aerials, capable of being fitted into aircraft, smaller warships, and battlefield vehicles, to achieve significant ranges and finer target discrimination against, for example, hostile aircraft, ground targets or submarine periscopes [1]. This was not achieved earlier, or even simultaneously, in Germany, to the shock of Germany's military leaders.

Academic comparative studies have focussed upon the pre-war development of radar itself [2], but rather less upon the wartime technological advance in microwave generation technique upon which the majority of post-war radar devices would be based [3].In this essay, we examine the relative progress in Britain and Germany of developing weapons of war using transmissions at centimetric wavelengths, against the considerations of changing military requirements, political constraints imposed on research, scientific advances and miscalculations, procurement processes, and the capabilities of existing technologies such as metric wavelength radar.

Background.

Significant progress in the development of radar technology was associated with the stable generation and amplification of electromagnetic oscillations, beginning with the invention of the electron tube in 1906, which over time displaced the 'spark' radio transmitter that had until then been commonplace. Following further development of vacuum tube (UK: radio valve) technology in the following decades, the short and ultra-short-wavelengths of less than 10 metres could increasingly be used [4], [5].

From the late 1920s onwards, employing the almost simultaneous development of radio technique in many countries including Britain and Germany, both these future combatants developed radar at metric wavelengths from 1935 up to the outbreak of war, for application on land, sea and in the air. The limitations of metric wavelength radar lay in its inability to distinguish targets, particularly those of small scale such as individual aircraft or small warships (surfaced submarines, motor torpedo boats) combined with its requirement for antennae of significant size – a frustration even for warships with limited space on their masts, but especially so on aircraft, where metric radar ranges were limited to aircraft height above the ground.

Further requirements for military service, though rarely stated in reviews of technological history, were robustness and compact size, The physical conditions under which war is fought demand robustness, as any device which is fragile rapidly becomes useless, while compact size, though less necessary in some ground-based applications such as long-range early warning, becomes important on-board ship and critical in airborne applications.

These factors drove British researchers to seek devices of high-power output at centimetric wavelengths. From the beginning of the 1930s, the development of magnetrons played an increasingly significant role, for this device was to become an essential component of powerful radar equipment. The magnetron had already been known in a wide variety of designs since at least the 1920s, e.g., on 27 November 1935 the German Hans Erich Hollmann filed a patent for a magnetron with multiple resonators. This patent was granted in America on 12 July 1938 [6].

In 1940, Britain achieved a series of technological breakthroughs. That January, Britain had none of the four capabilities needed for military radar equipment on centimetric wavelengths, these being a high-power, compact and robust microwave generator as transmitter; a detector of such microwaves, as the basis of a receiver; a low-power generator of microwaves, to mix with received radar signals so as to allow their amplification by existing technologies; and a means of using a single aerial for transmission and reception, to facilitate compact antennae for ships and aircraft. By December, Britain had all four capabilities; Germany, a much larger nation with greater scientific resources, did not, resulting in the technological superiority of Allied radar technology after 1943. A good overview of the development of tube technology in Germany with also a detailed description of the development and technical application of the magnetron can be found in Kern [7]. An excellent overview of tube development in Britain for the period from 1937 to 1944 can be found in Callick [8].

2. Germany

Strategically, pre-war Germany did not see long-range early warning or radar-based defensive fighter control as pressing priorities for radar. Radar applications were perceived as of value for anti-aircraft or coastal defence artillery, neither of which were priorities at the level of, for example, tanks, radio communications, secure cipher machines, or airborne radio-navigation and target-bombing technology.

Considering, therefore, the development of radar equipment for air surveillance, fighter guidance, and as anti-aircraft target location equipment, in Germany up to the middle of the Second World War, the wavelengths employed were predominantly in the ranges from a few metres, as for example with the Freya radar with a wavelength of 2.4 metres for air surveillance and fighter guidance, down to the decimetric range, of some 80centimetres in the case of the naval Seetakt radar, and the Luftwaffe Würzburg anti-aircraft radar at 53 cm. Indeed, even airborne German radars from 1941 onwards operated at decimetric wavelengths, in the range of half a metre.

Reliance on metric and decimetric wavelength technology did not disadvantage Germany strategically during the first three years of conflict. Luftwaffe fighters, relying on the metric Freya radar, inflicted such heavy losses on British daylight air raids [9] that the RAF was quickly forced to bomb by night, for which it was ill-prepared; its pitiful results until August 1941 were laid bare to the British by the Bensusan Butt report [10]. The result of the Battle of Britain was unlikely to have been reversed by microwave radar, and the accuracy of the Luftwaffe's Blitz bombing, reliant on the metric radio aids Knickebein, X- and Y-Verfahren [11], would not have been greatly enhanced, if at all. The Barbarossa campaign in Russia, extending over thousands of miles, revealed no obvious need which microwave radar could have fulfilled; likewise, the campaigns of the Afrika Corps. The winter of 1941/2, which saw Japan join the Axis and sweep through Eastern Asia and the Pacific islands, likewise offered very limited scope for microwave radar to have enhanced its military victories. At sea, the U-boat campaign could have benefitted from that technology only in a very few, extreme, cases.

However, as early as the end of 1942, 24 RAF aircraft were equipped with "H2S" radars, which operated at a wavelength of about 10cm [14]. This became known to Germany in February 1943, when a bomber so equipped was shot down near Rotterdam. The first German radars for aircraft in this wavelength range were produced by the Telefunken company as a copy of the British H2S radar, and by the end of 1943, such a radar, with the designation "Berlin A" (Funkgerät [FuG] 224), was released in Germany, operating at a wavelength of approx. 9cm [15].

To summarise German developments to that point, as early as autumn 1933 there had been concrete measurements in Germany at the Reichsmarine's (NVA) Nachrichtenmittel-Versuchsanstalt at wavelengths of 13.5cm, but the available transmitting power of 100mW was too low to receive reflected echo signals. After it became possible in the autumn of the same year to procure, from the Philips company in Eindhoven, Holland, six magnetrons, operating at a wavelength of 50 - 60cm and generating a continuous wave power of 40 watts, further experiments showed increasing success in increasing the achievable range at decimetric wavelengths and further research in the more challenging field of centimetric waves was not given any further priority.

There was a clear conviction among German scientists that centimetre waves were unsuitable for detection purposes, and in January 1939, due to the high level of development and reliable operation of existing, metric, devices, it was decided to continue to concentrate on existing wavelengths in the range of metre or decimetre waves, even though research in the field of decimetre and centimetre waves had a long tradition in Germany, going back to the 1920s, until the research work in the field of centimetre waves in particular was finally discontinued in November 1942.

In addition to the well-known industrial companies such as Siemens and Telefunken, and universities, other research institutes were also active in the field of radar technology. In this context, the "Drahtlostelegraphische und Luftelektrische Versuchsstation Gräfelfing (DVG)" founded as early as 1908, and the subsidiary institute "Flugfunk-Forschungsinstitut Oberpfaffenhofen (FFO)", founded in 1937, achieved great progress in the field of transmitter technology for radiolocation and navigation [16], [17], [18]. Tests were carried out in the wavelength range below 10cm and in 1937, for example, it was possible to carry out retro-reflection tests against an Ammersee steamer at a wavelength of 10cm at a distance of 1km and also to determine its speed by evaluating the Doppler effect. Further work of the FFO dealt e.g., with backscatter measurements on aircraft models to determine the frequency dependence of the radar backscatter cross section [19].

Hans Erich Hollmann

Fig. 3. Hans Erich Hollmann
Source: Hochfrequenztechnik und Elektroakustik, Feb. 1961, Bd. 70, Heft 1

Of three key German scientists, as early as 1928, Hans Erich Hollmann completed his doctorate at the TH Darmstadt on the generation of electrical oscillations of very high frequency [20]. He continued his work in this field first from 1928 to 1930 at the German Research Foundation (DFG) and then from 1930 at the Heinrich Hertz Institute for Vibration Research in Berlin. There, together with Gustav E. Leithäuser [21], who had been director of the institute since 1927, he succeeded in generating and receiving centimetre waves. In 1936, the second volume of the two-volume work "Physics and Technology with Ultra-short Waves" was published with the title "Ultra-short Waves in Technology" [22]. Here, he deals with „....the wide field of practical applications, as well as the phenomena whose control and knowledge is necessary for working with ultrashort waves" [23].

With the publication of this work, Hollmann was certainly ahead of his time, which had the consequence that the two books initially received little attention or distribution, and only later was great interest shown in his work, once the use of ultrashort to decimetre wavelengths was recognised for their importance for television and for emerging radar technology. In the second volume Hollmann deals with "seeing with electric waves" and describes the first findings that point in the direction of the radar principle [24]

Shortly afterwards, also in 1936, he founded his private "Laboratory for High Frequency Technology and Electromedicine" in Berlin; the aim of his work, carried out with his brother, being the application of physical-technical research to biological questions [25]. Additionally, as a consultant for the Tonographie company since mid-1933, and later as a scientific advisor and consultant to the newly-founded GEMA radar company from the beginning of 1934, Hollmann contributed significantly to the initial successes, such as the Freya radar, which made GEMA a national leader in this field [26].

In 1942, Hollmann was appointed scientific director of the newly founded Forschungsgesellschaft für Funk- und Tonfilmtechnik e.V. in Berlin and was able to pursue his scientific work in addition to supervising laboratories and research facilities [27], [28]. In doing so, he referred to works by GEMA. However, Hollmann had to submit his manuscript to the navy for censorship due to secrecy obligations and many text passages were deleted. This procedure greatly offended his desire for freedom of expression. The shareholders of the GEMA, Paul-Günter Erbslöh and Hans-Karl von Willisen, were horrified and had to realise that they had to submit to the regulations on secrecy in order to continue to work in the field of radar. Hollmann was not willing to bow to the rules of censorship and stopped cooperating with GEMA [29].

After the war, Hollmann would be appointed associate professor for special high-frequency technology at the University of Jena, but after a short time he moved to the USA, where he very successfully took on new scientific challenges in various functions [30].

Abraham Esau

Alongside Hollmann, Abraham Esau worked in the field of the development of ultra-short waves (VHF) and laid the foundation for later VHF technology with his work on the development of ultra-short-wave broadcasting in the mid-1920s [31]. Esau had studied mathematics and physics at the TH Berlin and received his doctorate in 1908 at the University of Berlin. In the year 1912 he joined the "Telefunkengesellschaft für drahtlose Telegraphie mbH" in Berlin.

After being sent by Telefunken to Togo in 1914 for the commissioning and experimental operation of a large-scale radio station in Kamina, he soon fell into French captivity and was taken to a distant prisoner-of-war camp in Central Africa and later to France. In 1917 and was then transferred to an internment camp in Switzerland until 1918 [32]. After his imprisonment, he returned to Berlin and quickly rose to become head of all Telefunken laboratories in 1919.

Fig.4. Abraham Esau
Source: Stiftung Deutsches Technikmuseum Berlin, Foto: Historisches Archiv (I.4.048 F 015-003)

In 1925, he followed a call to the University of Jena, where he first became associate professor and director of the Institute of Technical Physics and, from 1927, full professor of technical physics, devoting himself above all to the development of the ultra-short-wave range [33]. In 1939 he was appointed president of the Physikalisch-Technische Reichsanstalt (PTR) and held this office until 1945. In 1939, he was also appointed full professor for telecommunications technology at the TH Berlin and during the Second World War, he was Commissioner for High-Frequency Research, where he was particularly concerned with the development of wavelengths in the centimetre range. At the beginning of 1944, he succeeded Johannes Plendl as "Bevollmächtigter für die Hochfrequenzforschung" [34].

Postwar, after imprisonment in France, Germany, and the Netherlands until 1948 because of his role in National Socialism, he would receive an honorary professorship at the TH Aachen and in 1953 take over the management of the newly created Institute for High Frequency Technology of the German Aviation Research Institute (DVL) in Mühlheim/Ruhr [35].

In addition to his research work in the field of VHF technology, Esau also devoted himself to radar technology, here with the aim of generating wavelengths in the centimetre to millimetre range. As early as 1937, he conducted experiments in the wavelength range of 5mm with a magnetron.

On 10 December 1937, Esau reported on the generation of centimetre and millimetre waves in his lecture "Electric waves in the centimetre range" at the German Academy for Aeronautical Research [37], He had recognised early on that these wavelengths could be used to open up new areas of application, especially in the field of radar technology:
„I now turn to the wavelength range that forms the actual content of my lecture today, i.e. the waves that are smaller than 10cm. These are wavelengths which have not yet been used for technically significant problems, but which will be usable in the very near future, both for the actual transmission of messages and for certain problems which are of particular importance for aviation and the solution of which can perhaps be achieved with them: the localisation, altitude determination of aircraft and others" [38].

He further stated:
„As far as the attainable energies are concerned, it has been shown that a maximum of 3 watts can be produced at a wavelength of about 4 to 5cm, an amount which may seem very small compared with the powers attained in the decimetre range of several hundred watts, but in relation to the powers attainable with other methods can already be regarded as extraordinarily high" [39].

Later he added:
„The ranges obtained so far are more than 30km with a wavelength of about 4 to 4.5cm, 15km with a length of 2.5 to 3.5cm and about 6km with waves of 2.3cm. Waves of about 5 to 6cm should give ranges of more than 100km today if optical vision is available" [40].
Esau was also aware that research was also being carried out abroad in the field of centimetre wave technology, among other things, for locating aircraft, and that initial trials had already been undertaken, stating:
„The solution of these larger problems with the aid of centimetre waves has also been envisaged abroad, and certain experiments have already been made in this direction. On the one hand, stationary objects can be detected with centimetre waves by using the reflection effect, and on the other hand, moving objects such as aeroplanes can be detected. Even if the distance over which such experiments could be carried out must initially be regarded as very small - a little more than 1 km is given - it seems to be only a question of time before it will be possible to allow considerably greater distances" [41]
For Esau, it was thus only a question of time before greater distances could be achieved. He also saw other possible applications for this wavelength range, such as radio beacons for landing and directional purposes in fog, determining the ground speed of aircraft, determining altitude from aircraft using the echo method. He was convinced that this wavelength range was suitable for solving the challenges mentioned.

Esau states at the end of his lecture:
„The necessary preliminary work has been successfully carried out, as I have tried to show in my lecture. Now the second phase begins! The development of the devices and their application to the problems mentioned, which we hope will soon have a successful effect on aviation, both with regard to communications and for navigational purposes." [42].
This lecture was given on 10 December 1937. It would not be until much later, in 1943, that this second phase began, after parts of a radar operating at a wavelength of about 9cm were found in a British bomber shot down near Rotterdam.

Wilhelm Runge

Fig. 5. Wilhelm Runge
Source: Stiftung Deutsches Technikmuseum Berlin, Foto: Historisches Archiv (I.4.048 F 049-003)

The Head of Equipment Development at Telefunken, Wilhelm Runge [43], saw the practical application of wavelengths in the centimetre range as extremely important, even though technological mastery of receiver technology was still insufficient. For radio links with directional antennas, on the other hand, he saw the bridging of a distance of 100km as easily realisable, even with low transmission power, at wavelengths of a few decimetres. In his lecture "The Practical Significance of Microwaves", given simultaneously with Esau at the German Academy of Aeronautical Research on December 10, 1937, Runge saw the challenges to both transmitter and receiver technology down to the range of about 20cm as capable. However, he still saw need for considerable research and development in the area of even shorter wavelengths. He stated in his Lecture: „... From this point of view, microwaves down to about 20cm can be described today as technically fully developed and usable. Standards for the range down to a few centimetres are technically developed, while satisfactory solutions to the receiver problem in this band are only just being developed. How waves below a few centimetres will one day be received operationally, I believe no one knows today. It is a matter of free research, not of the economically and time-bound engineer, to find the way here." [44].
It should be noted that Telefunken had already been conducting experiments with magnetrons in the decimetre range since the end of 1933, especially in the field of radio relay technology, and thus had a great deal of practical experience. As early as 1934, Runge reported on the advantages of using decimetre waves below 1 metre for radio links with directional antennas in a lecture entitled "The Basics of Decimetre Wave Technology", which he gave on 21 March to the Academy for Aeronautical Research [45]. In December 1934, he reported in the Telefunken-Zeitung under the title "Generation of Oscillations with the Magnetron" on the technology and the mode of operation of various magnetrons [46].

Following the February 1940 'development stop order' by Goering, some centimetric research was discontinued, although this was mainly due to the military demand for diverting manpower into front-line troops, and the consequent lack of research personnel; only Telefunken worked on the research and development of centimetre waves in its laboratory in Berlin under the direction of Dr. Waldemar Illberg. This laboratory had been founded several years before the start of the war but achieved little success with its few available resources [47]. Runge also saw the problem in this wavelength range as being primarily the sensitivity of the receiver rather than power generation by the magnetron [48].

On 22 November 1942, Dr. Runge and Dr. Karl Rottgardt, a member of the Telefunken management, gave a demonstration of the latest development of the "Würzburg" radar device at a wavelength of 50 cm to military personnel at Telefunken in Berlin-Zehlendorf. It was repeatedly explained that centimetre waves were "reflected away" instead of being "reflected back" [49]. It must be emphasised that Prof. Dr. Esau of the Physikalisch-Technische Reichsanstalt had always denied this.

On 27 November 1942, General Erich Fellgiebel, General Plenipotentiary for Technical Means of Communication (GBN), visited Telefunken's production facilities together with Wolfgang Martini, General of Air Signals, and since November 1941 Special Representative for Radar Technology, and Admiral Maertens. In this context, Runge explained that the transmitters and receivers for centimetre waves developed in Dr. Illberg's laboratory would achieve only a small amount, combined with a higher financial expenditure for the development and production of centimetric equipment . As a consequence, Fellgiebel decided to discontinue work with centimetre waves in order not to fragment further the specialists, already few in number, and therefore, to concentrate on the further development of metric/ decimetric wavelengths, Dr Illberg's laboratory was disbanded [51].

Remaining outstanding was a request by Martini of 8 August 1942 to resume research in the field of centimetre waves; this was rejected by the Technical Office of the Reich Aviation Ministry (RLM) on 15 January 1943 on the grounds of increasingly limited chances of success and the continuing constrained personnel situation [52]. Almost at the same time, at the turn of the year 1942/43, a meeting with the military and technical officials responsible for the use of shorter wavelengths and convened by Plendl as "Plenipotentiary of High Frequency Research (BHF)", was held in Wannsee. They concluded that the effectiveness of the radiation reflected back to the transmitter increased with shorter wavelengths, but that it would drop off at 20cm, and consequently be useless at shorter wavelengths. This misjudgement also led to the total discontinuation of all work in Germany in the field of centimetre waves [53].

The story of the work under the 'Rotterdam Protokol' to rebuild a copy of H2S from the material recovered from a crashed Stirling bomber at Rotterdam, and to develop copies of centimetric radar, specifically various models of 'Naxos', for German use, are well-covered in the literature, but, as it is often implied that these were few in number, attention is specifically drawn to a paper by Arthur Bauer showing that several hundred sets of Naxos were produced. [54]

Centimetre Waves – Timeline Germany

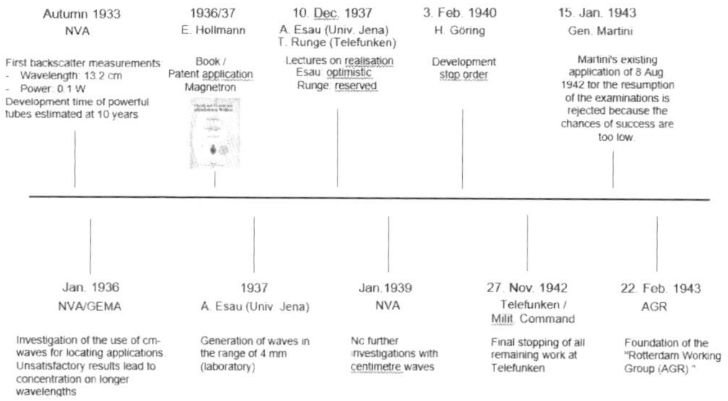

Centimetre Waves – Timeline Britain

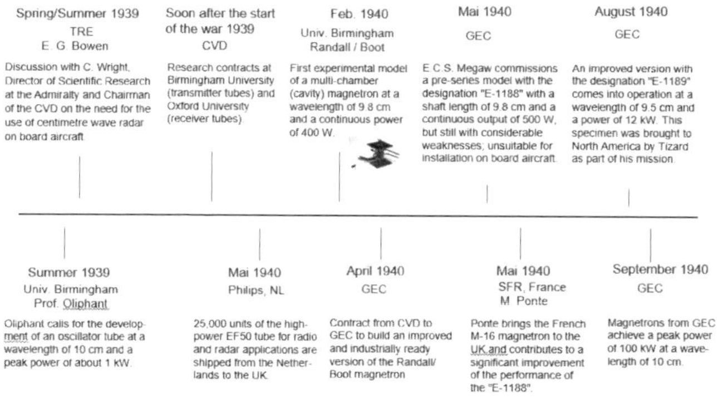

3. Britain.

The praise which has been lavished on the British development of the resonant multi-cavity magnetron and its transportation to the USA by the Tizard Mission in 1940 has been so showered with superlatives – "the most valuable cargo ever brought to these shores" [55], "the single most important piece of electronic equipment developed during the course of the war" [56] - that the historian may well wonder, with the Americans who received the magnetron, whether the appropriate response is simply to 'sit and gasp in amazement' rather than to attempt to assess its complicated genesis, and its path to making microwaves a weapon of war.

Analysis reveals a story with many twists and turns, a number of which do not reflect high credit on those involved. This paper now traces that path, but does not consider in detail American production following the Tizard Mission, as this has been well covered in other publications [57].

3.1. British Centimetric Radar before the Magnetron and the Tizard Mission.

In summer 1939, prior to the outbreak of war, the Air Ministry's EG 'Taffy' Bowen, responsible for developing airborne radar, had considered the need for compact high-power valves for centimetric wavelengths with Charles Horton of the Royal Navy's Signals School [58]; both aircraft and ships had little space for large aerials, and the use of microwaves was one answer.

The Royal Navy had previously made use of early low-powered and relatively fragile split-anode magnetrons in transmitters for microwave communications between larger warships [59], and the Navy, who had the longest military pedigree in radio valve (tube) research, managed Britain's Coordination of Valve Development (CVD) process, which coordinated the development of valve research in Britain on behalf of all three military services. Research groups, including those under Professor Mark Oliphant at the University of Birmingham, Brebis Bleaney at the Clarendon Laboratory in Oxford, and at Bristol University, had been contracted to develop powerful transmitting and receiving valves in the 10cm wavelength range [60].

The primary pre-war challenge anticipated by Britain was that of aerial onslaught from the German Air Force, the Luftwaffe. To meet this challenge, by day, Britain's metric-wavelength Chain Home early warning radar could locate Luftwaffe bombers at over 100 miles range, and direct RAF fighters to within 4 miles of their position [61]. On a clear day, 4 miles is a distance at which the human eye can discern a target such as a ship or aircraft, and, fortunately for Britain, the summer weather in 1940 was bright and clear. That summer's aerial Battle of Britain was hard-fought, and in autumn 1940 the Luftwaffe increasingly left the sky to the RAF by day - but continued their attack in the lengthening nights.

By night, and particularly autumn and winter nights of cloud, rain, fog and mist, the unaided human eye can only very rarely distinguish a blacked-out target at 4 miles range. Due to the restricted size of aerials which night-fighters could carry, Britain's metric air interception (AI) radar was effectively limited in range to the height of the fighter above the ground, some 10,000 to 15,000 feet (2-3 miles) [62]. There were two potential solutions:

- place a metric ground radar between the Chain Home long-range ground radar and the fighter's short-range air interception (AI) radar; this third radar, 'Ground Control of Interception' or 'GCI' radar, would guide the night-fighter closer to the German bomber to allow the fighter's radar operator to switch on the AI radar with a chance of acquiring the target; or
- develop a shorter wavelength, centimetric, AI radar, focussed like a torch beam by a mirrored reflector to generate a narrow radar beam unaffected by ground returns, with a range of more than 4 miles.

The first option, development of Ground Control of Interception (GCI) radar, was championed by WB "Ben" Lewis [63], newly arrived from Cambridge and rapidly to become deputy head of Air Ministry radar research, second only to AP Rowe. GCI, a metric system, would not enter service until November 1940, and was not truly effective until it was coupled with a redesigned and reliable AI radar in the spring of 1941; this was AI Mark 4, developed by Electrical and Musical Industries' (EMI's) world-class circuit genius, Alan Blumlein [64], the inventor of stereo, with some 80 patents to his name.

"Taffy" Bowen, the pre-war and early war team leader of AI research, saw the second option, a microwave radar, as the true solution, and he had persuaded Sir Henry Tizard, at that date the Scientific Adviser to the Chief of the Air Staff. Following discussions between Sir Henry Tizard and GEC's Director of Research, Sir Clifford Paterson on 6 November 1939, and later with Watson Watt as Director of Communications Development, a project had been awarded on 29 December to GEC to develop a 25cm AI radar [65] at GEC's Wembley research laboratories, well furnished with skilled engineers and test equipment to meet this challenge; the hope was for a transmitter power of 1kW yielding a 5-10 mile range, which would allow Chain Home to position a night fighter with a chance of making an interception. With Bowen acting as the Air Ministry liaison after 16 January [66], the GEC team (Marris, Espley, Edwards, Clayton and Cherry) focussed on using their E1130 "millimicropup" valve; and on 14 March 1940 they claimed a successful ground demonstration of a 25cm AI [67]. On 28th March, it was decided that AI systems operating at 25cm and '12cm or thereabouts' would be GEC's dual objective.

Priorities for Bowen's own small, overloaded, and frequently relocated, team had been those of fitting metric radar to fight the immediate war. Bowen's deputy, Gerald Touch, thought it essential to prioritise making metric radar operational [68], and devoted himself to that aim, essentially breaking up Bowen's team. GEC were researching centimetres but Bowen's own team, pressed to fit fighters with metric AI radar, and maritime patrol aircraft with metric ASV radar, carried out no major centimetric research before May 1940 [69], when RAF radar research moved to Worth Matravers, Swanage.

The move to Worth Matravers proved traumatic in many ways.
• Bowen's AI team was re-assigned and physically dispersed; Gerald Touch and many of that team relocated to RAE Farnborough to concentrate upon making metric airborne AI more effective.
• Bowen had clashed with AP Rowe, head of Air Ministry radar research, who had received an instruction to begin centimetric research [70], and who, apparently without advising Bowen or GEC, appointed an ex-university scientist, Herbert Skinner to head this research [71]; Bowen was left sidelined and alone. Skinner's small group of scientists at Worth working on centimetric AI was joined by a group of Cambridge University scientists, originally evacuated to Exeter, headed by the energetic, well-connected, disputatious, Philip Dee [72]. These researchers considered GEC's 25cm approach, employing scanning horn antennae, inferior to a 10cm wavelength using parabolic mirror reflectors to give tighter beams of high discrimination, rather than the wider beams resulting from horn antennae.
• Most importantly, at Birmingham, John Randall and Henry Boot had recently developed the 'resonant cavity magnetron', which had immediately proven capable of generating high power at a 10 centimetre wavelength, albeit at this stage their magnetron was very far from being small or robust.

3.2. The Randall and Boot Magnetron, and its results.

The physicists John T Randall and Henry AH Boot of Prof Mark Oliphant's group at Birmingham University had been working to develop a powerful klystron and experimenting with a laboratory model of a multi-cavity magnetron as a device to test the klystron. On 21 February 1940, their "cavity magnetron" generated an output of 400 watts at a wavelength of 9.8cm, and work was driven forward to increase the performance of the magnetron.

GEC's team under EGS Megaw had a great deal of experience in the field of conventional magnetrons, and so were contracted to deliver an improved and production engineered version. During this process, GEC had a quite exceptional stroke of good fortune; in June 1939, Megaw, together with Maurice Ponte of TSF, had visited Megaw's friend, Henri Gutton of SFR, who had a new magnetron, the M16, and suggested a thoriated tungsten spiral filament to boost output. Gutton had experimented further with a large oxide cathode, resulting in a tenfold increase in output. On 9 May 1940, Ponte brought samples of Gutton's M16 to GEC, stimulating a significant redesign with substantial increases in power. The magnetron was now optimised in design to a small block of copper, accurately machined with a series of interlocking cavities around a central cylindrical cavity for the large oxide cathode, and with the necessary magnet around it being a compact permanent magnet, rather than an electro-magnet with its demands for significant electrical power.

Fig.6. The first Randall and Boot magnetron, without electromagnet and water-cooling (Image credit: Science Museum, London)

The resulting laboratory model, based on the work of Boot and Randall, was of modest size and physically robust, designated "E1188", completed on 25 May 1940, and first put into service on 29 June 1940. The magnetron was initially operated in continuous wave mode, but after the pulse modulator was developed by EMI's Alan Dower Blumlein (1903 – 1942) as part of his earlier metric AI4 radar contract, a pulse power of 1 kilowatt at a wavelength of 10cm could be generated. The E1188, the first model that could be mass produced, quickly reached a peak power of 10kW in pulsed operation.

That success stimulated the Air Ministry to award a contract on 3 May to GEC for a 10cm (S-Band) AI, "AIS" [73]; in early June, GEC's parallel 25cm AI achieved ranges of 6 miles on ground targets [74]. Units and parts of both 25 and 10cm variants were demonstrated at an AI meeting on 8 July with Watson Watt present [75]. The work of GEC had the extra merit that, as its team were used to production engineering, this equipment was being produced with blueprints ready for mass production [76]. It is important to note that, at this time, the magnetron was not incorporated because its reduction in size to a volume easily capable of fitting into an aircraft was achieved only later, by July - by the same GEC laboratory as was working on the 25cm AI.

However, since May and in parallel with this work at GEC, Philip Dee at TRE Worth Matravers had been energetic, despite 'facilities' (an old stable and a large trailer) which were totally inadequate,. The challenge as described by Dee in his diary for 15 May, his first day of work, reads:
"Our main task is apparently to devise an AI equipment which will detect enemy bombers at the greatest possible range. Present 1½ metre AI equipment suffers from the defect that, radiating approx. uniformly in all directions, contacts cannot be obtained at ranges greater than one's own height owing to radiation reflected from the ground. The general idea therefore is to make a beamed system using the smallest possible wavelength and the largest possible parabolic mirror which can be put in an aircraft. This beam will have to scan in order to search. The key man upon AI at AMRE is a Dr Bowen who is in great disgrace .... Bowen has apparently had severe political rows with Mr Rowe. I think because Bowen exposed the lack of interest by Bowen and Lewis in the AI work which he, Bowen, and a group of people had been doing at St Athan. Bowen apparently was the person who put Tizard wise of this situation and probably inspired Tizard's visit

... a short while back when he, Tizard, beat up Rowe for his lack of interest in AI work. Neither Mr Rowe nor Lewis had ever visited St Athan and the people had to work in unheated hangars with no facilities and not even any subsistence. Work upon a cm AI has been in progress since January of last year when a contract was placed with the GEC. GEC has an overall contract to develop the equipment and appear to be the directing authority. GEC had a rough model of a scanning horn by March of this year but at that date emphasis appeared to be on 25cm and everyone was very worried about the possibilities of scanning such large mechanical assemblies. Hodgkin has been working on polar diagrams on 25cms and has established that a suitable horn would have to be very long. By the time of my arrival, opinions seem to have hardened at AMRE in favour of a system on wavelength of 8-10cm but GEC are pessimistic about the possibility of getting sufficient power on this wavelength and are pressing for a 25cm system. As an alternative to scanning, Espley of the GEC has a scheme with 4 transmitting horns to which the radiation is switched, in turn giving L-R, Up/Down indications".[77]

Fig. 7. The stables at Leeson House, Langton Matravers (Image credit: MS Dean)

The facilities are best described by Curran, a member of the team:
"There were so many mice in the stables that they had to be chased from the work benches, and trousers were often tied at the bottom. Some of the aspiring laboratory assistants found that their nice new notebooks had become a bundle of pages in the morning if they forgot to lock them in a metal cupboard: all the glue of the binding had been consumed in the night". [78]
Much inspiration came from Dee who had managed to shape a policy for centimetre wave research and somehow welded the group of individuals into a single project team.

Dee tried to obtain more people, but - this being the time of the Dunkirk evacuation – TRE was focussed upon immediate work on Chain Home and metric AI. Dee therefore went to visit Oliphant at Birmingham, who had a 200 watt klystron with a mirror reflector aerial working on his laboratory roof, the klystron being "not a sealed-off valve, but at that time only a collection of bits and pieces". Dee took the klystron drawing back to the Cavendish in Cambridge, and there asked Mond Laboratories to make one. Mond produced the klystron in a month, and GEC managed to make a 2kW magnetron by 20 June. From a previously unpublished series of file notes [79], it is apparent that Rowe had been following centimetric developments with more than passing interest, and at Rowe's bidding, from 12 June to mid-July Dee ran a series of tests to confirm that ground reflections would not be a problem at 10cm [80].

Fig.8. Philip Dee in his office at Malvern (Image credit: Group Captain Dudley Saward)

Dee also visited all the manufacturers involved in centimetric research. He found both EMI and STC 'working in the dark' owing to secrecy regulations, using equipment of 100mW power, unbriefed even on the aims of the project and unaware of what others were doing. [81] By contrast, due to their direct contact with Oliphant, GEC's 25cm AI team were well-briefed, and their magnetron team headed by EGS Megaw was developing 10cm magnetrons of increasing power, compactness and robustness; but GEC saw work under their 25cm AI contract as their own business, not to be shared with others, and with TRE restricted simply to testing for approval whatever equipment GEC produced.[82]

Dee therefore used his pre-war contacts with Sir Frank Smith, then Permanent Secretary of the Government's Department of Scientific and Industrial Research, now Controller of Telecommunications Equipment (CTE) in charge of all radar procurement. [83] Dee met Smith and persuaded him that the present situation 'could not be allowed to continue'. The result was a series of liaison meetings, chaired by Bowen, to ensure that all parties shared information fully, with GEC now to work under the 'guidance' of TRE.

Fig.9. Clifford Paterson, GEC Director of Research (Image credit: GEC)

On 16 July Rowe wrote a "Secret and Personal" letter [84] to GEC's Director of Research, Clifford Paterson:

"I want ... to tell you what I feel about the A.I.S. situation and to appeal to you once again for help, which I feel sure will be forthcoming. It remains true that the A.I. problem is about the most important work in the country. ... The limitations of the present A.I. are well-known to you, and we have awaited some favourable discontinuity in technical progress to start an intensive effort on A.I.S., using 10cm or less. This discontinuity is now being produced by you in the form of several kW from sealed off magnetrons operating at about

10cms. My personal view is that we must stop at nothing to push on with this work, and I have asked that it be given high priority here.
2. In this connection you can once again be of the very greatest help to us. It has become of the very greatest urgency for us to have samples of the new magnetrons for the short-wave developments here and I am wondering how we can most quickly obtain the development of specimen tubes to this establishment. ....
3. As you know, I have no authority whatever to ask you to spend money on behalf of the Air Ministry..... I am appealing to you personally simply because I am fearful of the delays involved in going through the normal channels, but having stated our desperate need, I am leaving it to you to decide how far you can help us. I may be going outside the limits of even your goodwill, but I most earnestly hope that this is not so. "

Rowe was advising Paterson that:
• Dee's small group of scientists, housed in a remote vermin-ridden stables in Dorset with no test facilities and mere weeks of expertise, had simply been awaiting an excuse to take over centimetric AI research, despite the facts that GEC had a formal contract, extensive laboratory and production facilities, and had demonstrated a production engineered AI, albeit on 25cm;
• GEC had provided such an excuse by their work on packaging the 10cm magnetron, so that it could now be used in AI; and
• since Rowe had no magnetrons, would Paterson please send some urgently, at GEC's own cost, so that Rowe's scientists could carry out the microwave research they had so brusquely taken over.

Paterson's reply of 18 July has, curiously, disappeared from the file [85], though the first powerful GEC magnetron was received at Worth Matravers on 20 July. The historian can only speculate on a 21 July file comment by Rowe [86] that "...we are up against a basic difficulty. Dr Paterson thinks he is developing the whole of 10cm AI whereas we feel firms should be doing specific parts. All this has been said to the Air Ministry and we are waiting for a reply. Our reply to Dr Paterson will therefore need to be rather non-committal".

Fig. 10. Sir Frank Smith (Image credit: Wikipedia)

In fact, Dee's direct contact with Sir Frank Smith had already achieved its end: a directive of 19 July confirmed both that TRE were to have charge of all centimetric developments, and that GEC had to pass its work across to them,. Professor Burns has well summarised GEC feelings [87] in their subsequent meetings with Dee and Skinner; from Paterson's perspective, TRE had "no radio engineering knowledge, no test equipment, no expertise, no team-working capability nor any articulate idea on the subject"; Dee was "seriously intolerant and entirely satisfied with his own ideas, which are often narrow and unsound", while Skinner "has no conscience in appropriating credit to himself and Swanage which should go to Wembley". Paterson's philippics include "I hope Rowe will manage to keep his wild men from doing too much harm" and "our academic-ridden radio effort is poisoned with intrigue, jealousy and uncharitableness. Industrial and commercial life seems clean in comparison..... Truly the Professor type needs to learn how to give honour where honour is due". The depth of ill-feeling between TRE and GEC, researcher and manufacturer is well-evidenced and did not augur well for speedy success.

The feelings of Bowen, the original liaison with GEC, can be imagined, but he was without power or support; Churchill having become Prime Minister in May had brought Lindemann, Tizard's opponent, as his scientific adviser, and Tizard had lost further credibility in late June, having been mistaken about the German bombing beams. He was sent to head a scientific mission to the USA, the 'Tizard Mission', whose terms of reference he himself drafted and to which he secured Churchill's agreement:
"To tell them what they want to know, to give all assistance I can on behalf of the British Government to enable the armed forces of the USA to reach the highest level of technical efficiency".[88]

Tizard insisted on Bowen's being a member of the Tizard Mission, which probably came as some relief to all parties, but which also gave the Mission a scientist who could speak with unique authority on airborne radar, both AI and ASV, and on the cavity magnetron he carried with him; Bowen had been a key figure in all these developments from their inception.

As Bowen was briefing American scientists and industry on the secrets of the magnetron, GEC continued to work on AI, focussing on a common transmit and receive antenna. But the effect of TRE and GEC competing against one another throughout July and August was corrosive – the urgent research appeared to observers to be slowing down; Burns reports [89] FS Barton, a Farnborough scientist, commenting that the "new arrangement seemed to have the only effect of slowing Wembley down to the pace of Swanage", while TRE's Batt [90] recounts the depression of the Government scientists after seeing how advanced the work of GEC was, and TRE's elation when TRE ideas showed results.

On 3 September, TRE and GEC met, with Dr Walmsley of the Ministry of Aircraft Production (MAP) as chair – after a reorganisation, MAP had taken over from the Air Ministry both the allocation of contracts for the RAF, and the management of their research. The forum was a new "Committee on 10cm wave AI" [91], which also included EMI, Oliphant's Birmingham researchers, and RAF Fighter Command in addition to GEC and TRE. At the first meeting [92], GEC emphasised that their equipments were already "drawn up and blueprinted"; Dee forcefully argued that Swanage should "formulate the technical character of the complete AI system"; GEC riposted that Swanage had "only recently been in the position to make any real contribution towards the solution of the AI centimetric problem". Walmsley, the MAP Chairman, pointed out that MAP had the power of control and supervision, and clearly had words outside the meeting, reporting back to the meeting of 28 September [93] that GEC's Paterson had agreed not to press the point, but "would like the idea of reciprocity to obtain as far as possible". Walmsley did not entirely take TRE's side, inquiring for example why it was necessary for TRE to build copies of GEC-designed equipment [94], since GEC could presumably provide these more easily;.

Rowe wanted to leave no room for doubt or ambiguity. His file note of 16 October [95] to his deputy Lewis says "the minutes (of Walmsley's Committee) ... give me no evidence that we are running the 10cm AIS. The arrangement promised ... whereby Dr Dee (and presumably his deputy) should give instructions to GEC and EMI does not seem to be working. ... I would like you to tell me whether we should again go into battle on this subject".

At this time, summer 1940 had faded into autumn and the Battle of Britain was drawing to its close; as the nights lengthened, attention turned to the urgency of combatting the night Blitz which would certainly follow. Rowe was facing a strong RAF challenge that centimetric radar 'was for the next war'; all effort should be focussed on metric radar which was both adequate and all that was available. Air Marshal Joubert, Assistant Chief of Air Staff for radar, visited TRE on 16 September [96] to emphasise that metric AI, with GCI radar under urgent construction, was all that was needed - the night 'Blitz' was now imminent. Dowding, Commander-in-Chief of Fighter Command, emphatically reinforced the point to the TRE scientists two days later [97].

Joubert recommended TRE to concentrate its centimetric radar researches on anti-aircraft gunnery [98], a simpler task than AI, and he was followed by Walmsley, in opening the meeting to which Rowe refers, who mentioned gun-laying as a likely application [99]. Rowe seized the initiative, and by the time of his October note, was attempting to take over all Army centimetric anti-aircraft gunnery research [100], while its head, John Cockcroft, was out of the country as part of the Tizard Mission – this being in addition to, not instead of, ignoring the instruction to focus upon metric radar and supporting his centimetric team of university scientists in every way.

Lewis responded to Rowe's October note that "Dee had now returned (he had been on sick leave suffering from pneumonia) and will soon be resuming his authority in the matter" [101]. The 26 November meeting's draft minutes are very heavily amended by Dee [102], indicative of a poor working relationship; interestingly, few of the actions from the meeting are taken by TRE.

From September onwards, centimetric AI was now being pursued in parallel by three groups - GEC; TRE, and from a standing start with pace and energy by the Americans, using the knowledge, techniques and technology gained from the Tizard mission. A summary of the developments in Britain may be summarised by the key points [103] that GEC worked more or less unhappily with TRE, providing the silicon for the TRE AI system's receiver crystal mixers in September [104] and the helical scanner in December; managing flight trials in February 1941; and refining and improving the offering until the prototype design, AI Mark 7, emerged in November 1941, and the production design, AI Mark 8, in early 1942. It is revealing that during this period the Americans had developed a centimetric AI radar, SCR-520, which they installed in a Boeing 247 aircraft and flew to Britain with 'Taffy' Bowen to demonstrate to TRE. TRE's response was, at best, flat; the transmitter performed better than the TRE version, the US receiver performed worse, but an order for 200 authorised by Dowding was cancelled on the grounds of excess bulk, weight and power consumption (and, one may suspect, 'not invented here'). The SCR-520 was used by the USA for ASV, and, once significantly reduced in size as the SCR-720, adopted widely by the RAF later in the war and post-war as AI Mark X, one of its main AI systems.

If one postulates a not-unreasonable estimate of six months' delay to service use of centimetric AI arising from fractious working relationships, such that a 25cm AI might have entered service early in 1941, a 10cm prototype by late Spring and the production variant by mid-year, what might the consequences have been? There had been, fortunately for the British, little negative impact in war-fighting terms from the delay; the Luftwaffe had since May 1941 been fully occupied in Operation Barbarossa, the German invasion of Russia, and air-raids on England were on a much smaller scale that during the 1940/41 Blitz, which in its final month, stages metric radar, GCI and Blumlein's airborne AI Mark 4, had been able to contain. Certainly, the grievous casualties and damage of the large raids on London of May 1941 might have been mitigated; but the impact in terms of total war, rather than human, terms, was not critical.

We now contrast the TRE's confrontational approach with that of the Royal Navy, who brought the first military centimetric radar speedily into operational service.

3.3. Britain's first operational centimetric radar: Royal Navy Type 271.

British disputes and dissensions might appear to have rendered worthless the nation's lead in centimetric radar. This was not so: the nation's first operational deployment of centimetric radar was, however, by the Royal Navy, not the RAF or the British Army, and for the most part their translation of centimetric discoveries into weapons of war did not depend on Rowe's TRE, even though the first steps certainly took place at Swanage, or more accurately, at Leeson House, Langton Matravers.

It was at Swanage at the end of October 1940 that a demonstration of the detection of small ships by an experimental 9cm radar mounted in a trailer was given to Cdr Fawcett and Lt Cdr Bayldon, Fawcett being responsible for Navy liaison with the RAF's Coastal Command, and both being in the Admiralty's Anti-submarine Warfare Division [105]. More senior Naval officers followed on 8 November, and further interest was generated by the tracking of a submarine, HMS Usk, on the 11th [106]. Within a week, the Admiralty despatched four scientists (Landale, Cochrane, Croney and Owen [107]) to Leeson House; they stayed for a month, working closely with TRE's Skinner, notably milder and more co-operative than Dee..

Only a fortnight later, the Admiralty party had constructed a microwave installation of their own, "Apparatus C", and then built it into a full system, including antenna, inside "the Admiralty trailer" [108]. Trials of range performance at low antenna (mast-head) heights were carried out against a small (92 ton) boat; TRE's Bernard Lovell then suggested to them the use of cylindrical paraboloid "cheese" aerials [109], later to become standard in most small boat installations, as a compensation for the natural pitch and roll of a smaller warship such as a destroyer or frigate at sea,. The 'cheeses' proved satisfactory in tests on 19 December, and very shortly afterwards, the "Admiralty trailer" departed to Eastney, Portsmouth.

Fig.11. Type 271 centimetric radar: the first in the world into operational service (Image credit: HMS Collingwood Naval Museum)

At Eastney, in the Admiralty laboratories, 12 sets were hand-built as a matter of utmost urgency; within two months, by the end of February 1941, the first was ready, 11 more almost ready, and parts ordered for a further 150 [110]. The Admiralty's industrial partners were Allen West Ltd of Brighton, whose drawing office was fully involved in the design stages of this equipment, christened Type 271X. The first 271X was installed on the new corvette HMS 'Orchis', and successfully trialled on 25 March 1941 [111]; the design was frozen, and Allen West proceeded at full speed onto the manufacture of the further 150.

Tizard, aware of both US and Naval progress, suggested that TRE and the Navy's Signal School should co-operate more, perhaps in an attempt to stimulate TRE to speed up. The response seems to have been an interesting memorandum from Lewis to Rowe [112] observing that the Navy, in a "vigorous programme for fitting corvettes", have fitted "10cm apparatus which is very far from ideal and which it is hoped soon will be obsolete". TRE's AIS, by contrast, has "not yet been able to accept anything but the best".

Near-obsolete or not, the Navy's Type 271 was at least operational at a critical time of the Battle of the Atlantic – TRE's AI would not be operational, even as the AI Mark 7 prototype, until November 1941, whereas by September 1941, 32 corvettes had already been fitted with Type 271, a modification to the antenna had allowed its fitting in destroyers also, as Type 272, and further development had produced a big-ship version, Type 273 [113]. Allen West's order had been expanded to 350 sets, and a redesign to optimise Type 271's mechanical engineering was commissioned from Metropolitan Vickers; 1,000 of this production design were ordered late in 1941 and completed before the end of 1942 [114].

The Royal Navy's smooth and rapid integration of its researchers with the equipment manufacturers, Allen West, the early fixing of the design, and the production of manufacturing drawings at a very early stage, contrast starkly with the dealings of the TRE scientists with GEC. It can, of course, only be speculation how much earlier the RAF could have had centimetric AI radar, but the delay postulated, of six months, seems reasonable in the light of the Royal Navy's achievements as well as those of the USA.

There were two further consequences for the RAF, to which we now turn.

3.4. Airborne Navigation and Blind-bombing Radar, H2S.

TRE's development of air interception radar on centimetric wavelengths (with the significant but little-acknowledged input from GEC) led to TRE's H2S, TRE's airborne navigation and blind-bombing radar. Urban and rural landscapes reflect H2S pulses as brighter and darker patches on the bomber navigator's display screen, providing him with a crude radar "picture" of the ground beneath the bomber. H2S would also lead to two further disputes between TRE and American scientists.

The Butt Report of 18 August 1941 starkly showed the widespread failure of RAF Bomber Command aircraft to hit their targets; photographic evidence revealed that only about 5% of bombers then setting out bombed within 75 square miles (a 5-mile radius) of the target.[115]. Lord Cherwell, Churchill's scientific adviser, who had initiated the report, told the TRE scientists on 26 October 1941 [116], that Bomber Command needed a radar bombing aid, self-contained within the bomber. Dee perhaps had already had the same idea, but – tellingly - had to obtain the approval of GEC's Dr O'Kane to flight-testing the idea, and it was O'Kane who worked on modifying a GEC AI scanner, which had been rejected by TRE, to look downwards and test the concept [117]. On 1 November - only 5 days after Cherwell's request – these tests showed that a town/country distinction was indeed possible to display, and would be useful as a bombing and navigational aid. It is revealing, as well as bizarre, that O'Kane's 'reward' was to be excluded by Rowe, the head of TRE, from succeeding meetings, on the grounds of "security" [118] – a amazing rationale, given that O'Kane would be responsible for air-testing the resulting equipment!

Within two months, Cherwell secured approval further to test this embryonic device, H2S, and to order 50 sets from EMI (not, it should be noted, GEC). Rowe instructed Lovell to take over this project [119], which at first appeared to be a modification of TRE's AI Mark 7 radar for which klystron power would be adequate; there were at this time major reservations about using the magnetron over enemy territory, for fear of revealing the magnetron to the Germans. In practice, the required modifications turned out to involve a complete redesign, for initial results were very poor, and in the middle of the test-flying, not only were the radar researchers significantly disrupted by being relocated from Swanage to Malvern; but also one of the test aircraft crashed, killing EMI's circuit genius Alan Blumlein and other key scientists, including two from GEC [120].

Churchill and his Cabinet had already been committed by Cherwell to a major campaign of area-bombing German cities and de-housing German workers [121], and at a meeting on 3 July 1942, Churchill personally instructed that 200 H2S sets had to be available by October [122], the 50 already ordered from EMI plus 150 from TRE's own Research Prototype Unit (RPU) – once more, it will be noted, trusting production to a new, untried group rather than to GEC; TRE had no expertise whatever in volume manufacture of electronic equipment and no skilled workforce with experience even of radio manufacture.

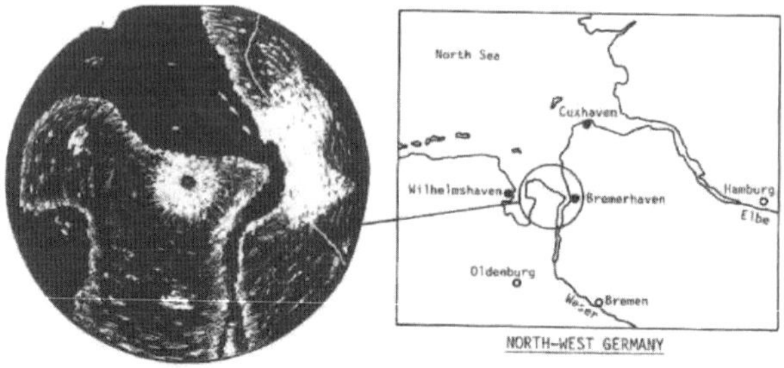

Fig. 12. The screen display of the H2S system with a map for comparison (Image credit: Crown copyright, believed expired)

The problems of H2S - which at this date, although better than nothing for the bomber crews, was truly a crude device with many faults - were known in the USA, where experiments convinced Bowen (who had remained in the USA after the Tizard Mission) that even cities of the size of Pittsburgh could not be seen on H2S [123]. Photographs of early H2S displays, together with the fact that that towns on the navigator's display could appear and disappear at differing heights and ranges, render such a view credible. Such lack of success, taken together with the fear that, if used over Germany, the magnetron could be revealed to the Germans from a shot-down aircraft (precisely as in fact happened at Rotterdam, as described above), stimulated a visit on 6/7 July 1942 – 3 days after Churchill's demand for 200 sets! - to Dee and Lovell at TRE Malvern by Isidor Rabi. Rabi was an influential Associate Director of the US radar research centre, the Radiation Laboratory, and he made a determined attempt to have H2S cancelled as "unscientific and unworkable" [124]. Hard words were exchanged, and Lovell writes [125] of "a good deal of antagonism being imposed by American opposition" to H2S.

The British saw the need for higher power and decided to solve by clearing the magnetron for flying over Germany, and there followed an intense programme of test-flying to improve results, the brunt of which was borne by GEC's O'Kane! As for production, by the end of 1942, whilst EMI had more than fulfilled its production target of 50 H2S sets, the TRE Research Prototype Unit had produced none at all [126] – totally predictably, given TRE's complete lack of production engineering or volume manufacture.

At the strategic level, on 17 August 1942 the USAAF had made its first raid on France, and quickly appreciated the difference between the clear skies of Arizona bombing ranges and operations over a cloudy and smoke-hazy Europe. Following the relative success during the winter of 1942/3 of H2S in RAF Bomber Command, and consequent demands from USAAF commanders for H2S, TRE was asked to instal H2S in USAAF aircraft by 15 March 1943, in addition to fitting those of Bomber Command [127].

By this date, it was recognised by both US and British researchers that still further improvements in H2S were needed, in particular that greater definition could be gained by changing frequency from 10cm to 3cm, but the TRE team were more than fully loaded with the work demanded urgently by both RAF and USAAF. It was a most unwelcome distraction when, late in May 1943, Lee DuBridge, Director of the Radiation Laboratory, visited TRE and proposed to Rowe, Dee and Lovell the British 10cm H2S be cancelled and replaced by the American 3cm development, H2X! [128] This origin of this proposal was apparently a demonstration given to Lord Cherwell in the USA; it was rejected only after urgent lobbying by the TRE scientists.

For the RAF, the price of this UK rejection of US manufacturing capability was that TRE had to modify 200 10cm H2S sets to 3cm by Christmas 1943 [129], a goal only partly achieved. The USAAF itself used British H2S for its first blind-bombing raid on 27 September; the US's own 3 cm H2X equipment was first used on 3 November over Wilhelmshaven. In operational trials, H2S and H2X achieved broadly similar results. Lovell has also stated [130] that the eventual success of UK 3cm H2S Mk III essentially depended upon US magnetrons, these being more powerful than the UK equivalents.

The question now to be posed is whether deliveries of H2S six months earlier, by say June 1942, might have had a significant military impact. It does not seem likely that major damage would have been caused to the German war economy in those months – part of which were summer 'light nights' so limited raids would have been flown, and Harris's Bomber Command was still of modest size. On the other hand, assuming a 'Rotterdam-type' loss of an H2S over enemy territory, H2S would have come into German hands around September 1942, rather than February 1943 – in fact at the time the whole question of centimetric wavelengths for radar was under active discussion in Germany, when a captured H2S would have disproved Runge's and Rottgardt's argument against microwave radar, vindicated Esau, probably saved Illberg's laboratory, and re-established microwave teams in other organisations. Given Bomber Command would almost be fought to a standstill by the German defences by the time of the Nuremberg Raid at end-March 1944; had the defences had a further six months to prepare, the results might have been even more negative for the RAF.

We consider next the potential impact on microwave ASV radar.

3.5 Centimetric ASV radar.

In 1942, the Battle of the Atlantic was going badly for the Allies. The introduction of the 4-rotor M4 Enigma and its 'SHARK' code rendered messages to U-boats unreadable by Bletchley Park [131], and the U-boats increasingly carried "Metox" [132], a radio receiver, which could detect the metric ASV Mk. II radar. Shipping losses increased rapidly, though many on the US East Coast were an unfortunate result of Admiral King's refusal to introduce coastal convoys until mid-year [133]. Centimetric radar offered the prospect of aircraft again being able to detect U-boats without themselves being detected, and, as a picture of the sea below an aircraft was as easy to generate as a picture of land, a modification of H2S to form centimetric ASV appeared achievable.

As described above, by the end of 1942, 24 bombers carried H2S. But there had been a separate centimetric ASV development by Ferranti working closely with RAF Coastal Command [134], radar tests against submarines having been started in April 1941 against HMS Sea Lion, and later in August, against HMS Sokol [135]. Development of this radar, known as ASVS, was slow, the first experimental version was not being test flown until December 1941. By summer 1942, ASVS radar was being production engineered by Ferranti – but on 30 September 1942, Ferranti was instructed to stop work, much to the dismay of Coastal Command, in order to speed up work on H2S, a variant of which was now proposed to replace ASVS [136]. H2S, intended for use over land at a flying height of 20,000ft, required redesign for operations over sea at 2,000ft. [137], introducing delay as well as the natural negativity following the cancellation of Coastal Command's own system. It was also apparent that there would now be conflicts with Bomber Command over the allocation of radar equipment between H2S and ASV [138], and given Cherwell was known to back Bomber Command, Coastal Command were less than enthusiastic in their support of H2S/ASV [139].

The capture of 4-rotor Enigma material from U-557 in October 1942 led to Bletchley Park's being able to resume regular deciphering of U-boat messages in December 1942 [140], with the last major clashes between U-boat wolfpacks and escorted convoys taking place in the following March [141]. The same month of March 1943 – two years after centimetric radar was first installed in warships - saw the first RAF centimetric radar air patrol over the Bay of Biscay [142]; by the end of that month, there had been 13 U-boat sightings [143], and by May, Coastal Command was detecting and attacking most U-boats in the Bay of Biscay [144]. Shipping losses decreased dramatically, from 400,000 to 100,000 tons per month [145]. Doenitz ordered his submarines to stay surfaced and fight it out with aircraft [146], but he lost 56 U-boats in April and May [147], and withdrew from the Atlantic; in Doenitz's own words, "The enemy has deprived the U-boat of its essential feature – namely the element of surprise – by means of radar. With this ... he has conquered the U-boat menace. The scientists who have created radar have been called the saviours of their country".

Assuming a delivery of microwave ASV some months earlier, around the time of the capture of U-557's Enigma M4 material, then the shipping losses of winter 1942/3 would have been reduced by perhaps 750,000 tons; this would certainly not have been trivial, but not, in the context of the entire war, crucial.

3.6 Centimetric Army AA radar.

In section 3.2 above, it was noted that Rowe had been advised to concentrate TRE's centimetric radar researches on anti-aircraft gunnery and had seized upon this to attempt to take over Army centimetric anti-aircraft gunnery research, while its head, John Cockcroft, was in America as part of the Tizard Mission. Some abortive work was carried out [148], but Cockcroft upon his return beat off Rowe's machinations [149]. Cockcroft's own research effort on centimetric radar was, however, continuously frustrated by his being deprived of the necessary electronic components and supplies in favour of the RAF[150]. The development of British AA radar is unlikely to have taken a different course had the RAF received centimetric radar earlier – indeed, the Army might have received even lower priority than it did.

4. Conclusions.

This paper has explored the reasons why Germany, unlike Britain, did not develop centimetric radar as a weapon of war. Germany certainly had the scientists and engineers, of the stature of Hollman and Esau, capable of so doing so. However, Germany chose to direct much of its technically-qualified manpower into front-line service, even though Goering's 'stop order' of February 1940 merely slowed, rather than stopped, radar research. More importantly, there was no pressing strategic need for high powered microwave radar; in 1940, the only military users of radar were the Kriegsmarine, in its few larger warships and for coastal gunnery, and the Luftwaffe, for antiaircraft artillery – the Luftwaffe had no radar-equipped fighters at that date, and indeed no airborne radar at all. However, the incursions of the RAF were, until 1942 of little more than nuisance value, and the night-fighting units under Kammhuber, formed in July 1940 and reliant upon metric and decimetric technologies, appeared quite sufficient to counter any such threat until at least the May 1942 Operation Millenium, the 1,000 bomber raid on Cologne. Perhaps most significant is the power of personalities - the mistake about the low reflectivity of centimetric waves, which was made and then strongly pressed, by prominent scientists such as Runge and Rottgardt. Their continuing dismissal of wavelengths below 20cm as useless, and their meetings and demonstrations to the military in November/December 1942, succeeded in closing down the few researches remaining. Only weeks later, in February 1943, the discovery of the British H2S/ Rotterdamgerät made their error starkly clear.

By contrast, the British centimetric endeavour met an obvious strategic need, of aerial threat detection and location to counter a clear and present danger of bombing, in particular by night. However, in Britain as in Germany, the military imperative was to concentrate on metric radar, to defeat the 1940/41 Blitz; 'centimetres were for the next war'. Nonetheless, three personalities, Bowen backed by Tizard and Watson Watt, ensured a commercial contract for GEC to create a centimetric AI. Separately, GEC were also entrusted with developing the basic resonant cavity magnetron of Randall and Boot into a usable device, through the engineering expertise of their own Eric Megaw and the important suggestions of French scientist Henri Gutton brought to London by his colleague Maurice Ponte.

The impact of a forceful and well-connected personality is shown by the arrival of Philip Dee in May 1940. A struggle developed between the Government's inexperienced and ill-equipped TRE university scientists, headed by Dee, and the well-equipped, skilled, and industrially experienced GEC team. Dee, backed by Rowe, the head of British radar research, used his pre-war contact with Sir Frank Smith to secure control of the project, supercharged by GEC's powerful, compact, and robust magnetron, but much energy was spent, and time wasted, in rivalry, dispute and reworking designs against the most obvious manufacturer of the microwave AI radar equipment, GEC.

Where such rivalry did not exist, as in the Royal Navy's working with Allen West Ltd, the speedy adoption of engineered equipment into urgently-needed naval use was dramatic; where there was rivalry, as in the Royal Air Force, the radar scientists' decisions to procure little from GEC, to reject US equipment, and instead to build up their own radar manufacturing capability from a zero base, slowed any chance of early adoption.

As it chances, the results of strategic decisions in other theatres – Barbarossa, the German invasion of Russia, and the Enigma 'black-out' of 1942 – meant that the impact of both Germany and Britain being delayed in their adoption of microwave technology in radar did not have a major overall impact on their war. Furthermore, counter-intuitively, had Britain had microwave radar without delays, the effect of an earlier loss to Germany might have been negative to the bombing campaign, though there would have been modest gains in other areas of war.

This essay has shown that German scientists were significantly hindered in the pursuit of microwave technology after 1940 by an absence of obvious need for centimetric radar (as opposed to developments on metric wavelengths) and by decisions to deploy skilled people into front-line service rather than research, but it is by no means certain that, had that not been the case, they would have made the scientific decisions which created the resonant cavity magnetron; German scientists in key positions were mistaken in their views that centimetric wavelengths were not suitable for radar purposes, a view which, forcibly expressed, lead to the demise of all such research in December 1942. The British scientists, not similarly constrained, were fortunate to discover and then develop, with important French input, a magnetron which was powerful, compact and robust, but then, in the case of the Air Ministry, wasted time and energy in bitter disputes with manufacturers; the Admiralty's choice of a co-operative approach led to speedier adoption into service use, a potential lesson for military procurement today.

REFERENCES

1. An excellent single-volume summary across all combatant nations and theatres of war is Louis Brown, 'A Radar History of World War 2: Technical and Military Imperatives', Institute of Physics Publishing, Bristol and Philadelphia, 1999.

2. Summaries of Origins across all combatant nations are given in S.S. Swords, 'Technical History of the Beginnings of Radar', London, Institution of Engineering and Technology,1986 chapter 3 pp.82-173, with an extended version of the British story in chapter 4 pp.174-257; in Raymond C Watson Jr, 'Radar Origins Worldwide: History of Its Evolution in 13 Nations Through World War II' Trafford Publishing, 2009, the British story is covered in chapter 2 pp. 37-110, and the German in chapter 5 pp. 227-272.

3. E.B.Callick, 'Metres to Microwaves British Development of Active Components for Radar Systems', 1937 to 1944 Peregrinus, 1990 covers the transition in part 2 pp. 17-106; the latest International assessments are the papers of the '2010 International Conference on the Origins and Evolution of the Cavity Magnetron' ('CAVMAG 2010'), 19-20 April 2010, Bournemouth, United Kingdom; ISBN: 9781424456093, published by the Institute of Electrical and Electronics Engineers (IEEE).

4. Frank Reuter, 'Funkmeß. Die Entwicklung und der Einsatz des RADAR-Verfahrens in Deutschland bis zum Ende des Zweiten Weltkrieges', wissenschaftliche Abhandlungen der Arbeitsgemeinschaft für Forschung des Landes Nordrhein-Westfalen, Wiesbaden: Springer, vol. 42, 1971, pp. 15 - 16.

5. Wolfgang Burkhardtsmaier, '75 Jahre Sendertechnik bei AEG-Telefunken', Ulm: AEG-Telefunken, 1984, pp. 13 - 38.
6. Hans E. Hollmann, US Patent 2,123,728, 'Magnetron', Telefunken Gesellschaft für Drahtlose Telegraphie m. b. H., Berlin, 12 July 1938.
7. Ulrich Kern, 'Die Entstehung des Radarverfahrens: Zur Geschichte der Radartechnik bis 1945', PhD thesis, Univ. Stuttgart, 1984, pp. 92 - 105.
8. E.B.Callick,'Metres to Microwaves. British Development of Active Components for Radar Systems', 1937 to 1944: Peregrinus, 1990.
9. Robin Holmes, 'The Battle of the Heligoland Bight 1939', London, Grub Street, 2009, chapters 4 & 5, pp. 56-125.
10. TNA/PRO AIR 14/1218; reproduced as Sir Charles Webster and Noble Frankland, 'The Strategic Air Offensive against Germany 1939-1945 Volume IV' Appendix 13, Report by Mr Butt to Bomber Command on his Examination of Night Photographs 18th August 1941..
11. TNA/PRO AIR 41/46 'Historical Report, No 80 Wing, RAF, 1939-45', Air Historical Branch. Laurie Brettingham, 'Royal Air Force Beam Benders: 80 (Signals) Wing 1940-1945', Ian Allan Publishing Ltd, 1997, particularly chapters 1&2, pp.19-26 ; R.V. Jones, 'Most Secret War: British Scientific Intelligence 1939-1945', London, Hamish Hamilton, 1978, particularly part 1, chapters 11-22 pages 92-185.
12. A. P. Rowe, 'One Story of Radar', Cambridge: Cambridge University Press, 1948, pp. 116 - 117.
13. E.B.Callick 'Metres to Microwaves, British Development of Active Components for Radar Systems, 1937 to 1944', Peregrinus, 1990, pp. 55 - 76 and pp. 107 - 123.
14. Sir Bernard Lovell, 'Echoes of War. The Story of H2S Radar', Bristol: Adam Hilger, 1991, p. XVII.
15. Fritz Trenkle, 'Die deutschen Funkmeßverfahren bis 1945', Stuttgart: Motorbuch, 1979, pp. 116 - 124.
16. Max Dieckmann, 'Beiträge zur Entwicklung der Funkortung', in: Leo Brandt (ed.): Zur Geschichte der Funkortung. Deutsche und britische Beiträge. Frankfurter Fachtagung 1953. Ausschuss für Funkortung. Dortmund: Verkehrs- u. Wirtschafts-Verlag, 1953, pp. 49 - 55.
17. Fritz Trenkle, 'Die deutschen Funkmeßverfahren bis 1945', Stuttgart: Motorbuch, 1979, pp. 21 - 23.
18. Wolfgang Keydel; Bernd Röde, 'Radar Research and Development in Oberpfaffenhofen During the 20th Century and Beyond', in: Deutsche Gesellschaft für Ortung und Navigation e.V. (DGON) (ed.): 100 Years of Radar. Special Issue - Contributions from Various Countries Describing the Individual Radar Development Processes, Bonn: DGON, 2005, pp. 201 - 225.
19. Bernd Röde, 'A Short Sketch of the History of the "Institut für Hochfrequenztechnik" in Oberpfaffenhofen', in: Deutsche Forschungsanstalt für Luft- und Raumfahrt (ed.): DLR-Nachrichten. Special Edition 60th anniversary of DLR´s "Institut für Hochfrequenztechnik", 1997, Issue 86. Mülheim: Thierbach, pp. 7–14.
20. H. Frühauf, 'H. E. Hollmann zum 60. Geburtstag', in: Hochfrequenztechnik und Elektroakustik (Ed.), Jahrbuch der drahtlosen Telegraphie und Telephonie, Bd. 68, Issue 5, 1959, pp. 141-143.
21. Kurt Jäger, Friedrich Heilbronner, 'Lexikon der Elektrotechniker', Berlin: VDE Verlag, 2010, pp. 260 - 261.
22. Hans E. Hollmann, 'Physik und Technik der ultrakurzen Wellen', vol. 2, ch. 4, 'Die ultrakurzen Wellen in der Technik', Berlin: Springer, 1936.
23. Ibid., Editorial
24. Hans E. Hollmann, 'Physik und Technik der ultrakurzen Wellen', vol. 2, ch. 4, 'Die ultrakurzen Wellen in der Technik', Berlin: Springer, 1936, pp. 160 - 161.
25. H. Frühauf, 'H.E. Hollmann zum 60. Geburtstag', in: Hochfrequenztechnik und Elektroakustik (Ed.), Jahrbuch der drahtlosen Telegraphie und Telephonie, vol. 68, Issue 5, 1959, p. 142.
26. H. Frühauf, 'H.E. Hollmann zum 60. Geburtstag', in: Hochfrequenztechnik und Elektroakustik (ed.). Jahrbuch der drahtlosen Telegraphie und Telephonie, Bd. 68, Issue 5, 1959, p. 142.
27. Hans E. Hollmann, 'Erzeugung und Verstärkung von Dezimeter- und Zentimeterwellen', in: Forschungsanstalt der Deutschen Reichspost (Ed.): Telegraphen-Fernsprech-Funk-und Fernsehtechnik, 1942, vol. 31, Issue 11.
28. Hollmann, Hans E., 'Erzeugung und Verstärkung von Dezimeter- und Zentimeterwellen (cont.)', in: Forschungsanstalt der Deutschen Reichspost (Ed.): Telegraphen-Fernsprech-Funk-und Fernsehtechnik, 1943, vol. 32, Issue 1.

29. Harry von Kroge, 'GEMA - Berlin. Geburtsstätte der deutschen aktiven Wasserschall- und Funkortungstechnik', Hamburg: Lühmanndruck, 1998, p. 51.
30. H. Frühauf, 'H.E. Hollmann zum 60. Geburtstag', in: Hochfrequenztechnik und Elektroakustik (ed.). Jahrbuch der drahtlosen Telegraphie und Telephonie, Bd. 68, Issue 5, 1959, p. 143.
31. Kurt Jäger, Friedrich Heilbronner, 'Lexikon der Elektrotechniker'. 2. Ed., Berlin: VDE Verlag, 2010, p.120.
32. Dieter Hoffmann, Rüdiger Stutz, 'Grenzgänger der Wissenschaft: Abraham Esau als Industriephysiker, Universitätsrektor und Forschungsmanager', in: U. Hoßfeld, et al. (ed.): "Kämpferische Wissenschaft". Studien zur Universität Jena im Nationalsozialismus. Köln: Böhlau Verlag, 2003, p. 141.
33. Kurt Jäger, Friedrich Heilbronner, 'Lexikon der Elektrotechniker'. 2. Ed., Berlin: VDE Verlag, 2010, p.120.
34. Ibid, p. 120.
35. Ibid, p. 120.
36. Dieter Hoffmann, Rüdiger Stutz, 'Grenzgänger der Wissenschaft: Abraham Esau als Industriephysiker, Universitätsrektor und Forschungsmanager', in: U. Hoßfeld et al. (ed.): "Kämpferische Wissenschaft". Studien zur Universität Jena im Nationalsozialismus. Köln: Böhlau Verlag, 2003, p. 143.
37. Abraham Esau, 'Elektrische Wellen im Zentimetergebiet', Lecture held at the 4th Scientific Session of Ordinary Members on 10 December 1937, Special reprint from the Yearbook of the German Academy of Aeronautical Research 1937/38, Berlin,1938, pp. 212 - 221.
38. Ibid, p. 214.
39. Ibid, p. 215.
40. Ibid, p. 219.
41. Ibid, p. 221.
42. Ibid, p. 221.
43. Kurt Jäger, Friedrich Heilbronner: 'Lexikon der Elektrotechniker'. 2. Ed., Berlin: VDE Verlag, 2010, pp. 371 - 372.
44. Wilhelm Runge, 'Die praktische Bedeutung der Mikrowellen', Lecture held at the 4th Scientific Session of Ordinary Members on 10 December 1937, Special reprint from the Yearbook of the German Academy of Aeronautical Research 1937/38, Berlin, 1938, p. 225.
45. Wilhelm Runge, 'Die Grundlagen der Dezimeterwellen-Technik', in: Telefunken-Zeitung (ed.), vol. 15, Issue 68, 1934, pp. 24 - 33.
46. Wilhelm Runge, 'Schwingungserzeugung mit dem Magnetron', in: Telefunken-Zeitung (ed.), vol. XV, No. 69, 1934, pp. 5 - 14.
47. Kai C. Handel, 'Anfänge der Halbleiterforschung und -entwicklung. Dargestellt an den Biographien von vier deutschen Halbleiterpionieren', PhD thesis, RWTH, Aachen, 1999, p. 50.
48. Frank Reuter, 'Funkmeß. Die Entwicklung und der Einsatz des RADAR-Verfahrens in Deutschland bis zum Ende des Zweiten Weltkrieges', wissenschaftliche Abhandlungen der Arbeitsgemeinschaft für Forschung des Landes Nordrhein-Westfalen, Wiesbaden: Springer, vol. 42, 1971, p. 197.
49. Fritz Trenkle, 'Die deutschen Funkmeßverfahren bis 1945', Stuttgart: Motorbuch, 1979, p. 43.
50. W. Runge 'Ich und Telefunken (Telefunken and I)', Memories of 40 years (1923 - 1963), Historical Archive of the German Museum of Technology Berlin: Unpublished manuscript,1971, p. 50.
51. Frank Reuter, 'Funkmeß. Die Entwicklung und der Einsatz des RADAR-Verfahrens in Deutschland bis zum Ende des Zweiten Weltkrieges', wissenschaftliche Abhandlungen der Arbeitsgemeinschaft für Forschung des Landes Nordrhein-Westfalen, Wiesbaden: Springer, vol. 42, 1971, p. 197.
52. Fritz Trenkle, 'Die deutschen Funkmeßverfahren bis 1945, Stuttgart: Motorbuch, 1979, p. 43.
53. Leo Brandt, 'Rückblick auf die deutsche Funkmeßtechnik', in: Leo Brandt (ed.): Forschen und Gestalten. Reden und Aufsätze von Leo Brandt 1930-1962. Köln: Westdeutscher Verlag, 1962, p. 69.
54. Arthur O. Bauer, 'The German Wartime Struggle to catch up with Allied Power Magnetron Technology', paper delivered at CAVMAG 2010 and available at www.cdvandt.org ; and also 'Naxos: A German Mobile Radar Direction Finder 1943-45' Defence Electronics History Society, e-Newsletter eDEN No 95 June 2020 Part I, p.105.
55. James Phinney Baxter III, 'Scientists against Time', Boston: Little, Brown, 1946, p.146.
56. David Zimmerman, 'Top Secret Exchange: the Tizard Mission and the Scientific War', Stroud: Alan Sutton Publishing, 1996, p.190.

57. Henry E. Guerlac, 'Radar in World War II', American Institute of Physics, 1987; Louis Brown, A Radar History of World War 2: Technical and Military Imperatives, IoP Bristol and Philadelphia, 1999.
58. RW Burns, 'The background to the development of the cavity magnetron', in RW Burns (ed.) Radar Development to 1945, London: Peter Peregrinus, 1988, p.268.
59. Public Record Office/The National Archives (TNA/PRO), Kew, London AVIA 15/648.
60. WE Willshaw, 'GEC's Wartime Contribution', in PM Rolph (ed.) 'Fifty Years of the Cavity Magnetron', Proceedings of a one-day symposium, School of Physics and Space Research, University of Birmingham, 1991, pp 61-2.
61. Phillip Edward Judkins, 'Making Vision into Power', PhD thesis, Cranfield University, November 2007, Chapter 4.
62. RW Burns, 'The early history of centimetric radar: the contributions of the General Electric Company', Proceedings of 25th IEE Weekend Meeting on the History of Electrical Engineering, University of Keele, 11-13 July, 1997, London: IET, 1998, Chapter 10, pp.10/3-4.
63. Ruth Fawcett, 'Nuclear Pursuits: the scientific biography of Wilfrid Bennett Lewis', Montreal: McGill/Queen's, 1994.
64. RW Burns, 'The Life and Times of Alan Dower Blumlein', London: IET, 1999.
65. RW Burns, 1998, op. cit, pp.10/9 - 10/10.
66. EG Bowen, 'Radar Days', Bristol: Adam Hilger, 1987, p.144.
67. RW Burns, 1998, op. cit, p.10/10.
68. Phillip Edward Judkins, 'Making Vision into Power', PhD thesis, Cranfield University, November 2007, Chapter 6 p.362: Professor Sir Bernard Lovell FRS, oral interview, October 2006.
69. Sir Bernard Lovell, 'Echoes of War', Bristol: Adam Hilger, 1991, Chapters 2 and 3.
70. JR Atkinson, 'Work at AMRE Worth Matravers and TRE Malvern', in PM Rolph, 1991, op. cit., p.24.
71. Ibid, p.25 and Footnote.
72. Sir Bernard Lovell, 1991, op. cit., p.29.
73. RW Burns, 1998, op. cit, p.10/10.
74. Ibid, p.10/11.
75. Ibid, p.10/11.
76. Ibid, p.10/13.
77. Diary note by Prof. Philip Dee provided to UK Official Historian KEB Jay and reproduced in Defence Electronics History Society, e-Newsletter eDEN 121, August 2022, p.158.
78. Samuel Crowe Curran, 'Philip Ivor Dee', Biographical Memoirs of Fellows of the Royal Society, London, November 1984, p.151.
79. TNA/PRO AVIA 7/137 Enc 4 to 11.
80. Ibid.
81. TNA/PRO AVIA 46/45 Enclosure 15a page 2 Interview with Professor Dee reprinted as Defence Electronics History Society, e-Newsletter eDEN No 121 August 2022, p.167.
82. TNA/PRO AVIA 7/137.
83. TNA/PRO AVIA 46/45 Enclosure 15a page 2 Interview with Professor Dee reprinted as Defence Electronics History Society, e-Newsletter eDEN No 121 August 2022, p.167.
84. TNA/PRO AVIA 7/137 Enc 13a.
85. Ibid, note on minute sheet ref Enc 14.
86. Ibid, Enc 15.
87. RW Burns, 1998, op. cit, p.10/12, quoting R Clayton and J Algar (eds.), 'A Scientist's War: the War Diary of Sir Clifford Paterson 1939 – 45', London: Peter Peregrinus, 1991, pp.72, 122, 168.
88. William Scott Farren, 'Henry Thomas Tizard' Biographical Memoirs of Fellows of the Royal Society, London, November 1961 p.337.
89. RW Burns, 1998, op. cit, p.10/13, quoting R Clayton and J Algar (eds.),1991, p. 63.
90. RGP Batt, 'Why Ten Centimetres?', in PM Rolph, 1991, op. cit., p.36.
91. TNA/PRO AVIA 7/251 Enc 2B.
92. Ibid., Minute 1.2 of First Meeting.
93. Ibid, Encl 12, Minute 1.6.

94. Ibid, Minute 4.3 of First Meeting.
95. TNA/PRO AVIA 7/251 Enc 13.
96. Sir Bernard Lovell, 1991, op. cit., p.45.
97. TNA/PRO AVIA 7/137 Enc 56.
98. Sir Bernard Lovell, 1991, op. cit., p.45.
99. TNA/PRO AVIA 7/251 Encl 12, p1.
100. Sir Bernard Lovell, 1991, op. cit., p.48.
101. TNA/PRO AVIA 7/251 Enc 15.
102. Ibid.
103. Ian White, 'The History of Air Intercept Radar & the British Nightfighter, 1935-1959', Pen and Sword Aviation, Barnsley UK, 2023.
104. TNA/PRO AVIA 7/251 Enc 15.
105. CA Cochrane, 'Development of Naval Warning & Tactical Radar Operating in the 10cm Band, 1940-5', Monograph 5 in FA Kingsley (ed.), 'Radar: The Developments of Equipments for the Royal Navy 1935-45', London: Macmillan, 1995, pp.189 -192; Sir Bernard Lovell, 1991, op. cit., pp.50ff.
106. Sir B Lovell, 1991, op. cit., pp. 51-2; CA Cochrane, 1995, op. cit, pp.192-3.
107. CA Cochrane, 1995, op. cit, p. 193; Sir B Lovell, 1991, op. cit., p.53.
108. CA Cochrane, 1995, op. cit, pp. 193 -4; Sir B Lovell, 1991, op. cit., pp.52-3.
109. Sir Bernard Lovell, 1991, op. cit., pp.52-3.
110. CA Cochrane, 1995, op. cit, pp.195-6.
111. TNA/PRO ADM 1/11063, 'New design of RDF sets (Type 271) Report on trials in HMS ORCHIS'.
112. TNA/PRO AVIA 7/137 Enc 103a.
113. CA Cochrane, 1995, op. cit, pp.197 - 206.
114. CA Cochrane, 1995, op. cit, pp.208-9.
115. TNA/PRO AIR 14/1218; reproduced as Sir Charles Webster and Noble Frankland, 'The Strategic Air Offensive against Germany 1939-1945 Volume IV' Appendix 13, Report by Mr Butt to Bomber Command on his Examination of Night Photographs 18th August 1941.
116. Sir Bernard Lovell, 1991, op. cit., pp.87 - 91.
117. Ibid, pp. 91-3.
118. Ibid, p. 94.
119. Ibid, p. 99.
120. Ibid, Chapter 13.
121. Ibid, pp.103-5.
122. Ibid, Chapter 14.
123. Ibid, pp.143 - 8.
124. Ibid, p.146.
125. Sir Bernard Lovell, 'H2S/ASV', in PM Rolph, 1991, op. cit., p.40.
126. Sir Bernard Lovell, 1991, op. cit., pp.149 — 150.
127. Ibid, Chapter 23.
128. Ibid, pp.183 - 4.
129. Ibid, p.184.
130. Ibid, p.185 footnote, quoting L Killip.
131. Hugh Sebag-Montefiore, 'Enigma: The Battle For The Code' Weidenfeld & Nicholson, 2017. ISBN 10: 1474608329.
132. David Pritchard, 'The Radar War -- Germany's Pioneering Achievement 1940-45' Patrick Stephens Limited, 1989.
133. Dean C Allard, A United States Overview' in ‚Battle of the Atlantic' Stephen Howarth and Derek Law (eds), Greenhill/Naval Institute Press, 1994, p 569.
134. Sir Bernard Lovell, 1991, op. cit., p.157.
135. Ibid.
136. Ibid, p.158. The Ferranti history, John F Wilson's 'Ferranti: A history', Carnegie Publishing, 1999, is extremely polite about the episode (p.276), although one imagines at the time it was rather less restrained.

137. Ibid, p.160.
138. Ibid, p.159.
139. Ibid.
140. Harper, Stephen 'Capturing Enigma: How HMS Petard Seized the German Naval Codes', Sutton Publishing 1999. ISBN 0-7509-3050-0.
141. Jürgen Rohwer, 'The critical convoy battles of March 1943' Naval Institute Press, 1977.
142. Sir Bernard Lovell, 1991, op. cit., p.161.
143. Ibid, p.116.
144. Ibid.
145. Ibid.
146. Ibid.
147. Ibid.
148. Brigadier A.P.Sayer, 'Army Radar', Official History, HMSO, London 1950 Chapter IX pp. 59-61.
149. Professor JD Cockcroft, 'Memories of Radar Research', IEE Proceedings, Vol. 132, Pt A, No. 6, October 1985, p.331, IET London, 1986
150. Ibid, p.332.

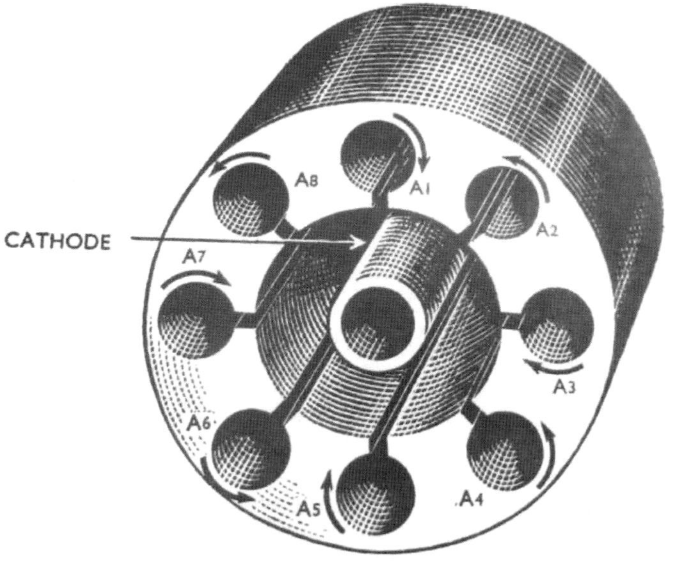

Fig.13. The structure of the resonant cavity magnetron.
(Image credit: 'Science at War', JG Crowther and R Whiddington, HMSO, 1947)

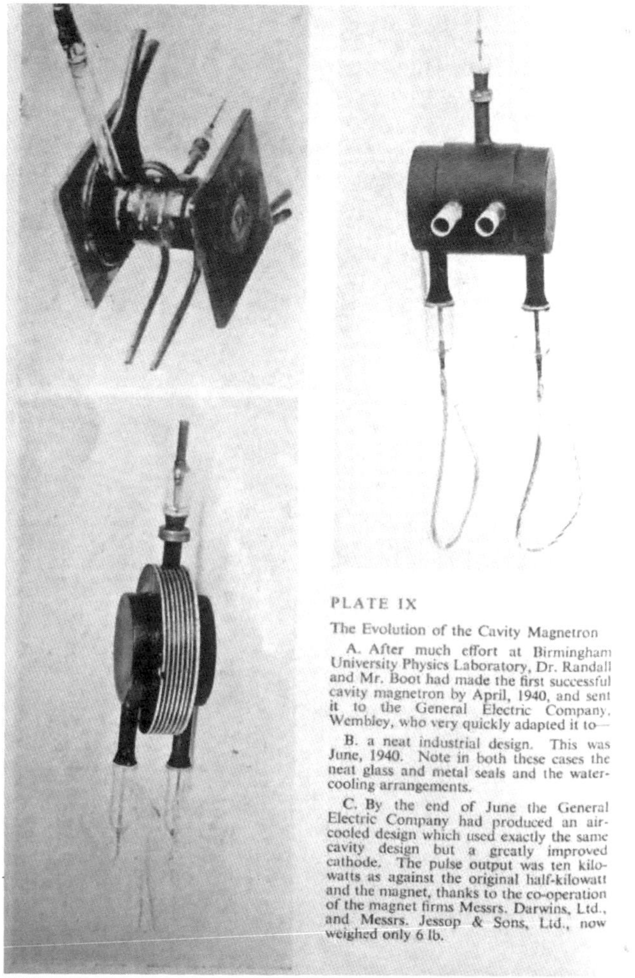

PLATE IX

The Evolution of the Cavity Magnetron

A. After much effort at Birmingham University Physics Laboratory, Dr. Randall and Mr. Boot had made the first successful cavity magnetron by April, 1940, and sent it to the General Electric Company, Wembley, who very quickly adapted it to—

B. a neat industrial design. This was June, 1940. Note in both these cases the neat glass and metal seals and the water-cooling arrangements.

C. By the end of June the General Electric Company had produced an air-cooled design which used exactly the same cavity design but a greatly improved cathode. The pulse output was ten kilowatts as against the original half-kilowatt and the magnet, thanks to the co-operation of the magnet firms Messrs. Darwins, Ltd., and Messrs. Jessop & Sons, Ltd., now weighed only 6 lb.

Fig.14. (Above and below) The British account of the evolution of the resonant cavity magnetron (Image credit: 'Science at War', JG Crowther and R Whiddington, HMSO, 1947)

D. The magnetron complete with magnet. These all operate at ten centimetres. Shorter waves than ten centimetres were on the way, and by 1942 a magnetron delivery power at three centimetres was available, mainly a joint effort by the Birmingham team and the British Thomson Houston Company at Rugby.

E. A late model; it is a "package" with magnetron and magnet complete with rectangular wave guide out of which the power pours. A continuation pipe of any length required can be bolted on to the flange shown at the bottom of the plate.

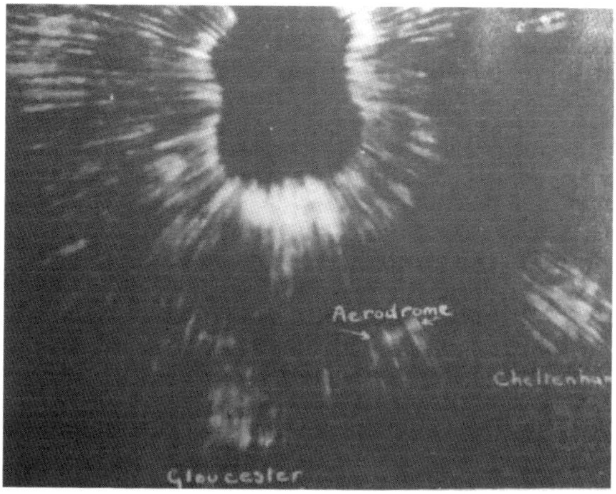

Fig 15. Contrasting screen displays. Above, September 1942: 10cm H2S image of the area north of Gloucester and Cheltenham; the 'eye of faoth' is needed for town identification. Below, December 1944. 1.25cm H2S image of the area around Orford in Suffolk. (Image credit: Sir Bernard Lovell, 'Echoes of War', Adam Hilger/IoP, 1991)

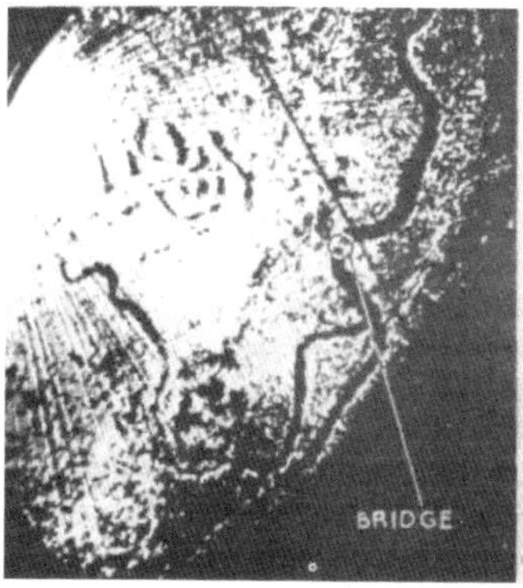

12 Abbreviations

AAC	Anti-Aircraft Command
ADEE	Air Defence Experimental Establishment
ADRC	Air Defence Research Sub-Committee of the C.I.D.
ADRDE	Air Defence Research and Development Establishment
ADRE	Air Defence Research Establishment
AEG	Allgemeine Elektricitäts-Gesellschaft
AGR	Arbeitsgemeinschaft Rotterdam
AI	Airborne Interception
AMES	Air Ministry Experimental Station
AMRE	Air Ministry Research Establishment
ARC	Aeronautical Research Committee
ASDIC	Anti Submarine Detection Investigation Committee
ASV	Air to Surface Vessel
BArch	Bundesarchiv Berlin-Lichterfelde
BArch-MA	Bundesarchiv-Militärarchiv Freiburg
BdU	Befehlshaber der U-Boote
BHF	Bevollmächtigter für die Hochfrequenzforschung
BRS	Bawdsey Research Station
C.A.S.	Chief of the Air Staff
CBE	Commander of the British Empire
CH	Chain Home
CHL	Chain Home Low
CHEL	Chain Home Extra Low
C.I.D.	Committee of Imperial Defence
COMINT	Communications Intelligence
CRT	Cathode Ray Tube
CSIR	Council for Scientific and Industrial Research
CSSAD	Committee for the Scientific Survey of Air Defence
CSSAO	Committee for the Scientific Survey of Air Offence
CSSAW	Committee for the Scientific Survey of Air Warfare
CVD	Committee for the Coordination of Valve Development

DASD	Deutschen Amateur-Sende- und Empfangsdienst
DeTe	Dezimeter-Telegraphie
DFG	Deutsche Forschungsgemeinschaft
DGON	Deutsche Gesellschaft für Ortung und Navigation e. V.
DLR	Deutsches Zentrum für Luft- und Raumfahrt
DNVP	Deutschnationale Volkspartei
DRC	Defence Requirements Sub-Committee
DRPC	Defence Research Policy Committee
DSIR	Department of Scientific and Industrial Research
DVG	Drahtlostelegraphische und Luftelektrische Versuchsstation Gräfelfing
DVL	Deutsche Versuchsanstalt für Luftfahrt
ECM	Electronic Countermeasures
EMI	Electric and Musical Industries
FAA	Fleet Air Arm
FdU	Führer der Unterseeboote
FFO	Flugfunk-Forschungsinstitut Oberpfaffenhofen
FLAK	Flugabwehrkanone
FLUM	Flugmeldung
FuMG, FuMO	Funkmeßgerät, Funkmeßortungsgerät
GBN	Generalbevollmächtigter für technische Nachrichtenmittel
GC&CS	Government Code & Cypher School
GCI	Ground-Controlled Interception
GEC	General Electric Company Ltd.
GEMA	Gesellschaft für elektroakustische und mechanische Apparate m.b.H.
GL	Generalluftzeugmeister
H_2S	High-resolution centimetric radar
HF/DF	High Frequency Direction Finder
HUMINT	Human Intelligence
HWA	Heereswaffenamt
IFF	Identification Friend or Foe
IRC	International Radiotelegraph Convention
IWM	Imperial War Museum
LORAN	Long Range Navigation
MAPRE	Ministry of Aircraft Production Research Establishment

MIT	Massachusetts Institute of Technology
MND	Marinenachrichtendienst
Natlab	Natuurkundig Laboratorium
NCR	National Cash Register Company
NDRC	National Defence Research Committee
NPL	National Physical Laboratory
NRL	Naval Research Laboratory
NSBDT	Nationalsozialistischer Bund Deutscher Technik
NSDAP	Nationalsozialistische Deutsche Arbeiterpartei
NVA	Nachrichtenmittel-Versuchsanstalt
NVK	Nachrichtenmittel-Versuchskommando
NVW	Nachrichten-Verbindungswesen
Ob.d.L.	Oberbefehlshaber der Luftwaffe
OKH	Oberkommando des Heeres
OKL	Oberkommando der Luftwaffe
OKM	Oberkommando der Marine
OKW	Oberkommando der Wehrmacht
OR	Operational Research; Operations Research
ORS	Operational Research Section
OSRD	Office of Scientific Research and Development
PTR	Physikalisch-Technische Reichsanstalt
RADAR	Radio Detection and Ranging
RadLab	MIT Radiation Laboratory
RAE	Royal Aircraft Establishment
RAF	Royal Air Force
RRB	Radio Research Board
RCAF	Royal Canadian Air Force
RCC	Research Construction Company
RDF	Radio Direction Finding
R.d.L.	Reichsminister der Luftfahrt
REM	Reichsministerium für Wissenschaft, Erziehung und Volksbildung
RFR	Reichsforschungsrat
RHF	Reichsstelle für Hochfrequenzforschung e. V.
RLE	Research Laboratory of Electronics
RLM	Reichsluftfahrtministerium
RMfBuM	Reichsministerium für Bewaffnung und Munition
RMfRuK	Reichsministerium für Rüstung und Kriegsproduktion

ROC	Royal Observer Corps
RRB	Radio Research Board
RRDE	Radar Research and Development Establishment
RRS	Radio Research Station
SAR	Synthetic Aperture Radar
SDF	Special Duties Flight
SFR	Société Française Radioélectrique
SIGINT	Signals Intelligence
SIS	Secret Intelligence Service
SKFM	Sonderkommission für Funkmeßtechnik
SLEE	Searchlight Experimental Establishment
Sonar	Sound Navigation and Ranging
SSR	Secondary Surveillance Radar
TFU	Telecommunications Flying Unit
TNA	The National Archives
TRE	Telecommunications Research Establishment
TVA	Torpedo-Versuchsanstalt
VHF	Very High Frequency
WAAF	Women's Auxiliary Air Force
WFG	Wehrforschungsgemeinschaft
WFM	Wissenschaftlicher Führungsstab beim Oberkommando der Kriegsmarine
WRNS	Women's Royal Naval Service

13 List of Illustrations

14 List of Tables

15 Sources

Bundesarchiv (BArch) Berlin-Lichterfelde

NS 19 / 2057, 7.1.1944, Hochfrequenzforschung.

NS 19 / 3912, 16.10.1944, Reichsverteidigung und Mobilmachung, Bd. 12.

R 26 III / 6, 1944. Forschungsaufträge: Bd. 4: Reichsforschungsrat (mit Bevollmächtigtem für Hochfrequenzforschung).

R 26 III / 91, 4.7.1944. Forschungsaufgaben der Kriegsmarine auf dem Nachrichten- und Ortungsgebiet.

R 26 III / 120, 28.12.1943. Denkschrift Osenberg.

R 26 III / 192, 12.5.1944, Erlaß betreffend Hauptkommission Elektrotechnik.

R 3 / 1739, 3.1.1944, Chronik der Dienststellen des Amtsbereichs Speer, Bd. 5.

R 3 / 1509, 1.1.-4.1. 1944, Führerbesprechungen, Bd. 7.

Bundesarchiv-Militärarchiv (BArch-MA) Freiburg

RL 2 V / 64, 11. Juli 1942, Taktisch-technische Forderungen für Funkmessgeräte.

RL 2 V / 75, 19. März 1943, Entwicklungsbesprechungen.

RL 2 V / 172, 30. Juli 1944, Die Funkmessgeräte der Luftwaffe.

RL 3 / 42, 24. März, 4. Mai 1944, Entwicklungsbesprechungen.

RM 7/ 106, Kriegstagebuch Teil B VI Nachrichtendienst (B-Dienst).

RM 7 / 1234, 11. Dez. 1943, Akte 1 Wehrwirtschaft.

RM 24 / M / 697 [E], Juni 1943 bis Dezember 1944, Forschungsaufgaben der Kriegsmarine.

Reichsgesetzblatt

Reichsgesetzblatt (1919): Gesetz über den Friedensschluß zwischen Deutschland und den alliierten und assoziierten Mächten vom 16. Juli 1919, Nr. 140, p. 687, p. 985.

Reichsgesetzblatt (1920): Verordnung über die Auslegung der Begriffe Friedensschluß und Kriegsende im Sinne rechtsgeschäftlicher Erklärungen vom 17. Februar 1920, Nr. 35, p. 237.

Reichsgesetzblatt (1933): Verordnung über den Reichskommissar für die Luftfahrt vom 2. Februar 1933, Teil I, Nr. 8, p. 35.

Reichsgesetzblatt (1933): Verordnung des Reichspräsidenten zum Schutz von Volk und Staat vom 28. Februar 1933, Teil I, Nr. 17, p. 83.

Reichsgesetzblatt (1933): Gesetz zur Behebung der Not von Volk und Reich (Ermächtigungsgesetz) vom 24. März 1933. Teil I, Nr. 25, p. 141.

Reichsgesetzblatt (1933): Gesetz zur Wiederherstellung des Berufsbeamtentums vom 7. April 1933. Teil I, Nr. 34, pp. 175–177.
Reichsgesetzblatt (1933): Verordnung über das Reichsluftfahrtministerium vom 5. Mai 1933. Teil I, Nr. 47, p. 241.
Reichgesetzblatt (1933): Gesetz gegen die Neubildung von Parteien vom 14. Juli 1933. Teil I, Nr. 81, p. 479
Reichsgesetzblatt (1934): Gesetz über das Staatsoberhaupt des Deutschen Reichs vom 1. August 1934. Teil I, Nr. 89, p. 747.
Reichsgesetzblatt (1935): Gesetz für den Aufbau der Wehrmacht vom 16. März 1935, Teil I, Nr. 28, p. 375.
Reichsgesetzblatt (1936): Verordnung zur Durchführung des Vierjahresplanes vom 18. Oktober 1936. Teil I, Nr. 96, p. 887.
Reichstagsprotokoll (1933): Friedensrede, 3. Sitzung vom 17. Mai 1933, pp. 47–54.
Hitler, Adolf (1936): Denkschrift über die Aufgaben eines Vierjahresplans vom August 1936. In: Bayerische Staatsbibliothek (Hg.): https://www.1000dokumente.de/Dokumente/Denkschrift_über_die_Aufgaben_eines_Vierjahresplans.

ADAP: Akten zur deutschen auswärtigen Politik

Hoesch von (1933): Telegramm des Botschafters in London von Hoesch an das Auswärtige Amt. Auswärtiges Amt (1971). In: ADAP. Akten zur Deutschen Auswärtigen Politik 1918–1945. Serie C: 1933–1937. Das Dritte Reich: Die ersten Jahre. Band I,1. 30. Januar bis 15. Mai 1933. Göttingen: Vandenhoeck & Ruprecht, Nr. 193, pp. 351–352.
Neurath (1933): Der Reichsminister des Auswärtigen Freiherr von Neurath an die Deutsche Delegation zur Abrüstungskonferenz (Genf). Auswärtiges Amt (1971). In: ADAP. Akten zur Deutschen Auswärtigen Politik 1918–1945. Serie C: 1933–1937. Das Dritte Reich: Die ersten Jahre. Band I,2. 16. Mai bis 14. Oktober 1933. Göttingen: Vandenhoeck & Ruprecht, Nr. 489, pp. 890–891.
unsigned (1935): Übersetzung des Sitzungsberichts vom 18. Juni 1935 im Foreign Office. Auswärtiges Amt (1975). In: ADAP. Akten zur Deutschen Auswärtigen Politik 1918–1945. Serie C: 1933–1937. Das Dritte Reich: Die ersten Jahre. Band IV,1. 1. April bis 13. September 1935. Göttingen: Vandenhoeck & Ruprecht, Nr. 156, pp. 315–320.
Hoßbach, Friedrich (1937): Niederschrift über die Besprechung in der Reichskanzlei am 5. November 1937. Auswärtiges Amt (1950). In: ADAP. Akten zur Deutschen Auswärtigen Politik 1918–1945. Serie D: 1937–1945. Von Neurath zu Ribbentrop. Band I. September 1937 – September 1938. Baden-Baden: Imprimerie Nationale, Nr. 19, pp. 25–32.

Sitzungsprotokolle der Arbeitsgemeinschaft Rotterdam (AGR)

(Available online at: https://www.cdvandt.org/agr_protocols.htm)

22./23.2.1943; 17.3.1943; 8.4.1943; 19.5.1943; 1.6.1943; 22.6.1943; 23.7.1943; 2.9.1943; 11.9.1943; 22.10.1943; 11.12.1943; 25.2.1944; 13.3.1944; 5.4.1944; 26.4.1944; 31.5.1944; 26.7.1944; 1.9.1944.

Personal Archive

AGR Bericht 1/44 – Die Arbeitsgemeinschaft "Rotterdam" in der Sonderkommission Funkmesstechnik. Jahrg. II, 1944/45, Heft 4.

Sitzungsprotokolle der Arbeitsgemeinschaft Rotterdam – Ausschuss für Funkortung Cl, gedrucktes Manuskript, Nov. 1952.

Würzburg-Fibel, Teil I, Oberkommando der Kriegsmarine, Ausschuss für Funkortung, Okt. 1943, Sonderbücherei der Funkortung.

Würzburg-Fibel, Teil II, Oberkommando der Kriegsmarine, Ausschuss für Funkortung, Okt. 1943, Sonderbücherei der Funkortung.

Freya-Fibel, Heft 8, Flugmeldedienst, Okt. 1943, Reichsdruckerei.

Berlin-Fibel, Oberkommando der Kriegsmarine, Amtsgruppe Technisches Nachrichtenwesen, Erste Ausgabe, Juli 1944.

Decknamenverzeichnis der deutschen Funkmess- und Ultrarot-Entwicklungen der Jahre 1944–1945, Ausschuss für Funkortung c2, edited by Leo Brandt.

16 References

Abelson, Philip (1996): Merle Antony Tuve. 1901–1982. In: National Academy of Sciences (Ed.): Biographical Memoirs. Vol. 70. Washington, D.C.: National Academies Press, pp. 407–422. Available online at: https://nap.nationalacademies.org/read/5406/chapter/20#407. Retrieved 01 September 2025.

Addison, P. (2018): Churchill, Sir Winston Leonard Spencer (1874–1965). Oxford Dictionary of National Biography. Available online at: https://doi.org/10.1093/ref:odnb/32413. Retrieved 01 September 2025.

AEG (1988): Die ersten 100 Jahre. Sonderdruck. Ulm: AEG.

Agar, J. (2008): Flowers, Thomas Harold (1905–1998). Oxford Dictionary of National Biography. Available online at: https://doi.org/10.1093/ref:odnb/71253. Retrieved 01 September 2025.

Air Ministry (Ed.) (1941): The Battle of Britain. An Air Ministry Account of the Great Days from 8th August – 31st October 1940. London: His Majesty's Stationery Office.

Air Ministry (Ed.) (1947): Introductory Survey of Radar. Part II. A non-technical account of selected airborne equipments. Air Publication. 1093 D. Vol. 1. First Ed. June 1946: Issued in reprint Oct. 1947. Available online at: http://vmarsmanuals.co.uk/archive/632_AP1093D_Pt2_Introductory_Survey_of_Radar_screenread.pdf. Retrieved 01 September 2025.

Air Ministry (1954): The Second World War 1939–1945. Royal Air Force. Signals. Volume VI. Radio In Maritime Warfare. Available online at: https://www.raf.mod.uk/what-we-do/our-history/air-historical-branch/second-world-war-thematic-studies/signals-vol-vi-radio-in-maritime-war/. Retrieved 01 September 2025.

Air Ministry (1956): The Second World War 1939–1945. Royal Air Force. Signals. Volume III. Aircraft Radio. Available online at: https://www.raf.mod.uk/what-we-do/our-history/air-historical-branch/second-world-war-thematic-studies/signals-vol-iii-aircraft-radio-chapters1-14/. Retrieved 01 September 2025.

Air Ministry (1963): The Origins and Development of Operational Research in the Royal Air Force. Air Publication 3368. London: Her Majesty's Stationery Office.

Air Publication 1093C (2016): The 'Rough Guide' to Marks of H2S. In: *e-newsletter of the Defence Electronics History Society (DEHS)* 50 (Part II, Sept. 2016), pp. 79–81. Available online at: https://www.dehs.org.uk. Retrieved 01 September 2025.

Air Publication 1093D (2016): Introduction H2S und ASV. In: *e-newsletter of the Defence Electronics History Society (DEHS)* 50 (Part I, Sept.

2016), pp. 6–28. Available online at: https://www.dehs.org.uk. Retrieved 01 September.

Alexander, Robert (2013): The Inventor of Stereo: The Life and Works of Alan Dower Blumlein. Burlington, MA: Focal Press.

Allibone, T. E. (2008): Cockcroft, Sir John Douglas (1897–1967). Oxford Dictionary of National Biography. Available online at: https://doi.org/10.1093/ref:odnb/32473. Retrieved 01 September 2025.

Allison, David (1981): New Eye for the Navy: The Origin of Radar at the Naval Research Laboratory. Washington, D.C.: Naval Research Laboratory (NRL Report 8466).

Alvarez, Luis (1980): Alfred Lee Loomis (November 4, 1887 – August, 11, 1975). In: National Academy of Sciences (Ed.): Biographical Memoir, Vol. 51. Washington, DC: National Academies Press, pp. 308–341. Available online at: https://nap.nationalacademies.org/read/574/chapter/13#340. Retrieved 01 September 2025.

Ambrose, Stephen (1981): Eisenhower and the Intelligence Community in World War II. In: *Journal of Contemporary History* 16 (1), pp. 153–166.

Arco, Georg von (1909a): Das neue Telefunken-System. In: *Elektrotechnische Zeitschrift* (23), pp. 535–540.

Arco, Georg von (1909b): Das neue Telefunken-System. In: *Elektrotechnische Zeitschrift* (24), pp. 561–567.

Austin, B. A. (1999): Precursors to Radar – The Watson-Watt Memorandum and the Daventry Experiment. In: *Int. J. Elect. Enging. Educ.* 36, pp. 365–372.

Baeumker, Adolf (1971): Ein Beitrag zur Geschichte der Führung der deutschen Luftfahrttechnik im ersten halben Jahrhundert 1900–1945. Bonn: ersetze durch: Deutsche Forschungs- und Versuchsanstalt für Luft- und Raumfahrt e.V.

Baldwin, Stanley (1999): 'The bomber will always get through'. London, 10 Nov. 1932. In: Brian MacArthur (Ed.): The Penguin Book of Twentieth-Century Speeches. 2nd rev. ed. London: Penguin, pp. 124–127.

Ball, Stuart (2019): Baldwin, Stanley, first Earl Baldwin of Bewdley (1867–1947). Oxford Dictionary of National Biography. Available online at: https://doi.org/10.1093/ref:odnb/30550. Retrieved 01 September 2025.

Bartel, Hans-Georg (2014): Ein Geheimrat im Militärdienst. Walther Nernst im Spannungsfeld von Kriegsforschung und Friedensbemühungen – aus Anlass seines 150. Geburtstages am 25. Juni 2014. In: *Physik Journal 13* (7), pp. 49–53.

Barton, David (1985): Radar System Analysis. 6. ed. Dedham, MA: Artech House.

Batey, Mavis (2006): Knox, (Alfred) Dillwyn (1884–1943). Oxford Dictionary of National Biography. Available online at: https://doi.org/10.1093/ref:odnb/37641. Retrieved 01 September 2025.

Bauer, Arthur (n.d.): Final Technical Report on the German RDF Equipment Captured at Bruneval on 28th February 1942. Transcript of T.R.E.

(TRE) report No. 6/R/25. Available online at: https://www.cdvandt.org/Bruneval TRE.pdf. Retrieved 01 September 2025.

Bauer, Arthur (1992): Deckname "Würzburg". Ein Beitrag zur Erhellung der Geschichte des geheimnisumwitterten deutschen Radargeräts 1937–1945. Herten: Verlag Historischer Technikerliteratur.

Bauer, Arthur (1997a): Funkpeilung als alliierte Waffe gegen deutsche U-Boote 1939–1945. Diemen, NL: Selbstverlag.

Bauer, Arthur (2004a): Christian Hülsmeyer and the Early Days of Radar, Sense and Nonsense – a Survey. In: Deutsche Gesellschaft für Ortung und Navigation e. V. (DGON) (Ed.): 100 Years of Radar. Special Issue. Contributions from Various Countries Describing the Individual Radar Development Processes. German Institute of Navigation. Bonn: DGON, pp. 13–57.

Bauer, Arthur (2004b): HF/DF an Allied Weapon against German U-Boats 1939–1945. Diemen, NL, pp. 1–23. Available online at: https://cdvandt.org/HFDF1998.pdf. Retrieved 01 September 2025.

Bauer, Arthur (2004c): Naxos, the History of a German Mobile Radar Direction Finder 1943–1945. Centre for German Communication and Related Technology. Diemen, NL, pp. 1–15. Available online at: https://www.cdvandt.org/Naxos95nw.pdf. Retrieved 01 September 2025.

Bauer, Arthur (2005a): Christian Hülsmeyer and about the early days of radar inventions, a survey. Diemen, NL, pp. 1–74. Available online at: https://aobauer.home.xs4all.nl/Huelspart1def.pdf. Retrieved 01 September 2025.

Bauer, Arthur (2005b): Christian Hülsmeyer and about the early days of radar, a survey. Part II. Diemen, NL, pp. 1–18. Available online at: https://www.cdvandt.org/Huelsmeyer part II.pdf. Retrieved 01 September 2025.

Bauer, Arthur (2006): Synopsis. Some Aspects of German Airborne Radar Technology, 1942 to 1945. Diemen, NL, pp. 1–32. Available online at: https://www.cdvandt.org/German airborne radar def.pdf. Retrieved 01 September 2025.

Bauer, Arthur (2008): Christian Hülsmeyer und die Erfindung des Radars. In: Horst Wessel (Ed.): Strom ohne Grenzen. Internationale Aspekte der Elektrotechnik. Tagung des VDE-Ausschusses "Geschichte der Elektrotechnik" am 23. Oktober 2006 in Aachen. Berlin: VDE Verlag, pp. 199–224.

Bauer, Arthur (2009): The German Wartime Struggle to catch up with Allied Power Magnetron Technology. Diemen, NL. Available online at: https://www.cdvandt.org/CAVMAG-2010-Wartime Struggle.pdf. Retrieved 01 September 2025.

Bauer, F. L. (2003): Codebreakers: Arne Beurling and the Swedish Crypto Program During World War II by Bengt Beckman. Book Review. In: *Notices of the American Mathematical Society* 50 (8), pp. 904–906.

Bauer, Friedrich (1991): Scherbius und die ENIGMA – Historische Notiz. In: *Informatik Spektrum* 14 (4), pp. 211–214.

Bauer, Friedrich (1997b): Entzifferte Geheimnisse. Methoden und Maximen der Kryptologie. 2. Aufl. Berlin: Springer.

Baxter 3rd, James (1968): Scientists Against Time. Cambridge, MA: MIT Press.

Beckh, Joachim (2005): Blitz & Anker. Informationstechnik – Geschichte, Hintergründe. Bd. II. Norderstedt: Books on Demand.

Beckman, Bengt (2002): Codebreakers. Arne Beurling and the Swedish Crypto Program During World War II. Providence, RI: American Math. Soc.

Beckman, Bengt (2006): Arne Beurling und Hitlers Geheimschreiber. Schwedische Entzifferungserfolge im 2. Weltkrieg. Berlin: Springer.

Beesly, Patrick (1976a): Das "Operational Intelligence Centre" der britischen Admiralität im Zweiten Weltkrieg. In: *Marine-Rundschau* 73 (3), pp. 147–164.

Beesly, Patrick (1976b): Das "Operational Intelligence Centre" der britischen Admiralität im Zweiten Weltkrieg (II). In: *Marine-Rundschau* 73 (6), pp. 368–383.

Beesly, Patrick (1976c): Das "Operational Intelligence Centre" der britischen Admiralität im Zweiten Weltkrieg (III). In: *Marine-Rundschau* 73 (11), pp. 698–707.

Beesly, Patrick (1978): Very Special Intelligence. Geheimdienstkrieg der britischen Admiralität 1939–1945. Frankfurt am Main: Ullstein.

Beesly, Patrick (1979): Das Operational Intelligence Centre der britischen Admiralität und die Schlacht im Atlantik. In: Jürgen Rohwer und Eberhard Jäckel (Eds.): Die Funkaufklärung und ihre Rolle im Zweiten Weltkrieg. Eine internationale Tagung in Bonn-Bad Godesberg und Stuttgart vom 15.–18. Nov. 1978. Stuttgart: Motorbuch, pp. 133–151.

Beesly, Patrick (2006): Very Special Intelligence. The Story of the Admiralty's Operational Intelligence Centre 1939–1945. Annapolis, MD: Naval Institute Press.

Bekker, Cajus (1980): Augen durch Nacht und Nebel. Die Radar Story. Herford: Mittler.

Bekker, Cajus (1991): Die deutsche Kriegsmarine 1939–1945. Augsburg: Weltbild.

Bell, Falko (2015). One of our Most Valuable Sources of Intelligence: British Intelligence and the Prisoner of War System in 1944. In: *Intelligence and National Security* 31 (4), pp. 556–578.

Bell, Falko (2016): Britische Feindaufklärung im Zweiten Weltkrieg. Stellenwert und Wirkung der »Human Intelligence« in der britischen Kriegsführung 1939–1945. Paderborn: Schöningh.

Benz, Wolfgang; Graml, Hermann; Weiß, Hermann (Eds.) (2001): Enzyklopädie des Nationalsozialismus. 4. Aufl. München: dtv.

Bertrand, Gustave (1973): Enigma ou la plus grande énigme de la guerre, 1939–1945. Paris: Plon.

Beyerchen, Alan (1980): Wissenschaftler unter Hitler. Physiker im Dritten Reich. Köln: Kiepenheuer & Witsch.

Beyrer, Klaus (1999): Mythos Enigma. Bletchley Park und die Funkentschlüsselung im Zweiten Weltkrieg. In: Klaus Beyrer (Ed.): Streng geheim. Die Welt der verschlüsselten Kommunikation. Frankfurt am Main: Museumsstiftung Post und Telekommunikation, pp. 113–126.

Biggs, Norman (2008): Tutte, William Thomas [Bill] (1917–2002). Oxford Dictionary of National Biography. Available online at: https://doi.org/10.1093/ref:odnb/76919. Retrieved 01 September 2025.

Birkenhead, Frederick (1961): The Prof in Two Worlds. The official life of Professor F. A. Lindemann, Viscount Cherwell. London: Collins.

Blackett, P.M.S (1960): Tizard and the Science of War. In: *Nature* (185), pp. 647–653.

Blake, Robert (2008): Lindemann, Frederick Alexander, Viscount Cherwell (1886–1957). Oxford Dictionary of National Biography. Available online at: https://doi.org/10.1093/ref:odnb/34533. Retrieved 01 September 2025.

Blanchard, Yves (2004): Le Radar 1904–2004. Histoire d'un siècle d'innovations techniques et opérationnelles. Paris: Ellipses.

Blanchard, Yves (2010): Maurice Ponte and Henri Gutton, pioneers of early French studies on Resonant Magnetron (1932–1940). In: IEEE, DEHS und IET (Eds.): *Proceedings of the 2010 International Conference on the Origins and Evolution of the Cavity Magnetron*. Bournemouth University, England. Piscataway, NJ: IEEE, pp. 5–10.

Blanchard, Yves; Galati, Gaspare; van Genderen, Piet (2013): The Cavity Magnetron: Not Just a British Invention. In: *IEEE Antennas and Propagation Magazine* 55 (5), pp. 244–254.

Blättel-Mink, Birgid; Ebner, Alexander (Eds.) (2020): Innovationssysteme. Technologie, Instutionen und die Dynamik der Wettbewerbsfähigkeit. 2. ed. Wiesbaden: Springer.

Bleaney, Brebis (2006): Oliphant, Marcus Laurence Elwin (1901–2000). Oxford Dictionary of National Biography. Available online at: https://doi.org/10.1093/ref:odnb/74397. Retrieved 01 September 2025.

Blumtritt, Oskar (1987): Lüschen, Friedrich Heinrich. Neue Deutsche Biographie 15. Available online at: https://www.deutsche-biographie.de/pnd140051325.html#ndbcontent. Retrieved 01 September 2025.

Blumtritt, Oskar (1988): Nachrichtentechnik. Sender – Empfänger – Übertragung – Vermittlung. München: Deutsches Museum.

Boemeke, Manfred; Feldman, Gerald; Glaser, Elisabeth (Eds.) (1998): The Treaty of Versailles. A Reassessment after 75 Years. Cambridge: Cambridge Univ. Press.

Boghardt, Thomas (2012): The Zimmermann Telegram. Intelligence, Diplomacy, and America's Entry into World War I. Annapolis, MD: Naval Institute Press.

Böhm, Martin (2015): Die Royal Air Force und der Luftkrieg 1922–1945. Personelle, kognitive und konzeptionelle Kontinuitäten und Entwicklungen. Paderborn: Schöningh.

Boog, Horst (1982): Mängel des Technikverständnisses. In: Militärgeschichtliches Forschungsamt (Ed.): Die deutsche Luftwaffenführung 1935–1945. Führungsprobleme, Spitzengliederung, Generalstabsausbildung. Stuttgart: Deutsche Verlagsanstalt, pp. 36–76.

Boot, Henry; Randall, John (1976): Historical Notes on the Cavity Magnetron. In: *IEEE Transactions on Electron Devises* 23 (7), pp. 724–729.

Borisova, Nina (2010): Cavity Magnetron in Russia. In: IEEE, DEHS und IET (Eds.): *Proceedings of the 2010 International Conference on the Origins and Evolution of the Cavity Magnetron*. Bournemouth University, England. Piscataway, NJ: IEEE, pp. 23–33.

Bowen, E. G. (1987): Radar Days. Bristol: Hilger.

Bowen, E. G. (1988a): The development of airborne radar in Great Britain 1935–1945. In: Russell Burns (Ed.): Radar Development to 1945. London: Peregrinus, pp. 177–188.

Bowen, E. G. (1988b): The Tizard Mission to the USA and Canada. In: Russell Burns (Ed.): Radar Development to 1945. London: Peregrinus, pp. 296–307.

Bradsher, Greg (2015). Hitler's final Words. His Political Testament, Personal Will, and Marriage Certificate: From the Bunker in Berlin to the National Archives. In: *Prologue*, pp. 16–25.

Bragg, Michael (1988): RDF 1. The Location of Aircraft by Radio Methods 1935–1945. Paisley: Hawkhead Publishing.

Brandt, Leo (Ed.) (n.d.). Decknamenverzeichnis der deutschen Funkmess- und Ultrarot-Entwicklungen der Jahre 1944–1945. Ausschuß für Funkortung. c 2.

Brandt, Leo (Ed.) (1952): Sitzungsprotokolle der Arbeitsgemeinschaft Rotterdam. Bücherei der Funkortung. Düsseldorf: Ausschuß für Funkortung C 1.

Brandt, Leo (Ed.) (1953a): Index zu den Sitzungsprotokollen der Arbeitsgemeinschaft Rotterdam. Sonderbücherreihe der Funkortung. Düsseldorf: Ausschuss für Funkortung. Available online at: https://www.cdvandt.org/AGR%20Index.pdf. Retrieved 01 September 2025.

Brandt, Leo (Ed.) (1953b): Zur Geschichte der Funkortung. Deutsche und britische Beiträge. Frankfurter Fachtagung 1953. Ausschuss für Funkortung. Bücherei der Funkortung. Band 2. Teil I. Dortmund: Verkehrs- u. Wirtschafts-Verlag.

Brandt, Leo (1962a): Der Stand der deutschen Zentimeterwellen-Technik am Ende des zweiten Weltkrieges. In: Leo Brandt (Ed.): Forschen und Gestalten. Reden und Aufsätze von Leo Brandt. 1930–1962. Köln: Westdeutscher Verlag, pp. 80–112.

Brandt, Leo (Ed.) (1962b): Forschen und Gestalten. Reden und Aufsätze von Leo Brandt. 1930–1962. Köln: Westdeutscher Verlag.

Brandt, Leo (1962c): Rückblick auf die deutsche Funkmeßtechnik. In: Leo Brandt (Ed.): Forschen und Gestalten. Reden und Aufsätze von Leo Brandt. 1930–1962. Köln: Westdeutscher Verlag, pp. 53–79.

Brandt, Leo (1968): Geschichte der Radartechnik. Rückblick auf die Leistungen der britischen und deutschen Forschung und Entwicklung. In: *VDI-Nachrichten* 19 (Teil III), p. 13.

Braun, F. (1901a): Ueber drahtlose Telegraphie. Das Princip der induktiven Sendererregung. In: *Elektrotechnische Zeitschrift* (12), pp. 258–260.

Braun, F. (1901b): Ueber einige Sendervarianten für drahtlose Telegraphie. In: *Elektrotechnische Zeitschrift* (23), pp. 469–470.

Braun, Ferdinand (1909): Electrical oscillations and wireless telegraphy. Nobel Lecture. Available online at: https://www.nobelprize.org/nobel_prizes/physics/laureates/1909/braun-lecture.html. Retrieved 01 September 2025.

Bredow, Hans (1928): Telefunken und der deutsche Rundfunk. In: Telefunken-Gesellschaft (Ed.): 25 Jahre Telefunken. Festschrift der Telefunkengesellschaft 1903–1928, pp. 22–26.

Bredow, Hans (1953): Rückblick auf 30 Jahre deutscher Rundfunk. In: *Telefunken-Zeitung* 26 (100), pp. 177–180.

Breit, Gregory; Tuve, Merle (1926): A Test of the Existence of the Conducting Layer. In: *Physical Review* 28, pp. 554–575.

Breitkopf, Klaus (2018): Radar-Entwicklung in Großbritannien 1934–1945. Tabellarische Chronik. In: *Funkgeschichte* 41 (242), pp. 251–257.

Brennecke, Jochen (1994): Die Wende im U-Boot-Krieg. Ursachen und Folgen 1939–1943. Augsburg: Weltbild.

Breuning, Ernst (1989): Mit LZ 130 Graf Zeppelin auf Funkhorch- und Funkortungsfahrt. In: Albert Sammt (Ed.): Mein Leben für den Zeppelin. 2. Aufl. Wahlwies: Pestalozzi Kinderdorf, pp. 156–170.

Broszat, Martin (2007): Der Staat Hitlers. Grundlegung und Entwicklung seiner inneren Verfassung. Wiesbaden: Marix Verlag.

Brown, Louis (1999): Technical and Military Imperatives. A Radar History of World War II. New York: Taylor & Francis.

Bruch, Walter (1983): Leithäuser, Gustav (1881–1969). In: Sigfrid von Weiher (Ed.): Männer der Funktechnik. Eine Sammlung von 70 Lebenswerken deutscher Pioniere der Funktechnik. (drahtlose Telegrafie, Radar, Rundfunk und Fernsehen). Berlin: VDE-Verlag, pp. 106–108.

Buderi, Robert (1997): The Invention that Changed the World. The Story of Radar from War to Peace. London: Little, Brown and Company.

Bührer, Werner (2001): Wirtschaft. In: Wolfgang Benz, Hermann Graml and Hermann Weiß (Eds.): Enzyklopädie des Nationalsozialismus. 4. ed. München: dtv, pp. 108–122.

Burcham, W. S. (2016): The Development of Centimetre AI. In: *e-Newsletter of the Defence Electronics History Society (DEHS)* 50 (Part I, Sept. 2026), pp. 34–48. Available online at: https://www.dehs.org.uk. Retrieved 01 September 2025.

Burcham, W.; Shearman, R. (2018): Fifty Years of the Cavity Magnetron 1940–1990. In: *e-Newsletter of the Defence Electronics History Society (DEHS)* 76 (Part I, Nov. 2018), pp. 42–55. Available online at: https://www.dehs.org.uk. Retrieved 01 September 2025.

Burns, R. W. (1988a): The background to the development of early radar, some naval questions. In: Russell Burns (Ed.): Radar Development to 1945. London: Peregrinus, pp. 1–28.

Burns, R. W. (1988b): The background to the development of the cavity magnetron. In: Russell Burns (Ed.): Radar Development to 1945. London: Peregrinus, pp. 259–283.

Burns, R. W. (2014): Blumlein, Alan Dower (1903–1942). Oxford Dictionary of National Biography. Available online at: https://doi.org/10.1093/ref:odnb/37201. Retrieved 01 September 2025.

Burns, Russell (2013): The Prof, Churchill & Science at War. The life of F A Lindemann to 1945. Reading: DEHS.

Bussey, Gordon (2000): Marconi's Atlantic Leap. Coventry: Marconi Communications.

Bussmann, Walter (1968): Zur Entstehung und Überlieferung der "Hoßbach-Niederschrift". In: *Vierteljahrshefte für Zeitgeschichte* 16 (4), pp. 373–384.

Butcher, Peter (2020): Wireless. In: *Transmission Lines – The Journal of the Defence Electronics History Society (DEHS)* 25 (4), pp. 1–20. Available online at: https://www.dehs.org.uk. Retrieved 01 September 2025.

Callick, E. B. (1990): Metres to Microwaves. British development of active componentes for radar systems 1937 to 1944. London: Peregrinus.

Calvocoressi, Peter (1979): Aufbau und Arbeitsweise des britischen Entzifferungsdienstes in Bletchley Park. In: Jürgen Rohwer und Eberhard Jäckel (Eds.): Die Funkaufklärung und ihre Rolle im Zweiten Weltkrieg. Eine internationale Tagung in Bonn-Bad Godesberg und Stuttgart vom 15.–18. Nov. 1978. Stuttgart: Motorbuch, pp. 88–99.

Chamberlain, Neville (1999a): *'Peace for our time'*. London, 30 September 1938. In: Brian MacArthur (Ed.): The Penguin Book of Twentieth-Century Speeches. 2. ed. London: Penguin, p. 170.

Chamberlain, Neville (1999b): *'This country is at war with Germany'*. London, 3 September 1939. In: Brian MacArthur (Ed.): The Penguin Book of Twentieth-Century Speeches. 2. ed. London: Penguin, pp. 179–180.

Chamberlain, Neville (2008): Radio Address by Neville Chamberlain, Prime Minister, September 3, 1939. Documents in Law, History and Diplomacy. Yale Law School (Ed.) The Avalon Project. New Haven, CT. Available online at: https://avalon.law.yale.edu/wwii/gb3.asp. Retrieved 01 September 2025.

Churchill, Winston (1950): Der Zweite Weltkrieg. Vierter Band. Erstes Buch. Kapitel XVI. Offensive im Äther. Zürich: Schweizer Verlagshaus.

Churchill, Winston (1954): Der Zweite Weltkrieg. Der Sturm zieht auf. Erster Band. Erstes Buch. 2. ed. Bern: Scherz & Goverts.

Churchill, Winston (1985): The Second World War. Volume II. Their Finest Hour. London: Penguin.
Churchill, Winston (1999a): *'A total and unmitigated defeat'*. London, 5 October 1938. In: Brian MacArthur (Ed.): The Penguin Book of Twentieth-Century Speeches. 2. ed. London: Penguin, pp. 172–175.
Churchill, Winston (1999b): *'Be ye men of valour'*. BBC Radio, 19 May 1940. In: Brian MacArthur (Ed.): The Penguin Book of Twentieth-Century Speeches. 2. ed. London: Penguin, pp. 188–190.
Churchill, Winston (1999c): *'I have nothing to offer but blood, toil, tears and sweat'*. London, 13 May 1940. In: Brian MacArthur (Ed.): The Penguin Book of Twentieth-Century Speeches. 2. ed. London: Penguin, pp. 187–188.
Churchill, Winston (1999d): *'This was their finest hour'*. London, 18 June 1940. In: Brian MacArthur (Ed.): The Penguin Book of Twentieth-Century Speeches. 2. ed. London: Penguin, pp. 190–191.
Churchill, Winston (2005): The Second World War. Volume I. The Gathering Storm. London: Penguin.
Churchill, Winston (2007): Der Zweite Weltkrieg. Frankfurt am Main: Fischer
CIOS (1946): Combined Intelligence Objectives Sub-Committee. Research Work undertaken by the German Universities and Technical High Schools for the Bevollmaechtigter fuer Hochfrequenztechnik; Independent Research in Associated Subjects. File Number XXXI-2. Item No. 1 & 7. London: H. M. Stationery Office. Available online at: https://www.cdvandt.org/CIOS-XXXI-2.pdf. Retrieved 01 September 2025.
Clark, Ronald (1965): Tizard. London: Methuen.
Clark, Ronald (1971): Sir Edward Appleton. Oxford: Pergamon Press.
Coales, J. F.; Rawlinson, J. D. S. (1988): The development of UK naval radar. In: Russell Burns (Ed.): Radar Development to 1945. London: Peregrinus, pp. 53–96.
Collar, A. R. (1967): Tizard and the Aeronautical Research Committee. In: *Journal of the Royal Aeronautical Society* 71 (680), pp. 529–538.
Cook, Alan (1999): Reginald Victor Jones, C.H., C.B., C.B.E. 29 September 1911–17 December 1997. In: *Biogr. Mems Fell. R. Soc.* 45, pp. 239–254.
Copeland, B. Jack (Ed.) (2010): Colossus. The Secrets of Bletchley Park's Codebreaking Computers. Oxford: Oxford University Press.
Corrigan, R. (2008): Airborne minefields and Fighter Command's information system. GikII. University of Edinburgh School of Law (Ed.). Available online at: https://marconiradarhistory.pbworks.com/f/BoBIFS.pdf. Retrieved 01 September 2025.
Crowther, J. G.; Whiddington, R. (1948): Science at War. New York: Philosophical Library.
Cunningham, W.; Freeman, D.; McCloskey, J. (1984): OR Forum – Of Radar and Operations Research: An Appreciation of A. P. Rowe (1898–1976). In: *Operations Research* 32 (4), pp. 958–967.

Darricau, Jaques; Blanchard, Yves (2002): La naissance du radar. In: *Pegase* 107, pp. 16–30.

Davies, Rodney; Graham-Smith, Francis; Lyne, Andrew (2016): Sir Alfred Charles Bernard Lovell OBE. 31 August 1913–6 August 2012. In: *Biogr. Mems Fell. R. Soc.* 62, pp. 323–344. Available online at: https://royalsocietypublishing.org/doi/10.1098/rsbm.2015.0026, Retrieved 01 September 2025.

Davis, John; Lovell, Bernard (2003): Robert Hanbury Brown. 31 August 1916-16 January 2002 Elected FRS 1960. In: *Biogr. Mems Fell. R. Soc.* 49, pp. 83–106.

Davis, Philip (2011): From Amateurs to Professionals: GC&CS and Institution-Building in Signals Intelligence. In: Ralph Erskine and Michael Smith (Ed.): The Bletchley Park Codebreakers. London: Biteback Publishing, pp. 341–354.

Deist, Wilhelm (1989): Die Aufrüstung der Wehrmacht. In: Wilhelm Deist; Manfred Messerschmidt; Hans-Erich Volkmann and Wolfram Wette (Eds.): Ursachen und Voraussetzungen des Zweiten Weltkrieges. Frankfurt am Main: Fischer Taschenbuch, pp. 437–637.

Dekker, Ronald (2014): Britische Radarentwicklung im Zweiten Weltkrieg: Der unsichtbare Vorhang. In: *Funkgeschichte* 37 (213), pp. 4–16.

Deutsch, Harold (1977): The Historical Impact of Revealing the Ultra Secret. In: *Parameters, Journal of the US Army War College* 7 (1), pp. 16–32.

Dieckmann, Max (1953): Beiträge zur Entwicklung der Funkortung. In: Leo Brandt (Ed.): Zur Geschichte der Funkortung. Deutsche und britische Beiträge. Frankfurter Fachtagung 1953. Ausschuss für Funkortung. Bücherei der Funkortung. Band 2. Teil I. Dortmund: Verkehrs- u. Wirtschafts-Verlag, pp. 49–55.

Diehl, H. (1953): Zur Geschichte der Funkmeßtechnik in Deutschland unter besonderer Berücksichtigung ihrer Anwendung für die Flugfunkortung. In: Leo Brandt (Ed.): Zur Geschichte der Funkortung. Deutsche und britische Beiträge. Frankfurter Fachtagung 1953. Ausschuss für Funkortung. Bücherei der Funkortung. Band 2. Teil I. Dortmund: Verkehrs- u. Wirtschafts-Verlag, pp. 56–75.

Dinçkal, Noyan; Dipper, Christof; Mares, Detlef (Eds.) (2010): Selbstmobilisierung der Wissenschaft. Technische Hochschulen im »Dritten Reich«. Darmstadt: WBG.

Dittmann, Frank et al. (2021): Technik – Innovation – Sicherheit. Der VDE Verband der Elektrotechnik, Elektronik und Informationstechnik als Expertennetzwerk in gesellschaftlicher Verantwortung 1893 bis 2018. Berlin: VDE Verlag.

DLR (n.d.): Deutsches Zentrum für Luft- und Raumfahrt. Institut für Hochfrequenztechnik und Radarsysteme. Available online at: https://www.dlr.de/de/hr/. Retrieved 01 September 2025.

Dobinson, Colin (2010): Building Radar. Forging Britain's early-warning chain, 1935–1945. London: Methuen.

Doherty, Julian (1972): Die Dominions und die britische Außenpolitik von München bis zum Kriegsausbruch 1939. In: *Vierteljahrshefte für Zeitgeschichte* 20 (3), pp. 209–234.

Dönitz, Karl (2018): Zehn Jahre und zwanzig Tage. Erinnerungen 1935–1945. 16. ed. Bad Neuenahr-Ahrweiler: Bernard & Graefe.

Douhet, Giulio; translated by Ferrari, Dino (2019): The Command of the Air. Maxwell Air Force Base, AL: Air University Press. Available online at: https://www.airuniversity.af.edu/Portals/10/AUPress/Books/B_0160_DOUHET_THE_COMMAND_OF_THE_AIR.PDF. Retrieved 01 September 2025.

Dowling, Michael; Kretschmer, Tobias (Eds.) (2018): Eberhard Witte. Sein Leben. Seine Forschung. Seine Wirkung. Norderstedt: Books on Demand.

Dragoni, Giorgio (2011): Marconi, Guglielmo (1874–1937). Oxford Dictionary of National Biography. Available online at: https://doi.org/10.1093/ref:odnb/37735. Retrieved 01 September 2025.

Dutton, D. J. (2013): Chamberlain, Sir (Joseph) Austen (1863–1937). Oxford Dictionary of National Biography. Available online at: https://doi.org/10.1093/ref:odnb/32351. Retrieved 01 September 2025.

E.A. (1897): Marconi's Telegraphensystem. In: *Elektrotechnische Zeitschrift* (33), pp. 505–507.

Eckert, Michael (2010): Heinrich Hertz. Hamburg: Ellert & Richter.

Edgerton, David (2005): Warfare State. Britain, 1920–1970. Cambridge: Cambridge University Press.

Eldridge, Chris (2000): Electronic eyes for the allies: Anglo-American cooperation on radar development during World War II. In: *History and Technology* 17 (1), pp. 1–20.

Elz, Wolfgang (2008): Versailles und Weimar. In: *APuZ: Aus Politik und Zeitgeschichte* (50–51), pp. 31–38.

Epkenhans, Michael (1997): Die kaiserliche Marine im Ersten Weltkrieg: Weltmacht oder Untergang? In: Wolfgang Michalka (Ed.): Der Erste Weltkrieg. Wirkung. Wahrnehmung. Analyse. Weyarn: Seehamer, pp. 319–340.

Epstein, Klaus (1976): Matthias Erzberger und das Dilemma der deutschen Demokratie. Frankfurt am Main: Ullstein.

Erskine, Ralph (2002): Der Krieg der Code-Brecher. In: Bayerische Akademie der Wissenschaften (Ed.): Akademie aktuell, Available online at: URL: https://badw.de/fileadmin/pub/akademieAktuell/2002/08/02_erskine.pdf. Retrieved 01 September 2025.

Erskine, Ralph (2009): Barry, Sir (Philip) Stuart Milner- (1906–1995). Oxford Dictionary of National Biography. Available online at: https://doi.org/10.1093/ref:odnb/58042. Retrieved 01 September 2025.

Erskine, Ralph; Smith, Michael (Eds.) (2011): The Bletchley Park Codebreakers. London: Biteback Publishing.

Esau, A.; Hahnemann, W. M. (1930): Report on Experiments with Electric Waves of about 3 Meters: Their Propagation and Use. In: *Proceedings of the Institute of Radio Engineers* 18 (3), pp. 471–489.

Esau, Abraham (1938): Elektrische Wellen im Zentimetergebiet. Vortrag gehalten am 10. Dezember 1937. In: Sonderabdruck aus dem Jahrbuch der Deutschen Akademie der Luftfahrtforschung 1937/38. Berlin, pp. 212–221.

Eschenburg, Theodor; Frank-Planitz, Ulrich (1978): Gustav Stresemann. Eine Bildbiographie. Stuttgart: Deutsche Verlags-Anstalt.

Escott, Beryl (1989): Women in Air Force Blue: Story of Women in the Royal Air Force from 1918 to the Present Day. Wellingborough: Patrick Stephens.

Eula, Michael (1986): Guilio Douhet and Strategic Air Force Operations. a study in the limitations of theoretical warfare. In: *Air University Review* XXXVI. (6), pp. 94–99.

Faensen, Hubert (2001): Hightech für Hitler. Die Hakeburg – Vom Forschungszentrum zur Kaderschmiede. Berlin: Ch. Links.

Farren, William (1961): Henry Thomas Tizard, 1885–1959. In: *Biogr. Mems Fell. R. Soc.* 7, pp. 313–348.

Federspiel, Ruth (2002): Mobilisierung der Rüstungsforschung? Werner Osenberg und das Planungsamt im Reichsforschungsrat 1943–1945. In: Helmut Maier (Ed.): Rüstungsforschung im Nationalsozialismus. Organisation, Mobilisierung und Entgrenzung der Technikwissenschaften. Göttingen: Wallstein, pp. 72–105.

Feiveson, H. A. (2018): Scientsists against Time. The Role of Scientists in World War II. Bloomington, In: Archway Publishing.

Fischer, Kurt (2011): Großadmiral Dr. phil. h. c. Erich Raeder. In: Gerd Ueberschär (Ed.): Hitlers militärische Elite. 68 Lebensläufe. 2. ed. Darmstadt: Primus, pp. 185–194.

Fisher, David (1989): A Race on the Edge of Time. Radar – The Decisive Weapon of World War II. New York: Paragon House.

Flachowsky, Sören (2005): Der Bevollmächtigte für Hochfrequenzforschung des Reichsforschungsrates und die Organisation der deutschen Radarforschung in der Endphase des Zweiten Weltkrieges 1942–1945. In: *Technikgeschichte* 72 (3), pp. 203–226.

Flachowsky, Sören (2008): Von der Notgemeinschaft zum Reichsforschungsrat. Wissenschaftspolitik im Kontext von Autarkie, Aufrüstung und Krieg. Stuttgart: Franz Steiner.

Fletcher, W. W. (2004): Curran, Joan Elizabeth, Lady Curran (1916–1999). Oxford Dictionary of National Biography. Available online at: https://doi.org/10.1093/ref:odnb/69524. Retrieved 01 September 2025.

Flugmeldedienst (1943): Freya-Fibel, Heft 8, Okt. 1943, Reichsdruckerei.

Folkerts, Liesa (2001): Promotoren in Innovationsprozessen. Empirische Untersuchung zur personellen Dynamik. Wiesbaden: Springer Fachmedien.

Fölsing, Albrecht (1997): Heinrich Hertz. Eine Biographie. Hamburg: Hoffmann und Campe.
Forbes, Nancy; Mahon, Basil (2014): Faraday, Maxwell, and the Electromagnetic Field. How Two Men Revolutionized Physics. Amherst, NY: Prometheus Books.
Forczyk, Robert (2016): We March Against England. Operation Sea Lion, 1940–41. Oxford: Osprey Publishing.
Foster, Richard (1986): Innovation. The Attackers Advantage. New York: Summit Books.
Foster, Richard (2006): Innovation. Die technologische Offensive. McKinseyClassics. Sonderausgabe. Bd. 6. Heidelberg: Redline Wirtschaft.
Fransecky, Tanja von (2003): Zwangsarbeit in der Berliner Metallindustrie 1939 bis 1945. Eine Firmenübersicht. Berlin: Otto Brenner Stiftung (Arbeitsheft Nr. 31).
Frei, Norbert (1983): "Machtergreifung". Anmerkungen zu einem historischen Begriff. In: *Vierteljahrshefte für Zeitgeschichte* 31 (1), pp. 136–145.
Frei, Norbert (2009): 1939 und wir. In: *APuZ: Aus Politik und Zeitgeschichte* (36–37), pp. 3–5.
Frei, Norbert (2013): Der Führerstaat. Nationalsozialistische Herrschaft 1933 bis 1945. München: Beck.
Friedewald, Michael (2001): Telefunken und der deutsche Schiffsfunk 1903–1914. In: *Zeitschrift für Unternehmensgeschichte* 46 (1), pp. 27–57.
Frühauf, H. (1959): H. E. Hollmann zum 60. Geburtstag. In: *Hochfrequenztechnik und Elektroakustik* 68 (5), pp. 141–143.
Fuchs, Margot (1998): Anfänge der drahtlosen Telegraphie im Deutschen Reich 1897–1918. In: Hans-Jürgen Teuteberg und Cornelius Neutsch (Eds.): Vom Flügeltelegraphen zum Internet. Geschichte der modernen Telekommunikation. Stuttgart: Franz Steiner, pp. 113–131.
Fuchs, Margot (2004): Georg von Arco (1869–1940) – Ingenieur, Pazifist, Technischer Direktor von Telefunken. Eine Erfinderbiographie. Berlin: GNT.
Funkschau (1943a): Hochfrequenzfachkräfte für die Luftwaffe. In: *Funkschau* 16 (6/7), p. 58.
Funkschau (1943b): Hochfrequenz-Fachkräfte für die Luftwaffe. In: *Funkschau* 16 (8/9), p. 73.
Galati, Gaspare (2016): 100 Years of Radar. Cham: Springer.
Gausemeier, Jürgen; Plass, Christoph; Wenzelmann, Christoph (2009): Zukunftsorientierte Unternehmensgestaltung. Strategien, Geschäftsprozesse und IT-Systeme für die Produktion von morgen. München: Hanser.
Gemünden, Hans Georg (2010): Das Scheitern von Promotoren. In: Harald Pechlaner; Brigitte Stechhammer und Hans Hinterhuber (Eds.): Scheitern: Die Schattenseite unternehmerischen Handelns. Die Chance zur Selbsterneuerung. Berlin: Erich Schmidt, pp. 86–107.

Gemünden, Hans Georg; Salomo, Sören; Hölzle, Katharina (2007): Role Models for Radical Innovations in Times of Open Innovation. In: *creativity and innovation management* 16 (4), pp. 408–421.

Gemünden, Hans Georg; Walter, Achim (1995): Der Beziehungspromotor. Schlüsselperson für inter-organisationale Innovationsprozesse. In: *Zeitschrift für Betriebswirtschaft* 65, pp. 971–986.

Gerlitzki, Werner (1994): An overview of radar development from 1950 to 1990: time frame and some comments on developments in Germany. In: Oskar Blumtritt, Hartmut Petzold and William Aspray (Eds.): Tracking the History of Radar. Piscataway, NJ: The IEEE-Rutgers Center for the History of Electrical Engineering & the Deutsches Museum, pp. 47–59.

Gerlitzki, Werner (1995): Ortung. Zur Geschichte der Telefunken-Radartechnik in Ulm 1954–1990. Ulm: Daimler-Benz Aerospace AG, Sensorsysteme.

Gerwin, Robert (1957): Marconi. Ein Erfinderleben unserer Zeit. Hannover: Theodor Oppermann.

Gießler, Helmuth (1958): Die Entwicklung der Funkmeßgeräte in Deutschland. Teil I: 1933–1939. In: *Wehrtechnische Monatshefte* 55 (3), pp. 108–114.

Gießler, Helmuth (1968): Die Revolutionierung des Seekrieges durch Funkmeß. Eine Analyse über Einsatz und Anwendung auf deutscher Seite. In: *Marine-Rundschau* 65 (3), pp. 181–191.

Gießler, Helmuth (1971): Der Marine-Nachrichten- und Ortungsdienst. Technische Entwicklung und Kriegserfahrungen. Wehrwissenschaftliche Berichte. Arbeitskreis für Wehrforschung. Bd. 10. München: J. F. Lehmanns.

Gießler, Helmuth (1977): RADAR für Marine und Schiffahrt. Entwicklung, Kriegserfahrungen und Nachkriegsanwendungen. In: *Schiff und Zeit* (6), pp. 37–47.

Göbel, Jürgen (2011): Radartechnik. Grundlagen und Anwendungen, 2. ed. Berlin: VDE Verlag.

Godt, Birgitta (2003): Aspekte der Radarentwicklung und -anwendung im Zweiten Weltkrieg. Dissertation. Universität Konstanz. Konstanz: Hartung-Gorre.

Goerth, Joachim (2010): Early magnetron development especially in Germany. In: IEEE, DEHS und IET (Eds.): Proceedings of the 2010 International Conference on the Origins and Evolution of the Cavity Magnetron. Bournemouth University, England. Piscataway, NJ: IEEE, pp. 17–22.

Goetzeler, Herbert (1983a): Braun, Karl Ferdinand (1850–1918). In: Sigfrid von Weiher (Ed.): Männer der Funktechnik. Eine Sammlung von 70 Lebenswerken deutscher Pioniere der Funktechnik (drahtlose Telegrafie, Radar, Rundfunk und Fernsehen). Berlin: VDE-Verlag, pp. 28–30.

Goetzeler, Herbert (1983b): Wien, Max (1866–1938). In: Sigfrid von Weiher (Ed.): Männer der Funktechnik. Eine Sammlung von 70 Lebenswerken deutscher Pioniere der Funktechnik (drahtlose Telegrafie, Radar, Rundfunk und Fernsehen). Berlin: VDE-Verlag, pp. 195–196.

Goetzeler, Herbert (1983c): Zennek, Jonathan (1871–1959). In: Sigfrid von Weiher (Ed.): Männer der Funktechnik. Eine Sammlung von 70 Lebenswerken deutscher Pioniere der Funktechnik (drahtlose Telegrafie, Radar, Rundfunk und Fernsehen). Berlin: VDE-Verlag, pp. 199–201.

Goodchild, James (2013): R. V. Jones and the Birth of Scientific Intelligence. PhD Thesis. Univ. of Exeter.

Grabau, Rudolf (1986): Funküberwachung und Elektronische Kampfführung. Grundlagen, Technik und Verfahren. Stuttgart: Franckh.

Grabau, Rudolf (1989): Technische Aufklärung. Sensoren, Systeme und Verfahren. Stuttgart: Franckh.

Graham-Smith, F. (2015): Brown, Robert Hanbury (1916–2002). Oxford Dictionary of National Biography. Available online at: https://doi.org/10.1093/ref:odnb/76626. Retrieved 01 September 2025.

Graml, Hermann (2010): Hitler und England. Ein Essay zur nationalsozialistischen Außenpolitik 1920 bis 1940. München: Oldenbourg.

Griffiths, Hugh; Austin, Brian (2016): The Daventry Experiment as a Bistatic Radar. In: *Transmission Lines – The Journal of the Defence Electronics History Society (DEHS)* 21 (2), pp. 3–8. Available online at: https://www.dehs.org.uk. Retrieved 01 September 2025.

Grüttner, Michael (2001): Wissenschaft. In: Wolfgang Benz, Hermann Graml und Hermann Weiß (Eds.): Enzyklopädie des Nationalsozialismus. 4. Aufl. München: dtv, pp. 135–153.

Grüttner, Michael (2004): Biographisches Lexikon zur nationalsozialistischen Wissenschaftspolitik. Heidelberg: Synchron.

Guerlac, Henry (1987): Radar in World War II. Sections A – C. Vol. 8. Los Angeles, CA: Tomash Publishers.

Gunston, Bill (1982): British Fighters of World War II. London: Hamlyn.

Hagenauer, Joachim; Pabst, Martin (2014): Anpassung, Unbotmäßigkeit und Widerstand. Karl Küpfmüller, Hans Piloty, Hans Ferdinand Mayer – Drei Wissenschaftler der Nachrichtentechnik im «Dritten Reich». Bayerische Akademie der Wissenschaften. München: Beck.

Haider, Carsten (2021): Führen wir diesen Krieg mit Waffen oder mit dem Rechenschieber? Blacketts Circus – britische Operationsforschung im Zweiten Weltkrieg. In: *Pallasch: Zeitschrift für Militärgeschichte,* 77, pp. 145–152.

Hammerstein, Notker (1999): Die Deutsche Forschungsgemeinschaft in der Weimarer Republik und im Dritten Reich. Wissenschaftspolitik in Republik und Diktatur 1920–1945. München: Beck.

Hanbury Brown, Robert (1991): Boffin. A Personal Storry of the Early Days of Radar, Radio Astronomy and Quantum Optics. Bristol: Hilger.

Hanbury Brown, Robert; Minnett, H. C.; White, Frederick (1992): Edward George Bowen. 14 January 1911–12 August 1991. In: *Biogr. Mems Fell. R.*

Soc. Vol. 38, pp. 41–65. Available online at: https://doi.org/10.1098/rsbm.1992.0003. Retrieved 01 September 2025.

Handel, Kai (1999): Anfänge der Halbleiterforschung und -entwicklung. Dargestellt an den Biographien von vier deutschen Halbleiterpionieren. Dissertation. RWTH Aachen.

Handel, Kai (2002): Die Arbeitsgemeinschaft Rotterdam und die Entwicklung von Halbleiterdetektoren. Hochfrequenzforschung in der militärischen Krise 1943–1945. In: Helmut Maier (Ed.): Rüstungsforschung im Nationalsozialismus. Organisation, Mobilisierung und Entgrenzung der Technikwissenschaften. Göttingen: Wallstein, pp. 250–270.

Harman, P. M. (2009): Maxwell, James Clerk (1831–1879). Oxford Dictionary of National Biography. Available online at: https://doi.org/10.1093/ref:odnb/5624. Retrieved 01 September 2025.

Harris, Arthur (1995): Despatch on War Operations. 23rd February, 1942 to 8th May, 1945. London: Cass.

Hars, Florian (1999): Ferdinand Braun (1850–1918). Ein wilhelminischer Physiker. Berlin: GNT.

Hauschildt, Jürgen; Chakrabarti, Alok K. (1988): Arbeitsteilung im Innovationsmanagement Forschungsergebnisse, Kriterien und Modelle. In: *Zeitschrift Führung + Organisation* 57, pp. 378–388.

Hauschildt, Jürgen; Salomo, Sören; Schultz, Carsten; Kock, Alexander (2016): Innovationsmanagement. 6. Aufl. München: Vahlen.

Hauser, Oswald (1972): England und das Dritte Reich. Erster Band. 1933 bis 1936. Stuttgart: Seewald.

Hauser, Oswald (1982): England und das Dritte Reich. Zweiter Band. 1936 bis 1938. Göttingen: Muster-Schmidt.

Heilbronner, Friedrich (n.d.): Marconi and the Germans. München: Deutsches Museum.

Heinemann, Winfrid (Ed.) (2011): Führung und Führungsmittel. Potsdamer Schriften zur Militärgeschichte. Militärgeschichtliches Forschungsamt. Bd. 14. Potsdam.

Heinrich-Hertz-Institut (Ed.) (1988): Heinrich Hertz. Festschrift anläßlich der Erforschung der elektromagnetischen Wellen vor 100 Jahren. Berlin: Heinrich-Hertz-Institut.

Henke, Josef (1973): England in Hitlers politischem Kalkül 1935–1939. Boppard am Rhein: Boldt.

Hentschel, Georg (1989): Die geheimen Konferenzen des Generalluftzeugmeisters. Ausgewählte und kommentierte Dokumente zur Geschichte der deutschen Luftrüstung und des Luftkrieges 1942–1944. Koblenz: Bernard & Graefe.

Herbst, Ludolf (2010): Speer, Albert. In: Neue Deutsche Biographie 24. Available online at: https://www.deutsche-biographie.de/pnd118615998.html#ndbcontent. Retrieved 01 September 2025.

Hertz, Gustav (1957): Die Entdeckungen von Heinrich Hertz und ihre Auswirkungen. In: Der Senat der Freien und Hansestadt Hamburg (Ed.):

Heinrich Hertz. Gedenkfeier aus Anlaß des 100. Geburtstages von Heinrich Hertz. Hamburg: Conrad Kayser, pp. 21–34.

Hertz, Heinrich (1923): Über die Beziehungen zwischen Licht und Elektrizität. Vortrag gehalten bei der 62. Versammlung deutscher Naturforscher und Ärzte in Heidelberg 1889. 13. Aufl. Leipzig: Kröner.

Hertz, Heinrich (2002a): Über elektrodynamische Wellen im Luftraume und deren Reflexion. In: Hans Wußing (Ed.): Über sehr schnelle elektrische Schwingungen. Vier Arbeiten (1887–1889). Ostwalds Klassiker der exakten Wissenschaften. Band 251. Frankfurt am Main: Harri Deutsch, pp. 64–79.

Hertz, Heinrich (2002b): Über Strahlen elektrischer Kraft. In: Hans Wußing (Ed.): Über sehr schnelle elektrische Schwingungen. Vier Arbeiten (1887–1889). Ostwalds Klassiker der exakten Wissenschaften. Band 251. Frankfurt am Main: Harri Deutsch, pp. 80–96.

Hildebrand, Klaus (2009): Das Dritte Reich. 7. Aufl. München: Oldenbourg.

Hillmann, Jörg (2022): Dönitz, Karl (1891–1980). In: NDB-online. Available online at: https://www.deutsche-biographie.de/118526251.html#dbocontent. Retrieved 01 September 2025.

Hinnen, Gijs (1987): Von Marconi bis Satellit. 1700 Daten aus d. internat. Entwicklung von Radio u. Fernsehen. Basel: Basilius.

Hinsley, F. H. revised by Erskine Ralph (2008): Denniston, Alexander Guthrie [Alastair] (1881–1961). Oxford Dictionary of National Biography. Available online at: https://doi.org/10.1093/ref:odnb/32783. Retrieved 01 September 2025.

Hitler, Adolf (1936): Denkschrift über die Aufgaben eines Vierjahresplanes. In: 100(0) Schlüsseldokumente der deutschen Geschichte im 20. Jahrhundert. Available online at: https://www.1000dokumente.de/Dokumente/Denkschrift_über_die_Aufgaben_eines_Vierjahresplans. Retrieved 01 September 2025.

Hitler, Adolf (1938): Mein Kampf. 2 Bände in 1 Band. Ungek. Ausg. 307–311. Tsd. München: Eher.

Hochfrequenztechnik und Elektroakustik (1943): Hochfrequenzfachkräfte für die Luftwaffe. In: *Hochfrequenztechnik und Elektroakustik* 61(5), p. 152.

Hodges, Andrew (2012): Alan Turing: The Enigma. London: Vintage.

Hodges, Andrew (2017): Turing, Alan Mathison (1912–1954). Oxford Dictionary of National Biography. Available online at: https://doi.org/10.1093/ref:odnb/36578. Retrieved 01 September 2025.

Hofer, Walther (Ed.) (2011): Der Nationalsozialismus. Dokumente 1933–1945. Frankfurt am Main: Fischer

Hofer, Walther (1979): Die Diktatur Hitlers bis zum Beginn des Zweiten Weltkrieges. In: Leo Just (Ed.): Handbuch der Deutschen Geschichte. Band IV, 2. Teil, Abschnitt IV. Wiesbaden: Akademische Verlagsgesellschaft Athenaion, pp. 1–257.

Hoffmann, Dieter; Stutz, Rüdiger (2003): Grenzgänger der Wissenschaft: Abraham Esau als Industriephysiker, Universitätsrektor und Forschungsmanager. In: Uwe Hoßfeld; Jürgen John; Oliver Lemuth und Rüdiger Stutz (Eds): »Kämpferische Wissenschaft«. Studien zur Universität Jena im Nationalsozialismus. Köln: Böhlau, pp. 136–179.

Hoffmann-Heyden, Adolf-Eckard (1954): Die Funkmessgeräte der deutschen Flakartillerie (1938–1945). In: Leo Brandt (Ed.): Ausschuss für Funkortung. Bücherei der Funkortung. Band 3. Dortmund: Verkehrs- u. Wirtschafts-Verlag.

Hoffmann, Karl Otto (1990): Martini, Wolfgang. In: Neue Deutsche Biographie 16. Available online at: https://www.deutsche-biographie.de/sfz58726.html. Retrieved 01 September 2025.

Hollmann, H. E. (1936): Physik und Technik der ultrakurzen Wellen. Zweiter Band. Die ultrakurzen Wellen in der Technik. Berlin: Springer.

Hollmann, H. E. (1942): Erzeugung und Verstärkung von Dezimeter- und Zentimeterwellen. In: *Telegraphen-Fernsprech-Funk-und Fernseh-Technik* 31 (11), pp. 281–293.

Hollmann, H. E. (1943a): Erzeugung und Verstärkung von Dezimeter- und Zentimeterwellen (Forts.). In: *Telegraphen-Fernsprech-Funk-und Fernseh-Technik* 32 (1), pp. 12–21.

Hollmann, H. E. (1943b): Erzeugung und Verstärkung von Dezimeter- und Zentimeterwellen (Forts.). In: *Telegraphen-Fernsprech-Funk-und Fernseh-Technik* 32 (2), pp. 37–44.

Holmes, Richard (2011): The Secret Headquarters at the Heart of Britain's Victory. London: Profile Books.

Holpp, Wolfgang (2004): The Century of Radar. In: Deutsche Gesellschaft für Ortung und Navigation e. V. (DGON) (Ed.): 100 Years of Radar. Special Issue. Contributions from Various Countries Describing the Individual Radar Development Processes. German Institute of Navigation. Bonn: DGON, pp. 171–200.

Holpp, Wolfgang (2014): Die Geschichte der Radarmesstechnik. In: Frank Dittmann und Martin Kahmann (Ed.): Geschichte der elektrischen Messtechnik. Messen mit und von Elektrizität. Beiträge der Veranstaltung des VDE-Ausschusses "Geschichte der Elektrotechnik" vom 02.11. bis 03.11.2009 im Technischen Museum Wien. Berlin: VDE Verlag, pp. 177–214.

Holpp, Wolfgang; Klausing, Helmut (2022): The Development of Radar Technology in Germany against the Background of Innovation Systems of Combatant Nations during World War II. In: *European Journal of Navigation* 22 (1), pp. 4–15.

Hölzle, Katharina; Gemünden, Hans Georg (2011): Schlüsselpersonen der Innovation. Champions und Promotoren. In: Sönke Albers und Gassmann Oliver (Eds.): Handbuch Technologie- und Innovationsmanagement. 2. ed. Wiesbaden: Gabler, pp. 495–512.

Hoßbach, Friedrich (1937): Niederschrift vom 10.11.1937 über die Besprechung in der Reichskanzlei am 5.11.1937. In: 100(0) Schlüsseldokumente der deutschen Geschichte im 20. Jahrhundert. Available online at: https://www.1000dokumente.de/Dokumente/Das_Hoßbach-Protokoll. Retrieved 01 September 2025.

Hoßbach, Friedrich (1965): Zwischen Wehrmacht und Hitler. 1934–1938. Göttingen: Vandenhoeck & Ruprecht.

Hovanessian, S. A. (1980): Synthetic Array and Imaging Radars. Dedham, MA: Artech.

Hovanessian, S. A. (1984): Radar System Design and Analysis. Dedham, MA: Artech House.

Howse, Derek (1993): Radar at Sea. The Royal Navy in World War 2. London: Macmillan Press.

Huder, Bernhard (1999): Einführung in die Radartechnik. Stuttgart: Teubner.

Hull, Albert (1921): The Magnetron. In: *Journal of the American Institute of Electrical Engineers* XL (9), pp. 715–723.

Hull, Mcallister (1998): Gregory Breit (July 14, 1899 – September 11, 1981). In: National Academy of Sciences (Ed.): Biographical Memoirs. Washington, D.C.: National Academy Press (74), pp. 26–56.

Hülsmeyer, Christian (1904): Patent: Verfahren, um entfernte metallische Gegenstände mittels elektrischer Wellen einem Beobachter zu melden. Veröffentlichungsnr: DRP 165546, 30.04.1904.

Hümmelchen, Gerhard (2011a): Generalfeldmarschall Erhard Milch. In: Gerd Ueberschär (Ed.): Hitlers militärische Elite. 68 Lebensläufe. 2. Aufl. Darmstadt: Primus, pp. 171–177.

Hümmelchen, Gerhard (2011b): Generaloberst Ernst Udet. In: Gerd Ueberschär (Ed.): Hitlers militärische Elite. 68 Lebensläufe. 2. ed. Darmstadt: Primus, pp. 258–264.

Hutchinson, Martin (2022): Most Secret. Boffins come to Malvern. The story of how Malvern became a key player in World War II. Malvern Radar and Technolgy History Society (MRATHS). Malvern.

Hüttenhain, Erich (1979): Erfolge und Mißerfolge der deutschen Chiffrierdienste im Zweiten Weltkrieg. In: Jürgen Rohwer und Eberhard Jäckel (Eds.): Die Funkaufklärung und ihre Rolle im Zweiten Weltkrieg. Eine internationale Tagung in Bonn-Bad Godesberg und Stuttgart vom 15.–18. Nov. 1978. Stuttgart: Motorbuch, pp. 100–109.

Hyland, L. A. (1988): A personal reminiscence: the beginnings of radar 1930–1934. In: Russell Burns (Ed.): Radar Development to 1945. London: Peregrinus, pp. 29–34.

IEEE (2000): Radar in the Twentieth Century. Jubilee Issue. In: *IEEE Aerospace & Electronic Systems Magazine*, pp. 27–44.

Isted, G. A. (1991a): Guglielmo Marconi and the History of Radio – Part I. In: *GEC Review* 7 (1), pp. 45–56.

Isted, G. A. (1991b): Guglielmo Marconi and the History of Radio – Part II. In: *GEC Review* 7 (2), pp. 110–122.

Jackson, A. (2008): Stewart, Charles Stewart Henry Vane-Tempest-, seventh marquess of Londonderry (1878–1949). Oxford Dictionary of National Biography. Available online at: https://doi.org/10.1093/ref:odnb/36627. Retrieved 01 September 2025.

Jacobsen, Hans-Adolf (1965): Der Zweite Weltkrieg. Grundzüge der Politik und Strategie in Dokumenten. Frankfurt am Main: Fischer.

Jacot, B. L.; Collier, D. M. B. (1937): Marconi. Beherrscher des Äthers. Berlin: Höger.

Jäger, Kurt; Heilbronner, Friedrich (2010): Lexikon der Elektrotechniker. 2. Aufl. Berlin: VDE Verlag.

James, F. (2011): Faraday, Michael (1791–1867). Oxford Dictionary of National Biography. Available online at: https://doi.org/10.1093/ref:odnb/9153. Retrieved 01 September 2025.

Jeffery, Keith (2011): MI6. The History of the Secret Intelligence Service, 1909–1949. London: Bloomsbury.

Jipp, August (1934): Moderne Telegraphie. Die Fernschreibetechnik mit der dazugehörigen Leitungs- und Nebentechnik. Berlin: Springer.

Johnson, Brian (1983): Streng geheim. Wissenschaft und Technik im 2. Weltkrieg. Stuttgart: Motorbuch.

Johnson, Brian (2004): The Secret War. Barnsley: Pen & Sword.

Johnson, Don (2008): The Oslo Person. The Biography of Hans Ferdinand Mayer: unpublished manuscript.

Jones, R. V. (2009): Most Secret War. London: Penguin Books.

Jones, R. V. (2017): The Wizard War. British Scientific Intelligence 1939–1945. Brattleboro, VT: Echo Point Books & Media.

Jones, R.V (1990): Reflections on Intelligence. London: Mandarin.

Jones, Reginald (1962): Lord Cherwell: 'The Prof.'. In: *Nature* 194, p. 1107.

Jones, Reginald (1978): Most Secret War. British Scientific Intelligence 1939–1945. London: Hamish Hamilton.

Jones, Reginald (1993): Der geheime wissenschaftliche Nachrichtendienst der Royal Air Force im Zweiten Weltkrieg. In: Horst Boog (Ed.): Luftkriegführung im Zweiten Weltkrieg. Ein internationaler Vergleich. Bd. 12. Herford: Mittler, pp. 667–684.

Judkins, Phil (2020): A Basic Aide-memoire on Enigma. In: *Transmission Lines – The Journal of the Defence Electronics History Society (DEHS)* 25 (3), pp. 18–24. Available online at: https://www.dehs.org.uk. Retrieved 01 September 2025.

Judkins, Phil (2022): The Power and the Glory. In: *Transmission Lines – The Journal of the Defence Electronics History Society (DEHS)* 27 (3), pp. 1, 3–28. Available online at: https://www.dehs.org.uk. Retrieved 01 September 2025.

Judkins, Phil (2023): Tommy Flowers, 1905–1998. Honouring People and Places. In: *e-Newsletter of the Defence Electronics History Society (DEHS)*

(No. 132. Part 1: Oct. 2023), pp. 1–3. Available online at: https://www.dehs.org.uk. Retrieved 01 September 2025.

Judkins, Phil; Bauer, Arthur (2014): CHESS – Electronic Warfare and Strategic Bomber Offensives in Europe; an Attempt at Balance. Defence Electronics History Society (DEHS): Cranfield University – DCMT, Shrivenham.

Judkins, Phillip (2007): Making Vision into Power. Britain's acquisition of the world's first radar-based integrated air defence system 1935–1941. PhD Thesis Cranfield University.

Kahn, David (1979): Fernmeldewesen, Chiffriertechniken und Nachrichtenaufklärung in den Kriegen des 20. Jahrhunderts. In: Jürgen Rohwer und Eberhard Jäckel (Eds.): Die Funkaufklärung und ihre Rolle im Zweiten Weltkrieg. Eine internationale Tagung in Bonn-Bad Godesberg und Stuttgart vom 15.–18. Nov. 1978. Stuttgart: Motorbuch, pp. 17–47.

Kahn, David (1996): The Codebreakers. The Story of Secret Writing. New York: Scribner.

Kaiser, Walter (1996a): A Case Study in the Relationship of History of Technology and of General History: British Radar Technology and Neville Chamberlain's Appeasement Policy. In: *ICON* (2), Available online at: https://www.jstor.org/stable/23790399. pp. 29–52. Retrieved 01 September 2025.

Kaiser, Walter (1996b): Die Bedeutung der Technikgeschichte für die politische Geschichte. Dargestellt am Beispiel der britischen Radartechnik und der Appeasement-Politik. In: *Blätter für Technikgeschichte* 1995/96 (57./58.), pp. 145–160.

Katz, Bernard revised by Quirke V. M. (2009): Hill, Archibald Vivian (1886–1977). Oxford Dictionary of National Biography. Available online at: https://doi.org/10.1093/ref:odnb/31230. Retrieved 01 September 2025.

Kern, Ulrich (1984): Die Entstehung des Radarverfahrens: Zur Geschichte der Radartechnik bis 1945. Dissertation. Universität Stuttgart.

Kershaw, Ian (1998): Hitler: 1889–1936: Hubris. London: Allen Lane.

Kershaw, Ian (2000): Hitler: 1936–1945: Nemesis. London: Allan Lane.

Kershaw, Ian (2004): Making Friends with Hitler. Lord Londonderry and Britain's Road to War. London: Allan Lane.

Kershaw, Ian (2005): Hitlers Freunde in England. Lord Londonderry und der Weg in den Krieg. München: Deutsche Verlags-Anstalt.

Kershaw, Ian (2009): Hitler 1889–1945. München: Pantheon.

Keydel, Wolfgang; Röde, Bernd (2004): Radar Research and Development in Oberpfaffenhofen During the 20th Century and Beyond. In: Deutsche Gesellschaft für Ortung und Navigation e. V. (DGON) (Ed.): 100 Years of Radar. Special Issue. Contributions from Various Countries Describing the Individual Radar Development Processes. German Institute of Navigation. Bonn: DGON, pp. 201–225.

Kind, Dieter; Mühe, Walter (2006a): Jonathan Zenneck. Pionier der Funktechnik und Erforscher der Eigenschaften elektrischer Schwingungen. In:

Stiftung Werner-von-Siemens-Ring (Ed.): Naturforscher und Gestalter der Technik. Die Träger des Werner-von-Siemens-Ringes und ihre Leistungen für Naturwissenschaft und Technik. Berlin: VDE Verlag, pp. 193–199.

Kind, Dieter; Mühe, Walter (2006b): Karl Küpfmüller. Grundlegende Arbeiten zur Theorie der elektrischen Nachrichtenübertragung und zur Systemtechnik. In: Stiftung Werner-von-Siemens-Ring (Ed.): Naturforscher und Gestalter der Technik. Die Träger des Werner-von-Siemens-Ringes und ihre Leistungen für Naturwissenschaft und Technik. Berlin: VDE Verlag, pp. 208–212.

Kingsley, F. A. (Ed.) (1995): The Development of Radar Equipments for the Royal Navy, 1935–1945. Basingstoke, UK: Macmillan Press.

Kinsey, Gordon (1983): Bawdsey – Birth of the Beam. The History of R.A.F. Stations Bawdsey and Woodbridge. Lavenham: Terence Dalton.

Kirby, M.; Capey, R. (1997): The Air Defence of Great Britain, 1920–1940: An Operational Research Perspective. In: *Journal of the Operational Research Society* 48 (6), pp. 555–568.

Kirby, Maurice (2003): Operational Research in War and Peace. The British Experience from the 1930s to 1970. London: Imperial College Press.

Klausing, Helmut (1989): Untersuchung der Realisierbarkeit eines Radars mit synthetischer Apertur durch rotierende Antennen. Dissertation Universität Karlsruhe. München: MBB-Publikation, MBB-UA-1150-89-Pub=OTN-029299.

Klausing, Helmut (1998): Grundlagen und Anwendungen der Radartechnik. Vorlesungsmanuskript: Berufsakademie Ravensburg.

Klausing, Helmut (2022): The Origin of the Innovation Systems in Germany and Great Britain and Their Impact on the Development of Radar Technology Before and During World War II. In: *e-Newsletter of the Defence Electronics History Society (DEHS)* 120 (Part I, July 2022), pp. 31–42.

Klausing, Helmut (2025): Politics and Innovation in the Development of Radar Technology in Germany and Great Britain up to 1945: In. *Proc. of the 2025 IEEE Radar Conference (RadarConf25).* Krakau, PL, pp. 48–53.

Klausing, Helmut; Holpp, Wolfgang (Eds.) (2000): Radar mit realer und synthetischer Apertur. Konzeption und Realisierung. München: Oldenbourg.

Klausing, Helmut; Holpp, Wolfgang (2022). A Contribution to the History of Radar Development in Europe and Overseas Besides Germany and Great Britain. In: *European Journal of Navigation* 22 (2), pp. 22–33.

Klausing, Helmut; Judkins, Phil (2023): Microwave Generation Into A Key Military Technology in World War 2: A Comparative Study. In: *Transmission Lines – The Journal of the Defence Electronics History Society (DEHS)* 28 (4), pp. 3–32. Available online at: https://www.dehs.org.uk. Retrieved 01 September 2025.

Klee, Ernst (2011): Das Personenlexikon zum Dritten Reich. Wer war was vor und nach 1945. Koblenz: Ed. Kramer.

Kniestedt, Joachim (1989): Heinrich Hertz. Die Entdeckung der elektromagnetischen Wellen vor 100 Jahren. In: *Archiv für das Post- und Fernmeldewesen* 1/89, pp. 41–57.

König, Wolfgang (2010): Ingenieure im Nationalsozialismus. In: Noyan Dinçkal; Christof Dipper und Detlef Mares (Eds.): Selbstmobilisierung der Wissenschaft. Technische Hochschulen im »Dritten Reich«. Darmstadt: WBG, pp. 63–81.

Kosin, Rüdiger (1990): Die Entwicklung der deutschen Jagdflugzeuge. Koblenz: Bernard & Graefe.

Kozaczuk, Władysław (1967): Bitwa o tajemnice;. Służby wywiadowcze Polski i Rzeszy Niemieckiej, 1922–1939. Warszawa: Książka i Wiedza.

Kozaczuk, Władysław (1979): W kręgu Enigmy. Warszawa: Robotnicza Spółdzielnia Wydawnicza.

Kozaczuk, Władysław (1987): Im Banne der Enigma. Ereignisse – Tatsachen – Zusammenhänge. Berlin: Militärverlag der Deutschen Demokratischen Republik.

Kozaczuk, Władysław; Straszak, Jerzy (2004): Enigma. How the Poles Broke the Nazi Code. New York: Hippocrene Books.

Kraus, Hans-Christof (2014): Versailles und die Folgen. Außenpolitik zwischen Revisionismus und Verständigung 1919–1933. Bonn: Bundeszentrale für politische Bildung.

Kraus, Herbert (2011): Großadmiral Dönitz. In: Gerd Ueberschär (Ed.): Hitlers militärische Elite. 68 Lebensläufe. 2. ed. Darmstadt: Primus, pp. 316–325.

Kroener, Bernhard (1988): Die personellen Ressourcen des Dritten Reiches im Spannungsfeld zwischen Wehrmacht, Bürokratie und Kriegswirtschaft 1939–1942. In: Militärgeschichtliches Forschungsamt (Ed.): Das Deutsche Reich und der Zweite Weltkrieg. Organisation und Mobilisierung des deutschen Machtbereichs. Band 5. Erster Halbband: Kriegsverwaltung, Wirtschaft und personelle Ressourcen 1939–1941. Stuttgart: Deutsche Verlags-Anstalt, pp. 693–1001.

Kroge, Harry von (1998): GEMA – Berlin. Geburtsstätte der deutschen aktiven Wasserschall- und Funkortungstechnik. Hamburg: Lühmanndruck.

Kroge, Harry von (2000): GEMA: Birthplace of German Radar and Sonar. Translated from the German and edited by Louis Brown. Bristol: Institute of Physics Publishing (IoP).

Krubasik, Edward (1982): Strategische Waffe. In: *Wirtschaftswoche* (25), pp. 28–33.

Krüger, Peter (1993): Versailles. Deutsche Außenpolitik zwischen Revisionismus und Friedenssicherung. 2. ed. München: dtv.

Kruh, Louis; Deavours, Cipher (2002): The Commercial Enigma: Beginnings of Machine Cryptography. In: *Cryptologia* 26 (1), pp. 1–16.

Kümmritz, Herbert (1988): German radar development up to 1945. In: Russell Burns (Ed.): Radar Development to 1945. London: Peregrinus, pp. 209–226.

Kümmritz, Herbert (1994): On the development of radar technologies in Germany up to 1945. In: Oskar Blumtritt, Hartmut Petzold und William Aspray (Eds.): Tracking the History of Radar. Piscataway, NJ: The IEEE-Rutgers Center for the History of Electrical Engineering & the Deutsches Museum, pp. 25–46.

Küpfmüller, Karl (1972): Hollmann, Hans Erich. In: Neue Deutsche Biographie 9. Available online at: https://www.deutsche-biographie.de/pnd125623852.html#ndbcontent. Retrieved 01 September 2025.

Kurylo, Friedrich (1965): Ferdinand Braun. Leben und Wirken des Erfinders der Braunschen Röhre. Nobelpreisträger 1909. München: Heinz Moos.

Landwehr, Dominik (2007): 10 000 Menschen und 1500 Elektronenröhren knackten die Nazi-Codes. In: *Neue Zürcher Zeitung*. Available online at: https://www.nzz.ch/articleeyv3k-ld.399994. Retrieved 01 September 2025.

Latham, Colin; Stobbs, Anne (1997): Radar. A Wartime Miracle. Stroud: Sutton Publishing.

Latham, Colin; Stobbs, Anne (1999): Pioneers of Radar. Stroud: Sutton Publishing.

Latham, Colin; Stobbs, Anne (Eds.) (2011): The Birth of British Radar. The memoirs of Arnold 'Skip' Wilkins OBE. 2. ed. Bedford / Wakefield: RSGB in association with DEHS.

Leconte, Marc (2010): A statistical study of magnetron patents in the early years of electronics between 1920 and 1945; heuristic focusing around the discovery of the cavity magnetron. In: IEEE, DEHS und IET (Eds.): *Proceedings of the 2010 International Conference on the Origins and Evolution of the Cavity Magnetron*. Bournemouth University, England. Piscataway, NJ: IEEE, pp. 11–16.

Lemke, Bernd (2001): Luftschutz in Großbritannien und Deutschland 1923–1939. Zivile Kriegsvorbereitungen als Ausdruck der staats- und gesellschaftspolitischen Grundlagen von Demokratie und Diktatur. Dissertation. Albert-Ludwigs-Universität zu Freiburg i. Br.

Lindemann, Frederick (1934): Science and Air Bombing. In: *The Times* 8. Aug. 1934.

Lisicki, Tadeusz (1979): Die Leistung des polnischen Entzifferungsdienstes bei der Lösung des Verfahrens der deutschen »Enigma« -Funkschlüsselmaschine. In: Jürgen Rohwer und Eberhard Jäckel (Eds.): Die Funkaufklärung und ihre Rolle im Zweiten Weltkrieg. Eine internationale Tagung in Bonn-Bad Godesberg und Stuttgart vom 15.–18. Nov. 1978. Stuttgart: Motorbuch, pp. 66–86.

Lohmann, Walter; Hildebrand, Hans (1956): Die deutsche Kriegsmarine 1939–1945. Band II. Hauptkapitel XVa. Kennziffer 151. Bad Nauheim: Podzun, pp. 1–6.

Lorenz AG (1955): 75 Jahre Lorenz. 1880–1955. Festschrift. Stuttgart.

Lotz, Wolfgang (1999): Ohnesorge, Wilhelm. In: Neue Deutsche Biographie 19. Available online at: https://www.deutsche-biographie.de/pnd117112356.html#ndbcontent. Retrieved 01 September 2025.

Lovell, B. (2004a): Rowe, Albert Percival (1898–1976). Oxford Dictionary of National Biography. Available online at: https://doi.org/10.1093/ref:odnb/51630. Retrieved 01 September 2025.

Lovell, Bernard (1988): Blackett in War and Peace. In: *Journal of the Operational Research Society* 39 (3), pp. 221–233.

Lovell, Bernard (1991): Echos of War. The Story of H_2S Radar. Bristol: Adam Hilger.

Lovell, Bernard (2004b): The cavity magnetron in World War II: was the secrecy justified? In: *Notes and Records of the Royal Society of London* 58 (3), pp. 283–294.

Lübbecke, Marco: Operations Research (OR). In: Gabler Wirtschaftslexikon online (Ed.). Available online at: https://wirtschaftslexikon.gabler.de/definition/operations-research-or-46854/version-270128. Retrieved 01 September 2025.

Ludloff, Albrecht (1993): Handbuch Radar und Radarsignalverarbeitung. Braunschweig: Vieweg.

Ludwig, Karl-Heinz (1974): Technik und Ingenieure im Dritten Reich. Düsseldorf: Droste.

Mache, Wolfgang (1989): The Siemens Cipher Teletype in the History of Telecommunications. In: *Cryptologia* 13 (2), pp. 97–117.

Mache, Wolfgang (1992): Der Siemens-Geheimschreiber – ein Beitrag zur Geschichte der Telekommunikation 1992: 60 Jahre Schlüsselfernschreibmaschine. In: *Archiv für deutsche Postgeschichte* (2), pp. 85–94.

Macksey, Kenneth; revised (2011): Pile, Sir Frederick Alfred, second baronet (1884–1976). Oxford Dictionary of National Biography. Available online at: https://doi.org/10.1093/ref:odnb/31549. Retrieved 01 September 2025.

MacMillan, Margaret (2003): Paris 1919. six months that changed the world. New York: Random House.

Maier, Helmut (Ed.) (2002): Rüstungsforschung im Nationalsozialismus. Organisation, Mobilisierung und Entgrenzung der Technikwissenschaften. Göttingen: Wallstein.

Maier, Helmut (2007a): Forschung als Waffe. Rüstungsforschung in der Kaiser-Wilhelm-Gesellschaft und das Kaiser-Wilhelm-Institut für Metallforschung 1900–1945/48. Band 2. Göttingen: Wallstein.

Maier, Helmut (2007b): Luftfahrtforschung im Nationalsozialismus. In: Helmuth Trischler und Kai-Uwe Schrogl (Eds.): Ein Jahrhundert im Flug. Luft- und Raumfahrtforschung in Deutschland 1907–2007. Frankfurt am Main: Campus, pp. 104–122.

Marconi, Guglielmo (1909): Wireless telegraphic communication. Nobel Lecture. Available online at: https://www.nobelprize.org/prizes/physics/1909/marconi/lecture. Retrieved 01 September 2025.

Marconi, Guglielmo (1922): Radio Telegraphy. In: *Proc. of the IRE* 10 (4), pp. 215–238.

Marquand, David (1997): Ramsay MacDonald. A Biography. London: Metro Books.

Marquand, David (2015): MacDonald, (James) Ramsay (1866–1937). Oxford Dictionary of National Biography. Available online at: https://doi.org/10.1093/ref:odnb/34704. Retrieved 01 September 2025.

Martin, G. (2011): Jones, Reginald Victor (1911–1997). Oxford Dictionary of National Biography. Available online at: https://doi.org/10.1093/ref:odnb/68827. Retrieved 01 September 2025.

Mauel, Kurt (1983a): Brandt, Leo (1908–1971). In: Sigfrid von Weiher (Ed.): Männer der Funktechnik. Eine Sammlung von 70 Lebenswerken deutscher Pioniere der Funktechnik (drahtlose Telegrafie, Radar, Rundfunk und Fernsehen). Berlin: VDE-Verlag, pp. 25–27.

Mauel, Kurt (1983b): Esau, Abraham (1884–1955). In: Sigfrid von Weiher (Ed.): Männer der Funktechnik. Eine Sammlung von 70 Lebenswerken deutscher Pioniere der Funktechnik (drahtlose Telegrafie, Radar, Rundfunk und Fernsehen). Berlin: VDE-Verlag, pp. 52–53.

Mauel, Kurt (1983c): Küpfmüller, Karl (1898–1977). In: Sigfrid von Weiher (Ed.): Männer der Funktechnik. Eine Sammlung von 70 Lebenswerken deutscher Pioniere der Funktechnik (drahtlose Telegrafie, Radar, Rundfunk und Fernsehen). Berlin: VDE-Verlag, pp. 101–102.

Mauel, Kurt (1983d): Slaby, Adolf (1849–1913). In: Sigfrid von Weiher (Ed.): Männer der Funktechnik. Eine Sammlung von 70 Lebenswerken deutscher Pioniere der Funktechnik. (drahtlose Telegrafie, Radar, Rundfunk und Fernsehen). Berlin: VDE-Verlag, pp. 176–177.

Maxwell, James (1954a): A Treatise on Electricity and Magnetism. Vol. 1. 3. ed. New York: Dover Publ.

Maxwell, James (1954b): A Treatise on Electricity and Magnetism. Vol. 2. 3. ed. New York: Dover Publ.

McCarthy, Jerry (2020): An Overview of Early Polish Methods for Breaking the 3-Rotor Enigma. In: *Transmission Lines – The Journal of the Defence Electronics History Society (DEHS)* 25 (3), pp. 1–17. Available online at: https://www.dehs.org.uk. Retrieved 01 September 2025.

McCloskey, Joseph (1987a): OR Forum: British Operational Research in World War II. In: *Operations Research* 35 (3), pp. 453–470.

McCloskey, Joseph (1987b): The Beginnings of Operations Research: 1934–1941. In: *Operations Research* 35 (1), pp. 143–152.

McCloskey, Joseph (1987c): U.S. Operations Research in World War II. In: *Operations Research* 35 (6), pp. 910–925.

McDonough, Frank (1998): Neville Chamberlain, appeasement and the British road to war. Manchester: Manchester Univ. Press.

McDonough, Frank (2002): Hitler, Chamberlain and appeasement. Cambridge: Cambridge Univ. Press.

McKay, Craig (2010): German Teleprinter Traffic and Swedish Wartime Intelligence. In: B. Jack Copeland (Ed.): Colossus. The Secrets of Bletchley Park's Codebreaking Computers. Oxford: Oxford University Press, pp. 328–333.

McKay, Sinclair (2011): The Secret Life of Bletchley Park. The WWII Codebreaking Centre and the Men and Women Who Worked There. London: Aurum Press.

McKercher, B. J. C. (2008): National Security and Imperial Defence: British Grand Strategy and Appeasement, 1930–1939. In: *Diplomacy & Statecraft* 19 (3), pp. 391–442.

McKinney, John (2006): Radar: A Reluctant Miracle. In: *IEEE Aerospace and Electronic Systems Magazine* 21 (8, Part II), pp. ix–xviii.

Mendelssohn, Kurt (1973): The World of Walther Nernst. The Rise and Fall of German Science. London: Macmillan.

Mendelssohn, Kurt (1976): Walther Nernst und seine Zeit. Aufstieg und Niedergang der deutschen Naturwissenschaften. Weinheim: Physik Verlag.

Messerschmidt, Manfred (1989): Außenpolitik und Kriegsvorbereitung. In: Wilhelm Deist; Manfred Messerschmidt; Hans-Erich Volkmann and Wolfram Wette (Eds.): Ursachen und Voraussetzungen des Zweiten Weltkrieges. Frankfurt am Main: Fischer Taschenbuch, pp. 638–850.

Metzmacher, Helmut (1966): Deutsch-Englische Ausgleichsbemühungen im Sommer 1939. In: *Vierteljahrshefte für Zeitgeschichte* 14 (4), pp. 369–412.

Michaelis, Herbert (1979): Der Zweite Weltkrieg 1939–1945. In: Leo Just (Ed.): Handbuch der Deutschen Geschichte. Band IV. 2. Teil, Abschnitt V. Wiesbaden: Akademische Verlagsgesellschaft Athenaion, pp. 1–393.

Michalka, Wolfgang (Ed.) (1985): Das Dritte Reich. Band 1: »Volksgemeinschaft« und Großmachtpolitik 1933–1939. München: dtv.

Michalka, Wolfgang (2002): Deutsche Geschichte 1933–1945. Dokumente zur Innen- und Außenpolitik. Frankfurt am Main: Fischer.

Michel, Andrée (1994): Hans Ferdinand Mayer. Pionier der elektrischen Nachrichtentechnik. In: Ernst Feldtkeller und Herbert Goetzeler (Eds.): Pioniere der Wissenschaft bei Siemens. Beruflicher Werdegang und wichtigste Ergebnisse. Erlangen: Publicis MCD, pp. 85–90.

Mielert, Helmut (1982): Küpfmüller, Karl. In: Neue Deutsche Biographie 13. Available online at: https://www.deutsche-biographie.de/pnd118778196.html#ndbcontent. Retrieved 01 September 2025.

Miller, Russell (2016): Trenchard. Father of the Royal Air Force. London: Weidenfeld & Nicolson.

Milner-Barry, Stuart (2001): Hut 6: Early days. In: F. H. Hinsley and Alan Stripp (Eds.): Codebreakers. The inside story of Bletchley Park. Oxford: University Press, pp. 89–99.

Milward, Alan (1966a): Die deutsche Kriegswirtschaft 1939–1945. Schriftenreihe der Vierteljahrshefte für Zeitgeschichte. Nr. 12. Stuttgart: Deutsche Verlags-Anstalt.

Milward, Alan (1966b): Fritz Todt als Rüstungsminister. In: *Vierteljahrshefte für Zeitgeschichte* 14(1), pp. 40–58.

Minnett, H. C. (2015): Bowen, Edward George (1911–1991). Oxford Dictionary of National Biography. Available online at: https://doi.org/10.1093/ref:odnb/49587. Retrieved 01 September 2025.

Mirow, Christoph; Hölzle, Katharina; Gemünden, Hans G. (2007): Systematisierung, Erklärungsbeiträge und Effekte von Innovationsbarrieren. In: *Journal für Betriebswirtschaft: management review quartely* 57 (2), pp. 101–134.

Mittermaier, Bernhard; Rusinek, Bernd-A. (Eds.) (2009): Leo Brandt (1908–1971) Ingenieur – Wissenschaftsförderer – Visionär. Wissenschaftliche Konferenz zum 100. Geburtstag des nordrhein-westfälischen Forschungspolitikers und Gründers des Forschungszentrums Jülich. Schriften des Forschungszentrums Jülich. Jülich: Forschungszentrum Jülich (Reihe Allgemeines. Bd. 6).

Moll, Martin (1997): "Führer-Erlasse" 1939–1945. Edition sämtlicher überlieferter, nicht im Reichsgesetzblatt abgedruckter, von Hitler während des Zweiten Weltkrieges schriftlich erteilter Direktiven aus den Bereichen Staat, Partei, Wirtschaft, Besatzungspolitik und Militärverwaltung. Stuttgart: Steiner.

Molyneux-Berry, R. B. (1988): Dr. Henri Gutton, French radar pioneer. In: Russell Burns (Ed.): Radar Development to 1945. London: Peregrinus, pp. 45–52.

Morse, Philip; Kimball, George (2003): Methods of operations Research. Mineola, NY: Dover Publ.

Mowry, David (2014): German Cipher Machines of World War II. Center for Cryptologic History. Fort George G. Meade, MD: National Security Agency.

Mühlhausen, Walter (2025): Friedrich Ebert 1871–1925. Reichspräsident der Weimarer Republik. Bonn: Dietz.

Müller, Gotthard (1981): Funkmeßgeräte-Entwicklung bei der Firma C. Lorenz AG im Zeitraum von 1935–1945. 2. erw. Fassung. Stuttgart: SEL, pp. 1–65.

Müller, Rolf-Dieter (1988): Die Mobilisierung der deutschen Wirtschaft für Hitlers Kriegsführung. In: Militärgeschichtliches Forschungsamt (Ed.): Das Deutsche Reich und der Zweite Weltkrieg. Organisation und Mobilisierung des deutschen Machtbereichs. Band 5. Erster Halbband: Kriegsverwaltung, Wirtschaft und personelle Ressourcen 1939–1941. Stuttgart: Deutsche Verlags-Anstalt, pp. 347–689.

Müller, Rolf-Dieter (1999): Albert Speer und die Rüstungspolitik im totalen Krieg. In: Militärgeschichtliches Forschungsamt (Ed.): Das Deutsche Reich und der Zweite Weltkrieg. Organisation und Mobilisierung des deutschen Machtbereichs. Band 5, Zweiter Halbband: Kriegsverwaltung, Wirtschaft und personelle Ressourcen 1942–1944/45. Stuttgart: Deutsche Verlags-Anstalt, pp. 273–773.

Müller, Werner (1992): Bodenfunkmessgeräte der deutschen Luftwaffe bis 1945. Friedberg/H.: Podzun-Pallas.

Müller, Werner (2002): Horchgeräte – Kommandogeräte und Scheinwerfer der schweren Flak. Wölfersheim-Berstadt: Podzun-Pallas.

Murdin, Paul (2016): Lovell, Sir (Alfred Charles) Bernard (1913–2012). Oxford Dictionary of National Biography. Available online at: https://doi.org/10.1093/ref:odnb/105432. Retrieved 01 September 2025.

Murray, Joan (2001): Hut 8 and naval Enigma, Part I. In: F. H. Hinsley and Alan Stripp (Eds.): Codebreakers. The inside story of Bletchley Park. Oxford: University Press, pp. 113–118.

Nassua, Martin (1992): "Gemeinsame Kriegführung, gemeinsamer Friedensschluß": das Zimmermann-Telegramm vom 13. Januar 1917 und der Eintritt der USA in den 1. Weltkrieg. Europäische Hochschulschriften: Reihe 3, Geschichte und ihre Hilfswissenschaften. Bd. 520. Frankfurt am Main: Lang.

Naylor, John (2008): Hankey, Maurice Pascal Alers, first Baron Hankey (1877–1963). Oxford Dictionary of National Biography. Available online at: https://doi.org/10.1093/ref:odnb/33683. Retrieved 01 September 2025.

Neale, B. T. (1985): CH – The First Operational Radar. In: *GEC Journal of Research* 3 (2), pp. 73–83.

Neale, B. T. (1988): CH – the first operational radar. In: Russell Burns (Ed.): Radar Development to 1945. London: Peregrinus, pp. 132–150.

Neiberg, Michael (2017): The Treaty of Versailles. A Concise History. New York, NY: Oxford University Press.

Neilson, Keith (2003): The Defence Requirements Sub-Committee, British Strategic Foreign Policy, Neville Chamberlain and the Path to Appeasement. In: *The English Historical Review* 118 (477), pp. 651–684.

Neitzel, Sönke (1995): Der Einsatz der deutschen Luftwaffe über dem Atlantik und der Nordsee 1939–1945. Bonn: Bernard & Graefe.

Neitzel, Sönke (2014): Zeitalter der Weltkriege. Wie die Weltkriege endeten. In: *Informationen zur politischen Bildung* (321), pp. 59–62.

Nestel, Werner (1983): Runge, Wilhelm T. (1895). In: Sigfrid von Weiher (Ed.): Männer der Funktechnik. Eine Sammlung von 70 Lebenswerken deutscher Pioniere der Funktechnik (drahtlose Telegrafie, Radar, Rundfunk und Fernsehen). Berlin: VDE-Verlag, pp. 151–153.

Neumann, Tanja (2021): Nicola Tesla – Dandy der Technikgeschichte? Leben, Erfindungen und Mythos. In: *Das Archiv – Magazin für Kommunikationsgeschichte* (2), pp. 50–53.

Nicholls, C. S. (2004): Winterbotham, Frederick William (1897–1990). Oxford Dictionary of National Biography. Available online at: https://doi.org/10.1093/ref:odnb/40178. Retrieved 01 September 2025.

Nicholls, C. S. (2011): Boot, Henry Albert Howard [Harry] (1917–1983). Oxford Dictionary of National Biography. Available online at: https://doi.org/10.1093/ref:odnb/30833. Retrieved 01 September 2025.

Nicoll, D. (2004): Tiltman, John Hessell (1894–1982). Oxford Dictionary of National Biography. Available online at: https://doi.org/10.1093/ref:odnb/38069. Retrieved 01 September 2025.

Noack, Karoline (1998): Alltag und Machtverhältnisse in einem Rüstungsbetrieb. Die »Gesellschaft für elektroakustische und mechanische Apparate m.b.H. (Gema)«. In: Leonore Scholze-Irrlitz und Karoline Noack (Eds.): Arbeit für den Feind. Zwangsarbeiter-Alltag in Berlin und Brandenburg (1939–1945). Berlin: be.bra, pp. 53–64.

Noskwith, Rolf (2001): Hut 8 and naval Enigma, Part II. In: F. H. Hinsley and Alan Stripp (Eds.): Codebreakers. The inside story of Bletchley Park. Oxford: University Press, pp. 119–122.

Noskwith, Rolf (2011): Hut 8 From the Inside. Ch. 12. In: Ralph Erskine and Michael Smith (Eds.): The Bletchley Park Codebreakers. London: Biteback Publishing, pp. 184–194.

Nye, M. (2011): Blackett, Patrick Maynard Stuart, Baron Blackett (1897–1974). Oxford Dictionary of National Biography. Available online at: https://doi.org/10.1093/ref:odnb/30822. Retrieved 01 September 2025.

Nye, Mary Jo (2004): Blackett. Physics, War, and Politics in the Twentieth Century. Cambridge, MA: Harvard University Press.

Oberkommando der Kriegsmarine (1943a): Würzburg-Fibel, Teil I. Leo Brandt (Ed.). Ausschuss für Funkortung. Sonderbücherei der Funkortung.

Oberkommando der Kriegsmarine (1943b): Würzburg-Fibel, Teil II. Leo Brandt (Ed.). Ausschuss für Funkortung. Sonderbücherei der Funkortung.

Oberkommando der Kriegsmarine (1944): Berlin-Fibel

O'Halpin, E. (2008): Hall, Sir (William) Reginald (1870–1943). Oxford Dictionary of National Biography. Available online at: https://doi.org/10.1093/ref:odnb/33657. Retrieved 01 September 2025.

Orange, V. (2008): Dowding, Hugh Caswall Tremenheere, first Baron Dowding (1882–1970). Oxford Dictionary of National Biography. Available online at: https://doi.org/10.1093/ref:odnb/32884. Retrieved 01 September 2025.

Orange, V. (2011): Trenchard, Hugh Montague, first Viscount Trenchard (1873–1956). Oxford Dictionary of National Biography. Available online at: https://doi.org/10.1093/ref:odnb/36552. Retrieved 01 September 2025.

Ormerod, Richard (2003): The Father of Operational Research. In: Peter Hore (Ed.): Patrick Blackett. Sailor, Scientist and Socialist. London: Cass, pp. 187–200.

Osenberg, Werner (1943): Die Kriegslage als Folge unzureichender Auslastung verfügbarer Kapazitäten der deutschen Forschung. Aktenzeichen: R 26 III / 120. Bundesarchiv Berlin-Lichterfelde.

Overy, Richard (2012): Goering. Hitler's Iron Knight. London: I. B. Tauris.

Penley, Bill (2011): TRE History. In: Penley Radar Archives (Ed.). Available online at: http://www.purbeckradar.org.uk/penleyradararchives/history/tre_history.htm. Retrieved 01 September 2025.

Peruzzi, Giulio (2000): Maxwell. Der Begründer der Elektrodynamik. In: *Spektrum der Wissenschaft.*

Petzina, Dieter (1968): Autarkiepolitik im Dritten Reich. Der nationalsozialistische Vierjahresplan. Schriftenreihe der Vierteljahrshefte für Zeitgeschichte. Nr. 16. Stuttgart: Deutsche Verlags-Anstalt.

Phelps, Stephen (2010): The Tizard Mission. The Top-Secret Operation That Changed the Course of World War II. Yardley, PA: Westholme Publishing.

Piller, G. (2008): Tizard, Sir Henry Thomas (1885–1959). Oxford Dictionary of National Biography. Available online at: https://doi.org/10.1093/ref:odnb/36528. Retrieved 01 September 2025.

Piper, Ernst (2021): Weimarer Republik. In: *Informationen zur politischen Bildung* (346), pp. 1–83.

Pöthe, R. (1905): Neue Seesignalapparate. In: *Prometheus. Illustrierte Wochenschrift über die Fortschritte in Gewerbe, Industrie und Wissenschaft* XVL (825), pp. 705–709.

Potempa, Harald (2016): Udet, Ernst. In: Neue Deutsche Biographie 26, pp. 514–515. Available online at: https://www.deutsche-biographie.de/sfz74986.html. Retrieved 01 September 2025.

Preece, William (1897): Die Telegraphie ohne Drähte. In: *Elektrotechnische Zeitschrift* (30), pp. 430–431.

Price, Alfred (1978): Blitz über England. Die Luftangriffe auf die Britischen Inseln 1939–1945. Stuttgart: Motorbuch.

Price, Alfred (2009): Blitz on Britain 1939–45. Stroud: The History Press.

Price, Alfred (2017): Instruments of Darkness. The History of Electronic Warfare, 1939–1945. Barnsley: Frontline Books.

Pritchard, David (1989): The Radar War. Germany's Pioneering Achievement 1904–45. Wellingborough: Stephens.

Pritchard, David (1992): Durch Raum und Nacht. Radarentwicklung und -einsatz 1904–1945. Stuttgart: Motorbuch.

Probert, Henry (1993): Die Führung der Royal Air Force im Zweiten Weltkrieg. In: Horst Boog (Ed.): Luftkriegführung im Zweiten Weltkrieg. Ein internationaler Vergleich. Bd. 12. Herford: Mittler, pp. 787–807.

Pröse, Michael (2004): Chiffriermaschinen und Entzifferungsgeräte im Zweiten Weltkrieg: Technikgeschichte und informatikhistorische Aspekte. Dissertation Techn. Univ. Chemnitz.

Putley, Ernest (2009): Science Comes to Malvern. TRE – A Story of Radar 1942–1953. Malvern: Aspect Design.

Putney, Diane (1987): ULTRA and the Army Air Forces in World War II. In: Richard Kohn and Joseph Harahan (Eds.): USAF Warrior Studies. Office of Air Force History. Washington D.C.

Pyta, Wolfram (2024): Hitler, Adolf (1889–1945). In: NDB-online. Available online at: https://www.deutsche-biographie.de/118551655.html#dbocontent. Retrieved 01 September 2025.

Raeder, Erich (1957): Mein Leben. Von 1935 bis Spandau 1955. Tübingen: Schlichtenmayer.

Rahn, Werner (1990): Einsatzbereitschaft und Kampfkraft deutscher U-Boote 1942. In: *Militärgeschichtliche Mitteilungen* 47 (1), pp. 73–132.

Rahn, Werner (2011): Der Einfluss der Funkaufklärung auf die deutsche Seekriegsführung im Ersten und Zweiten Weltkrieg. In: Winfrid Heinemann (Ed.): Führung und Führungsmittel. Potsdamer Schriften zur Militärgeschichte. Militärgeschichtliches Forschungsamt. Bd. 14. Potsdam, pp. 15–56.

Raithel, Thomas; Strenge, Irene (2000): Die Reichstagsbrandverordnung. Grundlegung der Diktatur mit den Instrumenten des Weimarer Ausnahmezustands. In: *Vierteljahrshefte für Zeitgeschichte* 48 (3), pp. 413–460.

Raju, N.V.S (2020): Operations Research. Theory and Practice. Boca Raton, FL: CRC Press.

Randall, J. (1946): The cavity magnetron. In: *Proceedings of the Physical Society* 58, pp. 247–252.

Rangnick, Doris (2005): Runge, Wilhelm. In: Neue Deutsche Biographie 22. Available online at: https://www.deutsche-biographie.de/pnd116707690.html#ndbcontent. Retrieved 01 September 2025.

Ratcliffe, J.; Hutchins, R. (2012): Appleton, Sir Edward Victor (1892–1965). Oxford Dictionary of National Biography. Available online at: https://doi.org/10.1093/ref:odnb/30426. Retrieved 01 September 2025.

Ratcliffe, John (1975): Robert Alexander Watson-Watt, 13 April 1892–5 December 1973. In: *Biogr. Mems Fell. R. Soc.* 21, pp. 548–568.

Rau, Erik (2005): Combat science: the emergence of Operational Research in World War II. In: *Endeavour* 29 (4), pp. 156–161.

Redhead, Paul (2001): The Invention of the Cavity Magnetron and Its Introduction into Canada and the U.S.A. In: *La Physique au Canada* (Nov./Dec. 2001), pp. 321–328.

Rennert, David (2021): Der Oslo-Report. Wie ein deutscher Physiker die geheimen Pläne der Nazis verriet. Salzburg – Wien: Residenz Verlag.

Reuter, Frank (1971): Funkmeß. Die Entwicklung und der Einsatz des RADAR-Verfahrens in Deutschland bis zum Ende des Zweiten Weltkrieges. Wissenschaftliche Abhandlungen der Arbeitsgemeinschaft für Forschung des Landes Nordrhein-Westfalen. Band 42. Wiesbaden: Springer.

Rieger, Tobias (2018a): Reichsluftfahrtministerium (RLM). Projekt: Beamte nationalsozialistischer Reichsministerien. Universität Heidelberg. Available online at: https://ns-reichsministerien.de/2018/03/16/reichsluftfahrtministerium-rlm/. Retrieved 01 September 2025.

Rieger, Tobias (2018b): Wolfgang Martini. Projekt: Beamte nationalsozialistischer Reichsministerien. Universität Heidelberg. Available online at:

https://ns-reichsministerien.de/2018/06/11/wolfgang-martini/. Retrieved 01 September 2025.
RLE currents (1991): The MIT Radiation Laboratory: RLE's Microwave Heritage. In: *RLE currents* 4 (2), pp. 1–32.
Robbins, K. (2006): Inskip, Thomas Walker Hobart, first Viscount Caldecote (1876–1947). Oxford Dictionary of National Biography. Available online at: https://doi.org/10.1093/ref:odnb/34107. Retrieved 01 September 2025.
Robbins, Keith (2008): Lister, Philip Cunliffe- [*formerly* Philip Lloyd-Greame], first earl of Swinton (1884–1972). Oxford Dictionary of National Biography. Available online at: https://doi.org/10.1093/ref:odnb/30990. Retrieved 01 September 2025.
Roberts, Andrew (2019): Churchill. Walking with Destiny. London: Penguin.
Roberts, Jerry (2018): Lorenz. Breaking Hitler's Top Secret Code at Bletchley Park. Stroud: The History Press.
Röde, Bernd (1997): A Short Sketch of the History of the "Institut für Hochfrequenztechnik" in Oberpfaffenhofen. Special Edition. 60th anniversary of DLR's "Institut für Hochfrequenztechnik". In: *DLR Nachrichten* (86), pp. 7–14.
Rogers, David (2014): Men Amidst the Madness. British Technology Development in World War II. Solihull: Helion & Company.
Rogers, David (2018): Defending Island Britain in the Second World War. Documentary Sources. Solihull, Uk: Helion & Company.
Rohling, Hermann (2004): Christian Hülsmeyer, eine Außergewöhnliche Biographie. In: Deutsche Gesellschaft für Ortung und Navigation e. V. (DGON) (Ed.): 100 Years of Radar. Special Issue. Contributions from Various Countries Describing the Individual Radar Development Processes. German Institute of Navigation. Bonn: DGON, pp. 227–237.
Rohwer, J.; Hümmelchen, G. (1968): Chronik des Seekrieges 1939–1945. Herrsching: Pawlak.
Rohwer, Jürgen (1969a): Die Funkführung der deutschen U-Boote im zweiten Weltkrieg. Teil I: Land-Peilung und Bordradar. In: *Wehrtechnik* (9), pp. 324–328.
Rohwer, Jürgen (1969b): Die Funkführung der deutschen U-Boote im zweiten Weltkrieg. Teil II: Funkpeilung von Bord. In: *Wehrtechnik* (10), pp. 360–364.
Rohwer, Jürgen; Jäckel, Eberhard (Eds.) (1979): Die Funkaufklärung und ihre Rolle im Zweiten Weltkrieg. Eine internationale Tagung in Bonn-Bad Godesberg und Stuttgart vom 15.–18. Nov. 1978. Stuttgart: Motorbuch.
Rohwer, Jürgen (1979): Der Einfluß der alliierten Funkaufklärung auf den Verlauf des Zweiten Weltkrieges. In: *Vierteljahrshefte für Zeitgeschichte* 27 (3), pp. 325–369.
Rohwer, Jürgen (1980): Nachwort. In: Cajus Bekker – Augen durch Nacht und Nebel. Die Radar Story. 3. Aufl. Herford: Mittler, pp. 275–283.
Rohwer, Jürgen; Jäckel, Eberhard (Eds.) (1984): Kriegswende Dezember 1941. Referate und Diskussionsbeiträge des internationalen his-

torischen Symposiums in Stuttgart vom 17. bis 19. September 1981. Koblenz: Bernard & Graefe.

Rowe, A. P. (2015): One Story of Radar. First published 1948. First paperback ed. 2015. Cambridge: Cambridge Univ. Press.

Rukop, H. (1953): Röhren, Rundfunk und kurze Wellen. In: *Telefunken-Zeitung* 26 (100), pp. 165–176.

Runge, Wilhelm (1934a): Die Grundlagen der Dezimeterwellen-Technik. In: *Telefunken-Zeitung* 15 (68), pp. 24–33.

Runge, Wilhelm (1934b): Schwingungserzeugung mit dem Magnetron. In: *Telefunken-Zeitung* XV. (69), pp. 5–14.

Runge, Wilhelm (1938): Die praktische Bedeutung der Mikrowellen. Vortrag gehalten am 10. Dezember 1937. In: Sonderabdruck aus dem Jahrbuch der Deutschen Akademie der Luftfahrtforschung 1937/38. Berlin, pp. 222–227.

Runge, Wilhelm (1941): Elektromagnetische Rückstrahlung, ihre Technik, Möglichkeiten und Grenzen. In: Deutsche Akademie der Luftfahrtforschung (Ed.). Berlin: Schriften der Deutschen Akademie der Luftfahrtforschung, pp. 1–37.

Runge, Wilhelm (1942): Ein räumlich peilendes Rückstrahlgerät als Zielgerät. In: Deutsche Akademie der Luftfahrtforschung (Ed.). Berlin: Schriften der Deutschen Akademie der Luftfahrtforschung, pp. 4–21.

Runge, Wilhelm (1953): Fernsehen, Richtstrecken und Funkortung. II. Richtstrecken. In: *Telefunken-Zeitung* 26 (Heft 100), pp. 182–185.

Runge, Wilhelm (1971): Ich und Telefunken. Erinnerungen aus 40 Jahren. Unveröffentlichtes Manuskript. Historisches Archiv des Deutschen Technikmuseums Berlin.

Runge, Wilhelm (1988): A personal reminiscence. In: Russell Burns (Ed.): Radar Development to 1945. London: Peregrinus, pp. 227–228.

Russer, Peter (2009): Ferdinand Braun – A Pioneer in Wireless Technology and Electronics. In: *Proc. of the 39th European Microwave Conference*, Rom, IT, pp. 547–554.

Ryan, Mary (2015): Hitler's final words – Pieces of history. In: *Prologue Magazine Blog. National Archives.* Available online at: https://prologue.blogs.archives.gov/2015/04/30/hitlers-final-words/. Retrieved 01 September 2025.

Sachs, Ernst (1941): Schwarzsenden ist Landesverrat. Jedem Schwarzsender droht die Todesstrafe. In: *Mitteilungen des deutschen Amateur-Sende- und Empfangsdienstes (DASD e. V.)* (7/8).

Salewski, Michael (2003): Raeder, Erich. In: Neue Deutsche Biographie 21. Available online at: https://www.deutsche-biographie.de/pnd118743511.html#ndbcontent. Retrieved 01 September 2025.

Sayer, A. P. (2022): Birth of British Radar. Official History of Army Radar. In: *The e-Newsletter of the Defence Electronics History Society (DEHS)* 120 (Part IV, July 2022), pp. 237–263. Available online at: https://www.dehs.org.uk. Retrieved 01 September 2025.

Scheffler, Wolfgang (1972): Himmler, Heinrich. In: Neue Deutsche Biographie 9. Available online at: https://www.deutsche-biographie.de/pnd11855123X.html#ndbcontent. Retrieved on 01 September 2025.

Scherbius, Arthur (1918): Chiffrierapparat. Patentschrift. Reichspatentamt. Angemeldet durch Gewerkschaft Securitas in Berlin. Veröffentlichungsnr: 416219. Available online at: https://www.cryptomuseum.com/crypto/enigma/patents/files/DE416219.pdf. Retrieved 01 September 2025.

Scherbius, Arthur (1923): "Enigma" Chiffriermaschine. In: *Elektrotechnische Zeitschrift* (47/48), pp. 1035–1036.

Scherner, Jonas (2007): Bericht zur deutschen Wirtschaftslage 1943/44. In: *Vierteljahrshefte für Zeitgeschichte* 55 (3), pp. 499–546.

Scherner, Jonas; Streb, Jochen (2006): Das Ende eines Mythos? Albert Speer und das so genannte Rüstungswunder. In: *VSWG: Vierteljahrschrift für Sozial- und Wirtschaftsgeschichte* 93 (2), pp. 172–196.

Schmucker, Georg (1992): Radartechnik in Großbritannien und in Deutschland von 1918 bis 1945. In: Armin Hermann und Hans-Peter Sang (Eds.): Technik und Staat. Bd. 9. Düsseldorf: VDI-Verlag, pp. 379–398.

Schoen, Lothar (1990): Mayer, Hans Ferdinand. In: Neue Deutsche Biographie 16. Available online at: https://www.deutsche-biographie.de/pnd139699961.html#ndbcontent. Retrieved 01 September 2025.

Schoen, Lothar (1994): Karl Küpfmüller (1897–1977). In: Ernst Feldtkeller und Herbert Goetzeler (Eds.): Pioniere der Wissenschaft bei Siemens. Beruflicher Werdegang und wichtigste Ergebnisse. Erlangen: Publicis MCD, pp. 96–102.

Schönborn, Franz (2003): Marconi, ein Leben für "Wireless Telegraphy". Leben und Wirken eines großen Visionärs. In: *telekom praxis 12/03*, pp. 13–35.

Schönbrunn, Günter (1979): Geschichte in Quellen. Weltkriege und Revolutionen 1914–1945. 3. ed. München: Bayerischer Schulbuch-Verlag.

Schon, Donald (1963): Champions for Radical New Inventions. In: *Harvard Business Review.* 41 (2), pp. 77–86.

Schramm, Percy (Ed.) (1961): Kriegstagebuch des Oberkommandos der Wehrmacht 1940–1941. Zweiter Halbband. Band 1. Im Auftrag des Arbeitskreises für Wehrforschung. Bonn: Bernard & Graefe.

Schramm, Percy (Ed.) (1963): Kriegstagebuch des Oberkommandos der Wehrmacht 1943. 1. Januar 1943-31. Dezember 1943. Zweiter Halbband. Band 3. Im Auftrag des Arbeitskreises für Wehrforschung. Bonn: Bernard & Graefe.

Schroeter, Wolfgang (2019): Albert Speer. Aufstieg und Fall eines Mythos. Paderborn: Schöningh.

Schröter, Fritz (1959): Esau, Abraham. In: Neue Deutsche Biographie 4. Available online at: https://www.deutsche-biographie.de/pnd128955856.html#ndbcontent. Retrieved 01 September 2025.

Schubert, Klaus; Klein, Martina (2021): Das Politiklexikon. Begriffe – Fakten- Zusammenhänge. Bonn: Dietz.

Schultes, Theodor (1953): Funkmeß-Übersichtsverfahren. In: Leo Brandt (Ed.). Die Funkortung der deutschen Flugsicherung. Frankfurter Fachtagung 1953. Ausschuss für Funkortung. Bücherei der Funkortung. Band 2. Teil III. Dortmund: Verkehrs- u. Wirtschafts-Verlag, pp. 56–94.

Schültz, Benjamin (2016): Prozesspromotoren in Innovationsprojekten. Theoretische Fundierung und erste Validierung eines psychometrischen Instruments zur Erfassung der Promotorenrolle. Dissertation. Universität der Bundeswehr, München.

Schulze, E. (1988): German anti-chaff measures. In: Russell Burns (Ed.): Radar Development to 1945. London: Peregrinus, pp. 405–409.

Schwab, Adolf; Fischer, Peter (1998): Maxwell, Hertz, and German Radio-Wave History. In: *Proc. of the IEEE* 86 (7), pp. 1312–1318.

Schwarz, Albert (1973): Die Weimarer Republik. Der Friede von Versailles. In: Leo Just (Ed.): Handbuch der Deutschen Geschichte. Band IV, 1. Teil, Abschnitt III. Frankfurt am Main: Akademische Verlagsgesellschaft Athenaion, pp. 11–26.

Seiler, Michael (2007): Kommandosache »Sonnengott«. Geschichte der deutschen Sonnenforschung im Dritten Reich und unter alliierter Besatzung. Frankfurt am Main: Harri Deutsch.

Senger und Etterlin von, Stefan (2021): Frauen in der Wehrmacht. 1939–1945. Stuttgart: Motorbuch

Serby, J.; Bradley, J. (2020): Wimperis, Harry Egerton (1876–1960). Oxford Dictionary of National Biography. Available online at: https://doi.org/10.1093/ref:odnb/36973. Retrieved 01 September 2025.

Shrader, Charles (2006): History of Operations Research in the United States Army. Volume I: 1942–1962. Washington, D.C.: Office of the Deputy under Secretary of the Army for Operations Research, United States Army.

Singh, Simon (2000): Geheime Botschaften. Die Kunst der Verschlüsselung von der Antike bis in die Zeiten des Internet. München: Hanser.

Sitzungsprotokolle der Arbeitsgemeinschaft Rotterdam (AGR) (n.d.): Ausschuß für Funkortung. *CDVandT.org* (Ed.). Available online at: https://www.cdvandt.org/agr_protocols.htm. Retrieved 01 September 2025.

Six, Alfred (1943): Die Hochfrequenztechnik im Kriege. In: *Funkschau* 16 (10/12), p. 89.

Skillen, Hugh (2024): The First Enigma Book: 'Enigma' by Gustave Bertrand. In: *e-Newsletter of the Defence Electronics History Society (DEHS)* 139 (Part II, Aug. 2024), pp. 62–148. Available online at: https://www.dehs.org.uk. Retrieved 01 September 2025.

Skolnik, Merrill (2002): Introduction to Radar Systems. 3. ed. Boston: McGraw-Hill.

Skolnik, Merrill (2008a): An Introduction and Overview of Radar. In: Merrill Skolnik (Ed.): Radar Handbook. 3. ed. New York: McGraw-Hill, Chapter 1.

Skolnik, Merrill (Ed.) (2008b): Radar Handbook. 3. ed. New York: McGraw-Hill.

Slaby, Adolf (1898): The new Telegraphy. Recent Experiments in Telegraphy with Sparks. In: *The Century Magazine*, pp. 867–874.

Slaby, Adolf (1902a): Die wissenschaftlichen Grundlagen der Funkentelegraphie. Erste Mittheilung. In: *Elektrotechnische Zeitschrift* (9), pp. 165–169.

Slaby, Adolf (1902b): Die wissenschaftlichen Grundlagen der Funkentelegraphie. Zweite Mittheilung. In: *Elektrotechnische Zeitschrift* (13), pp. 254–258.

Slaby, Adolf (1908): Glückliche Stunden. Entdeckungsfahrten in den elektrischen Ozean. Berlin: Simion.

Smith, C. (2011): Thomson, William, Baron Kelvin (1824–1907). Oxford Dictionary of National Biography. Available online at: https://doi.org/10.1093/ref:odnb/36507. Retrieved 01 September 2025.

Smith, Michael (2010): How It Began: Bletchley Park Goes to War. In: B. Jack Copeland (Ed.): Colossus. The Secrets of Bletchley Park's Codebreaking Computers. Oxford: Oxford University Press, pp. 18–35.

Smith, Michael (2011): The Secrets of Station X. How Bletchley Park Helped Win the War. London: Biteback Publ.

Smith, R. A.; Hanbury-Brown, R.; Mould, A. J.; Ward, A. G. (1985): ASV: the detection of surface vessels by airborne radar. In: *IEE Proceedings* 132 (6), pp. 359–384.

Snow, C. P. (1961a): Science and Government. London: Oxford University Press.

Snow, C. P. (1961b): Politik hinter verschlossenen Türen. Wissenschaft und Staatsführung. Stuttgart: Deutsche Verlags-Anstalt.

Snow, C. P. (1962): A Postscript to Science and Government. London: Oxford University Press.

Snow, C. P. (2013): Science and Government. Cambridge, MA: Harvard University Press.

Solari, Luigi (1942): Marconi. Im Privatleben und bei der Arbeit. Leipzig: List.

Speer, Albert (2013): Erinnerungen. 5. ed. Berlin: Ullstein.

Staal, Ir. M.; Weiller, J. L. C. (1988): Radar development in the Netherlands before the war. In: Russell Burns (Ed.): Radar Development to 1945. London: Peregrinus, pp. 235–242.

Stepp, Wilhelm (1947): Über die Reichweite drahtloser Anlagen im Wellengebiet von 1 cm bis 20 m. Dissertation. Technische Hochschule Darmstadt. Available online at: https://www.cdvandt.org/Stepp's-dissertation-1946.pdf. Retrieved 01 September 2025.

Stewart, Irvin (1948): Organizing Scientific Research for War. The Administrative History of the Office of Scientific Research and Development. Boston: Little, Brown and Company.

Stoltenberg, Gerhard (1964): Göring, Hermann. In: Neue Deutsche Biographie 6. Available online at: https://www.deutsche-biographie.de/pnd118540157.html#ndbcontent. Retrieved 01 September 2025.

Stone, Tessa (2008): Forbes, Dame (Katherine) Jane Trefusis (1899–1971). Oxford Dictionary of National Biography. Available online at: https://doi.org/10.1093/ref:odnb/51971. Retrieved 01 September 2025.

Stresemann, Gustav (1926): Rede zum deutschen Beitritt zum Völkerbund, 10. September 1926. In: Bayerische Staatsbibliothek (Ed.): 100(0) Schlüsseldokumente zur deutschen Geschichte im 20. Jahrhundert. Available online at: https://www.1000dokumente.de/index.html?c=dokument_de&dokument=0089_str&object=translation&l=de. Retrieved 01 September 2025.

Strunk, Peter (1999): Die AEG. Aufstieg und Niedergang einer Industrielegende. Berlin: Nicolaische Verlagsbuchhandlung.

Studt, Christoph (2002): Das Dritte Reich in Daten. München: Beck.

Sturm, Reinhard (2003): Weimarer Republik. Kampf um die Republik 1919–1923. In: *Informationen zur politischen Bildung* (261), pp. 18–31.

Suchenwirth, Richard (2012): Historical Turning Points in the German Air Force War Effort. USAF Historical Studies No. 189. Bennington VT: Merriam Press.

Suits, C. G.; Lafferty, J. M. (1970): Albert Wallace Hull (1880–1966). Biographical Memoir. Washington, D.C.: National Academy of Sciences, pp. 213–233.

Sullivan, Roger (2008): Synthetic Aperture Radar. In: Merrill Skolnik (Ed.): Radar Handbook. 3. ed. New York: McGraw-Hill, Chapter 17.

Surhone, Lambert et al (2010): Rudolf Kühnhold. Thuringia, Schwallungen: Betascript Publishing.

Süß, Dietmar (2011): Tod aus der Luft. Kriegsgesellschaft und Luftkrieg in Deutschland und England. Lizenzausgabe für die Bundeszentrale für politische Bildung. Band 1151. München: Siedler.

Susskind, C. (1988): Who invented radar? In: Russell Burns (Ed.): Radar Development to 1945. London: Peregrinus, pp. 506–517.

Swords, Sean (2008): Technical History of the Beginnings of Radar. London: The Institution of Engineering and Technology.

Taunt, Derek (2001): Hut 6: 1941–1945. In: F. H. Hinsley and Alan Stripp (Eds.): Codebreakers. The inside story of Bletchley Park. Oxford: University Press, pp. 100–112.

Taylor, A. H.; Young, L. E.; Hyland, L. A. (1934): System for Detecting Objects by Radio. United States Patent. No. 1,981,884.

Telefunken (Ed.) (1928): 25 Jahre Telefunken. Festschrift der Telefunken-Gesellschaft. 1903–1928. Berlin.

Tesla Memorial Society of New York (Ed.): Nicola Tesla's Interview in the New York Times about the "Death Rays". Available online at: https://www.teslasociety.com/deathray.jpg. Retrieved 01 September 2025.

Teuteberg, Hans-Jürgen; Neutsch, Cornelius (Eds.) (1998): Vom Flügeltelegraphen zum Internet. Geschichte der modernen Telekommunikation. Stuttgart: Franz Steiner.

Thamer, Hans-Ulrich (2003): Nationalsozialismus I. Von den Anfängen bis zur Festigung der Macht. In: *Informationen zur politischen Bildung* (251).

Thamer, Hans-Ulrich (2004): Nationalsozialismus II. Führerstaat und Vernichtungskrieg. In: *Informationen zur politischen Bildung* (266).

The Air Defense Debate, 1934–1939 (1985). In: *The Washington Quarterly* 8 (4), pp. 199–216.

The Manchester Guardian Weekly (1935). 8 March 1935, p. 183.

Thetford, Owen (1979): Aircraft of the Royal Air Force since 1918. London: Putnam.

Thiele, Erdmann (Ed.) (2003): Telefunken nach 100 Jahren. Das Erbe einer deutschen Weltmarke. Berlin: Nicolaische Verlagsbuchhandlung.

Thin, Jeremy (2008): The Pre-History of Royal Air Force Area Bombing, 1917–1942. Appendix 6: Trenchard Memorandum on the War Object of an Air Force. May 1928. M.A. in History. Univ. of Canterbury.

Thomas, William (2007): The heuristics of war: scientific method and the founders of operations research. In: *The British Journal for the History of Science* 40 (No. 2), pp. 251–274.

Thompson, F. C. (2016): H2S Mks I–III: Fundamentals of H2S. In: *e-newsletter of the Defence Electronics History Society (DEHS)* 50 (Part II, Sept. 2016), pp. 82–106. Available online at: https://www.dehs.org.uk. Retrieved 01 September 2025.

Thurn, H. (1913): Die Funkentelegraphie. 2. Aufl. Leipzig: Teubner.

Tidd, Joe; Bessant, John (2018): Managing Innovation. Integrating Technological, Market and Organizational Change. 6. ed. Hoboken, NJ: Wiley.

Tizard, Henry (1946): Science and the Services. In: *The Journal of the Royal United Service Institution* 91 (563), pp. 333–346.

Tofahrn, Klaus (2003): Chronologie des Dritten Reiches. Ereignisse, Personen, Begriffe. Darmstadt: Wiss. Buchges.

Tomlin, D. H. (1988): The origins and development of UK army radar to 1946. In: Russell Burns (Ed.): Radar Development to 1945. London: Peregrinus, pp. 284–295.

Traces of War (n.d.): Martini, Wolfgang (Generalleutnant). STIWOT (Stichting Informatie Wereldoorlog Twee) (Ed.) Available online at: https://www.tracesofwar.com/persons/35341/Martini-Wolfgang-Generalleutnant.htm. Retrieved 01 September 2025.

Tragner, Manfred (2004): Die frühen Hochspannungs- und Hochfrequenzversuche von Nikola Tesla. In: Horst Wessel (Ed.): Jahrhunderte der Hochspannung. Gemeinsame Tagung des Ungarischen Elektrotechnis-

chen Vereins und des VDE- Ausschusses "Geschichte der Elektrotechnik" vom 26. bis 27. September 2003 im Ungarischen Elektrotechnischen Museum Budapest. Berlin: VDE Verlag, pp. 217–245.
Trenkle, Fritz (1979): Die deutschen Funkmeßverfahren bis 1945. Stuttgart: Motorbuch.
Trenkle, Fritz (1982): Die deutschen Funkpeil- und Horch-Verfahren bis 1945. Ulm: AEG-Telefunken.
Trenkle, Fritz (1986): Die deutschen Funkmeßverfahren bis 1945. Heidelberg: Hüthig.
Trenkle, Fritz (1987): Die deutschen Funkführungsverfahren bis 1945. Heidelberg: Hüthig.
Trenkle, Fritz (1990): Die deutschen Funknachrichtenanlagen bis 1945. Band 2: "Der Zweite Weltkrieg". Heidelberg: Hüthig.
Treue, Wilhelm (1955): Hitlers Denkschrift zum Vierjahresplan 1936. In: *Vierteljahrshefte für Zeitgeschichte* 3 (2), pp. 184–210.
Tricker, R. A. R. (1975): Die Beiträge von Faraday und Maxwell zur Elektrodynamik. Wissenschaftliche Taschenbücher. Bd. 95. Berlin: De Gruyter.
Tuchman, Barbara (1982): Die Zimmermann Depesche. Bergisch-Gladbach: Bastei-Lübbe
Tuchman, Barbara (2014). The Zimmermann Telegram. America Enters the War, 1917–1918. Paperback. New York: Random House.
Tyrell, Albrecht (1987): Grossbritannien und die Deutschlandplanung der Alliierten 1941–1945. Frankfurt am Main: Metzner.
Ulaby, Fawwaz; Moore, Richard; Fung, Adrian (1981): Microwave Remote Sensing. Active and Passive. Volume I: Microwave Remote Sensing Fundamentals and Radiometry. Reading, MA: Addison-Wesley.
Ulbricht, Heinz (2005): Die Chiffriermaschine ENIGMA. Trügerische Sicherheit. Ein Beitrag zur Geschichte der Nachrichtendienste. Dissertation. Technische Universität Braunschweig.
Vahs, Dietmar; Brem, Alexander (2013): Innovationsmanagement. Von der Idee zur erfolgreichen Vermarktung. 4. ed. Stuttgart: Schäffer-Poeschel.
Völker, Karl-Heinz (1962): Planung und Maßnahmen zur Schaffung einer Fliegertruppe in der Reichswehr. In: Militärgeschichtliches Forschungsamt (Ed.): Die Entwicklung der militärischen Luftfahrt in Deutschland 1920–1933 Stuttgart: Deutsche Verlags-Anstalt, pp. 121–230.
Völker, Karl-Heinz (1967): Die Tarnzeit. Der Aufbau der Luftwaffe bis zum März 1935. In: Militärgeschichtliches Forschungsamt (Ed.): Die deutsche Luftwaffe 1933–1939. Aufbau, Führung und Rüstung der Luftwaffe sowie die Entwicklung der deutschen Luftkriegstheorie. Stuttgart: Deutsche Verlags-Anstalt, pp. 28–70.
Völker, Karl-Heinz (1968): Zur Entstehung und Organisation des Reichsluftfahrtministeriums (RLM) 1933–1939. In: Militärgeschichtliches Forschungsamt (Ed.): Dokumente und Dokumentarfotos zur Geschichte der deutschen Luftwaffe. Aus den Geheimakten des Reichswehrministeri-

ums 1919–1933 und des Reichsluftfahrtministeriums 1933–1939. Stuttgart: Deutsche Verlags-Anstalt, pp. 131–166.

Waddington, Conrad (1973): O.R. in World War 2. Operational Research against the U-boat. London: Elek Science.

Watson, Raymond (2009): Radar Origins Worldwide. History of Its Evolution in 13 Nations Through World War II: Trafford Publishing.

Watson-Watt, R. (1945): Radar in War and in Peace. In: *Nature* 156, pp. 319–324.

Watson-Watt, Robert (1946): The Evolution of Radiolocation. Part IIIA: Radiolocation. In: The Journal of the Institution of Electrical Engineers. 93(1), pp. 11–19.

Watson-Watt, Robert (1953): Resume of the Origins, Development and Applications of Radar in Great Britain. In: Leo Brandt (Ed.): Zur Geschichte der Funkortung. Deutsche und britische Beiträge. Frankfurter Fachtagung 1953. Ausschuss für Funkortung. Bücherei der Funkortung. Band 2. Teil I. Dortmund: Verkehrs- u. Wirtschafts-Verlag, pp. 32–39.

Watson-Watt, Robert (1957): Three Steps to Victory. London: Odhams Press.

Watt, Robert Alexander Watson-(1892–1973) (2017). Oxford Dictionary of National Biography. Available online at: https://doi.org/10.1093/ref:odnb/31811. Retrieved 01 September 2025.

Watts, Simon (2018a): Airborne Maritime Surveillance Radar. Volume 1. British ASV Radars in WW II 1939–1945. San Rafael, CA: Morgan & Claypool Publishers.

Watts, Simon (2018b): Airborne Maritime Surveillance Radar. Volume 2. Post-war British ASV Radars 1946–2000. San Rafael, CA: Morgan & Claypool Publishers.

Watts, Simon (2019): ASV Mark I, the First Airborne Maritime Surveillance Radar. In: *IEEE Aerospace and Electronic Systems Magazine* 34 (8), pp. 50–61.

Watts, Simon (2024): ASV Mk. XI, the first X-band radar for the FAA. In: *Transmission Lines – The Journal of the Defence Electronics History Society (DEHS)* 29 (2), pp. 3–18, 23–24. Available online at: https://www.dehs.org.uk. Retrieved 01 September 2025.

Watts, Simon (2025a): The Freya Air Warning Radar. In: *Transmission Lines – The Journal of the Defence Electronics History Society (DEHS)* 30 (2), pp. 1, 3–26, 28. Available online at: https://www.dehs.org.uk. Retrieved 01 September 2025.

Watts, Simon (2025b): The Birth of Airborne Radar. In: *Transmission Lines – The Journal of the Defence Electronics History Society (DEHS)* 30 (3), pp. 30–48. Available online at: https://www.dehs.org.uk. Retrieved 01 September 2025.

Watts, Simon; Klausing, Helmut (2025): History of German Radio Measurement Technology to 1936, Hans Karl Freiherr von Willisen. Translated by Simon Watts and Helmut Klausing. In: *e-Newsletter of the Defence*

Electronics History Society (DEHS) 147 (Aug. 2025), pp. 24–33. Available online at: https://www.dehs.org.uk. Retrieved 01 September 2025.

Webster, Charles; Frankland, Noble (2006): The Strategic Air Offensive Against Germany 1939–1945. Volume I. Preparation. Parts 1, 2 and 3. Uckfield, UK: Naval & Military Press.

Wehner, Donald (1987): High Resolution Radar. Norwood, MA: Artech House.

Weierud, Frode (2000): Sturgeon, The FISH BP Never Really Caught. In: David Joyner (Ed.): Coding Theory and Cryptography. From Enigma and Geheimschreiber to Quantum Theory. Conference Proceedings. Berlin: Springer, pp. 18–52.

Weierud, Frode (2010): Bletchley Park's Sturgeon – The Fish That Laid No Eggs. In: B. Jack Copeland (Ed.): Colossus. The Secrets of Bletchley Park's Codebreaking Computers. Oxford: Oxford University Press, pp. 307–327.

Weiher, Siegfrid von (1983a): Rindfleisch, Hans (1906). In: Sigfrid von Weiher (Ed.): Männer der Funktechnik. Eine Sammlung von 70 Lebenswerken deutscher Pioniere der Funktechnik (drahtlose Telegrafie, Radar, Rundfunk und Fernsehen). Berlin: VDE-Verlag, pp. 144–145.

Weiher, Siegfrid von (1983b): Schultes, Theodor Jakob Joseph (1901–1981). In: Sigfrid von Weiher (Ed.): Männer der Funktechnik. Eine Sammlung von 70 Lebenswerken deutscher Pioniere der Funktechnik. (drahtlose Telegrafie, Radar, Rundfunk und Fernsehen). Berlin: VDE-Verlag, pp. 171–173.

Weiher, Sigfrid von (1983c): Hülsmeyer, Christian (1881–1957). In: Sigfrid von Weiher (Ed.): Männer der Funktechnik. Eine Sammlung von 70 Lebenswerken deutscher Pioniere der Funktechnik (drahtlose Telegrafie, Radar, Rundfunk und Fernsehen). Berlin: VDE-Verlag, pp. 83–85.

Weinberg, Gerhard (1988): Hitler's Memorandum on the Four-Year Plan: A Note. In: *German Studies Review* 11 (1), pp. 133–135.

Weinreich, Gerhard (2001a): Deckname »Freya«. Aus dem Leben des HF-Technikers Theodor J. J. Schultes (1901–1981). In: *Post- und Telekommunikationsgeschichte* (1), pp. 106–108.

Weinreich, Gerhard (2001b): Die »Klingelei« an der Rheinbrücke. Aus dem Leben des Technikers Christian Hülsmeyer (1881–1957). In: *Post- und Telekommunikationsgeschichte* (1), pp. 102–103.

Weinreich, Gerhard (2001c): Erdtrabant im Funkvisier. Aus dem Leben des Physikers Edward Victor Appleton (1892–1965). In: *Post- und Telekommunikationsgeschichte* (1), pp. 104–105.

Weinreich, Gerhard (2001d): Funkmesstechnik kontra Bomber. Aus dem Leben des HF-Technikers Leo Brandt (1908–1971). In: *Post- und Telekommunikationsgeschichte* (1), pp. 109–110.

Weinreich, Gerhard (2001e): Kommandounternehmen Bruneval. Aus dem Leben des Physikers Wilhelm Tolmé Runge (1895–1987). In: *Post- und Telekommunikationsgeschichte* (1), pp. 105–106.

Weinreich, Gerhard (2001f): Radar zur Rettung Englands. Aus dem Leben des HF-Technikers Robert A. Wqatson-Watt (1892–1973). In: *Post- und Telekommunikationsgeschichte* (1), pp. 103–104.

Weinreich, Gerhard (2001g): Tarnbezeichnung DeTe-Gerät. Aus dem Leben des Physikers Rudolf Kühnhold (geb. 1903). In: *Post- und Telekommunikationsgeschichte* (1), pp. 108–109.

Weintraub, Stanley (2011): Snow, Charles Percy, Baron Snow (1905–1980). Oxford Dictionary of National Biography. Available online at: https://doi.org/10.1093/ref:odnb/31698. Retrieved 01 September 2025.

Weiss, Alexander von (1986): Die Maxwellsche Theorie, Auslöser der modernen Elektrotechnik. Ein historischer Rückblick. In: Horst Wessel (Ed.): Elektrotechnik im Wandel der Zeit. Geschichte der Elektrotechnik 4. Berlin: VDE-Verlag, pp. 71–82.

Weiss, Burghard (2002): Rüstungsforschung am Forschungsinstitut der Allgemeinen Electicitäts-Gesellschaft bis 1945. In: Helmut Maier (Ed.): Rüstungsforschung im Nationalsozialismus. Organisation, Mobilisierung und Entgrenzung der Technikwissenschaften. Göttingen: Wallstein, pp. 109–141.

Werther, Waldemar (1979): Die Entwicklung der deutschen Funkschlüsselmaschinen: Die »Enigma«. In: Jürgen Rohwer und Eberhard Jäckel (Eds.): Die Funkaufklärung und ihre Rolle im Zweiten Weltkrieg. Eine internationale Tagung in Bonn-Bad Godesberg und Stuttgart vom 15.–18. Nov. 1978. Stuttgart: Motorbuch, pp. 50–65.

Wessel, Horst (2004): Die Funkentelegraphie. In: Horst Wessel (Ed.): Jahrhunderte der Hochspannung. Gemeinsame Tagung des Ungarischen Elektrotechnischen Vereins und des VDE-Ausschusses "Geschichte der Elektrotechnik" vom 26. bis 27. September 2003 im Ungarischen Elektrotechnischen Museum Budapest. Berlin: VDE Verlag, pp. 173–216.

Wessel, Horst (2008): Grenzüberschreitende Kommunikation mittels Telegraf und Telefon. In: Horst Wessel (Ed.): Strom ohne Grenzen. Internationale Aspekte der Elektrotechnik. Tagung des VDE-Ausschusses "Geschichte der Elektrotechnik" am 23. Oktober 2006 in Aachen. Berlin: VDE Verlag, pp. 67–102.

Wette, Wolfram (1989): Ideologien, Propaganda und Innenpolitik als Voraussetzungen der Kriegspolitik des Dritten Reiches. In: Wilhelm Deist; Manfred Messerschmidt; Hans-Erich Volkmann and Wolfram Wette (Eds.): Ursachen und Voraussetzungen des Zweiten Weltkrieges. Frankfurt am Main: Fischer Taschenbuch, pp. 23–208.

White, Ian (2007a): A short history of air intercept radar and the British night-fighter, part one 1936–1945. Available online at: https://600sqnassociation.com/wp-content/uploads/2025/01/a_short_history_of_air_intercept_radar_and_the_british_pt_1.pdf. Retrieved 01 September 2025.

White, Ian (2007b): The History of Air Intercept (AI) Radar and the British Night-Fighter 1935–1959. Barnsley: Pen & Sword Aviation.

Wiesbeck, Werner (Ed.) (1988): 100 Jahre elektromagnetische Wellen. Vorträge d. Heinrich-Hertz-Symposiums vom 14. u. 15. März 1988 in Karlsruhe = 100 years electromagnetic waves. Berlin: VDE-Verlag.

Wildhagen, Karl Heinz (1970): Erich Fellgiebel. Meister operativer Nachrichtenverbindungen. Ein Beitrag zur Geschichte der Nachrichtentruppe. Wennigsen / Hannover: Selbstverlag.

Wildt, Michael (2012): Nationalsozialismus: Aufstieg und Herrschaft. In: *Informationen zur politischen Bildung* (314), pp. 22–27.

Wiley, Carl (1985): Synthetic Aperture Radars. In: *IEEE Transactions on Aerospace and Electronic Systems* AES-21 (3), pp. 440–443.

Wilkins, M. H. F. (2011): Randall, Sir John Turton (1905–1984). Oxford Dictionary of National Biography. Available online at: https://doi.org/10.1093/ref:odnb/31584. Retrieved 01 September 2025.

Wilkins, Maurice (1987): John Turton Randall. 23 March 1905–16 June 1984. In: *Biogr. Mems Fell. R. Soc.* 33, pp. 491–535.

Williams, Dennis (2018): Defford Airfield 1941–57. A pictorial history. Defford: Defford Airfield Heritage Group.

Willisen, Hans-Karl Freiherr von (ca. 1984): Geschichte der deutschen Funkmeßtechnik. Abschrift einer Tonbandaufzeichnung um 1952. Berlin-Köpenick: GEMA. pp. 1–27.

Willshaw, W. E. (1985): Microwave Magnetrons: A Brief History of Research and Development. In: *GEC Journal of Research* 3 (2), pp. 84–91.

Winterbotham, F. W. (1974): The Ultra Secret. London: Weidenfeld & Nicolson.

Winterbotham, F. W. (1999): The Ultra Secret. The Inside Story of Operation Ultra, Bletchley Park and Enigma. London: Weidenfeld & Nicolson.

Winterbotham, Frederick (1976): Aktion Ultra. Deutschlands Code-Maschine half den Alliierten siegen. Frankfurt am Main: Ullstein.

Wirsching, Andreas (2001): "Man kann nur Boden germanisieren". Eine neue Quelle zu Hitlers Rede vor den Spitzen der Reichswehr am 3. Februar 1933. In: *Vierteljahrshefte für Zeitgeschichte* 49 (3), pp. 517–550.

Wirsching, Andreas (2016): Hitlers Authentizität. Eine funktionalistische Deutung. In: *Vierteljahrshefte für Zeitgeschichte* 64 (3), pp. 387–418.

Wißner, Adolf (1961): Feddersen, Berend Wilhelm: Neue Deutsche Biographie 5. Available online at: https://www.deutsche-biographie.de/pnd11642589X.html#ndbcontent. Retrieved 01 September 2025.

Witte, Eberhard (1973a): Innovationsfähige Organisation. In: *Zeitschrift für Organisation* 42 (1), pp. 17–24.

Witte, Eberhard (1973b): Organisation für Innovationsentscheidungen. Das Promotoren-Modell. Schriften der Kommission für wirtschaftlichen und sozialen Wandel. 5. Aufl., Göttingen: Schwartz.

Zenneck, J.; Rukop H. (1925): Lehrbuch der drahtlosen Telegraphie. 5. ed, Stuttgart: Enke.

Zennek, Jonathan (1953): Arco, Georg Graf von. In: Neue Deutsche Biographie 1. Available online at: https://www.deutsche-biographie.de/sfz1190.html#ndbcontent. Retrieved 01 September 2025.

Zetzmann, Hans-Joachim (1983): Dieckmann, Max (1882–1960). In: Sigfrid von Weiher (Ed.): Männer der Funktechnik. Eine Sammlung von 70 Lebenswerken deutscher Pioniere der Funktechnik (drahtlose Telegrafie, Radar, Rundfunk und Fernsehen). Berlin: VDE-Verlag, pp. 46–48.

Zierold, Kurt (1968): Forschungsförderung in drei Epochen. Deutsche Forschungsgemeinschaft. Geschichte. Arbeitsweise. Kommentar. Wiesbaden: Steiner.

Zimmerman, David (1997): Tucker's Acoustical Mirrors: Aircraft Detection Before Radar. In: *War & Society* 15 (1), pp. 73–99.

Zimmerman, David (2003): Preparations for War. In: Peter Hore (Ed.): Patrick Blackett. Sailor, Scientist and Socialist. London: Cass Publ., pp. 110–125.

Zimmerman, David (2013): Radar. Britain's Shield and the Defeat of the Luftwaffe. Stroud: Amberley.

Zimmerman, David (2025): Radar and Sonar, 1917 to 1945: A Comparative Study of Military Technological Design. In: *Proc. of the 2025 IEEE Radar Conference (RadarConf25)*, Krakau, PL, pp. 36–41.

Zimmerman, David (1996): Top Secret Exchange. The Tizard Mission and the Scientific War. Stroud: Sutton Publ.

Zimmermann, Hans-Jürgen (2008): Operations Research. Methoden und Modelle. 2. ed. Wiesbaden: Vieweg.

Zimmermann, Peter (1992): Die Rolle der Technik im Ersten Weltkrieg. In: Armin Hermann und Hans-Peter Sang (Ed.): Technik und Staat. Bd. 9. Düsseldorf: VDI-Verlag, pp. 320–355.

17 Glossary of English Terms

Aeronautical Research Committee (ARC) Originally established as the Advisory Committee for Aeronautics (ACA) in 1909 and renamed in 1919, the ARC was a British research committee under the Air Ministry. It investigated scientific and technical problems in aviation and played a key role in advancing aircraft technologies, supporting the UK's aviation development in the interwar period.

Aeronautical Research Laboratory (ARL) Research facility established in Melbourne, Australia, following Harry Wimperis' advisory role to the Australian government, focusing on aeronautical and aviation research and development.

Air Defence Experimental Establishment (ADEE) Established in 1924, the ADEE was a British research facility responsible for testing and evaluating air defence technologies, including early radar systems. It played a key role in the development and assessment of equipment that would later be crucial for the UK's air defence capabilities.

Air Defence Research and Development Establishment (ADRDE) Formed in 1941 from the Air Defence Experimental Establishment (ADEE), the ADRDE was a British research establishment responsible for developing and testing air defence technologies, including radar systems, and contributed directly to the UK's air defence capabilities during World War II.

Air Defence Research Sub-Committee (ADRC) Established in spring 1935 as a sub-committee of the Committee of Imperial Defence (C.I.D.), the ADRC investigated scientific and technical solutions to strengthen Britain's air defence, playing a key role in coordinating research in the lead-up to World War II.

Air Ministry British government department responsible for the Royal Air Force (RAF) from 1918 to 1964, overseeing air strategy, procurement, organisation, and air defence efforts. It played a central role in planning and coordinating the UK's air operations and defence, particularly during the interwar period and throughout World War II.

Air Ministry Experimental Station (AMES) Series of British radar stations developed and operated by the Air Ministry from the 1930s, each assigned type numbers corresponding to specific radar systems and operational roles. The stations formed a key part of the UK's air defence network, supporting both research and operational deployment in the years leading up to and during World War II.

Air Ministry Research Establishment (AMRE) British research facility under the Air Ministry, responsible for developing and testing aviation technologies and air defence systems. It played a key role in advancing radar and other technical systems, contributing directly to the UK's air defence capabilities and supporting operational planning in the years leading up to and during World War II.

Air Ministry Wireless Station Facility operated by the British Air Ministry, responsible for transmitting and receiving wireless signals and supporting early radar development and air defence systems.

Air to Surface Vessel (ASV) radar Airborne radar system designed to detect and track ships and other surface vessels from aircraft, primarily used by maritime reconnaissance aircraft to locate enemy shipping and support naval operations.

Air Warning Radar Radar system designed to detect and track aircraft, primarily employed in air defence to provide advance warning of incoming aerial threats.

Airborne Interception (AI) radar Radar system primarily installed on night fighters during World War II, enabling them to detect, track, and engage enemy aircraft in darkness. Also referred to as Air Intercept Radar in some sources.

Anti-Aircraft Artillery Ground-based artillery systems designed to detect, track, and destroy enemy aircraft. Known in Germany during World War II as Flak, the term now broadly applies to any weapons or fire directed against airborne targets.

Anti-Aircraft Guns Rapid-fire, ground-based guns designed to target and destroy enemy aircraft. A subset of anti-aircraft artillery, they played a key role in air defence during World War II.

Anti-Aircraft Command (AAC) British Army organisation responsible for coordinating ground-based air defences in Great Britain during World War II.

Anti-Submarine Detection Investigation Committee (ASDIC) British research body tasked with developing methods for submarine detection. During World War II, ASDIC referred to the sonar technology developed by this committee. These devices allowed Allied ships to send out sound pulses to determine the range and direction of submarines.

Appeasement Policy British foreign policy in the 1930s, aimed at avoiding war by making concessions to Hitler's demands, particularly over territorial expansion in Europe. It was intended to maintain peace but ultimately failed to prevent the outbreak of World War II.

Armistice of Compiègne Ceasefire signed on 11 November 1918, between Germany and the Allies, marking the end of active fighting in World War I and paving the way for subsequent peace negotiations.

Atomic Energy Study Group at Chatham House Group within Chatham House examining the political, security, and scientific implications of atomic energy. Harry Wimperis was a member from 1946 to 1950, contributing his expertise on the technological and geopolitical aspects of nuclear energy.

Bawdsey Research Station (BRS) Research facility established in 1936 at Bawdsey Manor, where radar technology was developed and tested, forming the foundation of the UK's Chain Home early warning radar system and playing a key role in the development of British air defence prior to World War II.

Bentley Priory Headquarters of RAF Fighter Command during World War II, serving as the central hub for coordinating Britain's air defence, particularly during the Battle of Britain, and directing fighter operations under Air Chief Marshal Sir Hugh Dowding.

Biggin Hill Experiments Series of pioneering tests conducted at Biggin Hill that contributed to the development of the Dowding System, which integrated radar, radio communication, and centralised control to coordinate Britain's air defence and efficiently direct fighter operations during the Battle of Britain.

Blackett's Circus British team of scientists led by Patrick Blackett during World War II, conducting Operational Research. They analysed anti-submarine warfare, convoy tactics, and bombing operations, applying statistics and scientific methods to improve military effectiveness.

Boffins Informal British term for scientists and engineers, often highly dedicated and intensely focused on their technical work, particularly notable during World War II for their contributions to radar development.

Chain Home (CH) Radar stations forming the world's first integrated early warning system, developed and deployed by Great Britain from the mid-1930s. Operational by 1938, they detected incoming enemy aircraft at long range and played a key role in defending Great Britain during the Battle of Britain.

Chain Home Extra Low (CHEL) Radar system designed to detect very low-flying aircraft with high accuracy, supplementing the original Chain Home stations in areas where standard radars were less effective.

Chain Home Low (CHL) Radar system developed to detect low-flying aircraft, supplementing the original Chain Home stations by operating at higher frequencies and shorter wavelengths.

Clarendon Laboratory Research institution at the University of Oxford, founded in 1913, recognised for its contributions to experimental physics.

Commander-in-Chief (C-in-C) The highest-ranking officer with overall command of a nation's armed forces or a major military formation. In Britain, a formal title for large services or commands (e. g., RAF Air Officer Commanding-in-Chief). Germany had no direct equivalent.

Committee for the Coordination of Valve Development (CVD) British wartime committee established in 1940 to coordinate the research and production of specialised vacuum tubes for radar and other electronic systems

Committee for the Scientific Survey of Air Defence (CSSAD) British committee established in 1934 as a scientific advisory body to the Air Ministry, bringing together leading scientists and military advisers, playing a key role in the development of radar technology and laying the scientific foundation for Britain's air defence capabilities prior to World War II.

Committee for the Scientific Survey of Air Offence (CSSAO) Established in 1937 under the leadership of Henry Tizard, the CSSAO was a British research group serving as a scientific advisory body to the Air Ministry. It was tasked with studying air attack methods and assessing their potential impact on defence strategies, providing recommendations that informed the UK's air defence planning in the years leading up to World War II.

Committee for the Scientific Survey of Air Warfare (CSSAW) British committee formed in 1939 by merging the CSSAD and CSSAO, serving as a scientific advisory body to the Air Ministry and focused on advancing both defensive and offensive air technologies for Britain during World War II.

Committee of Imperial Defence (C.I.D.) British government committee established in 1904 to coordinate the strategic defence policy of the British Empire. It included senior political and military leaders and played a key advisory role in shaping defence decisions during the interwar period and the early years of World War II.

Council for Scientific and Industrial Research (CSIR) Australian government body founded in 1926 to promote scientific research. In 1938, Harry Wimperis advised on the establishment of the Aeronautical Research Division within the CSIR, focusing on aeronautical research.

Cyclometer Machine built in 1937 by mathematician and cryptologist Marian Rejewski and engineer Antoni Palluth to determine the daily ENIGMA key by calculating rotor cycles. It could deduce the key in around 20 minutes and was crucial for pre-war Polish cryptanalysis.

Defence Requirements Sub-Committee (DRC) British wartime committee responsible for defining and prioritising defence requirements within the Air Ministry and related organisations. It played an important role in guiding research, development, and production priorities for key technologies such as radar and aircraft during World War II.

Department of Scientific and Industrial Research (DSIR) British government agency founded 1916 to promote scientific research and innovation in industry. Henry Tizard served as its Permanent Secretary from 1927, overseeing the coordination of research projects and contributing to the development of technologies important for both national defence and industrial progress.

Directorate of Scientific Research of the Air Ministry Department within the British Air Ministry responsible for coordinating scientific research and technological development, particularly in areas vital to air defence, such as radar. It oversaw research programs and supported the implementation of technologies critical to Britain's air defence during World War II.

Dowding System Integrated air defence system developed by Air Chief Marshal Hugh Dowding during the late 1930s, combining radar, radio communication, and centralised control to coordinate Britain's air defences. It played a pivotal role in the Battle of Britain by enabling rapid detection, effective command, and efficient deployment of fighter aircraft against incoming Luftwaffe attacks.

Engineering Section of the British Association Division of the British Association for the Advancement of Science focused on the promotion and development of engineering sciences. Harry Wimperis served as its president in 1939, guiding discussions on advancements in engineering and their applications to industry and defence.

Fighter Command Branch of the Royal Air Force (RAF) responsible for controlling and coordinating British fighter aircraft during World War II. It played a key role in defending the UK during the Battle of Britain, directing fighter operations and integrating intelligence from radar and observation networks.

Filter Room Central facility that collected, processed, and verified data from Chain Home and other radar stations, as well as visual and radio reports, to provide a coherent picture of the air situation. It played a crucial role in coordinating timely fighter responses during World War II, forming an integral part of the Dowding System.

Fire-control radar Radar system that precisely tracks targets and provides accurate range, bearing, and altitude information to guide anti-aircraft guns. It was a key component of Britain's air defence during World War II, improving the accuracy and effectiveness of ground-based air defence systems.

Fleet Air Arm (FAA) Branch of the British Royal Navy responsible for operating naval aircraft. It carried out reconnaissance, air defence, and offensive operations from ships and shore bases, playing a crucial role in Britain's maritime operations during World War II.

Freya radar German air warning radar introduced in the late 1930s, comparable to Britain's Chain Home system but operating at a shorter wavelength with higher resolution and greater mobility. It provided early detection of incoming aircraft and formed a central component of Germany's air defence network during World War II.

Government Code and Cypher School (GC&CS) British government agency established in 1919, responsible for codebreaking and signals intelligence. It played a crucial role during World War II, particularly in decrypting German ENIGMA messages at Bletchley Park. In 1946, it was reorganised as the Government Communications Headquarters (GCHQ).

Ground-Controlled Interception (GCI) Radar-based system that enabled ground controllers to guide fighter aircraft directly to enemy targets via radio communication. Although not fully operational during the Battle of Britain, GCI later became a key component of British air defence, especially in directing night fighters with increased precision and effectiveness.

Group Operations Room Command centre coordinating fighter operations over a large area. It maintained map tables with real-time updates, directing Sector Operations Rooms and squadrons within its region, and played a vital role in the organisation and rapid deployment of fighter aircraft during World War II.

High-Frequency Direction Finder HF/DF, Huff-Duff Specialised form of radio direction finding for high-frequency signals, used by the Allies during World War II to locate German U-boat transmissions quickly and accurately. The nickname Huff-Duff comes from the colloquial pronunciation of HF/DF.

H_2S Radar British airborne radar system developed during World War II, first introduced in 1943. It was designed for ground mapping and navigation, allowing bomber crews to identify targets and navigate accurately at night or in poor visibility. H_2S played a key role in the effectiveness of the Royal Air Force's night bombing campaigns and represented one of the first operational uses of ground-mapping radar in combat.

Imperial College of Science and Technology British university in London, founded in 1907, recognised for its leading role in science, engineering, medicine, and technology, and its contributions to research and innovation.

Intelligence Service Military organisation responsible for gathering and analysing information about potential adversaries.

Ionospheric Research Station Cover name for the radar research facility at Orford Ness, involved in early radar development during the 1930s. It conducted experiments on radio wave propagation and aircraft detection, providing foundational knowledge for Britain's air defence radar systems.

Lord President of the Council Senior UK government position responsible for chairing important government meetings and coordinating policies, often involved in overseeing strategic decisions during times of national crisis.

Massachusetts Institute of Technology (MIT) Leading American research university in Cambridge, Massachusetts, which became a key centre for Allied radar research during World War II through the establishment of the Radiation Laboratory following the Tizard Mission.

Ministry of Aircraft Production Research Establishment (MAPRE) British radar research organisation established in 1940 under the Ministry of Aircraft Production. It conducted research on radar and aircraft electronics and is considered the predecessor of the Telecommunication Research Establishment (TRE), which played a central role in Britain's air defence during World War II.

Ministry of Supply British government department established in 1917 to oversee the procurement and supply of military equipment and materials. During World War II, it was also responsible for coordinating research establishments and supporting technological development for the war effort.

MIT Radiation Laboratory (Rad Lab) Research facility at the Massachusetts Institute of Technology (1940–1945) that played a central role in developing radar technology for the Allied forces during World War II.

National Cash Register Company (NCR) American manufacturer of cash registers; during WWII, some machines were used at Bletchley Park for administrative and data-processing tasks supporting codebreaking.

National Defence Research Committee (NDRC) United States government body established in 1940 to coordinate scientific research for military purposes during the Second World War, focusing on technologies such as radar, explosives, and early atomic research.

National Physical Laboratory (NPL) UK's national measurement institute, founded in 1900, specialising in scientific research and the development of measurement standards. It contributed to technological developments for Britain's defence and industry and continues today to support research and standardisation in science and engineering.

Naval Research Laboratory (NRL) United States Navy's principal research facility, officially established in July 1923 in Washington, D.C. It consolidated several earlier naval research units, including the Aviation Radio Laboratory in Anacostia. NRL became a leading centre for developments in radio, radar, and other naval technologies, and continues today to support research and innovation for the US Navy.

Office of Scientific Research and Development (OSRD) US government agency established in 1941 to coordinate scientific research and technological development for the war effort. The OSRD oversaw projects in radar, rocketry, medicine, and the Manhattan Project, playing a key role in advancing the United States' military capabilities during World War II.

Operation Biting British commando raid in February 1942 to capture parts of a German Würzburg radar in Bruneval, France. The operation provided vital intelligence for British radar countermeasures and demonstrated the effectiveness of small, well-planned special operations in gathering technical intelligence.

Operation Gomorrah Large-scale Allied bombing campaign against Hamburg from 24 July to 3 August 1943, conducted by the RAF and USAAF. It combined nighttime area bombing and daylight raids, using technologies such as H_2S radar and Window. The operation caused massive destruction and extensive civilian casualties, particularly during the firestorm on the night of 27–28 July.

Operational Research / Operations Research Scientific approach to military problem-solving developed in Britain during World War II. It applied mathematical analysis, statistics, and modelling to improve the effectiveness of operations such as anti-submarine warfare, radar deployment, and bombing strategies. After the war, the discipline evolved into a broader scientific field, known as Operations Research in the United States and Operational Research in the United Kingdom.

Operational Research Section (ORS) at the Admiralty Specialised unit established during World War II, responsible for applying scientific and mathematical methods to improve the efficiency and effectiveness of naval operations. It played a key role in optimising tactics, convoy protection, and the use of emerging technologies such as radar.

Operations Room (Bentley Priory) Central command centre of RAF Fighter Command during World War II, located at Bentley Priory near London. It coordinated the air defence of Great Britain by processing filtered radar and observer information, directing fighter squadrons, and implementing defensive measures against enemy air attacks.

Oslo Report Confidential intelligence document sent anonymously from Germany to the British in 1939, revealing detailed information on German military technology, including radar, aircraft, and naval equipment. The report provided early insights that helped the Allies understand and prepare for German technological capabilities, influencing operational planning and scientific efforts such as radar countermeasures and electronic warfare.

Paris Peace Conference International conference held from January 1919 to January 1920 to formally end World War I and establish a new world order. The main treaty negotiated was the Treaty of Versailles with Germany, which imposed harsh penalties including the war guilt clause. The conference was led by the major Allied powers: France, Great Britain, and the USA.

Pathfinder Aircraft Specialised bomber aircraft of the Royal Air Force used during World War II to locate and mark targets for the main bomber force, significantly improving the accuracy and effectiveness of Allied bombing missions.

Pipsqueak World War II British ground-based radio direction-finding system that tracked friendly aircraft by receiving periodic tone signals transmitted from the aircraft. It assisted in coordinating fighter interceptions and integrating aircraft positions into the air defence network.

Radar Research and Development Establishment (RRDE) British military research centre in Malvern focused on the development of Army radar. It was renamed from the Air Defence Research and Development Establishment (ADRDE) after its relocation in 1942 and contributed significantly to the advancement of radar technologies for military applications.

Radio Direction Finding (RDF) Technique used to determine the direction of a radio signal source. It was widely employed in early radar systems to locate aircraft and ships and formed a fundamental component of Britain's air and naval defence capabilities.

Radio Research Board (RRB) Division of the Department of Scientific and Industrial Research (DSIR) responsible for research and development in radio and radar technologies. It played an important role in advancing scientific knowledge and supporting military and air defence applications in Britain.

Radio Research Station (RRS) Research facility established in 1924 in Slough by the Department of Scientific and Industrial Research (DSIR), focused on radio technology and early radar development. It merged with the National Physical Laboratory (NPL) in 1927, providing a foundation for Britain's scientific and military advancements in radio and radar.

Research Construction Company (RCC) Company established in October 1941 by the RadLab to produce limited quantities of microwave radar systems and components not readily available from industry.

Research Laboratory of Electronics (RLE), MIT Research centre at the MIT, founded in 1946, specialising in electronics, radio-frequency technology, and early radar and semiconductor research

Royal Aeronautical Society UK-based professional society founded in 1866, dedicated to the advancement of aerospace science and engineering. Harry Wimperis served as its president from 1936 to 1938, during which he made significant contributions to the field.

Royal Aircraft Establishment (RAE) British research facility in Farnborough focused on developing aviation technologies, including aircraft design, radar, and wireless communication. It played a key role in advancing Britain's aeronautical research and military aviation capabilities.

Royal Aircraft Factory British research and production establishment founded in 1911 in Farnborough, focused on designing, developing, and testing military aircraft and aviation technologies during World War I.

Royal Flying Corps (RFC) British Army's air service during World War I, responsible for reconnaissance, artillery spotting, and air combat. It merged with the Royal Naval Air Service in 1918 to form the Royal Air Force (RAF).

Royal Observer Corps (ROC) British civilian organisation responsible for visually detecting and tracking aircraft over land during World War II. Integrated into the national air defence system, the ROC provided essential data to supplement radar coverage, particularly at low altitudes and inland, playing a key role in Britain's air defence.

Royal Society UK's national academy of sciences, founded in 1660, dedicated to promoting excellence in science and supporting research and innovation. It has played a central role in advancing scientific knowledge and advising on science and technology policy.

Searchlight Experimental Establishment (SLEE) British military research unit focused on developing searchlight technology for air defence during and after World War I. It was reorganised in 1924 into the Air Defence Experimental Establishment (ADEE), contributing to the advancement of Britain's air defence capabilities.

Secondary Surveillance Radar (SSR) Radar system that responds to interrogation signals by transmitting additional data such as identification codes. Early versions, like the World War II Identification Friend or Foe (IFF) systems, provided basic identification but did not transmit altitude information. Modern secondary radar systems, developed after World War II, include altitude and other data transmission and are widely used in air traffic control and military operations to improve target identification and tracking.

Secretary of State Senior British government minister responsible for overseeing a specific department, typically at the head of a major portfolio such as foreign affairs, defence, or home affairs.

Signals Intelligence (SIGINT) The collection and analysis of all enemy electronic signals, including communications and radar. Radio Reconnaissance is a sub-discipline of SIGINT, focused on monitoring, analysing, locating, and decrypting enemy radio transmissions, usually for tactical purposes.

Special Duties Flight (SDF) RAF unit during World War II responsible for test flights of airborne radar systems, supporting radar research and development. It initially operated at St Athan before relocating to Christchurch and becoming part of the Telecommunications Flying Unit.

Sunday Soviets Informal Sunday meetings at the Telecommunications Research Establishment (TRE) that encouraged open discussion, idea exchange, and innovation.

Synthetic Aperture Radar (SAR) Active radar system that uses the motion of the radar platform to synthesise a large antenna aperture, achieving much higher resolution than conventional radar. First developed in the 1950s, it is now widely used for earth observation, mapping, surveillance, and military applications, and operates effectively in all weather conditions, day and night.

Telecommunications Flying Unit (TFU) RAF unit established in 1941 to support the testing and development of airborne radar and other electronic systems. It evolved from the earlier Special Duties Flight and operated various aircraft to trial experimental equipment. The TFU played a key role in advancing airborne radar during World War II and continued its work in the post-war period.

Telecommunications Research Establishment (TRE) Main British radar research organisation during World War II, established in 1940 as the successor to the Ministry of Aircraft Production Research Establishment (MAPRE). Initially based at Worth Matravers and later relocated to Malvern, the TRE developed key systems such as Airborne Interception and H_2S radar, playing a central role in Britain's air defence and bomber operations.

Tizard Committee British research group officially established in 1935, chaired by Sir Henry Tizard, which played a pivotal role in the development of radar and Britain's air defence system. Its creation followed a 1934 memorandum by Harry Wimperis that highlighted the urgent need for scientific investigation into air defence. The committee's work laid the foundation for the Chain Home radar network and other technologies that proved decisive during World War II.

Tizard Mission British delegation to the USA, in 1940 that transferred critical military technologies, including the cavity magnetron for radar systems, to support Allied research and development. The mission laid the foundation for close UK-US collaboration in radar and electronic warfare during World War II.

Treaty of Versailles Peace agreement signed in 1919 that officially ended World War I. It imposed harsh penalties on Germany, including territorial losses, military restrictions, and reparations, and established the framework for the post-war international order.

Trenchard Doctrine British air power strategy of the 1920s, formulated by Hugh Trenchard. It stressed using long-range bombers against enemy industry, transport, and infrastructure to erode war capacity and morale, while affirming the RAF's independent strategic role.

Turing machine Theoretical device introduced by Alan Turing in 1936 to formalise computation. It manipulates symbols on an infinite tape according to rules, forming the foundation for modern computer science and the study of algorithms.

ULTRA Codename for intelligence from decrypted enemy communications, especially German cipher traffic in World War II, providing both tactical and strategic insights. Its existence was highly classified at the time, known only to a small number of Allied personnel.

War Office British government department responsible for the administration and logistics of the army, active from 1855 to 1964, when its functions were transferred to the Ministry of Defence.

Women's Auxiliary Air Force (WAAF) Women's branch of the Royal Air Force during World War II, responsible for radar operation, technical support, and administrative duties. The WAAF played a key role in maintaining Britain's air defence, operating radar stations and supporting RAF operations throughout the war.

Women's Royal Naval Service (WRNS) Women's branch of the Royal Navy during World War II, responsible for communications, technical support, and administrative duties. The WRNS played a key role in supporting naval operations, operating communication systems, assisting with technical tasks, and ensuring the effective functioning of the Royal Navy throughout the war.

18 Glossary of German Terms

Abteilung Technisches Nachrichtenwesen im Oberkommando der Kriegsmarine Technical Signals Department of the Naval High Command; responsible for the research, development, and coordination of naval communications and electronic detection systems.

Amtsgruppe Technisches Nachrichtenwesen im Oberkommando der Kriegsmarine Technical Communications of the Naval High Command; responsible for overseeing communications, radar technologies, and electronic warfare within the Kriegsmarine's High Command during World War II.

Arbeitsgemeinschaft Rotterdam (AGR) Rotterdam Working Group; founded in February 1943, the collaborative group brought together branches of the German armed forces, industry, and research institutions. Operating under industrial leadership, it focused on advancing radar technology, particularly in the centimetre-wave range, in an effort to address Germany's technological deficits.

Arier-Paragraph Aryan Paragraph; Nazi law excluding Jews and other so-called non-Aryans from public service and professions.

Außenminister Foreign Secretary/Foreign Minister; Foreign Secretary and Foreign Minister are titles for the government official responsible for a country's foreign policy. While Great Britain uses the title Foreign Secretary, most other countries such as Germany use Foreign Minister for this position.

Beauftragter für den Vierjahresplan Commissioner for the Four-Year Plan; Hermann Göring was appointed in 1936 to oversee the economic mobilisation of Germany for war, aiming for self-sufficiency in raw materials and increased armaments production by 1940.

Beauftragter für Elektrotechnik Commissioner for Electrical Engineering; Official appointed to oversee and coordinate electrical engineering projects within Albert Speer's ministry during World War II.

Befehlshaber der Unterseeboote (BdU) Commander of the German U-boat fleet during World War II, responsible for directing submarine operations.

Berlin Radar German radar devices developed during World War II, designed to operate primarily in the 10 cm wavelength band. While this technology promised improved detection accuracy and resolution, many of these systems remained experimental or in prototype stages and saw little to no practical deployment during the war due to technical and production challenges.

Bevollmächtigter für die Hochfrequenzforschung (BHF) Commissioner for High-Frequency Research; senior Nazi official appointed to coordinate and supervise high-frequency research across Germany's scientific, industrial, and military sectors. The BHF held overarching authority in radio, radar, communications, and electronic warfare technologies, reporting directly to Hermann Göring. The office played a central role in aligning scientific innovation with military priorities.

Braun'sche Röhre Braun tube; early type of cathode ray tube, invented by German physicist Karl Ferdinand Braun in 1897. It was used to visualize electrical signals and played a key role in the development of oscilloscopes and television screens. The Braun tube works by directing a beam of electrons onto a fluorescent screen, creating visible patterns or waveforms.

Denkschrift Memorandum; written document outlining proposals or recommendations, such as the 1939 memorandum by Osenberg that emphasised the importance of technological research for Germany's war effort.

DeTe German abbreviation for Dezimeter-Telegraphie; originally used to denote radar systems operating in the decimetre wavelength range. Over time, the term was more broadly adopted as a technical designation for various radar types developed by the German Air Force, regardless of wavelength.

DeTe I Technical designation for German naval radar (e.g. Seetakt), operating in the lower decimetre-wave band, used primarily for surface target detection.

DeTe II Technical designation for air reporting radar systems, primarily meter-wave early warning radars such as Freya, used for detecting and tracking aircraft.

Deutsch-Britisches Flottenabkommen Anglo-German Naval Agreement; a 1935 political agreement between Germany and Great Britain that established limits on the size of the German navy relative to the British Royal Navy.

Deutsche Akademie der Luftfahrtforschung German Academy of Aviation Research; Scientific institution that promotes, coordinates, and supports research and development in the field of aviation and aerospace.

Deutsche Forschungsgemeinschaft (DFG) German Research Foundation; central independent self-governing research funding organisation in Germany, supporting scientific research across all disciplines.

Deutsche Versuchsanstalt für Luftfahrt (DVL) German Experimental Institute for Aeronautics; German research institute focused on experimental studies and development in aeronautics and aviation technology.

Deutscher Amateur-Sende- und Empfangsdienst (DASD) German Amateur Radio Transmission and Reception Service; a Nazi-era organisation that regulated amateur radio activity in Germany, serving both as technical training and as a means of political and military oversight.

Dolchstoßlegende False narrative claiming that Germany had not lost World War I on the battlefield but was betrayed by politicians and civilians, later used by nationalists to undermine the Weimar Republic.

Drahtlostelegraphische und Luftelektrische Versuchsstation Gräfelfing (DVG) Wireless Telegraphy and Atmospheric Electricity Research Station Gräfelfing; founded in 1908 by Max Dieckmann, the DVG was a research facility near Munich that specialised in wireless telegraphy and atmospheric electricity. It contributed to early advances in radio technology, particularly in aviation communication and navigation.

Entwicklungsgruppe Funkmeß Radio Measurement Development Group; German research and development group specialising in radio measurement technology during World War II, responsible for advancements in radar and related technologies.

Entwicklungsstopp-Befehl Development Stop Order; Directive stating that all armament development projects expected to deliver results by 1940 or early 1941 must be given highest priority. Programs with an anticipated impact only after this timeframe should be postponed or suspended if economic or resource constraints require it. This was intended to ensure that resources and funding were focused on projects with more immediate benefits to meet the urgent needs of the war effort.

Ermächtigungsgesetz Enabling Act (formally titled Law to Remedy the Distress of People and the Reich); passed by the Reichstag on 23 March 1933. It conferred upon the German government, in practice Adolf Hitler and his cabinet, the authority to enact laws without parliamentary consent, including laws that deviated from the constitution. This legislation marked a decisive step in dismantling the Weimar Republic's democratic institutions and consolidating the Nazi dictatorship.

Flakartillerieschule Anti-Aircraft Artillery School; German military training institution responsible for educating and training anti-aircraft personnel during World War II.

FLAK-Batterie Flak Battery; Anti-Aircraft Battery; tactical unit equipped with anti-aircraft guns along with the necessary personnel and equipment, used to detect, engage, and defend against enemy aircraft.

Flakwaffe Contemporary German term referring to the anti-aircraft arm of the German armed forces. The term was used inconsistently and does not denote a formally distinct organisational branch.

Flakzielgerät Fire-control radar system; used during World War II to detect, track, and guide anti-aircraft artillery (flak) toward enemy aircraft. These devices provided precise target information to improve the accuracy and effectiveness of anti-aircraft guns.

Fliegertruppe Combat Aviation Units; operational combat units of the German Air Force in World War II, including fighter, bomber, and reconnaissance squadrons. They conducted offensive and defensive air missions, distinct from support units like the Luftnachrichtentruppe. The Flying Troops were central to Germany's air warfare strategy.

Flugabwehrartillerie Anti-Aircraft Artillery; broad term for all artillery systems designed to engage airborne targets, including anti-aircraft guns (FLAK) and other artillery types used in air defence.

Flugabwehrkanone (FLAK) General term for ground-based air defence weapons designed to target enemy aircraft. Historically, it primarily referred to rapid-fire anti-aircraft guns, but it can also denote the broader category of anti-aircraft artillery. The English equivalent is anti-aircraft guns or, more broadly, anti-aircraft artillery.

Flugabwehrwaffe Anti-Aircraft Weapons; weapons designed to detect, target, and destroy enemy aircraft, including guns, missiles, and related systems used in air defence.

Flugfunk-Forschungsinstitut Oberpfaffenhofen (FFO) Aviation Radio Research Institute Oberpfaffenhofen; German research institute specialising in aviation radio technology and radar development during World War II. It made significant contributions to the advancement of radio communication and radar systems for the Air Force.

Flugmeldedienst Air Reporting Service; German military unit responsible for detecting, tracking, and reporting aircraft movements. It provided early warnings to air defence forces during World War II, helping to coordinate defensive actions against enemy air attacks.

Flugmeldegerät (FLUM) Air Reporting Unit; German World War II ground station system responsible for collecting, processing, and integrating radar data from systems such as Freya and DeTe II, as well as visual observations. FLUM provided comprehensive air situation reports to the Luftwaffe command for operational decision-making.

Forschungsgesellschaft für Funk- und Tonfilmtechnik e. V. Research Society for Radio and Sound Film Technology; German research organisation focused on advancing radio communication and sound film technologies. It contributed to the development and improvement of radio and audio-visual equipment for both civilian and military use during the mid-20th century.

Freya Radar German Air Warning Radar; developed by GEMA before and during World War II. It was part of the German Air Reporting Service and primarily used for long-range detection and tracking of enemy aircraft, with an effective range of up to 120 kilometres, extendable to 160 kilometres under optimal conditions. Freya provided radar data to support fighter direction and coordinate anti-aircraft defences, serving both early warning and fighter control functions.

Friedensrede Peace Speech; Reichstag speech by Adolf Hitler on 17 May 1933, in which he proclaimed Germany's peaceful intentions and endorsed international disarmament. The speech was largely aimed at reassuring foreign powers while masking ongoing rearmament plans.

Funkmeßgerät (FMG) Radio Measurement Apparatus; primarily refers to radar systems, active radio measurement devices that emit radio waves and analyse their echoes to detect, locate and measure objects. Radar is a key example of an active radio measurement device. In technical literature, radio direction finding and related passive techniques are generally considered a separate category and use dedicated devices.

Funkmeßortungs-Gerät (FuMO) Radio Measurement and Detection Apparatus; term used by the German Navy during World War II. It refers specifically to naval radar systems designed for detecting and locating targets such as ships and aircraft. The designation FuMO was part of a broader system of radar nomenclature used to categorise equipment based on their function and service branch.

Funkmeß-Sondergerät (FuSe) Radio Measuring Special Apparatus; denotes special or experimental radar and radio measuring equipment used by the German Air Force (Luftwaffe) during World War II. These devices were often prototypes or custom systems designed for specific detection or targeting tasks, distinct from standard radar equipment.

Funkmeßtechnik Radio Measurement Technology; use of high-frequency electromagnetic waves for detection, measurement, and surveillance purposes, particularly in radar and related radio measurement systems.

Funkpeilung Radio Direction Finding (RDF); technique used to determine the direction of a radio signal's source by measuring the signal's arrival angle with specialised antennas. RDF was essential during World War II for locating enemy transmitters and aiding navigation.

Funk-Sendegerät (FuSE) Radio Transmit-Receive Apparatus; designation used by the German Luftwaffe for radio and radar equipment, especially early warning and fire-control systems. FuSE devices combined transmitter and receiver functions, often as integrated radar units.

GEMA Gesellschaft für elektroakustische und mechanische Apparate m.b.H.; German company specialising in radar and communication technology during World War II. As a pioneer in radar development, GEMA was responsible for designing and producing key radar systems such as the Freya early-warning radar. The company also developed other important radar devices, including the Seetakt naval radar system. GEMA played a crucial role in advancing German radar capabilities, which enhanced air defence and detection during the war.

Generalbevollmächtigter für technische Nachrichtenmittel (GBN) General Commissioner for Technical Communications; appointed in 1943 to coordinate the development and standardisation of military communications and radar systems across the Wehrmacht. The office aimed to reduce duplication, improve efficiency, and strengthen Germany's position in electronic warfare.

Generalinspekteur Inspector General; senior military officer responsible for overseeing inspections, readiness, and overall effectiveness of a branch or the entire armed forces.

Generalinspekteur der Luftwaffe Inspector General of the Luftwaffe; senior officer responsible for inspecting and overseeing the operational readiness, training, discipline, and organisational efficiency of Luftwaffe units and personnel. The office ensured that Luftwaffe forces met required standards and reported directly to the highest levels of command during World War II.

Generalluftzeugmeister General Master of Aircraft and Equipment; senior Luftwaffe official responsible for the development, procurement, maintenance, and supply of aircraft and technical air equipment. This role ensured that the Luftwaffe's logistical and technical requirements were met throughout World War II. He worked closely with the aviation industry to align production with operational demands and played a key role in managing wartime resource allocation and technical innovation.

Generalnachrichtenführer General of Communications; senior military officer responsible for overseeing all communication systems within the German armed forces during World War II, including coordination, standardisation, and operational readiness of signal units. The position reported directly to the top military leadership, ensuring communication effectiveness at the highest levels of command.

Gesetz für den Aufbau der Wehrmacht Law for the Establishment of the Wehrmacht; Legislation that legally established the Wehrmacht (the German armed forces) and reintroduced compulsory military service in Germany.

Gesetz gegen die Neubildung von Parteien Law Against the Formation of New Parties; enacted on 14 July 1933, this law prohibited the formation of any new political parties in Germany, effectively establishing the Nazi Party as the sole legal party and consolidating Hitler's one-party dictatorship.

Gesetz über das Staatsoberhaupt des Deutschen Reichs Law on the Head of State of the German Reich; passed on 1 August 1934 and enacted the following day after President Hindenburg's death, this law merged the offices of Reichspräsident and Reichskanzler, giving Adolf Hitler the combined title Führer und Reichskanzler and consolidating his totalitarian rule.

Gesetz über die Wiedervereinigung Österreichs mit dem Deutschen Reich Law on the Reunification of Austria with the German Reich; enacted on 13 March 1938, this law formalised the annexation (*Anschluss*) of Austria into Nazi Germany, abolished Austria's sovereignty, and integrated the country as a province of the German Reich.

Gesetz zur Wiederherstellung des Berufsbeamtentums Law for the Restoration of the Professional Civil Service; enacted on 7 April 1933, this law enabled the dismissal of civil servants considered politically unreliable or excluded based on Nazi racial criteria, allowing the regime to consolidate control over the public sector.

Hauptkommission Elektrotechnik Main Commission for Electrical Engineering; central body within Albert Speer's Ministry of Armaments and War Production, responsible for coordinating electrical engineering research and development in Germany during World War II.

Heer Army; the land forces branch of Germany's armed forces, responsible for ground-based military operations during World War II. It played a central role in the regime's campaigns of conquest and occupation. The Heer was the largest branch of the German military during the war.

Heereswaffenamt (HWA) Army Ordnance Office; central authority of the German Army tasked with the development, testing, and procurement of weapons and military equipment during the Nazi era. It oversaw the technical innovation and supply of armaments throughout World War II.

Heinrich-Hertz-Institut (HHI) Research institute based in Berlin named after physicist Heinrich Hertz, known for his work on electromagnetic waves. During the interwar years and World War II, the HHI conducted research in high-frequency technology and radar development, contributing to advances in telecommunications and military applications.

Hochseeflotte Main battle fleet of the German Navy before and during World War I.

Hoßbach-Protokoll Hoßbach Memorandum; record of a secret meeting held on 5 November 1937, in which Hitler outlined his plans for aggressive expansion and warned of an imminent war, providing evidence of Nazi Germany's expansionist intentions before World War II. The memorandum was written by Friedrich Hoßbach, Hitler's military adjutant at the time.

Inspekteur der Flakwaffe Inspector of Anti-Aircraft Artillery; senior officer of the German Air Force responsible for overseeing the training, organisation, and operational readiness of Germany's anti-aircraft artillery units during World War II.

Institut für Flugfunk und Mikrowellen Institute for Aeronautical Radio and Microwave Technology; German research institute specialising in the development of aviation radio communications and microwave technologies, primarily for military applications during World War II.

Jägerführung Fighter Control; coordination and direction of fighter aircraft during combat missions, including interception of enemy planes and air defence operations, typically managed through ground-based radar and communication systems.

Jagdschloß Radar Panoramic early-warning radar developed by GEMA, with patents registered as early as 1936/37 and deployed from 1943. It provided continuous 360-degree airspace surveillance, significantly enhancing the German air defence system's ability to detect and track incoming aircraft. The system integrated radar data with a central plotting system, enabling more effective coordination of fighter interceptions.

Julius Pintsch AG German industrial company founded in 1843, originally known for gas lighting systems such as gas lamps and gas meters. In the 20th century, the company expanded into electrical measurement technology and was involved in early radar technology developments, particularly in the area of Bremsfeld tubes for high-frequency applications.

Kathodenstrahlröhre Cathode Ray Tube (CRT); vacuum tube that uses an electron beam to display images on a fluorescent screen.

Knickebein German WWII radio navigation system that guided bombers to targets using intersecting radio beams; Allied jamming and deception operations reduce

Kommission für Nachrichtenmittel Commission for Communications Equipment; Nazi-era body active from at least 1942 under the Ministry for Armaments and Munitions (RMfBuM), responsible for co-ordinating the development and standardisation of military communication technologies, including radar.

Kriegsmarine German Navy; naval force of Nazi Germany responsible for all maritime military operations during World War II, including surface fleets, U-boat campaigns, and coastal defence. It succeeded the Reichsmarine following Germany's rearmament and naval expansion under the Nazi regime. The Kriegsmarine played a central role in naval warfare, notably in the Battle of the Atlantic, aiming to disrupt Allied shipping and assert naval dominance.

Kriegsschuldparagraph War Guilt Clause; Article 231 of the Treaty of Versailles (1919). This clause placed sole responsibility for the outbreak of World War I on Germany and its allies. It provided the legal and moral basis for demanding reparations, requiring Germany to compensate the Allied powers for the damage caused by the war.

Laboratorium für Hochfrequenztechnik und Elektromedizin Laboratory for High-Frequency Technology and Electromedicine; German research facility associated with physicist Hans Erich Hollmann, a pioneer in high-frequency electronics and radar. The laboratory focused on developing advanced high-frequency systems and exploring electromagnetic applications in both military and medical contexts during World War II.

Leigh Light British airborne searchlight in World War II, used together with radar on patrol aircraft. Radar located a U-boat at night, and the Leigh Light illuminated it during the final attack run, sharply increasing Allied anti-submarine effectiveness.

Lichtenstein Radar German airborne radar system first operational as Lichtenstein BC in 1942, developed by Telefunken. Used primarily on night fighters to detect enemy aircraft, it featured a distinctive but bulky antenna array mounted on the aircraft's nose, which was effective but created aerodynamic challenges.

Lorenz-Schlüsselmaschine Lorenz Cipher Machine; German rotor-based cipher device used during World War II to encrypt high-level military teleprinter communications. More complex than the ENIGMA machine, it was successfully deciphered by Allied cryptanalysts at Bletchley Park.

Luftnachrichtentruppe Air Force Signal Corps; military unit within the German Air Force responsible for signals intelligence, radar surveillance, and radio communication to support air operations. The Air Force Signal Corps played a crucial role in gathering intelligence, maintaining secure communications, and coordinating Luftwaffe missions throughout World War II.

Luftnachrichten-Versuchsregiment Air Signals Experimental Regiment; a specialised unit of the German Air Force responsible for testing, developing, and evaluating new communication technologies and radar systems to enhance operational effectiveness during World War II.

Luftwaffe German Air Force; aerial warfare branch of the German armed forces. The term most commonly refers to Nazi Germany's air force from 1935 to 1945. The term is also used for the modern German Air Force, re-established in 1956 as part of the Bundeswehr.

Mammut and Wassermann Radare Ground-based radar systems for air surveillance developed in Nazi Germany during World War II, both derived from the Freya radar. Wassermann, introduced in 1940, featured a rotating antenna for 360-degree surveillance. Mammut became operational in 1941, employing fixed antenna arrays to extend detection range and coverage. Both systems were integral components of Germany's integrated air defence network.

Marbach Radar German fire control radar introduced in 1945 during World War II, used for target acquisition and tracking to support anti-aircraft artillery. Operating at a wavelength of approximately 9 cm, its late deployment was limited by wartime constraints. The system complemented other Luftwaffe fire control radars, such as the Würzburg, within the integrated air defence network.

Marinenachrichtendienst (MND) Naval Intelligence Service; intelligence branch of the German Navy during World War II, responsible for naval reconnaissance, signals intelligence, cryptanalysis, and the interception and analysis of Allied communications. The MND supported naval operations and submarine warfare by providing critical intelligence.

Marinewaffenamt Naval Weapons Office; German Navy's authority responsible for developing and procuring naval weapons and equipment during World War II, including radar systems and fire control technology.

Metox German radar warning receiver introduced in 1940 on surface vessels and from August 1942 on U-boats to detect Allied 1.5-metre radar emissions from ships, aircraft, and land stations. It gave early warning and time to dive, but became useless once the Allies adopted centimetric radar.

Mitteilungen des Deutschen Amateur-Sende- und Empfangs-Dienstes (DASD) Messages of the German Amateur Radio Transmission and Reception Service; a publication and communication channel utilised by German amateur radio operators to exchange technical information, operational updates, and coordinate activities within the amateur radio community, primarily in the periods before and after World War II.

Münchener Abkommen Munich Agreement; political agreement signed on 29 September 1938 by Germany, Great Britain, France, and Italy, in which the Sudetenland, a predominantly German-speaking region of Czechoslovakia, was ceded to the German Reich. The agreement, part of the British and French policy of appeasement, was reached without Czechoslovak participation and is widely regarded as a failed attempt to prevent further German expansion.

Mythos Speer Speer Myth; post-war portrayal of Albert Speer, Adolf Hitler's chief architect and Minister of Armaments, as the 'good Nazi' who professed ignorance of the Holocaust and Nazi atrocities. Speer depicted himself as a technocrat focused on Germany's reconstruction, distancing himself from the regime's crimes. Subsequent historical research has thoroughly discredited this narrative, revealing Speer's significant involvement in war crimes and the exploitation of forced labour.

Nachrichtenmittel-Versuchsanstalt der Marine (NVA) Naval Signals Experimental Establishment; German naval research institution responsible for testing, developing, and evaluating communication and signalling technologies. From the early 1930s, under Rudolf Kühnhold, it conducted pioneering trials with electromagnetic wave reflection, marking some of the earliest German steps towards practical radar. The NVA's work included improving naval communications as well as detection and warning systems.

Nachrichtenmittel-Versuchskommando der Kriegsmarine (NVK) Communications Equipment Testing Command; Navy unit responsible for testing, developing, and evaluating communication and signalling technologies during World War II, including early radar experiments. It contributed to the advancement of naval communication systems and improved operational coordination.

Nachrichtenverbindungswesen (NVW) Communications and Signals Service; organisational and technical framework employed by the German military during World War II to establish and maintain reliable communication links between radar installations, anti-aircraft artillery units, and command headquarters. This system was essential for coordinating air defence operations and ensuring effective command and control.

Nah-Flugmeldegerät Short-range Air Reporting Unit; German ground-based system used during World War II for detecting and reporting nearby aircraft. As a short-range counterpart to larger air reporting units, it supported local air defence by providing immediate tactical information, often based on radar or visual signals

Nationalsozialistische Deutsche Arbeiterpartei (NSDAP) National Socialist German Workers' Party; political party in Germany from 1920 to 1945, commonly known as the Nazi Party. Founded in the aftermath of World War I, it promoted a nationalist, racist, and anti-Semitic ideology. Under Adolf Hitler's leadership, the NSDAP seized power in 1933, establishing a totalitarian dictatorship responsible for initiating World War II and perpetrating the Holocaust. The party was banned following Germany's defeat in 1945.

Nationalsozialistischer Bund Deutscher Technik (NSBDT) National Socialist Federation of German Technology; organisation in Nazi Germany that united engineers and technical experts. It aimed to promote technology aligned with Nazi ideology and support the regime's political and economic goals

Nationalversammlung Constituent Assembly; convened in Weimar in February 1919 to draft and adopt the constitution of the Weimar Republic. It approved the Treaty of Versailles under considerable pressure on 23 June 1919. Beyond the treaty, the Assembly also laid the foundation for Germany's democratic governance and legal framework. The Weimar Constitution was officially adopted on 11 August 1919.

Oberbefehlshaber der Luftwaffe (Ob.d.L.) Commander-in-Chief of the Luftwaffe; highest-ranking officer in the German Air Force during Nazi Germany. This position held supreme command over all Luftwaffe operations, including strategic planning, deployment of air forces, and coordination with other military branches. From 1935 to April 1945, the role was held by Hermann Göring.

Oberbefehlshaber der Marine Commander-in-Chief of the Navy; highest-ranking officer of the German Navy (Kriegsmarine) during Nazi Germany. The office held overall command over naval operations, including strategic planning, force deployment, and coordination with other military branches. Grand Admiral Erich Raeder served as Commander-in-Chief until 1943, when he was succeeded by Grand Admiral Karl Dönitz, who led the navy through the final phase of World War II and was later appointed Hitler's successor.

Oberkommando der Luftwaffe (OKL) High Command of the Air Force; central command authority of the German Air Force (Luftwaffe) during Nazi Germany. Responsible for strategic planning, coordination, and leadership of Luftwaffe operations, including air combat, reconnaissance, and support missions during World War II. Key figures included Hermann Göring, who served as both Commander-in-Chief of the Air Force and head of the OKL. The command played a crucial role in directing Germany's air warfare strategy and coordinating with other branches of the military.

Oberkommando der Marine (OKM) High Command of the Navy; central command authority of the German naval forces during Nazi Germany. Responsible for strategic planning, coordination, and overall leadership of all naval operations, including fleet deployments, submarine warfare, convoy protection, and maritime support missions throughout World War II. The OKM played a key role in shaping Germany's naval strategy and cooperating with other military branches. It was led by Grand Admiral Erich Raeder until 1943, followed by Grand Admiral Karl Dönitz until 1945.

Oberkommando der Wehrmacht (OKW) High Command of the German Armed Forces; supreme military command authority in Nazi Germany, tasked with strategic planning, coordination and command across all branches: Army (Heer), Navy (Kriegsmarine) and Air Force (Luftwaffe). It directed military operations and served as Adolf Hitler's main military staff. Under its Chief, Wilhelm Keitel, the OKW became an instrument of Hitler's personal control over the military, often bypassing traditional command structures.

Oberkommando des Heeres (OKH) High Command of the German Army; supreme leadership body of the German Army (Heer) from 1935 to 1945, tasked with planning and executing army operations, particularly on the Eastern Front. Though nominally subordinate to the High Command of the German Armed Forces, the OKH exercised considerable autonomy until Hitler increasingly centralised command authority from 1941 onward.

Operation Seelöwe Operation Sea Lion; planned German invasion of Great Britain in 1940. It involved a large-scale amphibious landing supported by the Luftwaffe and Kriegsmarine. The operation was never carried out because Germany failed to gain air superiority during the Battle of Britain. Its cancellation marked a significant turning point by halting Germany's plans to invade Western Europe by sea and forcing a strategic shift in the war.

Organisation der deutschen Schwingungsforschung für den U-Boot-Krieg German Organisation for Acoustic and Vibration Research in Submarine Warfare; interdisciplinary group established during World War II to improve the effectiveness and stealth of German submarines. It brought together experts to study vibrations, sound propagation, and signal behaviour, focusing on acoustic camouflage, jamming, deception, and detection to counter Allied anti-submarine efforts.

Physikalisch-Technische Reichsanstalt (PTR) Physical-Technical Institute of the German Reich; prominent Nazi-era research institute dedicated to advancing physical and technical sciences, with a strong emphasis on military technology development and standardisation. The PTR contributed to Germany's war efforts through cutting-edge scientific research, technological innovation, and providing technical expertise across various defence projects.

Planungsamt im Reichsforschungsrat Planning Office within the Reich Research Council; central administrative unit responsible for the coordination, planning, and supervision of scientific research initiatives in Nazi Germany. The office played a pivotal role in directing research efforts to align with state priorities, particularly in advancing military technology and supporting the war effort during World War II.

Reichsakademie der Forschung Reich Academy of Research; planned as a central institution in Nazi Germany to coordinate and control scientific research in line with the regime's ideological and political goals. Established plans in 1934 were never realised due to internal power struggles and rivalries within the Nazi leadership.

Reichsbahn German Reich Railway; the national railway company of Germany during the Nazi era. It was responsible for civilian transportation and played a crucial role in military logistics, including the movement of troops and the deportation of victims during the Holocaust.

Reichsdozentenführung Reich Lecturers' Leadership; Nazi-era organisation responsible for overseeing university lecturers and academic staff throughout Germany. It enforced National Socialist ideology by controlling appointments, curricula, and research, ensuring political conformity and loyalty. The Reichsdozentenführung played a central role in the Nazification and control of German higher education.

Reichsforschungsamt: Reich Research Office; established by the Nazi regime in 1936 to coordinate scientific research and development, focusing on technologies for the war effort. Its work included radar, rocket technology, and other military innovations supporting Germany's objectives during World War II.

Reichsforschungsrat (RFR) Reich Research Council; established by the Nazi regime in 1937 to coordinate and supervise scientific research and development across Germany. Its goal was to centralise research efforts. Despite its official mandate, the RFR often faced challenges due to overlapping authority and competition among Nazi organisations, which limited its overall effectiveness. Nevertheless, it played a significant role in advancing military research during World War II.

Reichsführer-SS Highest rank in the SS (Schutzstaffel), held exclusively by Heinrich Himmler during the Nazi era. As Reichsführer SS, Himmler was the supreme commander of the SS and oversaw all SS operations, including the security and police forces, the concentration camp system, and numerous internal intelligence and terror apparatuses. His role was central to the enforcement of Nazi policies and the regime's system of repression and genocide, making him one of the most powerful men in the Third Reich.

Reichsgesetzblatt Official Legal Gazette of the German Reich; central publication medium for all laws, decrees, and official announcements issued by the Nazi regime. Legal acts became binding only upon publication in the Reichsgesetzblatt, making it a key instrument in enforcing National Socialist policy.

Reichskabinett Reich Cabinet; the highest executive body of the German government during the Nazi era, composed of senior ministers tasked with advising Adolf Hitler and implementing government policies. Although it formally held legislative and administrative authority, its power rapidly diminished after the Enabling Act of 1933, which allowed Hitler to bypass the cabinet and rule by decree. The Reich Cabinet became largely symbolic, serving to ratify decisions already made by Hitler.

Reichskanzler Reich Chancellor; head of government in Germany during the Weimar Republic and Nazi era, appointed by the President and initially reliant on parliamentary support. After Adolf Hitler's appointment in 1933, the office was transformed into a central instrument of dictatorship. Through the use of emergency decrees and the Enabling Act, Hitler bypassed democratic structures and concentrated legislative and executive powers in his hands, making the position synonymous with total authoritarian rule.

Reichskommissar für den Luftverkehr Reich Commissioner for Air Transport; position created by Adolf Hitler in January 1933 and assigned to Hermann Göring, who had simultaneously been appointed Minister without Portfolio in Hitler's cabinet. The office served to oversee civil and military aviation, consolidate control over the sector, and laid the groundwork for the Reichsluftfahrtministerium, RLM, established on 5 May 1933 with Göring as its head. It marked an early step in Germany's rearmament in violation of the Treaty of Versailles.

Reichsleiter für Rüstung und Kriegsproduktion Senior Nazi Party official responsible for coordinating and overseeing armaments and war production within the party hierarchy.

Reichsluftfahrtministerium (RLM) Reich Ministry of Aviation; central governmental authority in Nazi Germany responsible for the organisation, administration, and oversight of both military and civil aviation. Its remit encompassed aircraft production, aeronautical research, regulation of the aviation industry, and air traffic control. The ministry played a pivotal role in Germany's rearmament efforts and the advancement of air warfare capabilities.

Reichsmarine German Navy during the interwar period from 1919 to 1935. Limited in size and armament by the Treaty of Versailles, it was primarily tasked with defending German coastal waters and maritime interests. The Reichsmarine focused on naval training, coastal protection, and maintaining Germany's naval presence within treaty limits. After the Anglo-German Naval Agreement of 1935, it began expanding and modernising before being renamed the Kriegsmarine.

Reichsmarschall Highest military rank in Nazi Germany, established in 1940. The Reichsmarschall held supreme authority over the armed forces, ranking above all generals, and was responsible for coordinating military strategy and overseeing key operations.

Reichsminister Reich Minister; ministerial position in Nazi Germany, typically held by influential government officials. The title was used from the early 1930s to designate heads of various government departments reporting directly to the Reich Chancellor.

Reichsminister der Luftfahrt (R.d.L.) Reich Minister of Aviation; official head of the RLM and the Air Force, held by Hermann Göring from 1933 to 1945. Responsible for Germany's civil and military aviation, including research, production, logistics, and operational command.

Reichsminister ohne Geschäftsbereich German government minister without a specific departmental responsibility. Such ministers were part of the cabinet and often assigned special tasks or supervisory roles, as in the case of Hermann Göring, who was appointed Reichsminister ohne Geschäftsbereich on 30 January 1933 while also becoming Reichskommissar für den Luftverkehr.

Reichsministerium für Bewaffnung und Munition (RMfBuM) Reich Ministry for Armaments and Munitions; Nazi government ministry established on 17 March 1940 to coordinate and expand the production of weapons and ammunition. From 1942, it was led by Albert Speer, who implemented extensive reforms to increase wartime industrial output. In September 1943, the ministry was reorganised and expanded into the Reich Ministry for Armaments and War Production.

Reichsministerium für Rüstung und Kriegsproduktion (RMfRuK) Reich Ministry for Armaments and War Production; successor to the Reich Ministry for Armaments and Munitions, established in 1943 and headed by Albert Speer. It was responsible for centralising Germany's entire war production, overseeing arms manufacturing and military technology research, including radar and high-frequency systems. It played a key role in sustaining the German war economy in the later stages of World War II.

Reichsministerium für Wissenschaft, Erziehung und Volksbildung (REM) Reich Ministry of Science, Education and Culture; Nazi government ministry responsible for science, education, and cultural cultural/ideological indoctrination. It oversaw universities and research institutes, aligning academic work with regime ideology and, during the war, increasingly directing science and education towards military objectives.

Reichsparteitag Nuremberg Rally; annual mass assemblies of the Nazi Party held in Nuremberg between 1927 and 1938, with some interruptions. These highly choreographed events served as a central platform for propagating Nazi ideology, demonstrating party unity and discipline, and announcing key political, social, and economic policies. The rallies featured large-scale parades, speeches by Adolf Hitler and other senior officials, and theatrical displays designed to evoke loyalty and awe. The rallies played a central role in consolidating the regime's power and shaping domestic and international perceptions.

Reichsparteitag der Ehre Rally of Honour (1936 Nuremberg Rally); unique Nazi Party congress held in Nuremberg on 9 September 1936. At this rally, the second Vierjahresplan was announced, marking a key moment for the regime to present its economic policies focused on rearmament and self-sufficiency.

Reichspost German Reich Postal and Telecommunications Authority; responsible for mail delivery, telephone and telegraph services, and the operation of communication infrastructure. It formed an integral part of the Nazi regime's communications network and supported propaganda and military coordination.

Reichspostforschungsanstalt Reichspost Research Institute; scientific institute of the German Reich Postal and Telecommunications Authority, responsible for research in telecommunications, postal systems, and broadcasting. It contributed to both civilian infrastructure and military developments during the Nazi period. The institute played a significant role in advancing communication technologies under the regime.

Reichspostminister Reich Minister of Posts; government official responsible for managing and overseeing the German Reich Postal and Telecommunications Authority, including regulation, infrastructure development, and coordination of communication policies. The position was integral to maintaining the Reich's communication network and supporting state objectives.

Reichspostzentralamt Central administrative authority of the German Imperial Post during the early 20th century. It was responsible for managing and regulating postal services, telecommunication networks, and related technical developments across Germany. The Reichspostzentralamt played a crucial role in coordinating communication infrastructure and overseeing technological innovations within the German postal and telecommunication system.

Reichspräsident Reich President; head of state of the Weimar Republic (1919–1934), elected by popular mandate and endowed with extensive constitutional powers, including the appointment of the Reich Chancellor, dissolution of the Reichstag (Article 25), and issuance of emergency decrees (Article 48). Paul von Hindenburg held the office from 1925 until 1934 and played a pivotal role in the appointment of Adolf Hitler as Chancellor in 1933. Following Hindenburg's death, the office was merged with that of the Reich Chancellor, thereby consolidating executive authority under Hitler.

Reichsprotektorat Böhmen und Mähren Protectorate of Bohemia and Moravia;
established by Nazi Germany on 16 March 1939 after the dissolution of Czechoslovakia, this territory included the Czech regions of Bohemia and Moravia. Under German control, a limited Czech administration continued to operate. The protectorate was a key part of Nazi expansion policy.

Reichsstelle für Hochfrequenzforschung (RHF) Reich Office for High Frequency Research; Nazi-era government agency responsible for coordinating, funding, and overseeing high-frequency research, particularly in support of military applications such as radar, communications, and electronic countermeasures for the Navy and Air Force.

Reichstagsbrandverordnung President's Decree for the Protection of the People and State; emergency decree issued on 28 February 1933 by Reichspräsident Hindenburg immediately following the Reichstag fire. It suspended key civil liberties such as freedom of speech, press, assembly, and association, enabling the Nazi government to persecute political opponents, ban opposition parties, and effectively dismantle democratic institutions.

Reichsverteidigungsgesetz Reich Defence Law; German law enacted on 21 May 1935 that granted the government extensive powers to prioritise resources and coordinate arms production, facilitating rapid militarisation of the economy under the Nazi regime.

Reichswehr Reichswehr was the official name of the German armed forces from 1919 to 1935, consisting of the Army (Heer) and Navy (Reichsmarine). It was established after World War I as a limited military force under the restrictions of the Treaty of Versailles. The Reichswehr laid the foundation for the later Wehrmacht, which was formed in 1935 with the establishment of an official Air Force (Luftwaffe) as Germany began rearmament. Throughout its existence, the Reichswehr operated within the constraints imposed by the Allied powers but also secretly prepared for future expansion

Reichswirtschaftsminister Reich Minister of Economics; head of the German Ministry of Economics during the Nazi era, responsible for economic policy, including industrial production, trade, and resource allocation. The office was central to implementing the regime's policies of economic mobilisation and autarky, working closely with other ministries and the military.

Rüstungsforschung im National-sozialismus Armaments Research under National Socialism; scientific and technological research aimed at developing and improving weapons and military equipment, closely linked to the Nazi regime's war preparations and conducted in cooperation with industry, universities, and the armed forces.

Rüstungsministerium Armaments Ministry; government department responsible for coordinating weapons production and managing the wartime economy in Germany, especially under the leadership of Albert Speer, who increased armaments output through centralised control and forced labour.

Rüstungswunder Armaments Miracle; rapid and significant increase in German weapons production during World War II under Albert Speer's leadership, achieved through improved organisation, centralised control, and the extensive use of forced labour.

Rundsuchradar 360-degree search radar; radar system capable of scanning the entire surrounding area to detect and track targets from all directions. Introduced and widely used during World War II, it played a crucial role in early warning and air defence by providing continuous surveillance of airspace and sea approaches.

Schlüsselkraft-Befehl Key Personnel Order; directive issued on 19 February 1942 to prevent the conscription of essential skilled workers from key sectors such as armaments production, railway expansion, and the energy industry, in order to safeguard critical functions of the war economy.

Schlüsselzusatz (G-Zusatz) SZ 40/42 Key Supplement (G-Supplement) SZ40/SZ42; component of the Lorenz cipher system (developed by Lorenz AG for the German Army), the key supplement SZ40 (later improved as SZ42) generated pseudo-random key streams for encrypted high-level military communications. It was used together with the Lorenz Lo15 teleprinter.

Seetakt Radar German naval radar system used during World War II to detect surface ships and coastal targets. Operating primarily in the decimetre wave band (80 cm wavelength), it provided detection ranges typically between 10 and 25 kilometres but could detect targets up to approximately 40 kilometres under optimal conditions. Factors such as target size, weather, sea state, and antenna height influenced its effective range. Seetakt played a crucial role in naval engagements by enhancing situational awareness and targeting accuracy.

Sekundär-Radar Secondary Surveillance Radar (SSR); radar system that responds to interrogation signals by transmitting additional data, such as identification codes. Early versions, including World War II Identification Friend or Foe (IFF) systems, provided basic identification but no altitude information. Modern secondary radar systems, developed after World War II, transmit altitude and other data and are widely used in air traffic control and military operations to enhance target identification and tracking.

S-Gerät / S-Anlage Abbreviation of Sondergerät / Sonderanlage (special device / special installation); these terms were used by the German Navy and GEMA to designate specialised sound-location equipment developed alongside radar systems before and during World War II. Initially, they referred to experimental or non-standard acoustic detection devices without formal code names.

Siemens-Geheimschreiber T52 Siemens T52 cipher teleprinter; German cipher machine developed by Siemens & Halske and used during World War II. Unlike the ENIGMA, it was based on teleprinter technology and employed primarily for high-level communications, especially by the Luftwaffe and the German Foreign Office.

Sonderbeauftragter für Funkmeßgeräte Special Commissioner for Radar Equipment; Nazi-era official responsible for overseeing the development, production, and deployment of radar and related radio measurement technologies. Held by Martin Martini, who played a key role in coordinating radar innovations and their military applications during World War II.

Sonderkommission für Funkmeßtechnik (SKFM) Special Commission for Radio Measurement Technology; commission established by the German High Command in 1944 to centralise and accelerate the development, coordination, and deployment of radar and radio measurement technologies. It aimed to improve the effectiveness of German radar systems in the final phase of World War II.

Staatsrat State Councillor; honorary political-administrative title awarded in Nazi Germany, particularly within the Prussian State Council under Hermann Göring. The position was largely symbolic and served to integrate influential figures from science, industry, and the party into the regime's institutional structure. Physicist Abraham Esau, who played a central role in coordinating military research, was one such appointee.

Staatssekretär State Secretary; highest-ranking civil servant in a government ministry in Germany, subordinate only to the minister. In Nazi Germany, State Secretaries held significant administrative power and were often key figures in implementing government policy. Their role was crucial in coordinating between political leadership and ministerial bureaucracy.

Staatssekretär der Luftfahrt State Secretary for Aviation; high-ranking official in Nazi Germany responsible for overseeing aviation matters, including the development and expansion of the Luftwaffe, as well as coordinating civil and military aviation activities. Notable holders, such as Erhard Milch, played key roles in managing aircraft production and the aviation industry during World War II.

Sudetenkrise Sudeten Crisis; political and diplomatic crisis in 1938 caused by Adolf Hitler's demands for the annexation of the Sudetenland, a region of Czechoslovakia with a large ethnic German population. The crisis culminated in the Munich Agreement, in which Britain, France, Italy, and Germany agreed to cede the Sudetenland to Nazi Germany, marking a key step in the lead-up to World War II.

Tag der Machtergreifung Day of the Takeover of Power; the date 30 January 1933 marks Adolf Hitler's appointment as Chancellor of Germany, leading to the Nazi Party's rapid consolidation of control over the government and the establishment of a dictatorship. Historians emphasise that it was not a violent takeover but rather a legal transfer of power, as Hitler was appointed through constitutional means.

Tannenbaumstrahler Christmas tree antenna; type of antenna with staggered elements resembling a Christmas tree. The German term Strahler referred to the whole antenna, not just the radiating element.

Technische Hochschule (TH) Technical University; higher education institution specialising in engineering, applied sciences, and technology.

Technisches Amt des RLM Technical Office of the RLM; responsible for the development, design, and production oversight of aircraft and aviation technology in Nazi Germany during World War II. The office coordinated technical specifications, supervised manufacturers, and played a central role in advancing the German Air Force's capabilities.

Telemobiloskop Early obstacle detection device developed by Christian Hülsmeyer in 1904. It used radio waves to detect metallic objects, such as ships, and is regarded as an early forerunner of radar. Although technically advanced, it was not commercially successful and was poorly understood at the time.

Torpedo-Kommission des Rüstungsministeriums Torpedo Commission of the Ministry of Armaments; commission within the Ministry of Armaments responsible for overseeing the development, production, and deployment of torpedoes for the German military during World War II. It coordinated technical requirements and supervised manufacturing to support naval warfare capabilities.

Torpedo- und Artillerie-Waffenamt Torpedo and Artillery Weapons Office; department responsible for the development, testing, and production oversight of torpedoes and artillery weapons during World War II. It coordinated technical specifications and supervised manufacturing to meet operational needs for naval and land forces.

Torpedo-Versuchsanstalt (TVA) Torpedo Testing Establishment; German Navy institution dedicated to the research, development, and testing of torpedo technologies. The TVA improved torpedo performance, guidance, and reliability, contributing directly to the operational effectiveness of the Kriegsmarine. Its work encompassed experimental designs, propulsion systems, and control mechanisms to enhance accuracy and range in combat..

Vierjahresplan Four-Year Plan; economic programme initiated by the Nazi regime in 1936 under Hermann Göring to prepare Germany for war within four years. It focused on achieving economic self-sufficiency, boosting raw material and energy production, and rapidly expanding armament manufacturing to support military rearmament and wartime needs.

Wehrforschungsgemeinschaft (WFG) German Military Research Community; established in 1944, this organisation was created to coordinate and advance military research and development efforts in Nazi Germany during World War II. It focused especially on weaponry, radar, and communications technology.

Wehrgesetz Military Service Law; German military law enacted on 21 May 1935 that reintroduced general conscription and regulated military service obligations, marking a key step in Germany's remilitarisation under the Nazi regime.

Wehrmacht German Armed Forces; unified military forces of Nazi Germany from 1935 to 1945, consisting of the Army (Heer), Navy (Kriegsmarine), and Air Force (Luftwaffe). The Wehrmacht was responsible for conducting Germany's military campaigns during World War II and played a crucial role in implementing Nazi expansionist policies and warfare strategies.

Wehrstaat Nazi concept of a defence state in which all aspects of society including politics, economy, science, education and culture were subordinated to military needs and the population prepared for war.

Weingut I Code name for a late war German project to move armaments production to underground facilities near Mühldorf in Bavaria. It was planned to be operational by July 1945 to protect war manufacturing from Allied air raids. The project was only partly completed and never fully operational before the war ended.

Wismar/Vollwismar Camouflage designations for German radar countermeasures in World War II. Wismar introduced additional frequency ranges for Freya and Würzburg radars, while Vollwismar aimed to equip them with multiple alternative frequencies to resist Allied jamming.

Wissenschaftlicher Führungsstab beim Oberkommando der Kriegsmarine (WFM) Scientific Command Staff of the Navy High Command

(WFM); specialised group within the German Navy during World War II responsible for coordinating and overseeing scientific and technological research and development. The WFM played a key role in advancing naval technologies and supporting operational effectiveness.

Wolfsrudeltaktik Wolfpack tactic; German U-boat strategy in World War II where groups of submarines attacked Allied convoys together to maximise impact and minimise risk.

Würzburg-Radar Fire-control radar used by Germany during World War II for target detection and directing anti-aircraft artillery. The Würzburg system improved the accuracy of gunfire against enemy aircraft and was a key component of the German air defence network, providing real-time tracking and fire coordination. It was widely deployed at critical defence sites and contributed to prolonging resistance against Allied air raids.

Zentralinstitut für Hochfrequenz- und Fernmeldetechnik, Reichsführer SS Central Institute for High Frequency and Telecommunications Technology, Reichsführer SS; wartime research institute under Heinrich Himmler's direct authority, focused on developing high-frequency and telecommunications technologies for military use. It worked on radar, radio communications, and signal intelligence, enhancing the SS's and German military's technological capabilities during World War II.

19 Index of Persons